Asymmetric Synthesis

Volume 4

THE CHIRAL CARBON POOL AND CHIRAL SULFUR, NITROGEN, PHOSPHORUS, AND SILICON CENTERS

Asymmetric Synthesis

Volume 4

THE CHIRAL CARBON POOL AND CHIRAL SULFUR, NITROGEN, PHOSPHORUS, AND SILICON CENTERS

Edited by

James D. Morrison

Department of Chemistry
University of New Hampshire
Durham, New Hampshire

John W. Scott

Roche Research Center
Hoffmann–La Roche Inc.
Nutley, New Jersey

1984

ACADEMIC PRESS, INC.

(Harcourt Brace Jovanovich, Publishers)

Orlando San Diego New York London
Toronto Montreal Sydney Tokyo

ACADEMIC PRESS, INC.
Orlando, Florida 32887

United Kingdom Edition published by
ACADEMIC PRESS, INC. (LONDON) LTD.
24/28 Oval Road, London NW1 7DX

Library of Congress Cataloging in Publication Data

Main entry under title:

Asymmetric synthesis.

Vol. 4 edited by James D. Morrison and John W. Scott.
Includes bibliographical references and indexes.
Contents: v. 1. Analytical methods -- -- v. 4 The chiral carbon pool and chiral sulfur, nitrogen, phosphorus, and silicon centers.
1. Stereochemistry. 2. Chirality. 3. Optical rotation. 4. Addition reactions. 5. Chemistry, Organic--Synthesis. I. Morrison, James D., Date . II. Scott, John W.
QD481.A78 1983 541.3'93 83-4620
ISBN 0-12-507704-1 (v. 4 ; alk. paper)

PRINTED IN THE UNITED STATES OF AMERICA

84 85 86 87 9 8 7 6 5 4 3 2 1

Dedication*

Each volume of this treatise is dedicated to an asymmetric synthesis pioneer. Volume 4 is dedicated to Professor Shun-ichi Yamada, Professor Emeritus of the University of Tokyo and Professor of the Faculty of Pharmaceutical Sciences, Josai University.

Professor Yamada was born August 1915, at Osaka, Japan. He graduated from the Pharmaceutical Institute, Faculty of Medicine, Tokyo Imperial University (now the Faculty of Pharmaceutical Sciences, University of Tokyo) in 1940. After five years of military service, he joined Tanabe Seiyaku Co., Ltd., where he developed a method for the production of optically active α-amino acids on an industrial scale using an *N*-acylase from *Aspergillus oryzae*. In 1959 he resigned as director of Tanabe Seiyaku Research Laboratories (Tokyo branch) to become a professor at his alma mater.

At Tokyo University, Professor Yamada developed novel methods for the synthesis of many chiral natural products and biologically active compounds, often using optically active amino acids as chiral sources. His other contributions include a general synthesis of optically active α-substituted carbonyl compounds via the alkylation of proline-derived enamines, a biogenetic type of asymmetric synthesis of alkaloids, biomimetic asymmetric reductive transaminations and imine alkylations, catalytic asymmetric epoxidations of allylic alcohols, new coupling reagents for peptide synthesis, and numerous absolute configuration determinations that have clarified the stereochemical and mechanistic features of many carbon–carbon and carbon–nitrogen bond-breaking and bond-making reactions.

Professor Yamada has received the Academic Prize of the Pharmaceutical Society of Japan (1975) and the Naito Foundation Research Prize (1978) for his outstanding contributions to the chemistry of optically active compounds. His protégés include many eminent contemporary Japanese chemists.

*The assistance of Professors T. Mukaiyama and K. Koga in collecting information for this dedication is gratefully acknowledged.

Contents

Contributors

Numbers in parentheses indicate the pages on which the authors' contributions begin.

MICHAEL R. BARBACHYN (227), Department of Chemistry, Wayne State University, Detroit, Michigan 48202

FRANKLIN A. DAVIS (313), Department of Chemistry, Drexel University, Philadelphia, Pennsylvania 19104

ROBERT H. JENKINS, JR. (313), Texaco Research Center, Beacon, New York 12508

CARL R. JOHNSON (227), Department of Chemistry, Wayne State University, Detroit, Michigan 48202

CYNTHIA A. MARYANOFF (355), McNeil Pharmaceutical, Spring House, Pennsylvania 19477

BRUCE E. MARYANOFF (355), McNeil Pharmaceutical, Spring House, Pennsylvania 19477

JOHN W. SCOTT (1), Roche Research Center, Hoffmann–La Roche Inc., Nutley, New Jersey 07110

DONALD VALENTINE, JR.[1] (263), Santa Clara, California

[1]Present address: Discovery Research, American Cyanamid, Stamford, Connecticut 06904.

Preface

Asymmetric synthesis has evolved from an academic curiosity to one of the most intensely studied synthetic methodologies. Leading research groups in both industrial and academic laboratories are now concentrating great effort on the subject.

This is the fourth volume of a multivolume treatise that reviews progress in the field since the 1971 publication of *Asymmetric Organic Reactions* by J. D. Morrison and H. S. Mosher. Volume 1 covers the major analytical methods used to determine enantiomer ratios; Volumes 2 and 3 are concerned with asymmetric addition reactions; Volume 5 treats asymmetric catalysis.

This volume is the first book to review the chemistry of centrodissymmetric chiral compounds containing carbon, sulfur, phosphorus, nitrogen, and silicon asymmetric centers, as it relates to chiral synthesis. Relatively simple chiral fragments, obtained by a variety of means, can trigger asymmetric synthetic sequences of considerable scope and diversity. The chapters in this volume show how one can obtain chiral fragments, with emphasis in most instances on the more practical methods, and they also illustrate how the initial chirality can be transmitted and extended.

Chapter 1 describes the so-called chiral carbon pool, readily available chiral carbon fragments used as building blocks in synthesis. A carefully organized tabulation that lists 375 chiral building blocks, along with their commercial sources, approximate prices, and methods of synthesis with literature citations, provides a valuable ready reference. In schemes involving almost 1300 structures, there is detailed the use of chiral building blocks to prepare alkaloids, arachidonic acid metabolites, compounds with chemotherapeutic (β-lactams, macrolides, polyether antibiotics, etc.) and pharmacological (carbinolamines) activity, insect pheromones, sugars, vitamins, terpenoids, and miscellaneous compounds. The schemes are supported by an extensively referenced text.

The remaining chapters provide a similar overview of the chemistry of compounds chiral at sulfur (Chapter 2), phosphorus (Chapter 3), nitrogen (Chapter 4), and silicon (Chapter 5).

All synthetic chemists, especially those engaged in pharmaceutical research and specialty chemicals production, will benefit from the focused treatment provided by this volume. This treatise seeks to enrich and clarify both the scientific and artistic facets of modern chiral synthesis.

1

Readily Available Chiral Carbon Fragments and Their Use in Synthesis

John W. Scott
Roche Research Center
Hoffmann–La Roche Inc.
Nutley, New Jersey

I. The Method

A. Introduction

In recent years, the once desirable preparation of a chiral compound in enantiomerically pure form has become a virtual necessity. The develop-

ISBN 0-12-507704-1

ment of this higher order criterion for a successful synthesis has esthetic as well as very practical foundations [e.g., the dependence of pharmacological (*1*) or pheromonal activity (*2*) on absolute configuration].

Seebach (*3, 4*) has coined the term EPC (enantiomerically pure compound) synthesis to cover the manifold of methods by which a chiral molecule can be obtained. The methodology has three basic forms, some of the advantages and disadvantages of which can be mentioned briefly:

1. *Resolution.* Despite significant advances (*5, 6*) in recent years from both theoretical and operational viewpoints, resolution remains largely an empirical science for the practicing organic chemist. The attendant risks in planning a synthesis on the basis of a resolution are thus evident. Unless a recycling of the undesired enantiomer is available, resolution is inefficient and for practical reasons has to occur early in the synthetic sequence. On the other hand, many compounds can be, and have been, resolved in a very straightforward way to give materials of high enantiomeric purity. Also, resolution usually gives rapid access to both enantiomers of a molecule, an important consideration in, for example, biological studies.

2. *Asymmetric synthesis.* The multivolume treatise of which this contribution is a part attests to the tremendous surge of interest in recent years in the field of asymmetric synthesis. Asymmetric synthesis (in the broadest form, including enzymic transformations) is, at its best, a very effective method for producing chiral molecules. In cases where the chiral auxiliary can be used catalytically, asymmetric synthesis provides by far the most efficient use of one chiral material to prepare another. There are still only a few asymmetric syntheses, however, that can be relied upon to routinely produce products of high enantiomeric excess (ee). Even then, a high enantiomeric excess may not be sufficient to guarantee the success of an asymmetric synthesis. From a synthetic viewpoint, substrate synthesis and product enantiomeric enrichment (since virtually no asymmetric syntheses give ≥99% ee) must be major concerns (7).

3. *Chiral building blocks.* All approaches to chiral target molecules ultimately rely on other naturally produced chiral entities. In the resolutions, one chiral molecule is used to "extract" another from a mixture with its enantiomer. Asymmetric synthesis employs chiral auxiliaries either catalytically or stoichiometrically (with or without the possibility of recovery of the auxiliary) to create new chiral molecules. The third use of the naturally occurring chiral molecules is as starting points,[1] themselves,

[1] Occasionally, one is in the fortunate position of requiring, as a final product, a compound that is a chiral natural product and additionally an article of commerce. In this special, limiting case, synthesis is not required, and one must only apply the FRN (Federal Reserve note) method (see Morrison, Volume 1, Chapter 1).

for syntheses. Presuming (see below) that these compounds are enantiomerically pure, their use projects the least risk of obtaining an enantiomerically impure product. If the subsequent reactions are designed and executed in a manner to avoid adventitious racemization, one can be reasonably confident that a "chirally safe" route to the target molecule is in hand. The major disadvantages of the method are that the original chiral molecule is, of necessity, consumed in the synthesis and that the range of easily accessed compounds is relatively limited. This latter fact may require that substantial chemistry be done to convert a chiral starting material to an enantiomerically pure intermediate, the synthesis of which in racemic form is a trivial matter.

Unless one is wedded to a particular synthetic methodology, retrosynthetic analysis of a chiral target should take into consideration all of the facets of EPC synthesis. The considerations in making a choice between, for example, a resolution and the use of (*S*)-malic acid as a starting material are so numerous as to make generalizations impossible. What is an acceptable methodology for a synthesis as a structure proof can be entirely inappropriate for industrial preparation of the same compound.

The chiral building block approach to EPC synthesis has been reviewed (*4*) a number of times, both from the viewpoint of target classes (*2, 8–19*) and specific starting materials (*20–31*). The aim of this chapter is to take a somewhat broader view, encompassing as many as possible of the relevant starting materials and the major product classes. Specifically, the remainder of Section I will discuss the considerations in using and the scope and limitations of the method. Section II contains a tabulation of the chiral building blocks, and Section III gives, by target-molecule class, an overview of syntheses in which chiral building blocks have been employed.

B. When Are Chiral Building Blocks Used?

Virtually every modern synthetic organic chemist is a disciple of the "synthon approach" (*32*) in the preparation of desired compounds. The retrosynthetic analysis and synthetic planning thus used for a chiral molecule must include provision for introduction of that chirality. As noted above, the choice of means for effecting the introduction of chirality may be the use of a chiral starting material. While there are often tangential factors that impinge on this decision, some of the major considerations might be:

1. *Cost and bulk availability of the synthon.*
2. *Need for both enantiomers.* For a number of materials both enantiomers are available, albeit often at substantially different costs. There

are often, however, alternatives in these cases and in situations where only one enantiomer is available. For example, D-rhamnose is not commercially available, but its L enantiomer is relatively inexpensive. On the

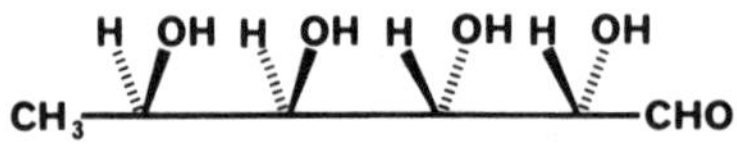

D-Rhamnose

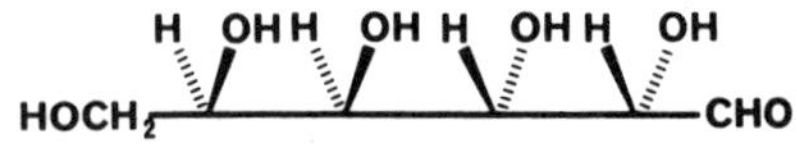

D-Mannose

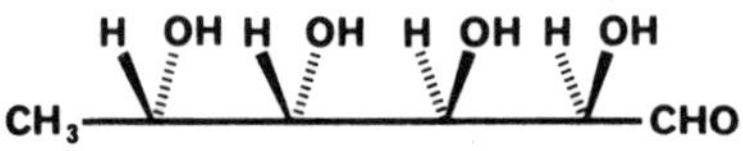

L-Rhamnose

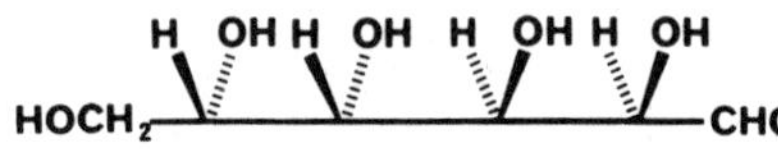

L-Mannose

other hand, D-mannose is substantially less expensive than its L isomer. Comparison of the projection formulas of these compounds clearly indicates the configurational relationships. It is not difficult to see that, for many synthetic purposes, rhamnose would be an equivalent of mannose and vice versa. In these cases, enantiomeric products would be most cheaply prepared from D-mannose and L-rhamnose.

A second means of circumventing the unavailability of one enantiomer of a chiral compound is the use of a variation of the "meso trick" (*4*). In this modification, one employs a chiral material that is asymmetric only by virtue of differentiation of oxidation level around a central carbon atom. Subsequent manipulation of these oxidation levels can formally invert the center atom. A good illustration of this principle is provided by the work of Cohen (*33*). Microbial oxidation of isobutyric acid provided (*S*)-β-hydroxyisobutyric acid. Manipulation of the molecule gave the enantiomeric four-carbon synthons **A** and **B** (*34*). Proper combination of

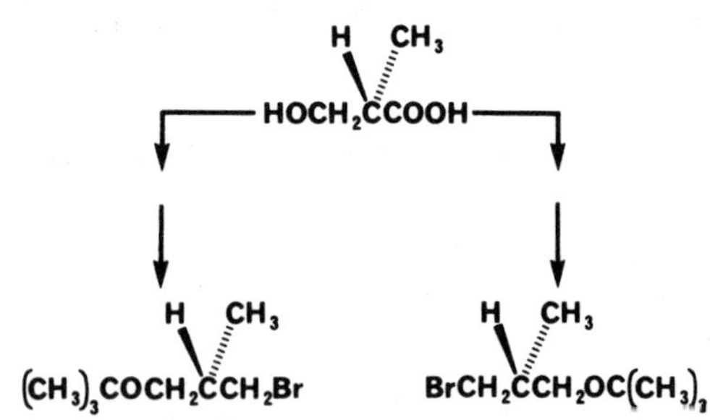

Synthon **A** Synthon **B**

these units allowed the preparation of all eight diastereomers of α-tocopherol (vitamin E) (*33*).

3. *Enantiomeric integrity of the synthesis*. The considerations here are independent of the source of the chirality and are virtually self-evident. A synthetic design that allows the possibility of racemization (e.g., Wittig

reaction on isopropylidene glyceraldehyde) should obviously be avoided if possible. Also to be avoided are routes in which introduction of new chiral centers occurs with low stereoselectivity, resulting in major diastereomer separations and lower yields.

The thought processes involved in a retrosynthetic analysis are a major contributor to the "art" of organic synthesis. Indeed, the quality of this analysis is often in direct proportion to the quality of the resultant work. The conceptualization or "inspiration" step in connecting particular starting materials and products [e.g., (+)-camphor to cobyric acid; Section III,G,2] can often be analyzed only in retrospect.

Some attempts have been made, particularly with carbohydrates as educts, to define and analyze the method more closely. Hanessian (*26*) preferred to consider carbohydrates as "chiral templates" and has attempted to describe the means by which one retrosynthetically decodes the product stereochemistry and transposes the result to a chiral template. Two types of synthetic targets are defined: those having readily apparent carbohydrate portions (primarily tetrahydrofuran and pyran-containing compounds, as well as their nitrogen analogs) and those containing a "hidden" carbohydrate. Fraser-Reid (*27*), on the other hand, preferred definitions involving the manner in which the carbohydrate is transformed into the product: by acyclic transfer, cyclic transfer, or transcription.

As will be seen in Section III, many types of chiral molecules can be synthesized from the appropriate building blocks. There are a few situations, however, in which consideration of the chiral building block method is particularly appropriate:

1. *Epoxides*. The ready availability of 1,2-diol units (e.g., carbohydrates, tartaric acid, etc.), along with the existence of excellent methodology for their stereospecific conversion to epoxides, finds no match in resolution or asymmetric synthesis. An exception is the epoxides derived from allylic alcohols (see Volume 5 of this treatise). The value of epoxides as synthetic intermediates, as well as an end in themselves, emphasizes the need and desirability of obtaining these compounds in a high state of enantiomeric purity.

2. *Secondary methyl groups* [$R^1CH_2CH(CH_3)CH_2R^2$]. Because of the flexible nature of the chain containing this unit, resolution is often difficult and may not be complete. The methodologies for asymmetric synthesis of a chiral secondary methyl group are limited; self-immolative sigmatropic rearrangements (Volume 3, Chapter 8) are the only ones for which one can confidently predict a high enantiomeric excess. On the other hand, a variety of terpenoids and related molecules possess this functional group-

ing. Although one often has to be concerned with the enantiomeric purity of terpenes, their ready availability strongly recommends them as chiral educts.

3. *Polyhydroxylated and/or methylated chains.* These functional units appear in a variety of compounds generated biosynthetically by the acetate–propionate–butyrate pathways. The macrolide and polyether antibiotics are prime examples of these types of compounds. Because of the large ring size of the macrolides and the special nature of the polyethers, syntheses of these compounds are an exercise in acyclic stereoselection. While the tremendous recent advances in aldol methodology (Volume 3, Chapter 2) offer one excellent method for control of this acyclic stereochemistry, there are two highly competitive means of obtaining the same result using chiral building blocks. The simplest of these methods is to dissect the target into very small chiral fragments that are then synthesized and joined at centers that are achiral. The second approach is to use the constrained pyranose or furanose form of a carbohydrate, manipulate the existing chirality, and then liberate the acyclic form. Examples illustrating both techniques are given in Section III.

Although not often considered, it is occasionally possible to use the biosynthetic precursor of a desired target as an educt for its chemical synthesis.

II. The Chiral Building Blocks

A. Sources of Chiral Carbon Fragments

Two major groups of chiral building blocks exist: those that can be purchased directly and those reported in the literature as having been prepared from the first group. In constructing the tabulation of such compounds that concludes this section, initial emphasis was placed on commercially available materials. To this end, catalogs of a number of supply houses were perused for listings of relevant compounds. This proved to be an exceedingly difficult task since the nomenclature employed was often a mixture of trivial and nonspecific names proceeded (in most cases) by equally confusing descriptors D, L, *d, l,* (+), or (−). In addition, structural formulas were usually missing or, when provided, were stereochemically ambiguous, thus making the task even more difficult. Three sources were particularly useful in converting the provided information to the data presented in the tabulation:

Klyne, W. and Buckingham, J. (1978). "Atlas of Stereochemistry," 2nd Ed. Oxford University Press, New York.

Buckingham, J. (ed.) (1982). "Dictionary of Organic Compounds," 5th Ed. Chapman and Hall, New York.

Jacques, J., Gros, C., and Bourcier, S. (1977). Absolute Configurations of 6000 Selected Compounds with One Asymmetric Carbon Atom. *In* "Stereochemistry, Fundamentals and Methods" (H. B. Kagan, ed.), Vol. 4. Thieme, Stuttgart.

These sources can be recommended as extremely reliable in providing leading references and accurate stereochemical information for many chiral compounds.

On the more positive side, several major suppliers have recognized the importance of chiral carbon fragments to the synthetic chemist and are making deliberate efforts to expand their offerings of these compounds.

In addition to those chiral building blocks that can be purchased directly, there is a substantial group of synthetically useful compounds derived by chemical modification of the purchased materials. Most of these compounds arose as the result of specific, target-oriented syntheses. Of more general significance, however, are recent efforts to provide a "manifold" of such compounds as an end in itself. Leading the way in this endeavor is Seebach, who has reviewed the method (*4*) and provided a number of specific examples (*35–40*). Many compounds arising from the work of Seebach and others are included in the tabulation.

In a review of asymmetric synthesis, we (*7*) made the following observations concerning use of optical rotations as an indication of enantiomeric purity:

> In this article, we use the term enantiomeric excess (e.e.) instead of the operational term optical purity. Many e.e. values we report actually were estimated from optical purity measurements and some have been "corrected" for the use of a chiral reagent supposed not to be a pure enantiomer. Estimation of the e.e. of a chiral substance from its optical purity may give erroneous results. Optical rotations may be very sensitive to experimental conditions and affected by unsuspected impurities. Attempts to remove impurities may lead to enantiomeric fractionation. Even if the rotation of crude product is correctly determined, the optical purity may still be incorrect because the optical rotation of [the] pure enantiomer may have been determined or reported incorrectly. Many of these problems are avoided when the e.e. of a crude reaction product is determined by a direct method which does not rely on optical rotation measurements. It is now possible to determine by direct methods the e.e. of many typical asymmetric synthesis products, including acids, alcohols, amines, amino acids, esters, etc. Using these methods, the mixture of enantiomers of the chiral substance to be analyzed is allowed to react with an excess of a second, enantiomerically pure substance. The resulting diastereomeric compounds or complexes are analyzed by N.M.R. or analytically separated by G.L.C. or H.P.L.C. Direct methods are less sensitive than optical rotations to adventitious impurities and hence are better suited to analysis of crude reaction products. These methods also require only the racemate for calibration and may be used on substrates not pre-

viously prepared in enantiomerically pure form. Examples of the use of direct methods for e.e. determination are given throughout this article. We think these methods should replace determinations of optical purities whenever possible. In our view, authors should report and editors should require the raw data on crude reaction products including isolation method, e.e. determination method, chemical yield, enantiomer [(*R*) or (*S*)] produced in excess and the enantiomeric excess.

Most of the sentiments expressed regarding asymmetric syntheses have a direct translation in the chemistry of chiral building blocks. Unfortunately, no supplier of these materials yet provides any information regarding the enantiomeric purity of his products. This situation exists even with compounds that have been long known to be mixtures of enantiomers, for example, the pinenes (*41*) and citronellal (*42*). What conclusion is one then to draw regarding the enantiomeric homogeneity of compounds represented by the following catalog entries?

(−)-Fenchone, ~97% (GC), $[\alpha]_D^{20}$ $-52 \pm 2°$
(+)-Fenchone, >98% (GC), $[\alpha]_D^{20}$ $+60 \pm 3°$
(+)-Limonene, 97%, $[\alpha]^{24}$ $+106°$
(−)-Limonene, 92%, $[\alpha]^{19}$ $-100°$

The problem is, regrettably, not limited to the terpenes. Ethyl (*S*)-lactate used in a synthesis of (*S*)-2-methyloxetane was found (*43*) to be only 96.6% ee.

One has the general feeling that commercial lots of sugars and amino acids are of high enantiomeric purity, that small molecules such as lactic acid or esters of 3-hydroxybutyric acid (*44*) are somewhat less reliably so, and that terpenes should always be viewed with suspicion. These, however, are only generalizations and *the only safe operating principle is to assume, until shown otherwise by a methodology other than optical rotation, that the chiral building blocks are* ***not*** *enantiomerically pure.*

The problems, real or potential, with the enantiomeric purity of commercial materials carry over to the chiral carbon synthons derived from them. Even at this late date, many authors continue to ignore this point and rely on optical rotation data alone as criteria for the enantiomeric excesses of starting materials and products. Since some of the reaction sequences involve at least the possibility of partial or even complete racemization, the italicized statement above is even more appropriate when applied to literature syntheses. All compounds in the tabulation must be viewed in this light. Some groups (*43, 45–51*) are aware of the problem, however, and provide experimental descriptions that could well serve as models.

B. Chiral Building Blocks

In choosing compounds for the tabulation, we were guided primarily by a desire to include as wide a variety of structural types as possible. Some of the compounds thus selected will perhaps be more familiar to the reader in another context, for example, as resolving agents or chiral auxiliaries for asymmetric syntheses. This in no way should mitigate their use as chiral building blocks.

Some compounds were deliberately excluded from the tabulation. Most simple derivatives or protected forms (esters of acids or amino acids and N-blocked amino acids) and compounds with more than eight carbon atoms (steroids, the quinine alkaloids, and peptides) fall into this category. However, many selectively protected sugars are included. The tabulation is intended to be indicative of the chiral building blocks available and not to be all-inclusive. With the ever widening range of such compounds, a literature and/or catalog search is certainly merited for any desired substance not found here.

The compounds are organized in the tabulation primarily on the basis of the length of the carbon chain containing the chiral center. Aromatic and heteroaromatic groups were ignored in determining chain length; thus, both alanine and phenylalanine are found in the three-carbon-chain group. A secondary form of organization is the nature of the substituent group at the chiral center, either heteroatom (O, N, S, halogen) or carbon. Within these subgroups, the compounds are further ordered by the number of asymmetric carbon atoms in the molecule and, finally, alphabetically. Some liberties have been taken, however, with this last ordering so that closely related compounds could be grouped together.

The following information is provided, as appropriate, for each compound:

1. *Compound number*
2. *Current systematic nomenclature*
3. *Empirical formula*
4. *Chemical structure with Cahn–Ingold–Prelog (R,S) descriptors for each chiral center.* Some of the commercial materials are sold as hydrates, salts, etc. Only the basic structure is indicated.
5. *Pricing information.* Pricing ranges (U.S. dollars per gram):

*****	<0.10
****	0.10–0.25
***	0.25–1.00
**	1.00–2.50
*	>2.50

The absence of a pricing code implies that the compound was not commercially available at the beginning of 1983. In cases of multiple suppliers, the lowest price is reported. Prices are given for the largest catalog unit; negotiation of bulk prices can often lower these numbers. In some cases, the price differential between the base compound and some of its derivatives can be substantial. For example, the esters of (*S*)-lactic acid are less than half the cost of the water-soluble parent.

6. *Catalog or literature name*. The inconsistencies and ambiguities of the names in common use are apparent to the most casual observer. In some cases we have appended several additional configurational descriptors commonly used for the compounds as an aid in their identification.
7. *Method of synthesis and literature citation*. This information is provided for all of the commercially unavailable materials and also for some of the most useful, yet most expensive, substances that can be bought. In some cases a yield determination was not possible. In others only modest assumptions were required to obtain an "estimated" yield. The citations are intended as leads and do not necessarily refer to the original preparation of a compound.

Chiral Building Blocks

Two-Carbon Chain: Heteroatom Substituted

1. (*R*)-4-[1-Hydroxy-2-(methylamino)ethyl]-1,2-benzenediol ($C_9H_{13}NO_3$)

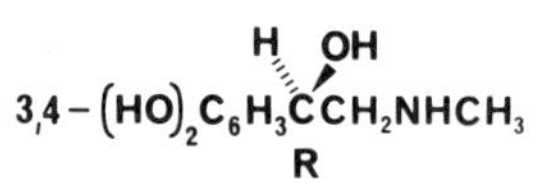

l-Epinephrine [L-(−)-adrenaline] **

2. (*R*)-4-[1-Hydroxy-2-[(1-methylethyl)amino]ethyl-1,2-benzenediol ($C_{11}H_{17}NO_3$)

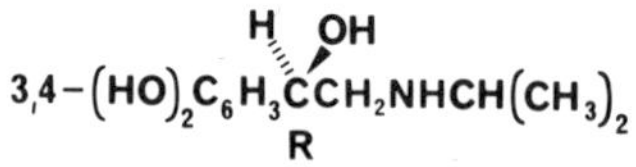

l-Isoproterenol *

3. (*S*)-4-[1-Hydroxy-2-[(1-methylethyl)amino]ethyl] 1,2 benzenediol ($C_{11}H_{17}NO_3$)

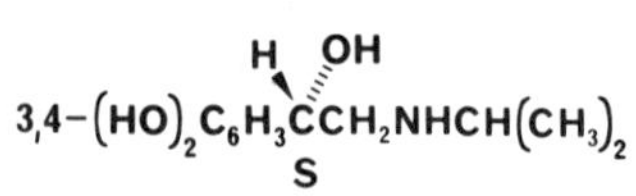

d-Isoproterenol *

4. (*R*)-α-Hydroxybenzeneacetic acid ($C_8H_8O_3$)

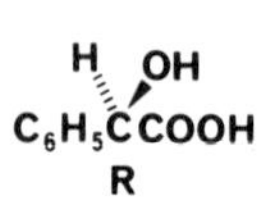

l-Mandelic acid [D-(−)] ***

5. (*S*)-α-Hydroxybenzeneacetic acid ($C_8H_8O_3$)

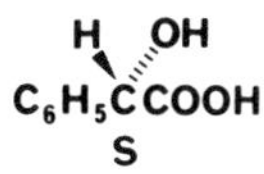

d-Mandelic acid [L-(+)] ***

6. (*R*)-4-(2-Amino-1-hydroxyethyl)-1,2-benzenediol ($C_8H_{11}NO_3$)

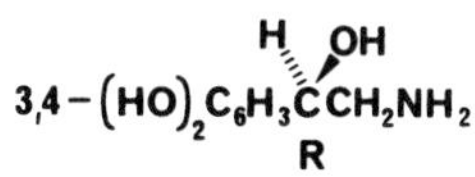

l-Norepinephrine [L-noradrenaline] *

7. (*R*)-α-Methylbenzenamine ($C_8H_{11}N$)

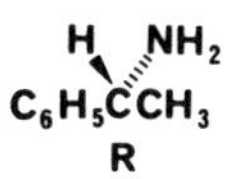

d-(+)-α-Methylbenzylamine ***

8. (*S*)-α-Methylbenzenamine ($C_8H_{11}N$)

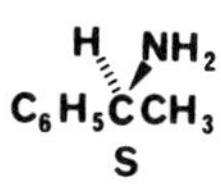

l-(−)-α-Methylbenzylamine ***

9. (*R*)-4-Hydroxy-α-[(methylamino)methyl]benzenemethanol ($C_9H_{13}NO_2$)

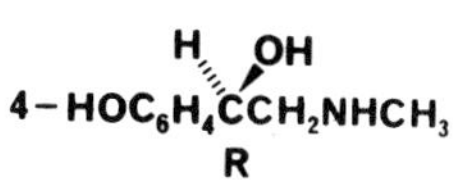

l-Phenylephrine **

10. (*R*)-α-Methylbenzenemethanol ($C_8H_{10}O$)

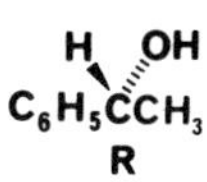

(*R*)-(+)-*sec*-Phenethyl alcohol *

11. (*S*)-α-Methylbenzenemethanol ($C_8H_{10}O$)

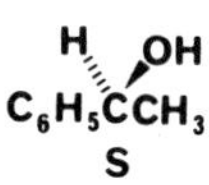

(*S*)-(−)-*sec*-Phenethyl alcohol *

12. (*R*)-α-Aminobenzeneacetic acid ($C_8H_9NO_2$)

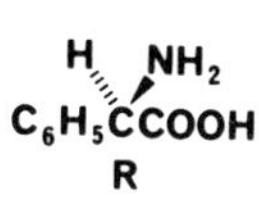

D-(−)-α-Phenylglycine ****

13. (*S*)-α-Aminobenzeneacetic acid ($C_8H_9NO_2$)

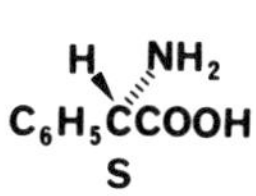

L-(+)-Phenylglycine **

Three-Carbon Chain: Heteroatom Substituted

14. (*R*)-2-Aminopropanoic acid ($C_3H_7NO_2$)

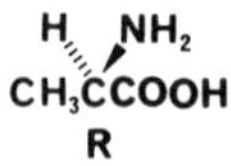

D-(−)-Alanine **

15. (*S*)-2-Aminopropanoic acid ($C_3H_7NO_2$)

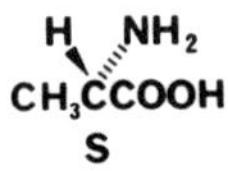

L-(+)-Alanine ****

16. (*S*)-*β*-Aminobenzenepropanol ($C_9H_{13}NO$)

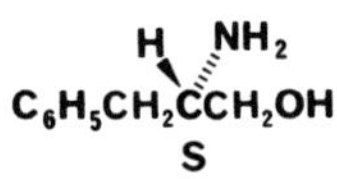

L-2-Amino-3-phenyl-1-propanol *

17. (*R*)-1-Amino-2-propanol (C_3H_9NO)

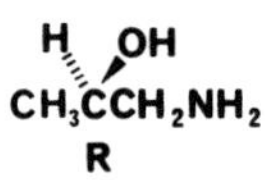

(*R*)-(−)-1-Amino-2-propanol *

18. (*S*)-1-Amino-2-propanol (C_3H_9NO)

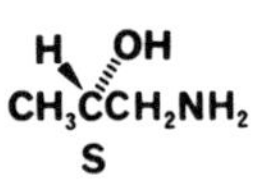

(*S*)-(+)-1-Amino-2-propanol *

19. (*S*)-2-Amino-1-propanol (C_3H_9NO)

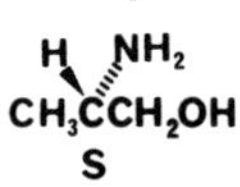

L-2-Amino-1-propanol
[L-(+)-alaninol] *

20. (*S*)-*α*-Methylbenzeneethanamine ($C_9H_{13}N$)

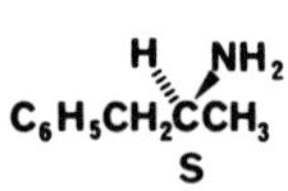

D-Amphetamine ***

21. (*R*)-[(Phenylmethoxy)methyl]oxirane ($C_{10}H_{12}O_2$)

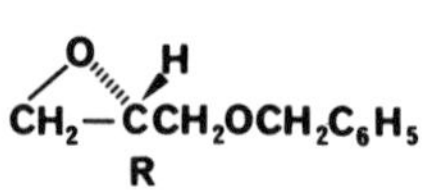

(*R*)-Benzyl-2,3-epoxypropyl ether

8 steps (~12%) ex D-mannitol (*52*)
4 steps (51%) ex (*R*)-1-benzylglycerol (*554*)

22. (*S*)-[(Phenylmethoxy)methyl]oxirane ($C_{10}H_{12}O_2$)

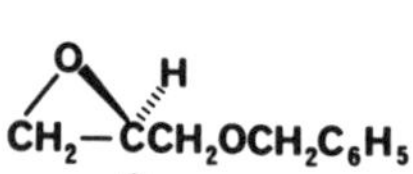

(*S*)-Benzyl-2,3-epoxypropyl ether

6 steps (51%) ex D-mannitol bisacetonide (*53*)
7 steps (~12%) ex D-mannitol (*52*)

23. (*R*)-3-(Phenylmethoxy)-1,2-propanediol ($C_{10}H_{14}O_3$)

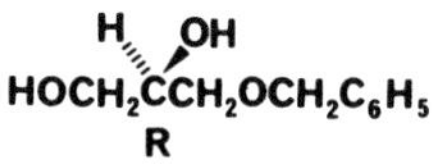

L-α-*O*-Benzylglycerol *
[3-*O*-benzyl-*sn*-glycerol]

3 steps [67% (*54*); 68% (*53*)] ex D-mannitol bisacetonide

24. (*S*)-3-(Phenylmethoxy)-1,2-propanediol ($C_{10}H_{14}O_3$)

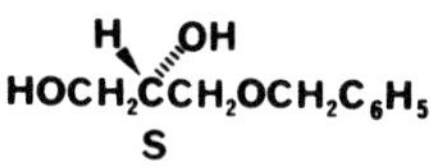

1-*O*-Benzyl-*sn*-glycerol *

6 steps (21%) ex D-mannitol bisacetonide (*55*)
3 steps (83%) ex (*R*)-1-benzylglycerol (*554*)

25. (*S*)-2,2-Dimethyl-4-[(phenylmethoxy)methyl]-1,3-dioxolane ($C_{13}H_{18}O_3$)

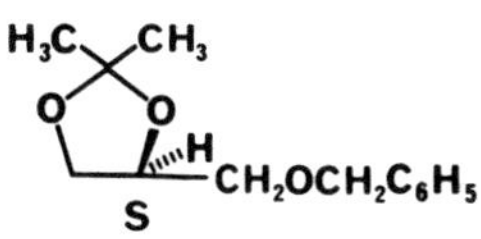

1,2-Isopropylidene-3-*O*-benzyl-*sn*-glycerol *

26. (*R*)-2,2-Dimethyl-1,3-dioxolane-4-methanol ($C_6H_{12}O_3$)

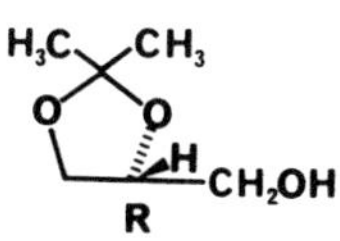

(*R*)-(−)-2,2-Dimethyl-1,3-dioxolane-4-methanol *
[L-2,3-isopropylidene-*sn*-glycerol]

4 steps (46%) ex L-serine (*56*)
5 steps (40–50%) ex ascorbic acid (*57, 58*)

27. (*S*)-2,2-Dimethyl-1,3-dioxolane-4-methanol ($C_6H_{12}O_3$)

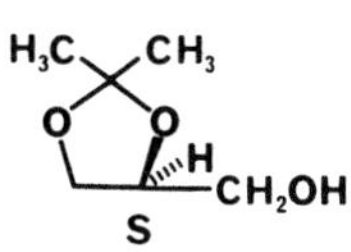

(*S*)-(+)-2,2-Dimethyl-1,3-dioxolane-4-methanol *
[L-1,2-isopropylidene-*sn*-glycerol]

3 steps (≥40%) ex D-mannitol (*45, 53, 59, 60*)
4 steps ex D-arabinose (*61*)
4 steps ex D-serine (*56*)

28. (*R*)-2,2-Dimethyl-1,3-dioxolane-4-methanol-4-methylbenzenesulfonate ($C_{13}H_{18}O_5S$)

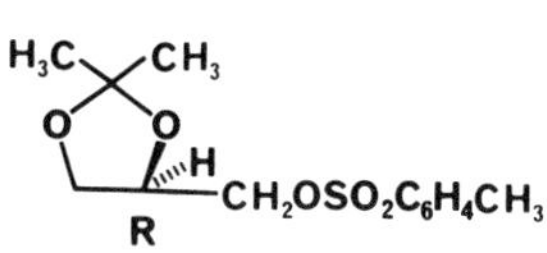

(2*R*)-3-Tosyloxy-1,2-propanediol acetonide

4 steps (30%) ex D-mannitol (*62*)

29. (*S*)-2,2-Dimethyl-1,3-dioxolane-4-methanol-4-methylbenzenesulfonate ($C_{13}H_{18}O_5S$)

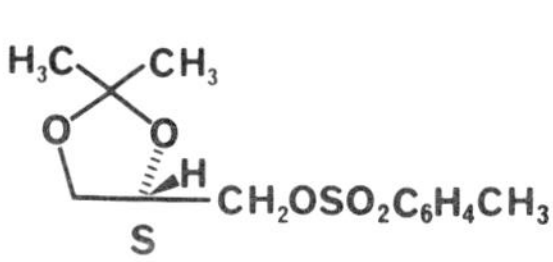

(2*S*)-3-Tosyloxy-1,2-propanediol acetonide

8 steps (4%) ex D-mannitol (*62*)

30. (*R*)-2-(Phenylmethoxy)propanal ($C_{10}H_{12}O_2$)

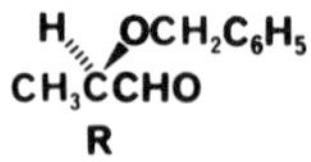

(2*R*)-2-(Benzyloxy)propanal

6 steps (~11%) ex D-mannose (*63*)

31. (*S*)-2-(Phenylmethoxy)propanal ($C_{10}H_{12}O_2$)

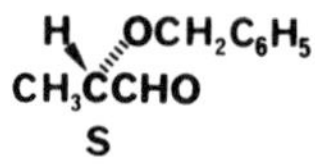

(2*S*)-2-(Benzyloxy)propanal

5 steps (~32%) ex L-rhamnose (*63*)

32. (*R*)-4-Amino-3-isoxazolidinone ($C_3H_6N_2O_2$)

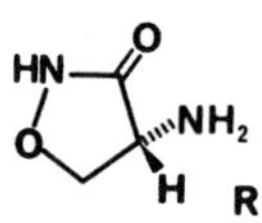

D-Cycloserine **

33. (*S*)-4-Amino-3-isoxazolidinone ($C_3H_6N_2O_2$)

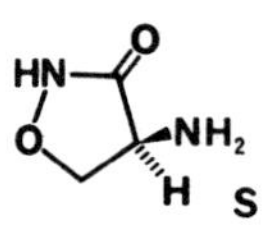

L-Cycloserine *

34. (*R*)-2-Amino-3-mercaptopropanoic acid ($C_3H_7NO_2S$)

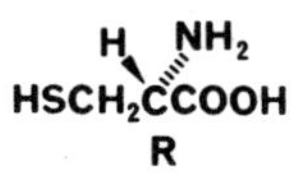

L-(+)-Cysteine *****

35. (*S*)-2-Amino-3-mercaptopropanoic acid ($C_3H_7NO_2S$)

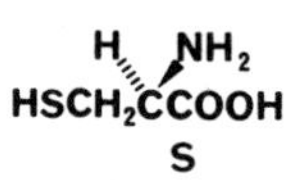

D-(−)-Cysteine *

36. [*R*-(*R**,*R**)]-3,3′-Dithiobis(2-aminopropanoic acid) ($C_6H_{12}N_2O_4S_2$)

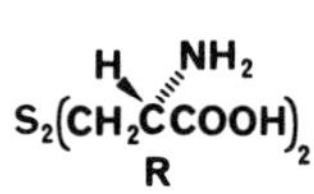

L-Cystine *****

37. [*S*-(*R**,*R**)]-3,3′-Dithiobis(2-aminopropanoic acid) ($C_6H_{12}N_2O_4S_2$)

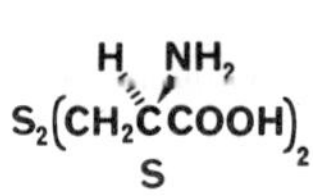

D-Cystine *

38. (*R*)-Chloromethyloxirane (C_3H_5ClO)

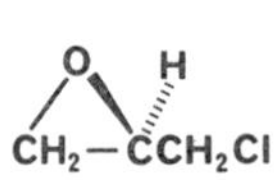

(*R*)-Epichlorohydrin

6 steps (50%) ex D-mannitol bisacetonide (*45*)

39. (*S*)-Chloromethyloxirane (C_3H_5ClO)

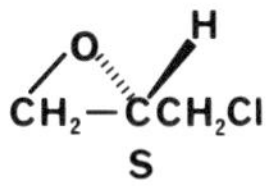

(*S*)-Epichlorohydrin

4 steps (56%) ex D-mannitol bisacetonide (*45*)

40. (*R*)-2,3-Dihydroxypropanal ($C_3H_6O_3$)

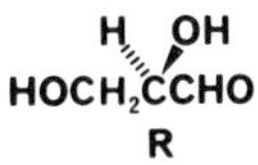

D-(+)-Glyceraldehyde *

2 steps (86%) ex D-mannitol bisacetonide (*64*)

41. (*S*)-2,3-Dihydroxypropanal ($C_3H_6O_3$)

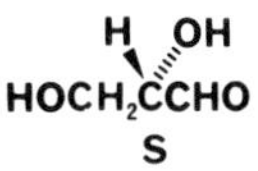

L-(−)-Glyceraldehyde ["unnatural"] *

42. (*R*)-2,3-Dihydroxypropanoic acid ($C_3H_6O_4$)

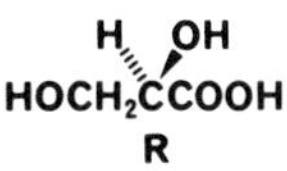

D-(+)-Glyceric acid *

43. (*S*)-2,3-Dihydroxypropanoic acid ($C_3H_6O_4$)

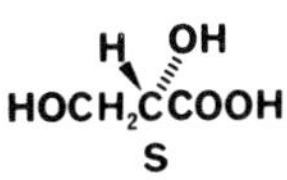

L-(−)-Glyceric acid *

44. (*R*)-α-Amino-1*H*-imidazole-4-propanoic acid ($C_6H_9N_3O_2$)

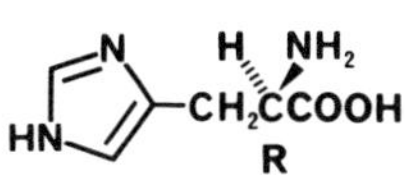

D-Histidine **

45. (*S*)-α-Amino-1*H*-imidazole-4-propanoic acid ($C_6H_9N_3O_2$)

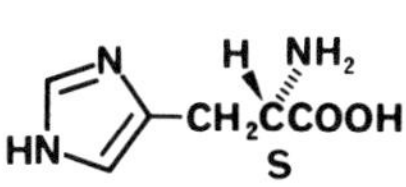

L-Histidine ****

46. (*R*)-2-Hydroxypropanoic acid ($C_3H_6O_3$)

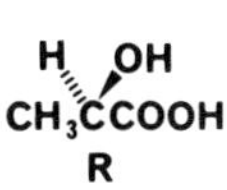

D-(−)-Lactic acid *

47. (*S*)-2-Hydroxypropanoic acid ($C_3H_6O_3$)

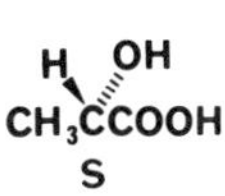

L-(+)-Lactic acid *****

48. (*S*)-2-Chloropropanoic acid methyl ester ($C_4H_7ClO_2$)

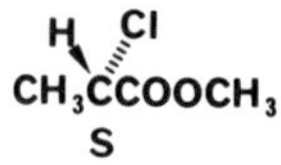

(*S*)-(−)-Methyl 2-chloropropionate *

49. (*R*)-2-Methylaziridine (C_3H_7N)

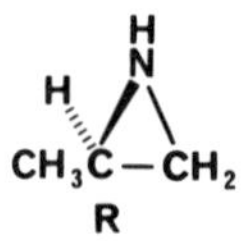

(*R*)-2-Methylethyleneimine

1 step (72%) ex (*S*)-1-amino-2-propanol (*65*)
3 steps ex (*R*)-alanine (*66*)

50. (*S*)-2-Methylaziridine (C_3H_7N)

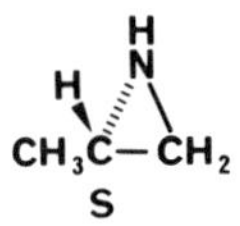

(*S*)-2-Methylethyleneimine

1 step (67%) ex (*R*)-1-amino-2-propanol (*65*)
3 steps ex (*S*)-alanine (*66*)

51. (*R*)-2,2-Dimethyl-1,3-dioxolane-4-carboxylic acid methyl ester ($C_7H_{12}O_4$)

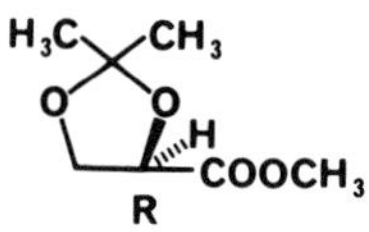

(+)-Methyl *O*-isopropylidene-D-glycerate

3 steps ex D-serine (*56*)

52. (*S*)-2,2-Dimethyl-1,3-dioxolane-4-carboxylic acid methyl ester ($C_7H_{12}O_4$)

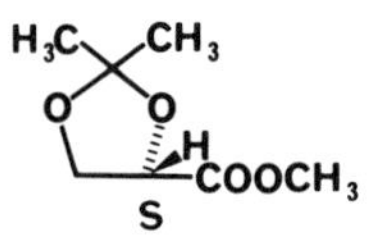

(−)-Methyl *O*-isopropylidene-L-glycerate

3 steps (57%) ex L-serine (*56*)

53. (*S*)-Thiiranecarboxylic acid methyl ester ($C_4H_6O_2S$)

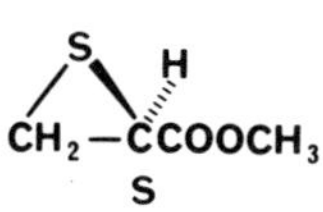

(*S*)-Methyl thiiranecarboxylate

1 step (47%) ex methyl (*R*)-cysteinate (*67*)

54. (*R*)-α-Aminobenzenepropanoic acid ($C_9H_{11}NO_2$)

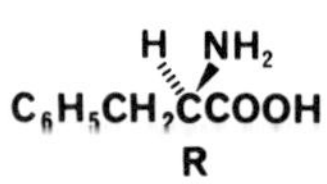

D-(+)-Phenylalanine ****

55. (*S*)-α-Aminobenzenepropanoic acid ($C_9H_{11}NO_2$)

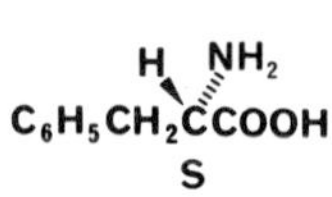

L-(−)-Phenylalanine ****

56. (*S*)-α-Hydroxybenzenepropanoic acid ($C_9H_{10}O_3$)

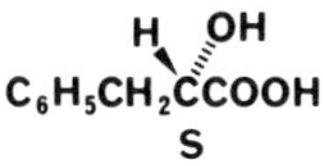

L-(−)-3-Phenyllactic acid *

57. (*S*)-1,2-Propanediol ($C_3H_8O_2$)

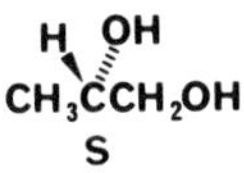

(*S*)-1,2-Propanediol *

58. (*S*)-1,2-Propanediol-1-(4-methylbenzenesulfonate) ($C_{10}H_{14}O_4S$)

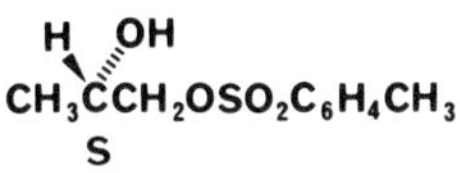

(*S*)-1-Tosyloxypropan-2-ol

8 steps (27%) ex L-rhamnose (*68*)

59. (*R*)-1,2-Propanediol-1-(4-methylbenzenesulfonate) ($C_{10}H_{14}O_4S$)

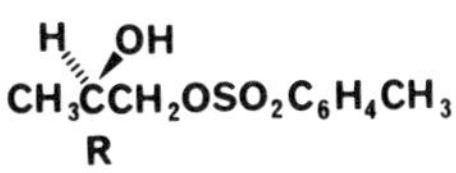

(*R*)-1-Tosyloxypropan-2-ol

9 steps (~25%) ex D-mannitol (*68*)

60. (*R*)-2-Methyloxirane (C_3H_6O)

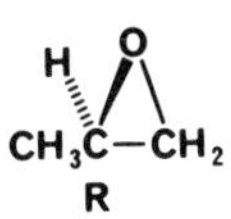

(*R*)-(+)-Propylene oxide *

3 steps [71% (*69*); 50% (*70*)] ex ethyl (*S*)-lactate

3 steps (30%) ex (*S*)-alanine (*50*); general method for (*R*)-oxiranes ex (*S*)-amino acids (*51*)

61. (*S*)-2-Methyloxirane (C_3H_6O)

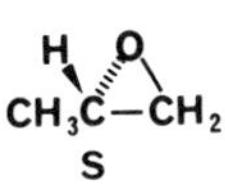

(*S*)-(−)-Propylene oxide *

3 steps (54%) (*71*); 4 steps (46%) (*38*); 5 steps (44%) (*72*) ex ethyl (*S*)-lactate (*50*)

62. (*R*)-2-Amino-3-hydroxypropanoic acid ($C_3H_7NO_3$)

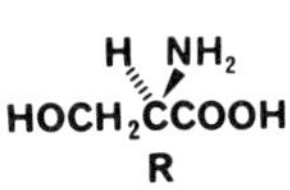

D-Serine **

63. (*S*)-2-Amino-3-hydroxypropanoic acid ($C_3H_7NO_3$)

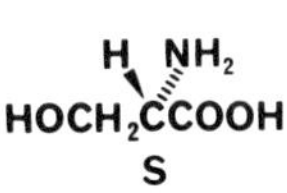

L-Serine ****

64. (*R*)-Thiazolidine-4-carboxylic acid ($C_4H_7NO_2S$)

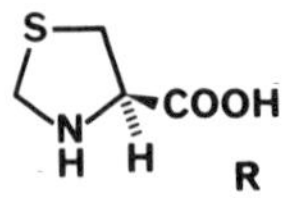

L-Thiazolidine-4-carboxylic acid [thioproline] ***

65. (*R*)-α-Amino-1*H*-indole-3-propanoic acid ($C_{11}H_{12}N_2O_2$)

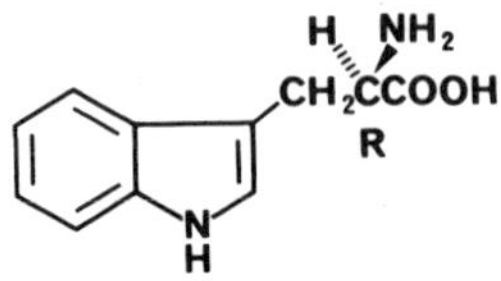

D-(+)-Tryptophan ***

66. (*S*)-α-Amino-1*H*-indole-3-propanoic acid ($C_{11}H_{12}N_2O_2$)

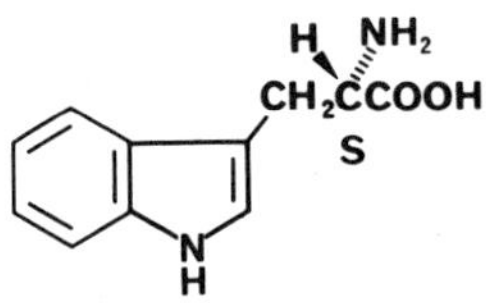

L-(−)-Tryptophan ****

67. (*R*)-α-Amino-4-hydroxybenzenepropanoic acid ($C_9H_{11}NO_3$)

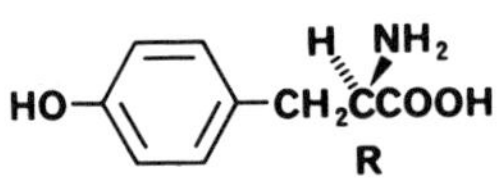

D-(+)-Tyrosine *

68. (*S*)-α-Amino-4-hydroxybenzenepropanoic acid ($C_9H_{11}NO_3$)

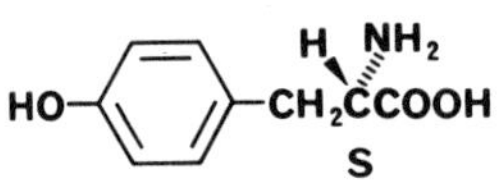

L-(−)-Tyrosine *****

69. (4*S*)-*cis*-2,2-Dimethyl-4-phenyl-1,3-dioxan-5-amine ($C_{12}H_{17}NO_2$)

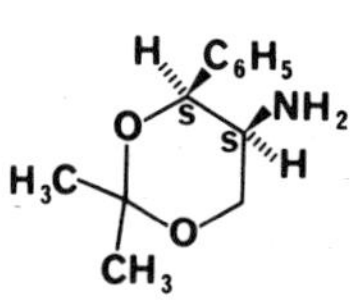

(4*S*,5*S*)-(+)-5-Amino-2,2-dimethyl-4-phenyl-1,3-dioxane ****

70. [*S*-(*R**,*R**)]-2-Amino-1-phenyl-1,3-propanediol ($C_9H_{13}NO_2$)

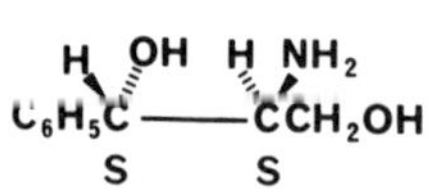

(+)-2-Amino-1-phenyl-1,3-propanediol [L-(+)-threo] ****

71. [*S*-(*R**,*S**)]-α-[1-(Methylamino)ethyl]benzenemethanol ($C_{10}H_{15}NO$)

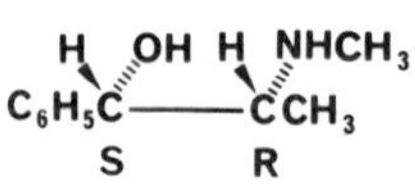

d-Ephedrine [D-(+)] ****

72. [*R*-(*R**,*S**)]-α-[1-(Methylamino)ethyl]benzenemethanol ($C_{10}H_{15}NO$)

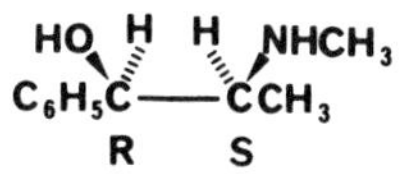

l-Ephedrine [L-(−)] ***

73. [*R*-(*R**,*S**)]-α-[1-(Dimethylamino)ethyl]benzenemethanol ($C_{11}H_{17}NO$)

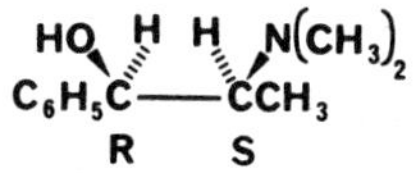

(−)-*N*-Methylephedrine **

74. [*S*-(*R**,*S**)]-α-(1-Aminoethyl)benzenemethanol ($C_9H_{13}NO$)

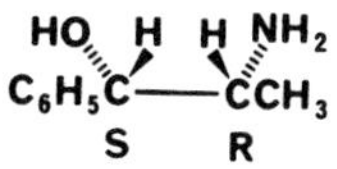

(+)-Norephedrine ***

75. [*R*-(*R**,*R**)]-α-(1-Aminoethyl)benzenemethanol ($C_9H_{13}NO$)

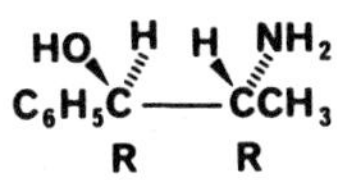

(−)-Norpseudoephedrine ***

76. [*S*-(*R**,*R**)]-α-[1-(Methylamino)ethyl]benzenemethanol ($C_{10}H_{15}NO$)

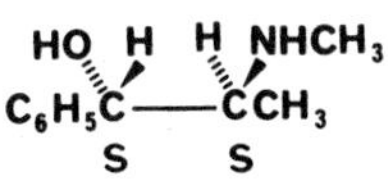

D-(+)-Pseudoephedrine ****

77. [*R*-(*R**,*R**)]-α-[1-Methylamino)ethyl]benzenemethanol ($C_{10}H_{15}NO$)

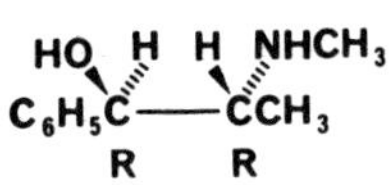

L-(−)-Pseudoephedrine ****

78. (4*S*)-*trans*-4,5-Dihydro-2-methyl-5-phenyl-4-oxazolemethanol ($C_{11}H_{13}NO_2$)

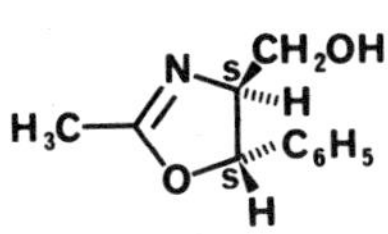

(4*S*,5*S*)-(−)-2-Methyl-5-phenyl-2-oxazoline-4-methanol **

Three-Carbon Chain: Carbon Substituted

79. (*R*)-3-(Acetylthio)-2-methylpropanoic acid ($C_6H_{10}O_3S$)

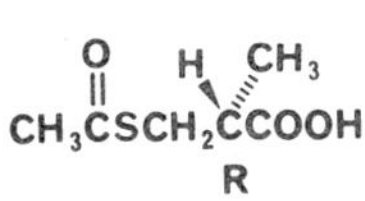

L-(+)-*S*-Acetyl-β-mercaptoisobutyric acid *

80. (*S*)-3-(Acetylthio)-2-methylpropanoic acid ($C_6H_{10}O_3S$)

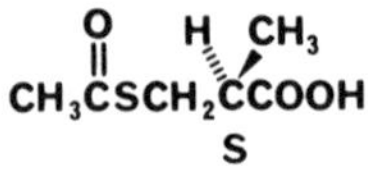

D-(−)-*S*-Acetyl-*β*-mercaptoisobutyric acid *

81. (*R*)-1-Bromo-3-(1,1-dimethylethoxy)-2-methylpropane ($C_8H_{17}BrO$)

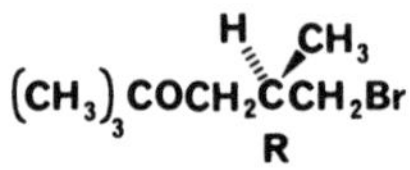

(*R*)-(−)-3-*tert*-Butoxy-2-methyl-1-bromopropane

5 steps (~35%) ex (*S*)-3-hydroxy-2-methylpropanoic acid (*33*)

82. (*S*)-1-Bromo-3-(1,1-dimethylethoxy)-2-methylpropane ($C_8H_{17}BrO$)

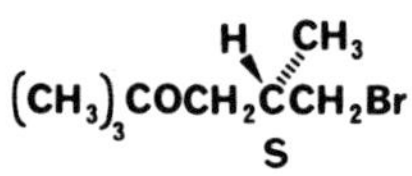

(*S*)-(+)-3-*tert*-Butoxy-2-methyl-1-bromopropane

3 steps (50%) ex (*S*)-3-hydroxy-2-methylpropanoic acid

83. (*R*)-3-(1,1-Dimethylethoxy)-2-methyl-1-propanol ($C_8H_{18}O_2$)

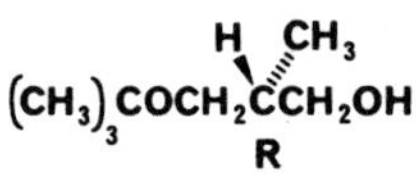

(*R*)-(+)-3-*tert*-Butoxy-2-methyl-1-propanol

2 steps (71%) ex (*S*)-3-hydroxy-2-methylpropanoic acid (*34*)

84. (*S*)-3-(1,1-Dimethylethoxy)-2-methyl-1-propanol ($C_8H_{18}O_2$)

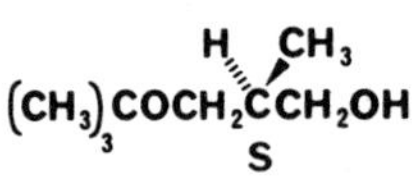

(*S*)-(−)-3-*tert*-Butoxy-2-methyl-1-propanol

4 steps (49%) ex (*S*)-3-hydroxy-2-methylpropanoic acid (*34*)

85. (*S*)-3-Hydroxy-2-methylpropanoic acid ($C_4H_8O_3$)

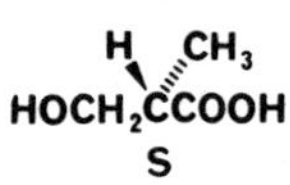

L-*β*-Hydroxyisobutyric acid *

86. (*R*)-*α*-Methylbenzeneacetic acid ($C_9H_{10}O_2$)

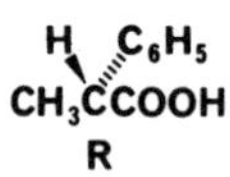

(−)-2-Phenylpropionic acid [hydratropic acid] *

87. (*S*)-*α*-Methylbenzeneacetic acid ($C_9H_{10}O_2$)

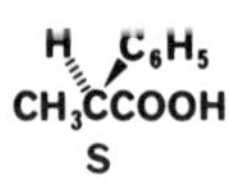

(+)-2-Phenylpropionic acid [hydratropic acid] *

88. [*S*-(*R**,*S**)]-*α*-[1-[(Dimethylamino)methyl]ethyl]-*α*-phenylbenzeneethanol ($C_{19}H_{25}NO$)

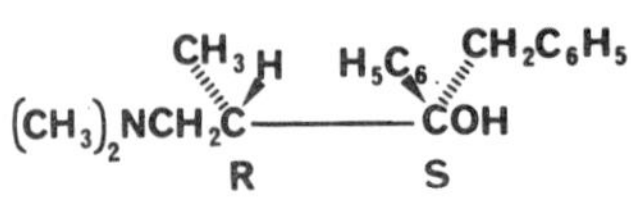

(2*S*,3*R*)-(+)-4-Dimethylamino-1,2-diphenyl-3-methyl-2-butanol ***

FOUR-CARBON CHAIN: HETEROATOM SUBSTITUTED

89. (*R*)-2-Butanamine ($C_4H_{11}N$)

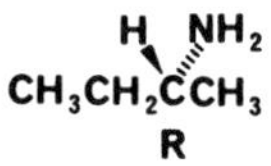

(−)-2-Aminobutane *

90. (*S*)-2-Butanamine ($C_4H_{11}N$)

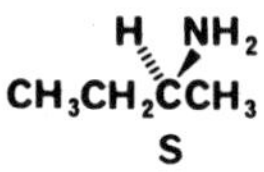

(+)-2-Aminobutane *

91. (*R*)-2-Amino-1-butanol ($C_4H_{11}NO$)

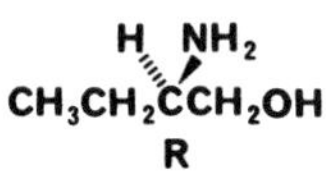

L-(−)-2-Amino-1-butanol ****

92. (*S*)-2-Amino-1-butanol ($C_4H_{11}NO$)

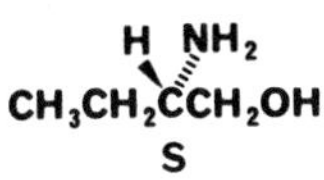

D-(+)-2-Amino-1-butanol *

93. (*R*)-2-Aminobutanoic acid ($C_4H_9NO_2$)

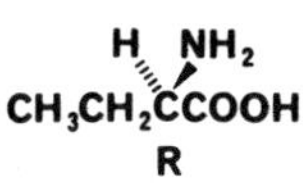

D-(−)-2-Aminobutyric acid *

94. (*S*)-2-Aminobutanoic acid ($C_4H_9NO_2$)

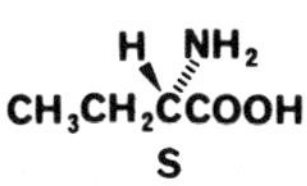

L-(+)-2-Aminobutyric acid *

95. (*R*)-α-Aminobenzenebutanoic acid ($C_{10}H_{13}NO_2$)

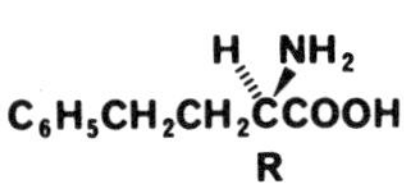

(+)-α-Amino-4-phenylbutyric acid *

96. (*S*)-α-Aminobenzenebutanoic acid ($C_{10}H_{18}NO_2$)

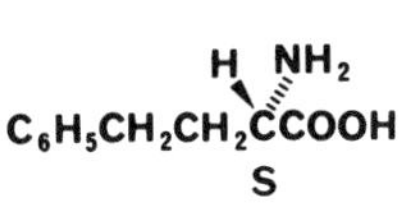

(−)-α-Amino-4-phenylbutyric acid *

97. (*R*)-2,4-Diamino-4-oxobutanoic acid ($C_4H_8N_2O_3$)

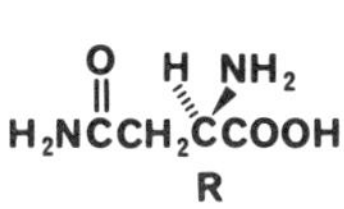

D-Asparagine ****

98. (*S*)-2,4-Diamino-4-oxobutanoic acid ($C_4H_8N_2O_3$)

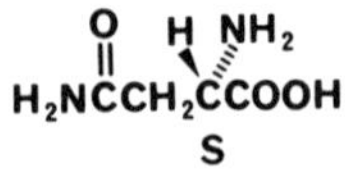

L-Asparagine *****

99. (*R*)-α-Aminobutanedioic acid ($C_4H_7NO_4$)

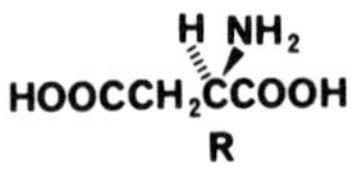

D-(−)-Aspartic acid ***

100. (*S*)-α-Aminobutanedioic acid ($C_4H_7NO_4$)

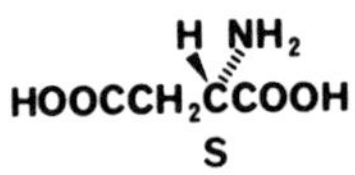

L-(+)-Aspartic acid *****

101. (*S*)-2-Azetidinecarboxylic acid ($C_4H_7NO_2$)

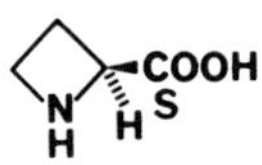

L-2-Azetidinecarboxylic acid *

102. (*S*)-Dihydro-4-(phenylmethoxy)-2(3*H*)-furanone ($C_{11}H_{10}O_3$)

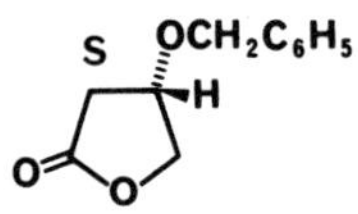

(*S*)-3-Benzyloxybutyrolactone

4 steps (51%) ex diethyl (*S*)-malate (*74*)

103. (*R*)-(2-Bromoethyl)oxirane (C_4H_7BrO)

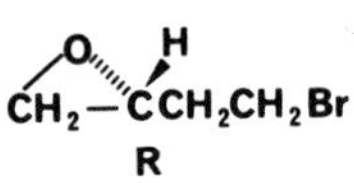

(*R*)-4-Bromo-1,2-epoxybutane

9 steps (20%) ex dimethyl (2*R*,3*R*)-tartrate (*37*)

104. (*S*)-(2-Bromoethyl)oxirane ($C_4H_{17}BrO$)

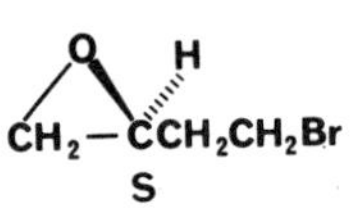

(*S*)-4-Bromo-1,2-epoxybutane

5 steps (56%) (*75*); 6 steps (57%) (*38*) ex diethyl (*S*)-malate

105. (*R*)-(2-Iodoethyl)oxirane (C_4H_7IO)

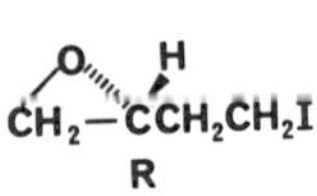

(*R*)-(+)-4-Iodo-1,2-epoxybutane

7 steps (32%) ex (*S*)-malic acid (*76*)

106. (*S*)-(2-Iodoethyl)oxirane (C_4H_7IO)

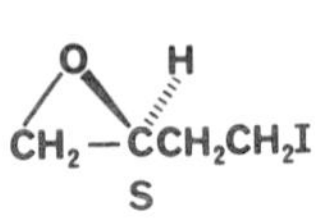

(*S*)-(−)-4-Iodo-1,2-epoxybutane

8 steps (20%) ex (*S*)-malic acid (*76*)

107. (*R*)-1,3-Butanediol ($C_4H_{10}O_2$)

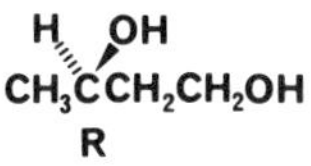

(*R*)-(−)-1,3-Butanediol *

7 steps (29%) ex diethyl (*S*)-malate (*37*)
5 steps (28%) ex (*S*)-malic acid (*77*)

108. (*S*)-1,3-Butanediol ($C_4H_{10}O_2$)

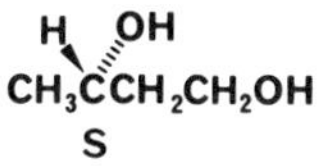

(*S*)-(+)-1,3-Butanediol *

1 step (37%) ex methyl (*S*)-3-hydroxybutyrate (*78*)

109. (*R*)-Butane-1,2,4-triol ($C_4H_{10}O_3$)

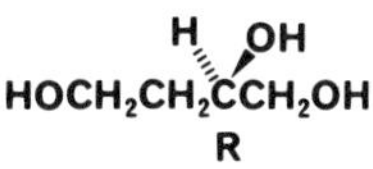

(*R*)-Butane-1,2,4-triol

4 steps (69%) ex (*R*)-malic acid (*79*)
7 steps (~31%) ex dimethyl (2*R*,3*R*)-tartrate (*37*)

110. (*S*)-Butane-1,2,4-triol ($C_4H_{10}O_3$)

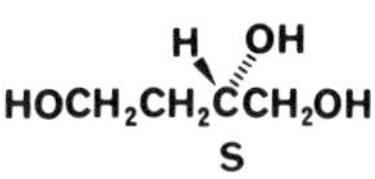

(*S*)-Butane-1,2,4-triol

2 steps (58%) ex (*S*)-malic acid (*77*)
3 steps (66%) ex diethyl (*S*)-malate (*37*)

111. (*R*)-2,2-Dimethyl-1,3-dioxolane-4-ethanol ($C_7H_{14}O_3$)

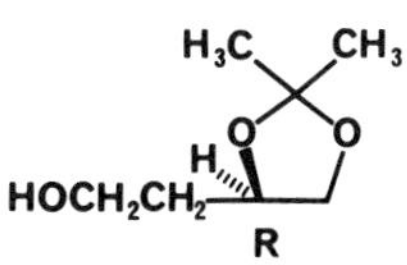

(R)-(+)-1,2-Isopropylidenebutane-1,2,4-triol

5 steps (53%) ex (*R*)-malic acid (*79*)
[see ref. *80* re chemical purity]

112. (*S*)-2,2-Dimethyl-1,3-dioxolane-4-ethanol ($C_7H_{14}O_3$)

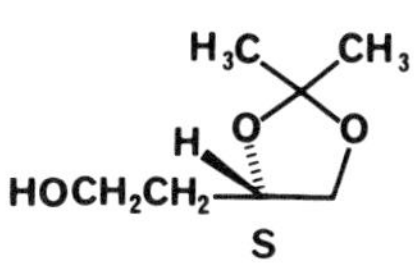

(*S*)-(−)-1,2-Isopropylidenebutane-1,2,4-triol

3 steps (76%) ex (*S*)-malic acid (*81, 82*)
[see ref. *80* re chemical purity]

113. (*R*)-2-Butanol ($C_4H_{10}O$)

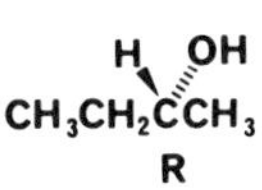

(*R*)-(−)-2-Butanol *

114. (*S*)-2-Butanol ($C_4H_{10}O$)

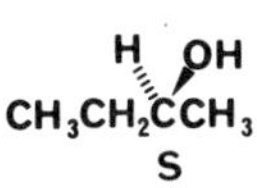

(*S*)-(+)-2-Butanol *

115. (*R*)-2-Chlorobutanedioic acid ($C_4H_5ClO_4$)

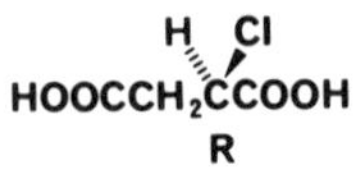

D-2-Chlorosuccinic acid *

116. (*S*)-2-Chlorobutanedioic acid ($C_4H_5ClO_4$)

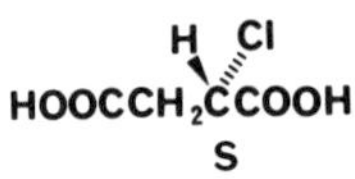

L-2-Chlorosuccinic acid *

117. (*S*)-2,4-Diaminobutanoic acid ($C_4H_{10}N_2O_2$)

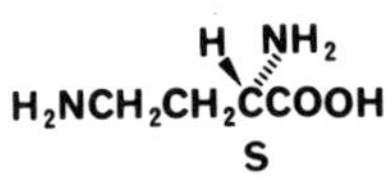

L-2,4-Diaminobutyric acid *

118. (*R*)-Ethyloxirane (C_4H_8O)

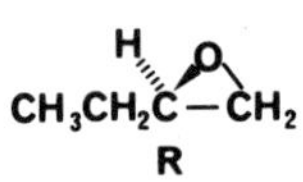

(*R*)-(+)-1,2-Epoxybutane

7 steps (6%) ex (*R*)-α-amino-*n*-butyric acid (*83*)

119. (*S*)-Ethyloxirane (C_4H_8O)

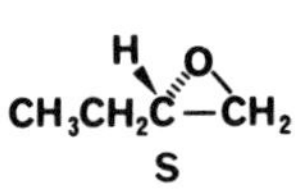

(*S*)-(−)-1,2-Epoxybutane

6 steps (51%) ex diethyl (*S*)-malate (*38*)
7 steps (9%) ex (*S*)-α-amino-*n*-butyric acid (*83*)
6 steps ex D-mannitol (*84*); general method for homologs

120. (*S*)-3-Hydroxybutanoic acid ethyl ester ($C_6H_{12}O_3$)

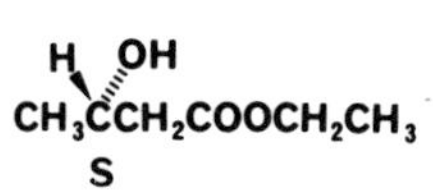

(*S*)-(+)-Ethyl 3-hydroxybutanoate

5 steps (18%) ex ethyl (*S*)-lactate (*43*)

121. (*S*)-3-Hydroxybutanoic acid methyl ester ($C_5H_{10}O_3$)

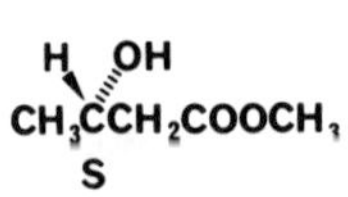

(*S*)-(+)-Methyl 3-hydroxybutyrate *

122. (*R*)-3-Hydroxybutanoic acid methyl ester ($C_5H_{10}O_3$)

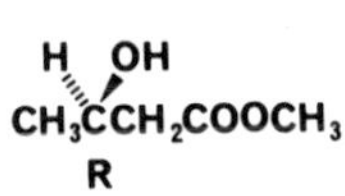

(*R*)-(−)-Methyl 3-hydroxybutyrate *

123. (*R*)-Dihydro-4-hydroxy-2(3*H*)-furanone ($C_4H_6O_3$)

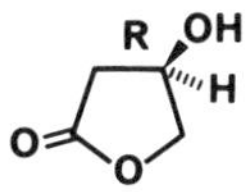

(*R*)-(+)-3-Hydroxybutyrolactone

7 steps (19%) ex (*R*)-malic acid (*79*)

124. (*S*)-Dihydro-4-hydroxy-2(3*H*)-furanone ($C_4H_6O_3$)

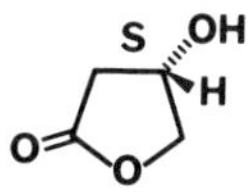

(*S*)-(−)-3-Hydroxybutyrolactone

4 steps (23%) ex (*S*)-malic acid (*85*)

125. (*S*)-Dihydro-3-[(phenylmethoxy)methoxy]-2(3*H*)-furanone ($C_{12}H_{14}O_4$)

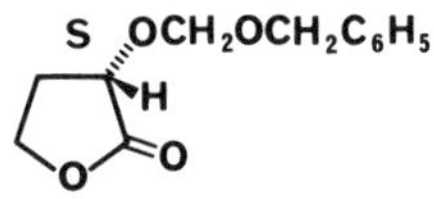

(*S*)-2-(Benzyloxy)methoxybutyrolactone

3 steps (63%) ex (*S*)-malic acid (*86*)

126. (*R*)-Hydroxybutanedioic acid ($C_4H_6O_5$)

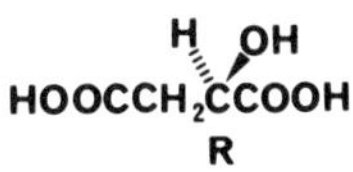

D-(+)-Malic acid *

3 steps (56%) (*87*); 4 steps (47%) (*37*) ex diethyl (2*R*,3*R*)-tartrate

127. (*S*)-Hydroxybutanedioic acid ($C_4H_6O_5$)

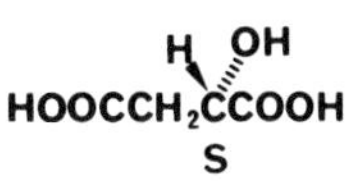

L-(−)-Malic acid *****

128. (*R*)-2-Amino-4-(methylthio)butanoic acid ($C_5H_{11}NO_2S$)

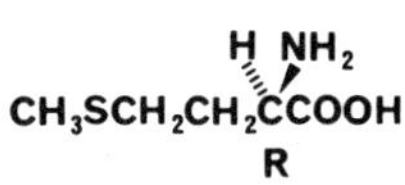

D-Methionine ***

129. (*S*)-2-Amino-4-(methylthio)butanoic acid ($C_5H_{11}NO_2S$)

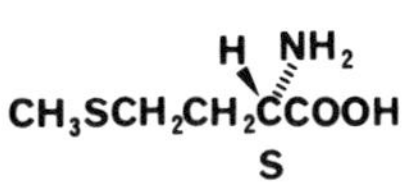

L-Methionine *****

130. (*R*)-3-Methyl-2-butanol ($C_5H_{12}O$)

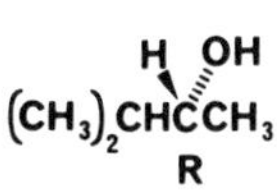

(−)-3-Methyl-2-butanol *

131. (*S*)-3-Methyl-2-butanol ($C_5H_{12}O$)

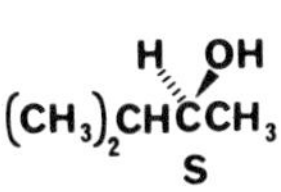

(+)-3-Methyl-2-butanol *

132. (*S*)-3-(1-Methyl-2-pyrrolidinyl)pyridine ($C_{10}H_{14}N_2$)

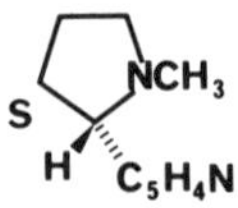

l-Nicotine ****

133. (*R*)-Dihydro-3-hydroxy-4,4-dimethyl-2(3*H*)-furanone ($C_6H_{10}O_3$)

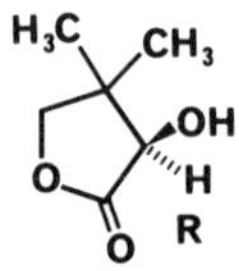

D-(−)-Pantolactone ***

134. (*R*)-2-Amino-3-methylbutanoic acid ($C_5H_{11}NO_2$)

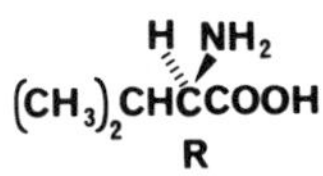

D-Valine ***

135. (*S*)-2-Amino-3-methylbutanoic acid ($C_5H_{11}NO_2$)

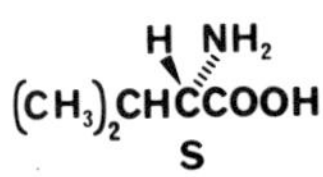

L-Valine *****

136. [*R*-(*R**,*R**)]-2,2-Dimethyl-β-(phenylmethoxy)-1,3-dioxolane-4-ethanol ($C_{14}H_{20}O_4$)

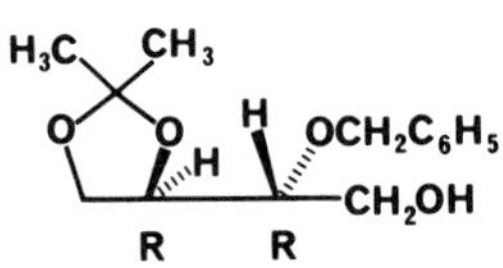

2-*O*-Benzyl-3,4-*O*-isopropylidene-D-threitol

3 steps (68%) ex (2*S*,3*S*)-tartaric acid (*88*)

137. [*S*-(*R**,*R**)]-2,2-Dimethyl-β-(phenylmethoxy)-1,3-dioxolane-4-ethanol ($C_{14}H_{20}O_4$)

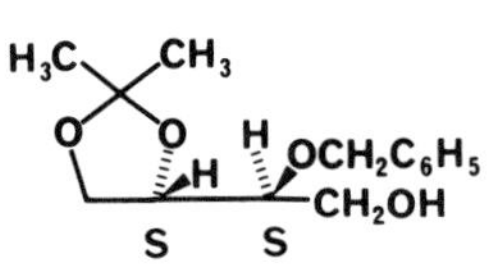

2-*O*-Benzyl-3,4-*O*-isopropylidene-L-threitol

3 steps (68%) ex (2*R*,3*R*)-tartaric acid (*88*)

138. (*S*)-*cis*-2-Methyl-3-[(phenylmethoxy)methyl]oxirane ($C_{11}H_{14}O_2$)

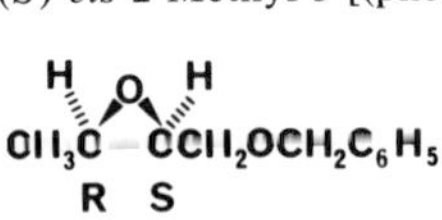

(2*S*,3*R*)-1-Benzyloxy-2,3-epoxybutane

9 steps (20%) ex dimethyl (2*R*,3*R*)-tartrate (*35, 37*)

139. (*S*)-*trans*-2-Methyl-3-[(phenylmethoxy)methyl]oxirane ($C_{11}H_{14}O_2$)

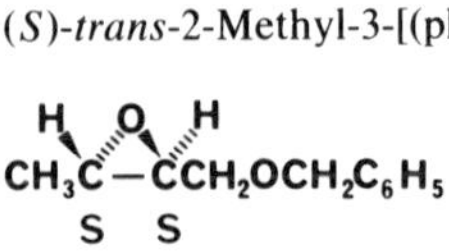

(2*S*,3*S*)-1-Benzyloxy-2,3-epoxybutane

11 steps (16%) ex dimethyl (2*R*,3*R*)-tartrate (*35, 37*)

140. (*R*)-*trans*-3-[(Phenylmethoxy)methyl]oxiranemethanol ($C_{11}H_{14}O_3$)

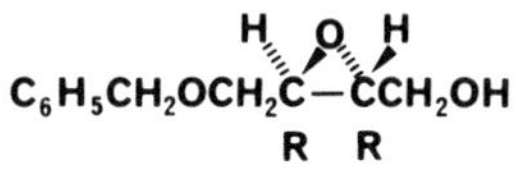

(2*R*,3*R*)-4-Benzyloxy-2,3-epoxy-1-butanol

6 steps (22%) ex dimethyl (2*R*,3*R*)-tartrate (*36*)

141. (*S*)-*cis*-3-[(Phenylmethoxy)methyl]oxiranemethanol ($C_{11}H_{14}O_3$)

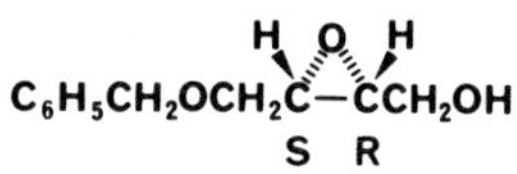

(2*R*,3*S*)-4-Benzyloxy-2,3-epoxy-1-butanol

6 steps (19%) ex dimethyl (2*S*,3*S*)-tartrate (*36*)

142. (*R*)-*cis*-3-[(Phenylmethoxy)methyl]oxiranemethanol ($C_{11}H_{14}O_3$)

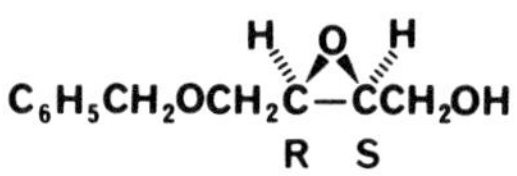

(2*S*,3*R*)-4-Benzyloxy-2,3-epoxy-1-butanol

6 steps (31%) ex dimethyl (2*R*,3*R*)-tartrate (*36*)

143. (*S*)-*trans*-3-[(Phenylmethoxy)methyl]oxiranemethanol ($C_{11}H_{14}O_3$)

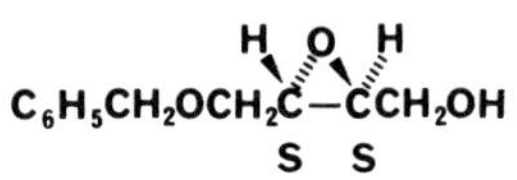

(2*S*,3*S*)-4-Benzyloxy-2,3-epoxy-1-butanol

6 steps (17%) ex dimethyl (2*S*,3*S*)-tartrate (*36*)

144. [*R*-(*R**,*R**)]-α-[(Phenylmethoxy)methyl]oxiranemethanol ($C_{11}H_{14}O_3$)

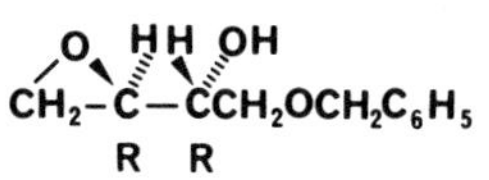

(2*R*,3*R*)-1-Benzyloxy-3,4-epoxy-2-butanol

4 steps (41%) ex diethyl (2*S*,3*S*)-tartrate (*35, 36*)

145. [*S*-(*R**,*S**)]-α-[Phenylmethoxy)methyl]oxiranemethanol ($C_{11}H_{14}O_3$)

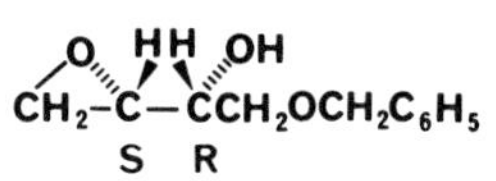

(2*R*,3*S*)-1-Benzyloxy-3,4-epoxy-2-butanol

6 steps (21%) ex dimethyl (2*R*,3*R*)-tartrate (*35, 36*)

146. [*R*-(*R**,*S**)]-α-[(Phenylmethoxy)methyl]oxiranemethanol ($C_{11}H_{14}O_3$)

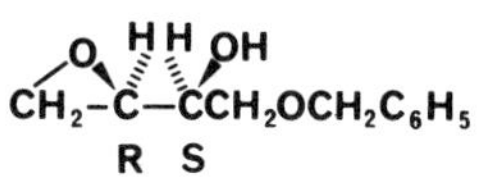

(2*S*,3*R*)-1-Benzyloxy-3,4-epoxy-2-butanol

6 steps (24%) ex diethyl (2*S*,3*S*)-tartrate (*36*)

147. [*S*-(*R**,*R**)]-α-[(Phenylmethoxy)methyl]oxiranemethanol ($C_{11}H_{14}O_3$)

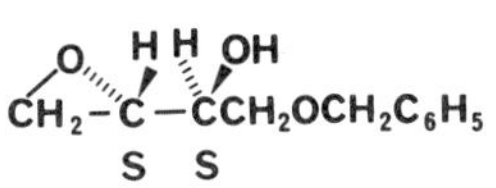

(2*S*,3*S*)-1-Benzyloxy-3,4-epoxy-2-butanol

4 steps (36%) ex dimethyl (2*R*,3*R*)-tartrate (*35, 36*)

148. (*R*)-*cis*-2,3-bis[(Phenylmethoxy)methyl]oxirane ($C_{18}H_{20}O_3$)

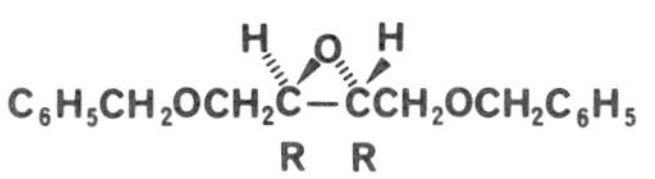

(2*R*,3*R*)-1,4-Bis(benzyloxy)butane-2,3-epoxide

5 steps (70–75%) ex diethyl (2*S*,3*S*)-tartrate (*89*)

149. [*R*-(*R**,*R**)]-2,3-Butanediol ($C_4H_{10}O_2$)

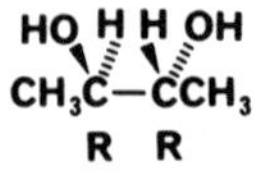

(2*R*,3*R*)-(−)-2,3-Butanediol

7 steps ex (2*S*,3*S*)-tartaric acid (*90*)

150. [*S*-(*R**,*R**)]-2,3-Butanediol ($C_4H_{10}O_2$)

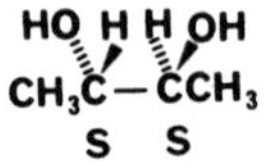

(2*S*,3*S*)-(+)-2,3-Butanediol

5 steps (55%) (*49*); 7 steps (*90*) ex (2*R*,3*R*)-tartaric acid

151. (*S*)-*cis*-3-Formyl-1,4-dioxaspirodecane-2-carboxylic acid ($C_{10}H_{14}O_5$)

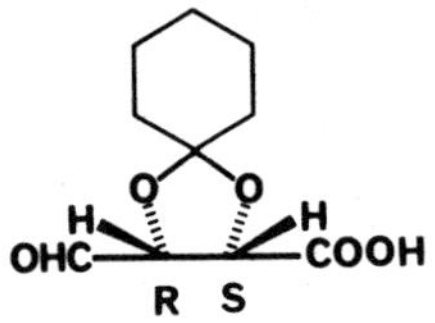

2,3-*O*-Cyclohexylidene-L-(−)-erythruronic acid *

152. (*S*)-*cis*-5-Formyl-2,2-dimethyl-1,3-dioxolane-4-carboxylic acid ($C_7H_{10}O_5$)

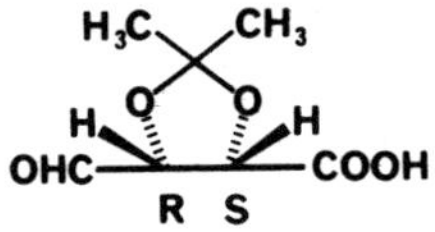

2,3-*O*-Isopropylidene-L-erythruronic acid *

153. (*R*)-*trans*-2,3-Oxiranedicarboxylic acid diethyl ester ($C_8H_{12}O_5$)

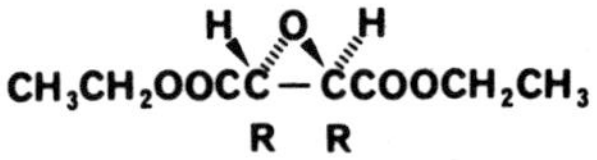

Diethyl (2*R*,3*R*)-epoxysuccinate

3 steps (51%) ex diethyl (2*R*,3*R*)-tartrate (*91*)

154. (*S*)-*trans*-2,3-Oxiranedicarboxylic acid diethyl ester ($C_8H_{12}O_5$)

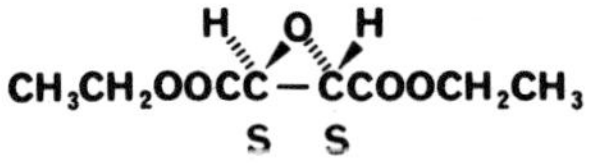

Diethyl (2*S*,3*S*)-epoxysuccinate

3 steps (56%) ex diethyl (2*S*,3*S*)-tartrate (*91*)

155. (*S*)-*trans*-2,3-Oxiranedicarboxylic acid ($C_4H_4O_5$)

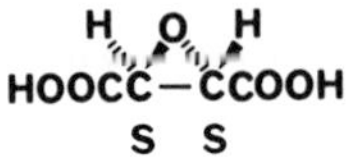

(2*S*,3*S*)-2,3-Epoxysuccinic acid

2 steps (~50%) ex diethyl (*S*)-malate (*39*)

156. (*R*)-*trans*-2,3-Dimethyloxirane (C_4H_8O)

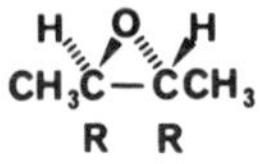

(2*R*,3*R*)-2,3-Epoxybutane

9 steps ex (2*S*,3*S*)-tartaric acid (*90*)

157. (*S*)-*trans*-2,3-Dimethyloxirane (C_4H_8O)

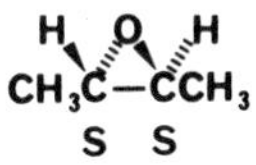

(2*S*,3*S*)-2,3-Epoxybutane

8 steps (39%) (*49*); 9 steps (*90*) ex (2*R*,3*R*)-tartaric acid

158. [*R*-(*R**,*R**)]-2,3,4-Trihydroxybutanal ($C_4H_8O_4$)

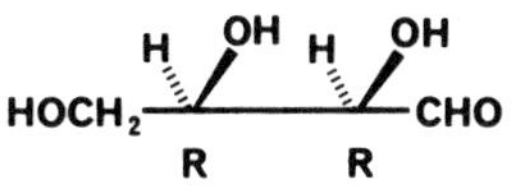

D-(−)-Erythrose *

159. [*R*-(*R**,*R**)]-2,3-Dihydroxybutanedioic acid ($C_4H_6O_6$)

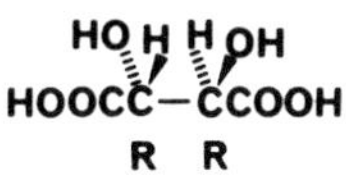

d-L-Tartaric acid ["natural"] *****

160. [*S*-(*R**,*R**)]-2,3-Dihydroxybutanedioic acid ($C_4H_6O_6$)

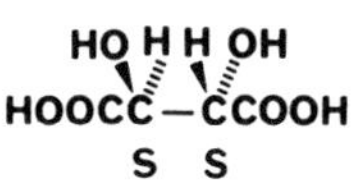

l-D-Tartaric Acid ["unnatural"] ****

161. [*S*-(*R**,*R**)]-1,2,3,4-Butanetetrol ($C_4H_{10}O_4$)

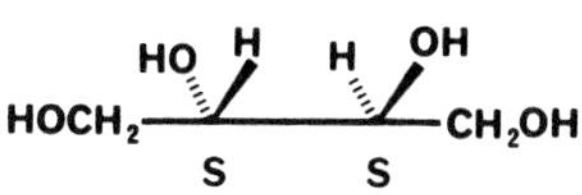

L-Threitol *

162. (*S*)-*trans*-2,2-Dimethyl-1,3-dioxolane-4,5-dimethanol ($C_7H_{14}O_4$)

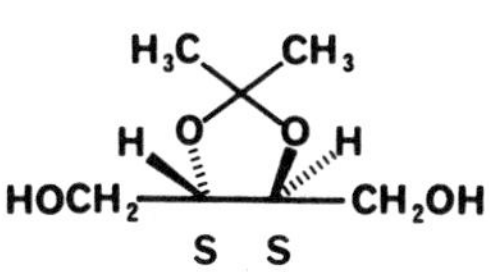

(+)-2,3-*O*-Isopropylidene-L-threitol *

3 steps (95%) ex (2*R*,3*R*)-tartaric acid (*36*)

163. (*R*)-*trans*-2,2-Dimethyl-1,3-dioxolane-4,5-dimethanol ($C_7H_{14}O_4$)

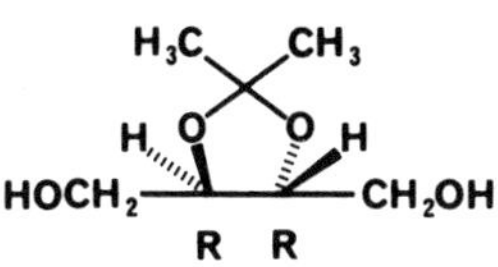

(−)-2,3-*O*-Isopropylidene-D-threitol *

3 steps (75%) ex (2*S*,3*S*)-tartaric acid (*36*)

164. (*R*)-*cis*-2-Phenyl-1,3-dioxane-4-methanol 4-methylbenzenesulfonate ($C_{18}H_{20}O_5S$)

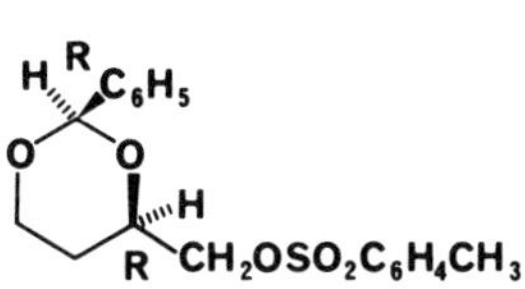

(2*R*,4*R*)-2-Phenyl-4-tosyloxymethyl-1,3-dioxane

9 steps (~17%) ex dimethyl (2*R*,3*R*)-tartrate (*37*)

165. (*S*)-*cis*-2-Phenyl-1,3-dioxane-4-methanol 4-methylbenzenesulfonate ($C_{18}H_{20}O_5S$)

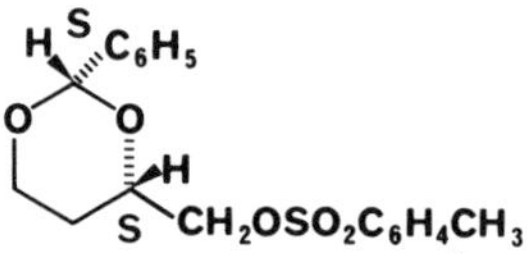

(2*S*,4*S*)-2-Phenyl-4-tosyloxymethyl-1,3-dioxane

5 steps (37%) ex diethyl (*S*)-malate (*37*)

166. [*R*-(*R**,*R**)]-2-Amino-3-hydroxybutanoic acid ($C_4H_9NO_3$)

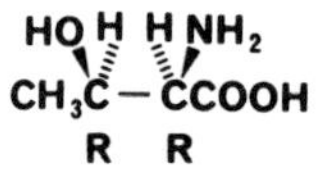

D-(−)-allo-Threonine *

167. [*S*-(*R**,*S**)]-2-Amino-3-hydroxybutanoic acid ($C_4H_9NO_3$)

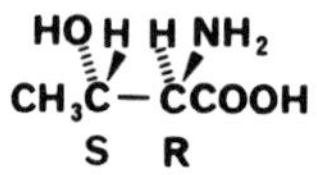

D-Threonine ****

168. [*R*-(*R**,*S**)]-2-Amino-3-hydroxybutanoic acid ($C_4H_9NO_3$)

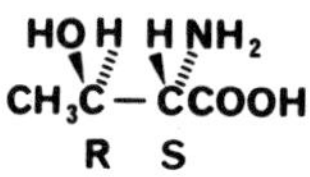

L-Threonine ***

169. [*S*-(*R**,*R**)]-2-Amino-3-hydroxybutanoic acid ($C_4H_9NO_3$)

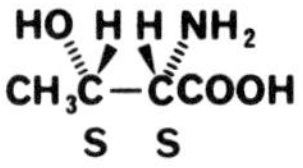

L-(+)-allo-Threonine *

4 steps (74%) ex L-threonine (*43*)

Four-Carbon Chain: Carbon Substituted

170. (*S*)-1-Bromo-2-methylbutane ($C_5H_{11}Br$)

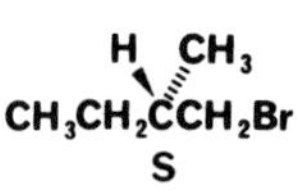

(*S*)-(+)-1-Bromo-2-methylbutane *

171. (*R*)-2-Hydroxy-2-methylbutanedioic acid ($C_5H_8O_5$)

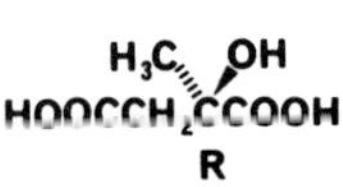

D-(−)-Citramalic acid *

172. (*S*)-2-Hydroxy-2-methylbutanedioic acid ($C_5H_8O_5$)

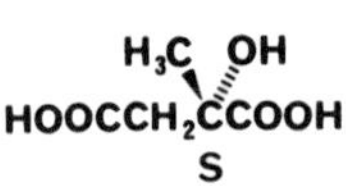

L-(+)-Citramalic acid *

173. (*S*)-4-Bromo-3-methylbutanoic acid ethyl ester ($C_7H_{13}BrO_2$)

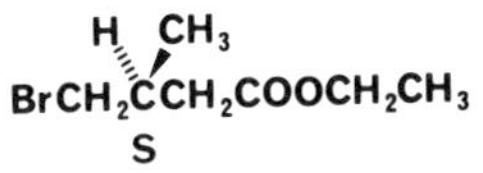

Ethyl (*S*)-4-bromo-3-methylbutyrate

1 step (87%) ex (*S*)-methyl-γ-butyrolactone (*92*)

174. (*S*)-2-Methylbutanal ($C_5H_{10}O$)

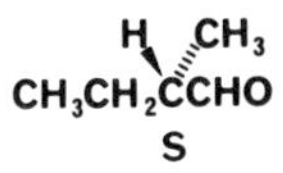

(*S*)-2-Methylbutanal

1 step (71%) ex L-isoleucine (*93*)

175. (*S*)-2-Methyl-1-butanol ($C_5H_{12}O$)

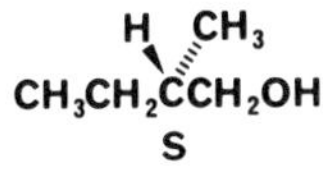

(−)-2-Methyl-1-butanol ****
[“primary active amyl alcohol”]

176. (*S*)-2-Methyl-1-butanamine ($C_5H_{13}N$)

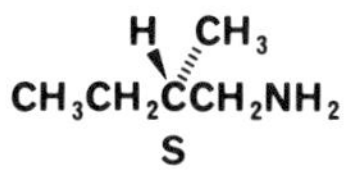

(*S*)-(−)-Methylbutylamine *

177. (*S*)-2-Methylbutanoic acid ($C_5H_{10}O_2$)

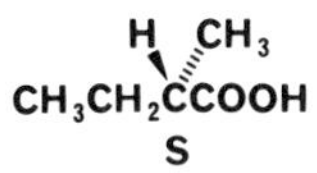

(*S*)-(+)-Methylbutyric acid *

178. (*R*)-2,2,4-Trimethyl-1,3-dioxolane-4-ethanol ($C_8H_{16}O_3$)

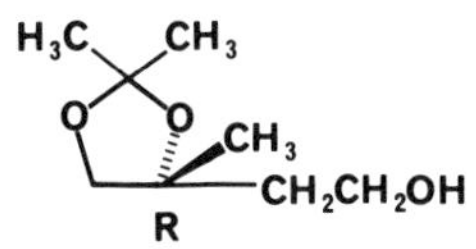

(*R*)-3-Methyl-3,4-*O*-isopropylidenebutane-1,3,4-triol

2 steps ex (*R*)-citramalic acid (*94*)

179. (*S*)-2,2,4-Trimethyl-1,3-dioxolane-4-ethanol ($C_8H_{16}O_3$)

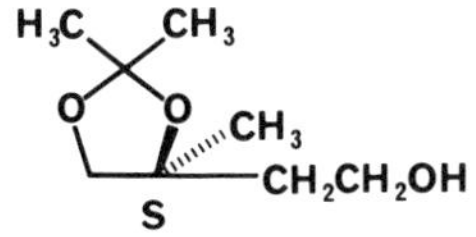

(*S*)-3-Methyl-3,4-*O*-isopropylidenebutane-1,3,4-triol

2 steps (65%) ex dimethyl (*S*)-citramalate (*95, 96*)
2 steps (48%) ex (*S*)-citramalic acid (*97*)

180. (*S*)-Dihydro-4-methyl-2(3*H*)-furanone ($C_5H_8O_2$)

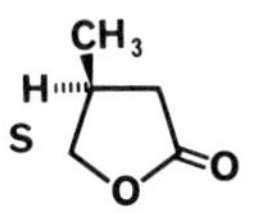

(*S*)-β-Methyl-γ-butyrolactone *

181. (*R*)-α-Ethylbenzeneacetic acid ($C_{10}H_{12}O_2$)

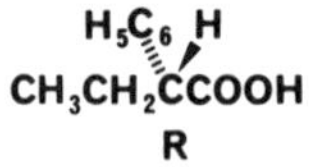

(−)-2-Phenylbutyric acid *

182. (*S*)-α-Ethylbenzeneacetic acid ($C_{10}H_{12}O_2$)

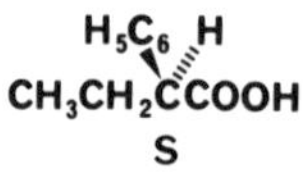

(+)-2-Phenylbutyric acid *

183. [*R*-(*R**,*S**)]-2-Hydroxy-3-methylbutanedioic acid diethyl ester ($C_9H_{16}O_5$)

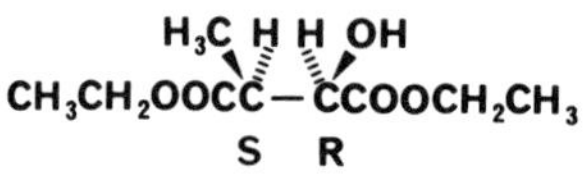

Diethyl (2*R*,3*S*)-3-methylmalic acid

4 steps (33%) ex diethyl (2*R*,3*R*)-tartrate (*91*)

184. [*S*-(*R**,*S**)]-2-Hydroxy-3-methylbutanedioic acid diethyl ester ($C_9H_{16}O_5$)

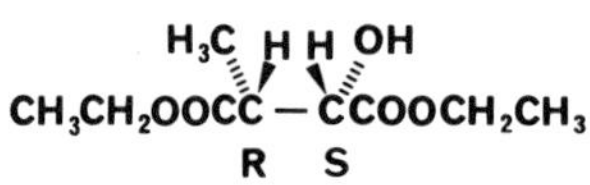

Diethyl (2*S*,3*R*)-3-methylmalic acid

4 steps (44%) ex diethyl (2*S*,3*S*)-tartrate (*91*)

3 steps (~40%) ex diethyl (*S*)-malate (*39*)

185. [*S*-(*R**,*R**)]-3-Methylbutane-1,2,4-triol ($C_5H_{12}O_3$)

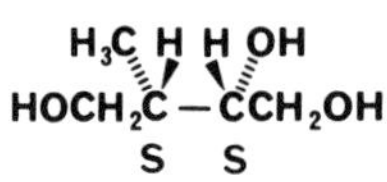

(2*S*,3*S*)-3-Methylbutane-1,2,4-triol

7 steps (~70%) ex diethyl (2*S*,3*S*)-tartrate (*89*)

186. (3*S*)-*cis*-3-Ethyldihydro-4-[(1-methyl-1*H*-imidazol-5-yl)ethyl]-2(3*H*)-furanone ($C_{12}H_{18}N_2O_2$)

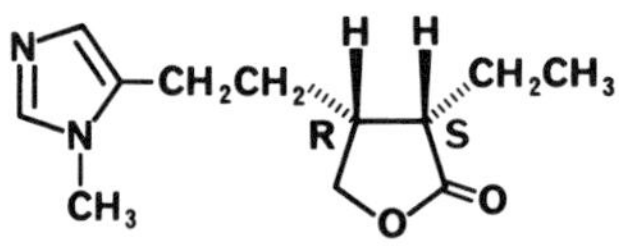

(+)-Pilocarpine **

Five-Carbon Chain: Heteroatom Substituted

187. (*R*)-2-Amino-5-[(aminoiminomethyl)amino]pentanoic acid ($C_6H_{14}N_4O_2$)

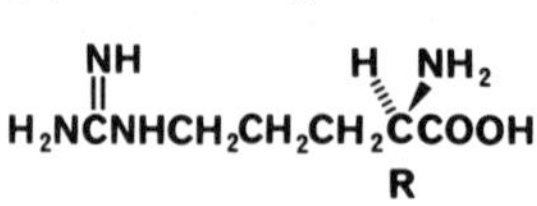

D-(−)-Arginine *

188. (*S*)-2-Amino-5-[(aminoiminomethyl)amino]pentanoic acid ($C_6H_{14}N_4O_2$)

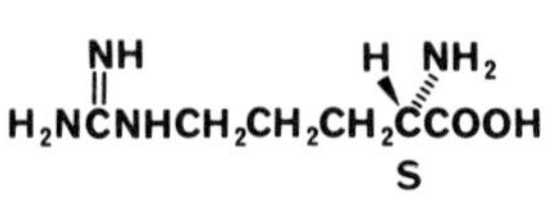

L-(+)-Arginine *****

189. (*R*)-2-Aminopentanedioic acid ($C_5H_9NO_4$)

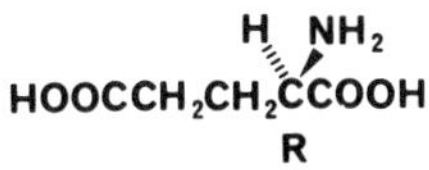

D-(−)-Glutamic acid **

190. (*S*)-2-Aminopentanedioic acid ($C_5H_9NO_4$)

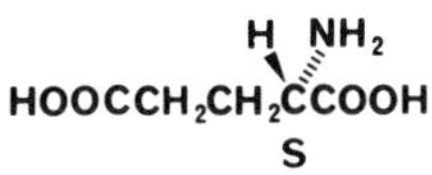

L-(+)-Glutamic acid *****

191. (*R*)-2,5-Diamino-5-oxopentanoic acid ($C_5H_{10}N_2O_3$)

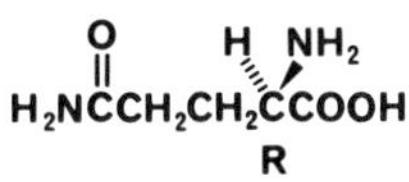

D-Glutamine *

192. (*S*)-2,5-Diamino-5-oxopentanoic acid ($C_5H_{10}N_2O_3$)

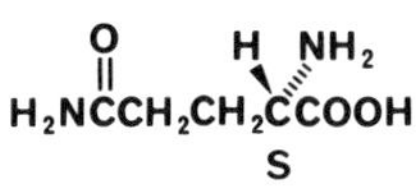

L-(+)-Glutamine ****

193. (*R*)-4-Hydroxy-2-cyclopenten-1-one ($C_5H_6O_2$)

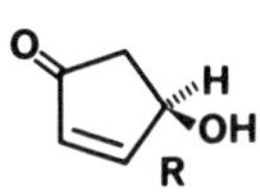

(*R*)-4-Hydroxy-2-cyclopentenone

6 steps (22%) ex (2*S*,3*S*)-tartaric acid (*98*)

194. (*S*)-4-Hydroxy-2-cyclopenten-1-one ($C_5H_6O_2$)

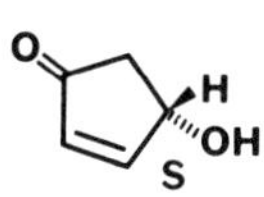

(*S*)-4-Hydroxy-2-cyclopentenone

6 steps (~30%) ex diethyl (2*R*,3*R*)-tartrate (*98*)

195. (*R*)-2-Hydroxypentanedioic acid ($C_5H_8O_5$)

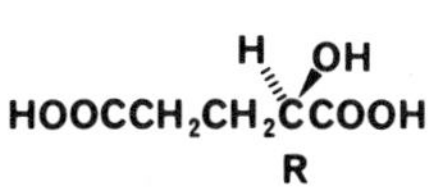

D-α-Hydroxyglutaric acid *

196. (*S*)-2-Hydroxypentanedioic acid ($C_5H_8O_5$)

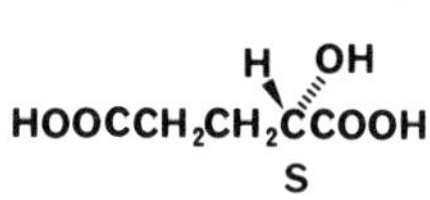

L-α-Hydroxyglutaric acid *

197. (*R*)-Dihydro-5-(hydroxymethyl)-2(3*H*)-furanone ($C_5H_8O_3$)

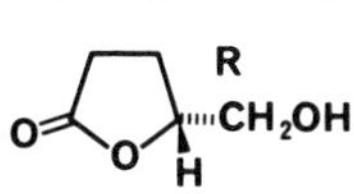

(*R*)-(−)-γ-Hydroxymethyl-γ-butyrolactone

3 steps (62%) ex (*R*)-glutamic acid (*99*)

198. (S)-Dihydro-5-(hydroxymethyl)-2(3H)-furanone ($C_5H_8O_3$)

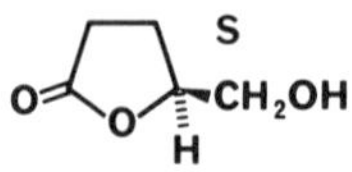

(S)-(+)-γ-Hydroxymethyl-γ-butyrolactone

3 steps (49%) ex (S)-glutamic aid (*100*)
6 steps (37%) ex D-mannitol bisacetonide (*101*)

199. (R)-Dihydro-5-[(phenylmethoxy)methyl]-2(3H)-furanone ($C_{12}H_{14}O_3$)

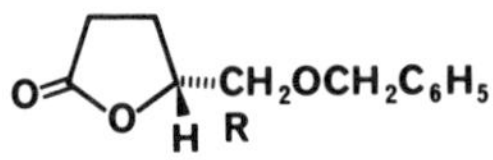

(R)-(−)-γ-Benzyloxymethyl-γ-butyrolactone

8 steps (46%) ex D-mannitol bisacetonide (*53*)

200. (S)-Dihydro-5-[(phenylmethoxy)methyl]-2(3H)-furanone ($C_{12}H_{14}O_3$)

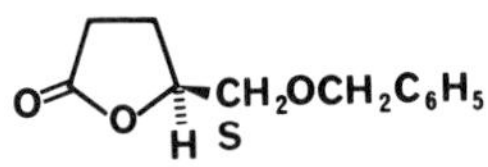

(S)-(+)-γ-Benzyloxymethyl-γ-butyrolactone

4 steps (39%) ex (S)-glutamic acid (*100*)

201. (R)-Dihydro-5-[(triphenylmethoxy)methyl]-2(3H)-furanone ($C_{24}H_{22}O_3$)

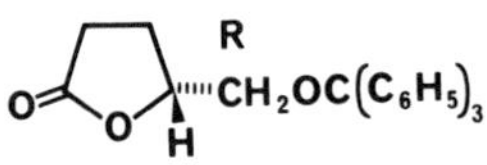

(R)-(−)-γ-Trityloxymethyl-γ-butyrolactone *

5 steps (16%) ex (S)-glutamic acid (*102*)

202. (S)-Dihydro-5-[(triphenylmethoxy)methyl]-2(3H)-furanone ($C_{24}H_{22}O_3$)

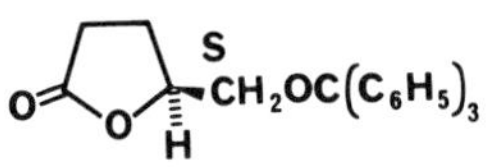

(S)-(+)-γ-Trityloxymethyl-γ-butyrolactone *

4 steps (31%) ex (S)-glutamic acid (*102*)

203. (S)-5-(Hydroxymethyl)-2(5H)-furanone ($C_5H_6O_3$)

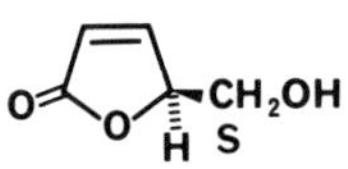

(S)-5-Hydroxymethyl-5H-furan-2-one

2 steps (52%) ex D-ribonolactone (*103*)

204. (S)-5-[(Phenylmethoxy)methyl]-2(5H)-furanone ($C_{12}H_{12}O_3$)

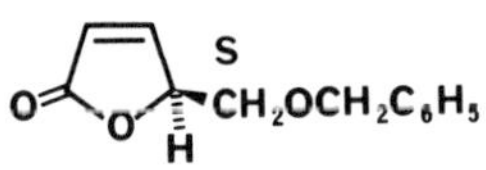

(S)-5-Benzyloxymethyl-5H-furan-2-one

3 steps (38%) ex D-ribonolactone (*103*)
6 steps (20%) ex (S)-glutamic acid (*104*)

205. (S)-5-[(Triphenylmethoxy)methyl]-2(5H)-furanone ($C_{24}H_{20}O_3$)

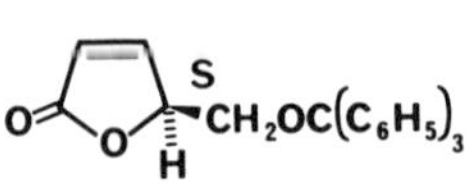

(S)-5-Trityloxymethyl-5H-furan-2-one

3 steps (34%) ex D-ribonolactone (*103*)
6 steps (20%) ex (S)-glutamic acid (*105*)

206. (S)-3-Piperidinol ($C_5H_{11}NO$)

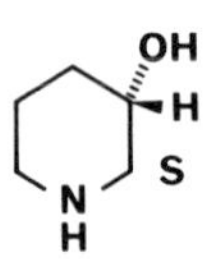

(S)-3-Hydroxypiperidine

11 steps ex D-mannitol (*106*)

207. (*S*)-5-Hydroxy-2-piperidinone ($C_5H_9NO_2$)

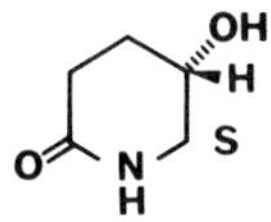

(*S*)-5-Hydroxypiperid-2-one
10 steps ex D-mannitol (*106*)

208. (*R*)-2-Amino-4-methylpentanoic acid ($C_6H_{13}NO_2$)

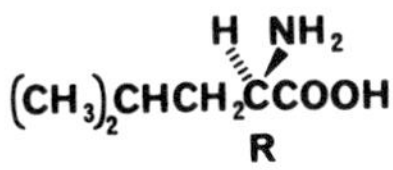

D-Leucine *

209. (*S*)-2-Amino-4-methylpentanoic acid ($C_6H_{13}NO_2$)

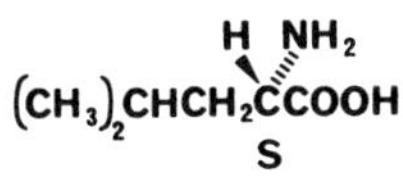

L-Leucine *****

210. (*R*)-Dihydro-5-methyl-2(3*H*)-furanone ($C_5H_8O_2$)

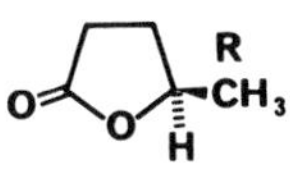

(*R*)-(+)-γ-Methyl-γ-butyrolactone
6 steps (~15%) ex (*S*)-glutamic acid (*107*)

211. (*S*)-Dihydro-5-methyl-2(3*H*)-furanone ($C_5H_8O_2$)

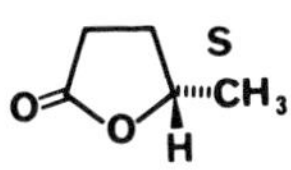

(*S*)-(−)-γ-Methyl-γ-butyrolactone
6 steps (15%) ex (*R*)-glutamic acid (*107*)

212. (*S*)-Oxiranepropanoic acid methyl ester ($C_6H_{10}O_3$)

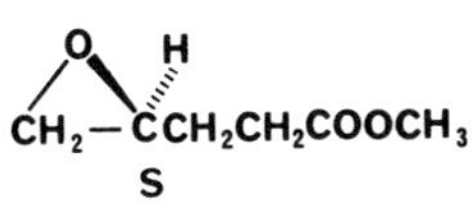

(4*S*)-Methyl-4,5-epoxyvalerate
5 steps (35%) ex (*S*)-glutamic acid (*108*)

213. (*R*)-2-Aminopentanoic acid ($C_5H_{11}NO_2$)

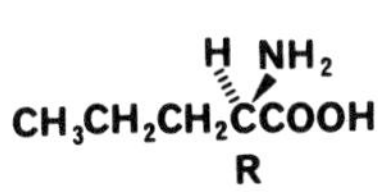

D-(−)-Norvaline *

214. (*S*)-2-Aminopentanoic acid ($C_5H_{11}NO_2$)

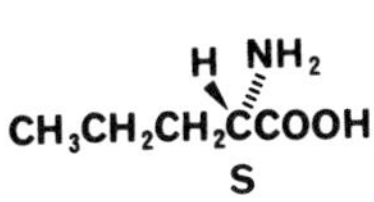

L-(+)-Norvaline **

215. (*R*)-2,5-Diaminopentanoic acid ($C_5H_{12}N_2O_2$)

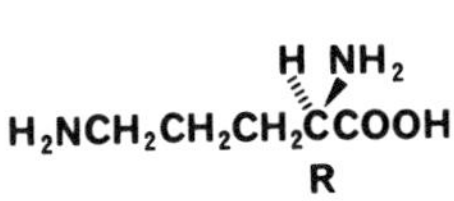

D-(−)-Ornithine *

216. (*S*)-2,5-Diaminopentanoic acid ($C_5H_{12}N_2O_2$)

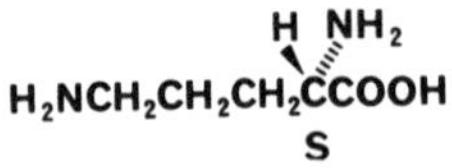

L-(+)-Ornithine *****

217. (*S*)-Pentane-1,4-diol ($C_5H_{12}O_2$)

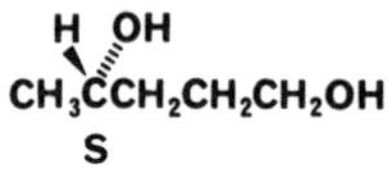

(4*S*)-(+)-Pentane-1,4-diol

5 steps (23%) ex D-glutamic acid (*99, 107, 109*)

218. (*S*)-Pentane-1,2,5-triol ($C_5H_{12}O_3$)

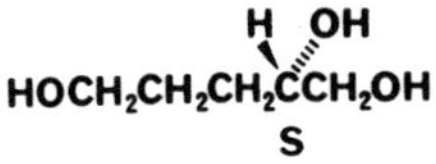

(*S*)-(−)-1,2,5-Pentanetriol

3 steps ex (*S*)-glutamic acid (*110*)

219. (*R*)-2-Pyrrolidinecarboxylic acid ($C_5H_9NO_2$)

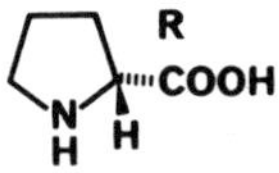

D-(+)-Proline *

220. (*S*)-2-Pyrrolidinecarboxylic acid ($C_5H_9NO_2$)

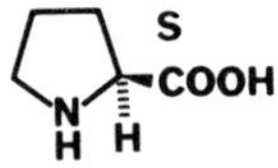

L-(−)-Proline ****

221. (*R*)-5-Oxo-2-pyrrolidinecarboxylic acid ($C_5H_7NO_3$)

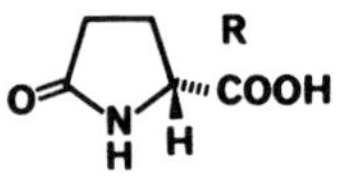

D-2-Pyrrolidone-5-carboxylic acid [D-pyroglutamic acid] **

222. (*S*)-5-Oxo-2-pyrrolidinecarboxylic acid ($C_5H_7NO_3$)

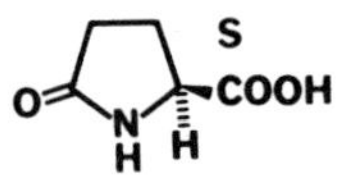

L-2-Pyrrolidone-5-carboxylic acid [L-pyroglutamic acid] *****

223. (2*R*)-*cis*-4-Hydroxy-2-pyrrolidinecarboxylic acid ($C_5H_9NO_3$)

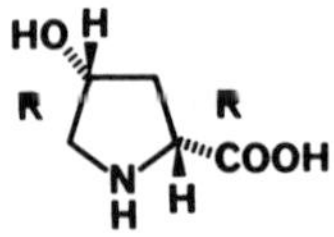

cis-4-Hydroxy-D-proline *

224. (2*S*)-*trans*-4-Hydroxy-2-pyrrolidinecarboxylic acid ($C_5H_9NO_3$)

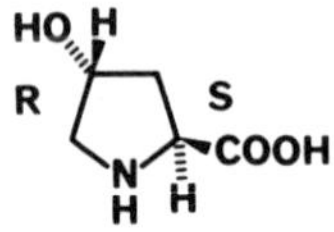

trans-4-Hydroxy-L-proline ***

225. (2*S*)-*cis*-4-Hydroxy-2-pyrrolidinecarboxylic acid ($C_5H_9NO_3$)

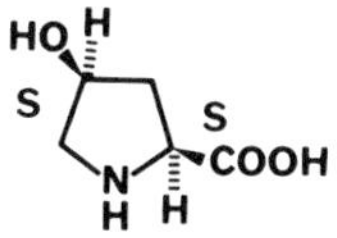

cis-4-Hydroxy-L-proline *

226. [*R*-(*R**,*R**)]-Pentane-2,4-diol ($C_5H_{12}O_2$)

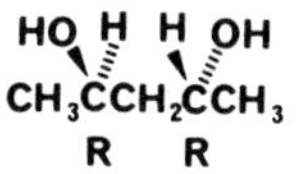

(2*R*,4*R*)-(−)-Pentanediol *

227. D-Arabinose ($C_5H_{10}O_5$)

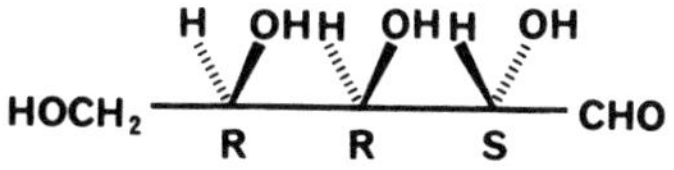

D-(−)-Arabinose ****

228. L-Arabinose ($C_5H_{10}O_5$)

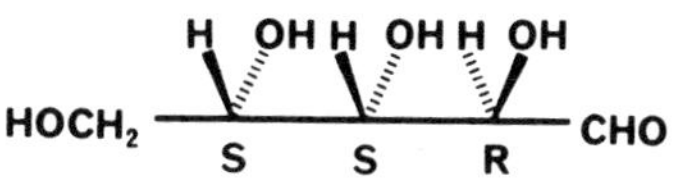

L-(+)-Arabinose *****

229. D-Arabinitol ($C_5H_{12}O_5$)

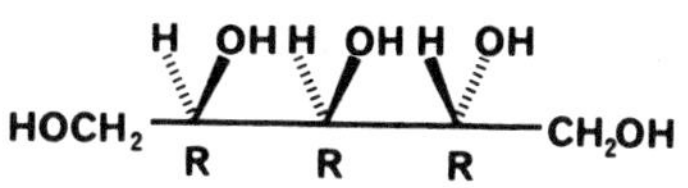

D-(+)-Arabitol **

230. L-Arabinitol ($C_5H_{12}O_5$)

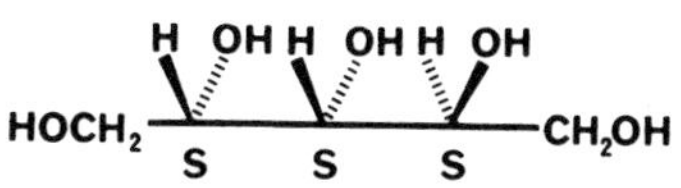

L-(−)-Arabitol **

231. 5-Deoxy-L-arabinose ($C_5H_{10}O_4$)

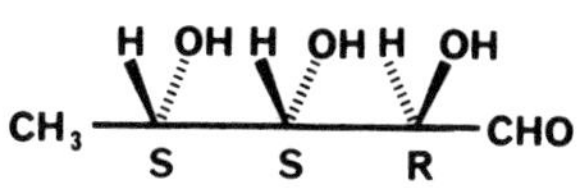

5-Deoxy-L-arabinose *

232. L-Lyxose ($C_5H_{10}O_5$)

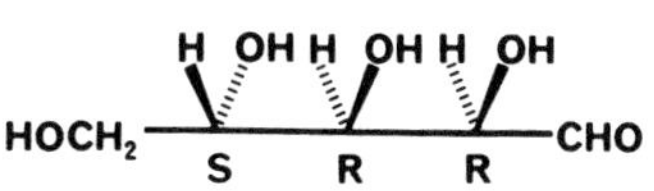

L-Lyxose *

233. D-Lyxose ($C_5H_{10}O_5$)

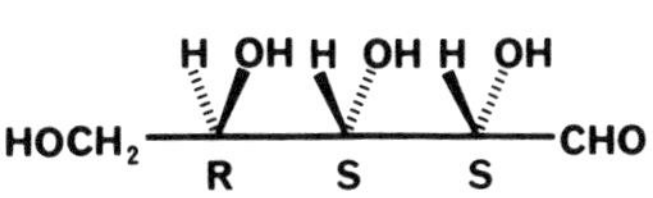

D-Lyxose *

234. 2-Deoxy-D-erythropentose ($C_5H_{10}O_4$)

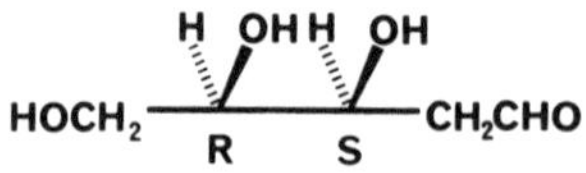

2-Deoxy-D-ribose **

235. D-Ribose ($C_5H_{10}O_5$)

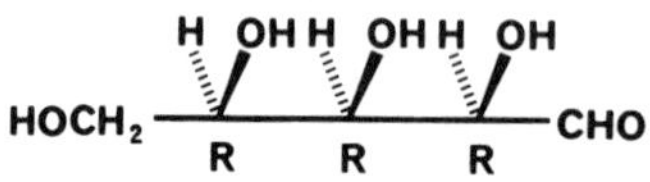

D-(−)-Ribose ****

236. L-Ribose ($C_5H_{10}O_5$)

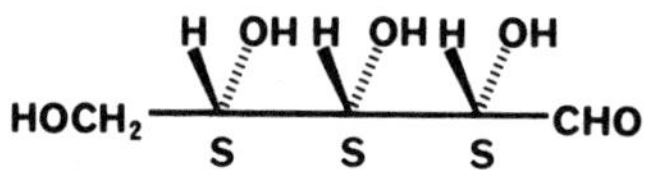

L-(+)-Ribose *

237. Methyl-2,3-*O*-(1-methylethylidene)-β-D-ribofuranoside ($C_9H_{16}O_5$)

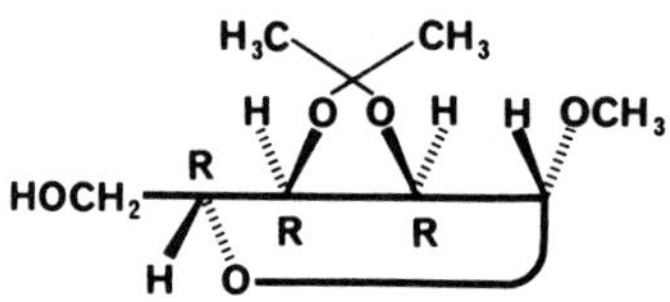

Methyl-2,3-*O*-Isopropylidene-D-ribofuranoside ***

238. Methyl-2,3-*O*-(1-methylethylidene)-β-D-ribofuranoside 4-methylbenzenesulfonate ($C_{16}H_{22}O_7S$)

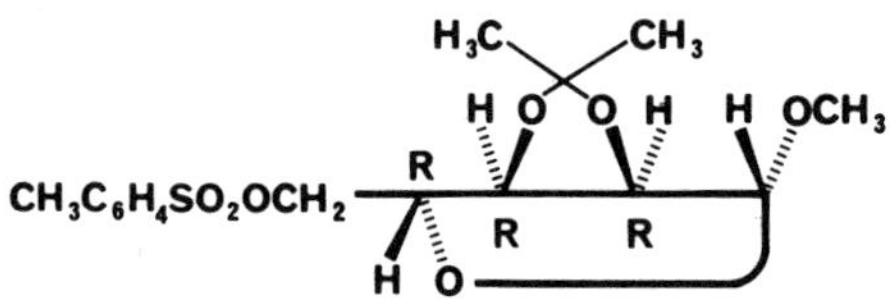

1-*O*-Methyl-2,3-*O*-isopropylidene-5-*O*-*p*-(tolylsulfonyl)-β-D-ribofuranoside **

239. D-Ribonic acid γ-lactone ($C_5H_8O_5$)

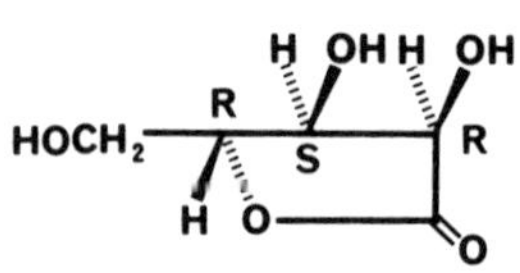

D-(+)-Ribonic acid γ-lactone [D ribonolactone] ****

240. D-Xylose ($C_5H_{10}O_5$)

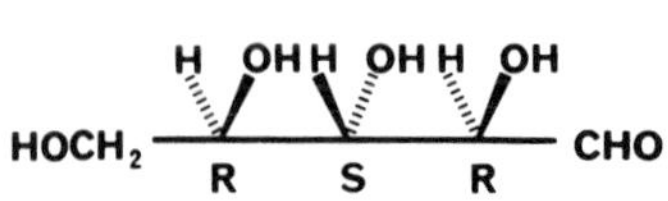

D-(+)-Xylose *****

241. L-Xylose ($C_5H_{10}O_5$)

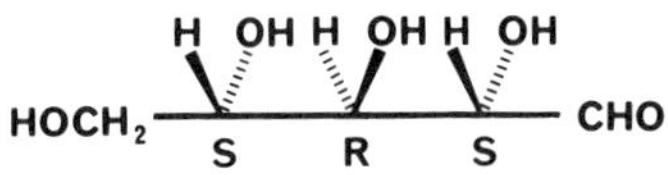

L-(−)-Xylose **

242. 1,2-*O*-(1-Methylethylidene)-α-D-xylofuranose-5-methylcarbonate ($C_{10}H_{16}O_7$)

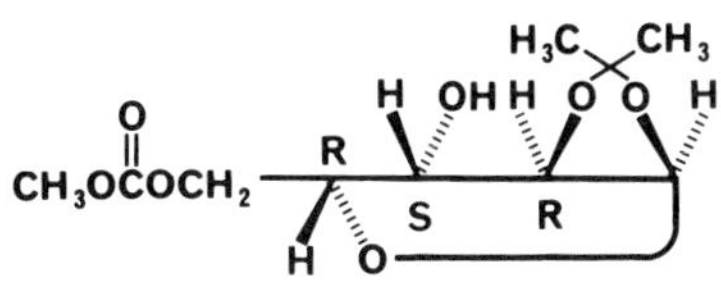

5-*O*-Carbomethoxy-1,2-*O*-isopropylidene-D-xylofuranose ***

243. 1,2-*O*-(1-Methylethylidene)-α-D-xylofuranose-5-methylcarbonate-3-(4-methylbenzenesulfonate) ($C_{17}H_{22}O_9S$)

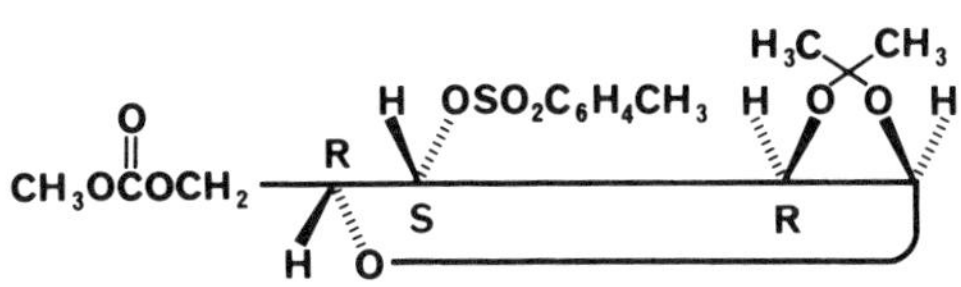

5-*O*-Carbomethoxy-1,2-*O*-isopropylidene-3-(*p*-tosyl)-D-xylofuranose ***

244. 1,2 : 3,5-Bis-*O*-(1-methylethylidene)-α-D-xylofuranose ($C_{11}H_{18}O_5$)

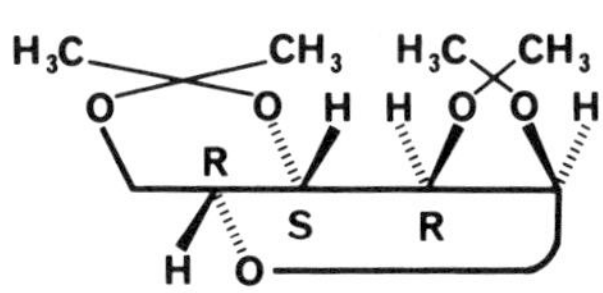

1,2 : 3,5-di-*O*-Isopropylidene-D-xylofuranose [diacetone-D-xylose] ***

245. 1,2-*O*-(1-Methylethylidene)-α-D-xylofuranose ($C_8H_{14}O_5$)

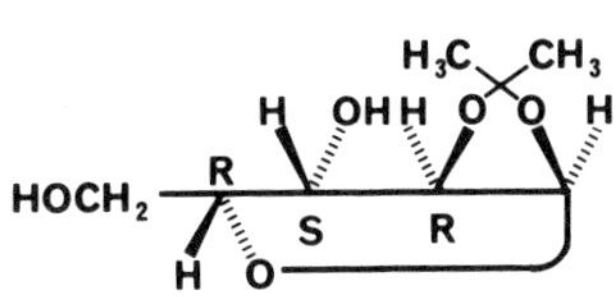

1,2-*O*-Isopropylidene-D-xylofuranose ***

Five-Carbon Chain: Carbon Substituted

246. (*R*)-5-(Phenylmethoxy)-2-methyl-1-pentanol ($C_{13}H_{20}O_2$)

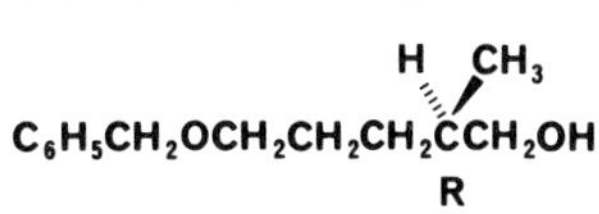

(*R*)-5-Benzyloxy-2-methylpentanol

9 steps (79%) ex (*R*)-citronellol (*111*)

247. (*R*)-5-Bromo-3-methyl-1-pentene ($C_6H_{11}Br$)

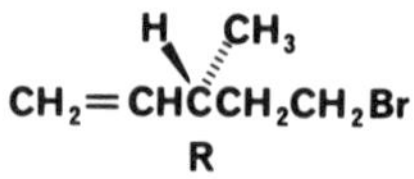

(*R*)-1-Bromo-3-methyl-4-pentene

7 steps (28%) ex (*R*)-(−)-citronellene (*112*)

248. (*S*)-5-Bromo-3-methyl-1-pentene ($C_6H_{11}Br$)

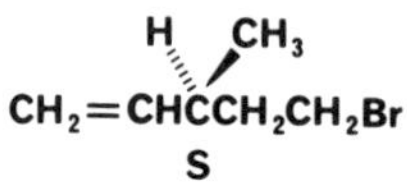

(*S*)-1-Bromo-3-methyl-4-pentene

6 steps (26%) ex (*R*)-(−)-citronellene (*112*)

249. (*R*)-3-Methylcyclopentanone ($C_6H_{10}O$)

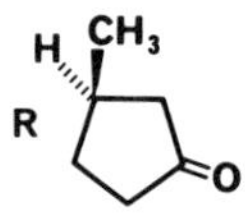

(+)-3-Methylcyclopentanone *

250. (*R*)-3-Methyl-5-oxopentanoic acid methyl ester ($C_7H_{12}O_3$)

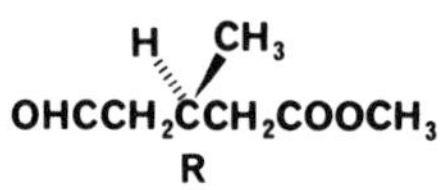

(3*R*)-Methyl-5-oxopentanoate

4 steps (47%) ex (3*R*)-methylcyclopentanone or

10 steps (<8%) ex *l*-menthone (*113*)

251. (*R*)-Tetrahydro-4-methyl-2*H*-pyran-2-one ($C_6H_{10}O_2$)

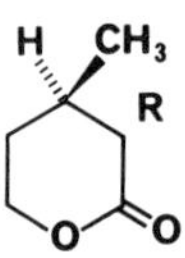

(*R*)-3-Methylvalerolactone

5 steps (39%) ex (3*R*)-methylcyclopentanone (*113*)

252. (4*S*)-*cis*-Dihydro-5-(hydroxymethyl)-4-methyl-2(3*H*)-furanone ($C_6H_{10}O_3$)

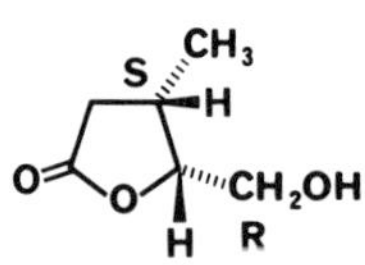

(3*S*,4*R*)-4-Hydroxymethyl-3-methylbutyrolactone

6 steps (54%) ex benzyl (*R*)-citronellate (*86*)

253. [*R*-(*R**,*R**)]-2-Amino-3-methylpentanoic acid ($C_6H_{13}NO_2$)

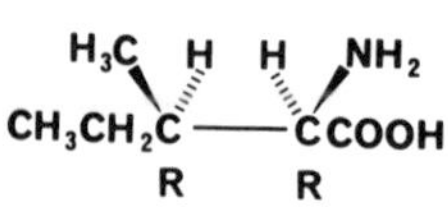

D-Isoleucine *

254. [*S*-(*R**,*R**)]-2-Amino-3-methylpentanoic acid ($C_6H_{13}NO_2$)

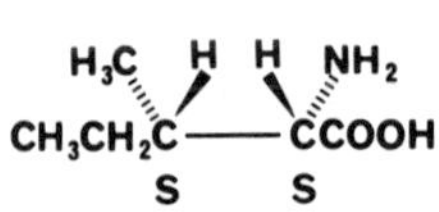

L-Isoleucine ****

255. [*R*-(*R**,*S**)]-2-Amino-3-methylpentanoic acid ($C_6H_{13}NO_2$)

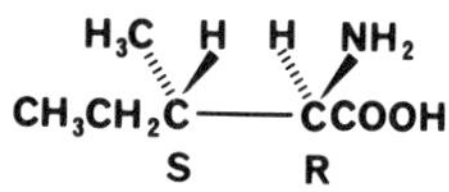

D-allo-Isoleucine *

256. [*S*-(*R**,*S**)]-2-Amino-3-methylpentanoic acid ($C_6H_{13}NO_2$)

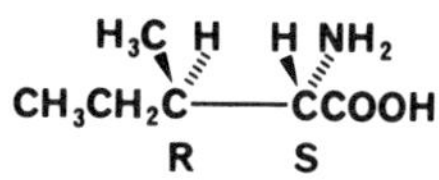

L-allo-Isoleucine *

Six-Carbon Chain: Heteroatom Substituted

257. (*R*)-2-Aminohexanedioic acid ($C_6H_{11}NO_4$)

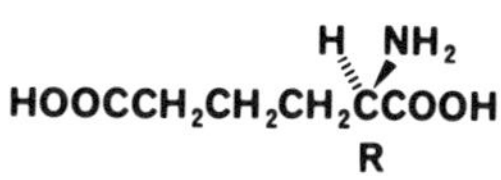

D-α-Aminoadipic acid *

258. (*S*)-2-Aminohexanedioic acid ($C_6H_{11}NO_4$)

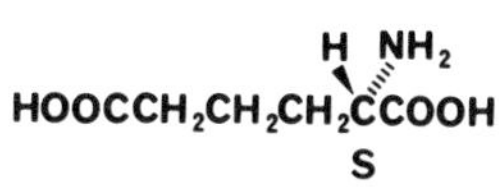

L-α-Aminoadipic acid *

259. (*R*)-2-Hexanol ($C_6H_{14}O$)

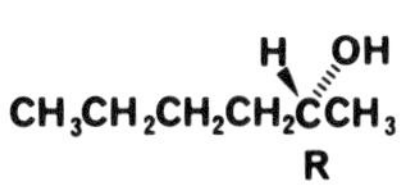

(−)-2-Hexanol *

260. (*S*)-2-Hexanol ($C_6H_{14}O$)

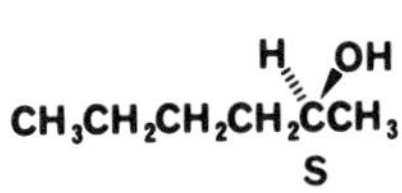

(+)-2-Hexanol *

261. (*R*)-2,6-Diaminohexanoic acid ($C_6H_{14}N_2O_2$)

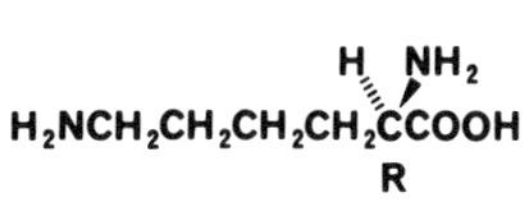

D-(−)-Lysine *

262. (*S*)-2,6-Diaminohexanoic acid ($C_6H_{14}N_2O_2$)

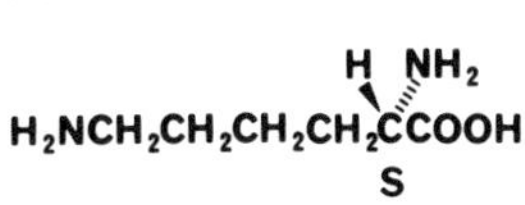

L-(+)-Lysine ****

263. (*R*)-2-Aminohexanoic acid ($C_6H_{13}NO_2$)

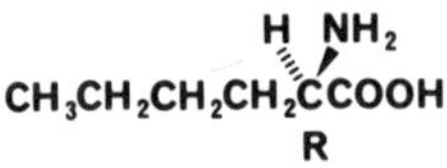

D-(−)-Norleucine *

264. (*S*)-2-Aminohexanoic acid ($C_6H_{13}NO_2$)

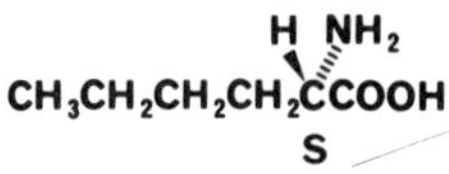

L-(+)-Norleucine *

265. (*R*)-2-Piperidinecarboxylic acid ($C_6H_{11}NO_2$)

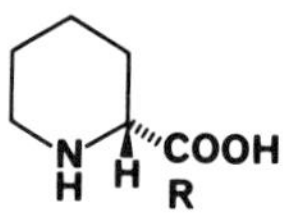

D-Pipecolinic acid *

266. (*S*)-2-Piperidinecarboxylic acid ($C_6H_{11}NO_2$)

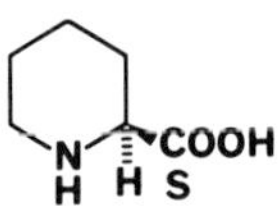

L-Pipecolinic acid *

1 step (39%) ex L-lysine (*114*)

267. L-*threo*-Hex-2-enonic acid γ-lactone ($C_6H_8O_6$)

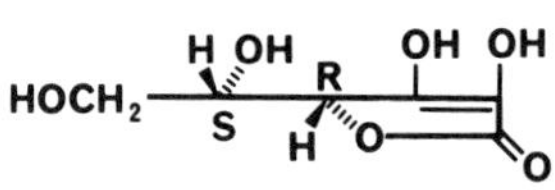

L-(+)-Ascorbic acid [vitamin C] *****

268. D-*erythro*-Hex-2-enonic acid γ-lactone ($C_6H_8O_6$)

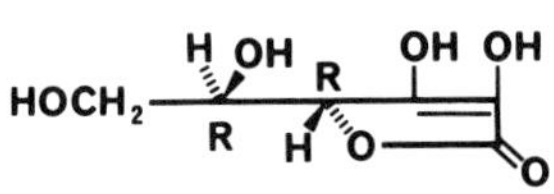

D-(−)-Isoascorbic acid [D-(−)-araboascorbic acid] *****

269. 1,5-Anhydro-2,6-dideoxy-L-arabinohex-1-enitol diacetate ($C_{10}H_{14}O_5$)

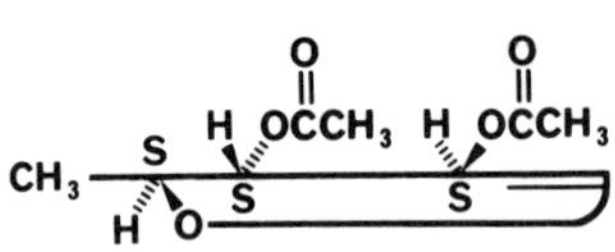

3,4-Di-*O*-acetyl-6-deoxy-L-glucal [di-*O*-acetyl-L-rhamnal] **

270. 1,5-Anhydro-2-deoxy-D-arabinohex-1-enitol triacetate ($C_{12}H_{16}O_7$)

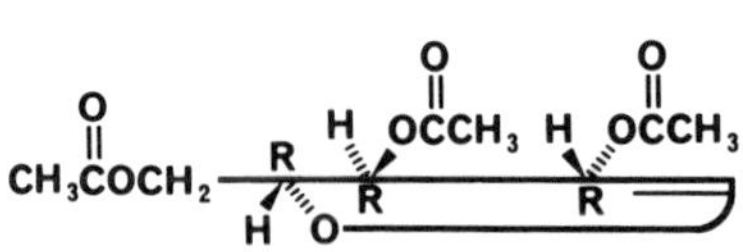

Tri-*O*-acetyl-D-glucal ***

271. 2-Deoxy-D-ribohexose ($C_6H_{12}O_5$)

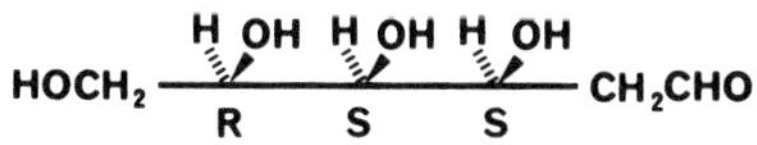

D-(+)-Digitoxose *

272. D-Fructose ($C_6H_{12}O_6$)

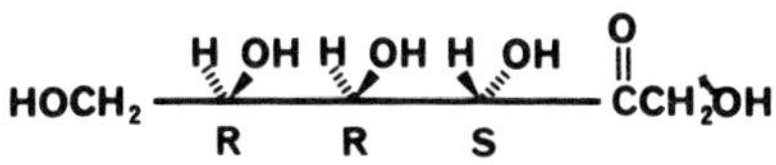

D-(−)-Fructose *****

273. 2,3 : 4,5-Bis-*O*-(1-methylethylidene)-β-D-fructopyranose ($C_{12}H_{20}O_6$)

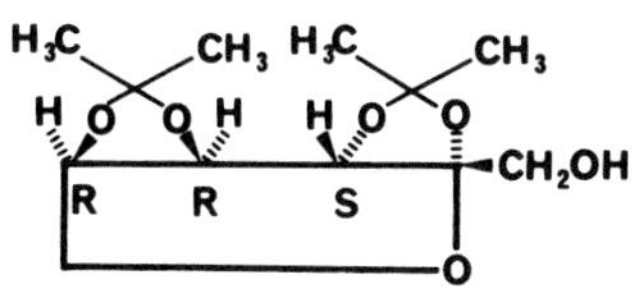

2,3 : 4,5-Di-*O*-isopropylidene-β-D-fructopyranose *

274. D-Arabino-2-hexulosonic acid ($C_6H_{10}O_7$)

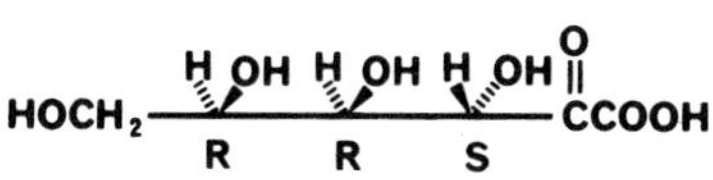

2-Keto-D-gluconic acid *

275. D-Xylo-5-hexulosonic acid ($C_6H_{10}O_7$)

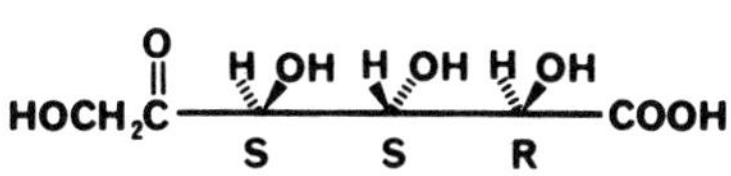

5-Keto-D-gluconic acid *

276. D-Sorbose ($C_6H_{12}O_6$)

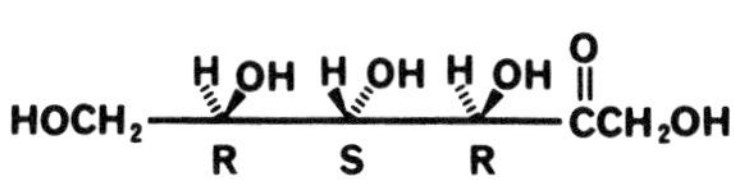

D-(+)-Sorbose *

277. L-Sorbose ($C_6H_{12}O_6$)

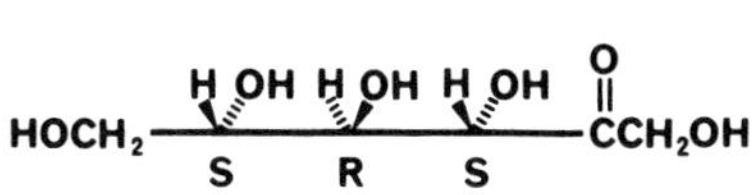

L-(−)-Sorbose *****

278. D-Tagatose ($C_6H_{12}O_6$)

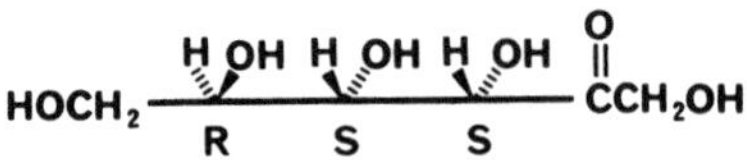

D-(−)-Tagatose *

279. D-Allose ($C_6H_{12}O_6$)

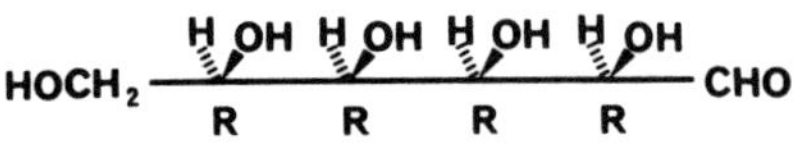

D-Allose *

280. 6-Deoxy-D-galactose ($C_6H_{12}O_5$)

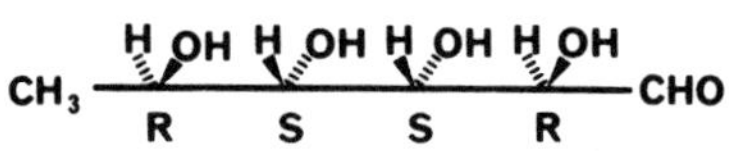

D-(+)-Fucose [rhodeose] *

281. L-Fucose ($C_6H_{12}O_5$)

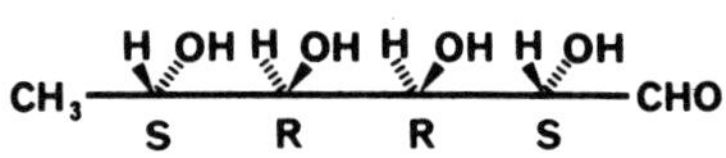

L-(−)-Fucose *

282. D-Galactonic acid ($C_6H_{12}O_7$)

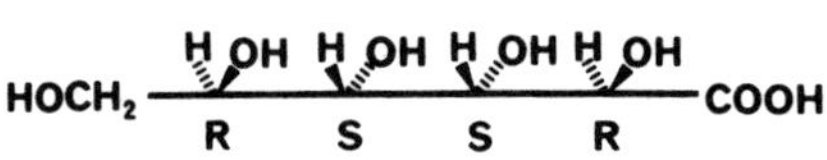

D-Galactonic acid *****

283. D-Galactonic acid γ-lactone ($C_6H_{10}O_6$)

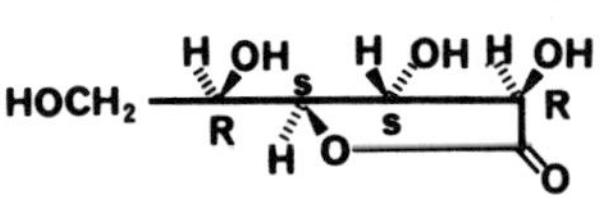

D-Galactonic acid γ-lactone [D-galactono-1,4-lactone] ****

284. L-Galactonic acid γ-lactone ($C_6H_{10}O_6$)

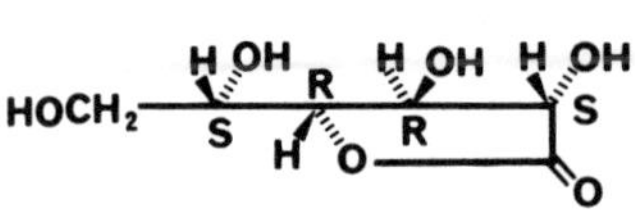

L-Galactonic acid γ-lactone [L-galactono-1,4-lactone] **

285. D-Galactose ($C_6H_{12}O_6$)

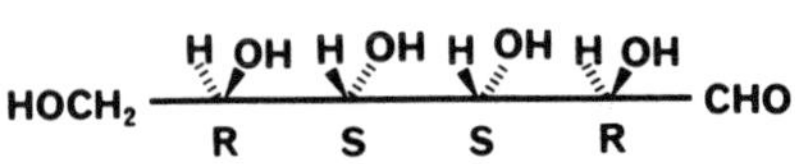

D-(+)-Galactose *****

286. L-Galactose ($C_6H_{12}O_6$)

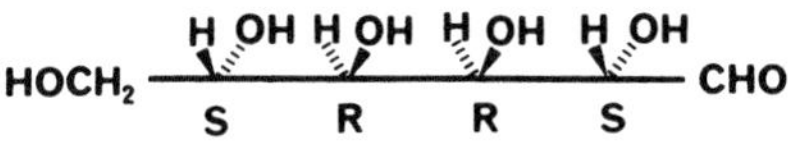

L-Galactose *

287. 2-Deoxy-D-lyxohexose ($C_6H_{12}O_5$)

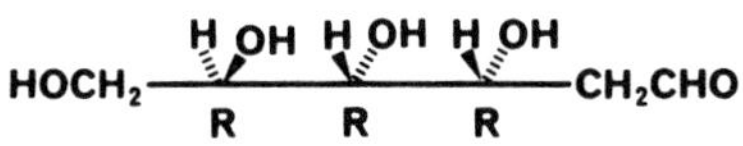

2-Deoxy-D-galactose *

288. 1,6-Anhydro-3,4-*O*-(1-methylethylidene)-β-D-galactopyranose 2-(4-methylbenzenesulfonate) ($C_{16}H_{20}O_7S$)

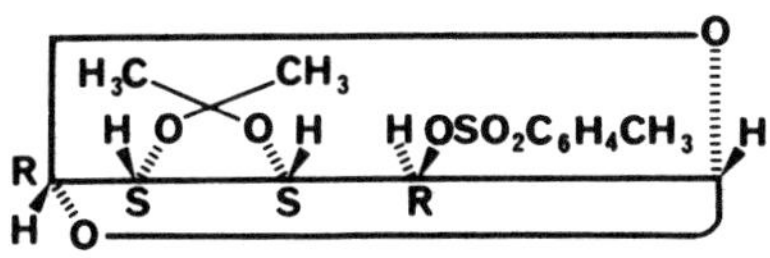

1,6-Anhydro-3,4-*O*-isopropylidene-2-tosyl-D-galactose *

289. 6-Deoxy-6-iodo-1,2 : 3,4-bis-*O*-(1-methylethylidene)-α-D-galactopyranose ($C_{12}H_{19}IO_5$)

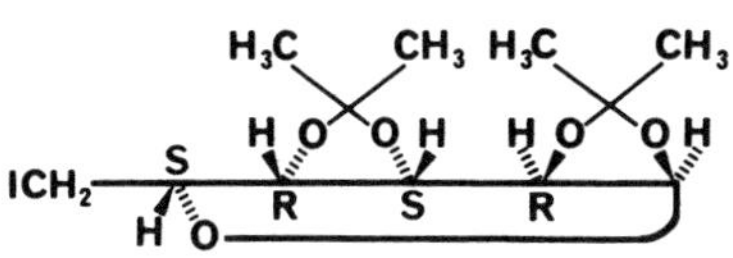

6-Deoxy-1,2 : 3,4-di-*O*-isopropylidene-6-iodo-α-D-galactopyranose **

290. 1,2 : 3,4-Bis-*O*-(1-methylethylidene)-α-D-galactopyranose ($C_{12}H_{20}O_6$)

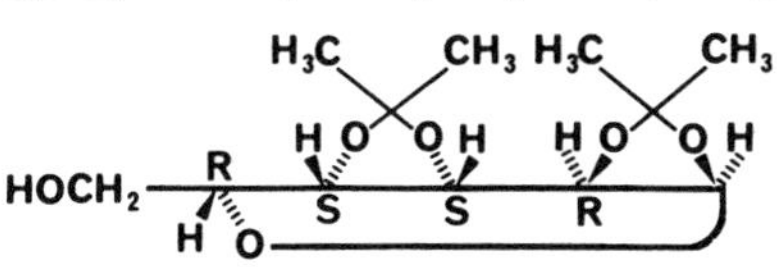

1,2 : 3,4-Di-*O*-isopropylidene-D-galactopyranose **

291. D-Galacturonic acid ($C_6H_{10}O_7$)

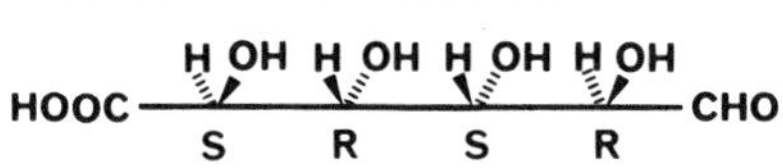

D-(+)-Galacturonic acid [D-galactopyranuronic acid] ***

292. D-Gluconic acid ($C_6H_{12}O_7$)

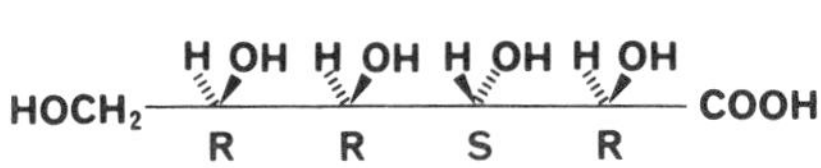

D-Gluconic acid *****

293. D-Gluconic acid δ-lactone ($C_6H_{10}O_6$)

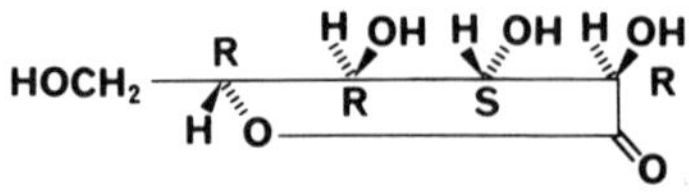

D-Gluconic acid δ-lactone [D-glucono-1,5-lactone] *****

294. L-Gluconic acid δ-lactone ($C_6H_{10}O_6$)

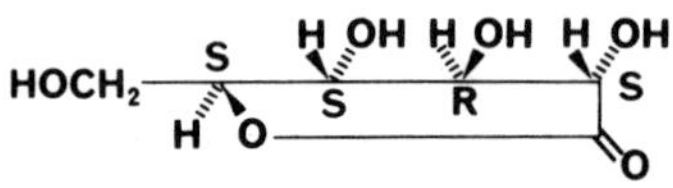

L-Glucono-1,5-lactone *

295. 2-Amino-2-deoxy-D-glucose ($C_6H_{13}NO_5$)

D-(+)-Glucosamine *****

296. D-Glucose ($C_6H_{12}O_6$)

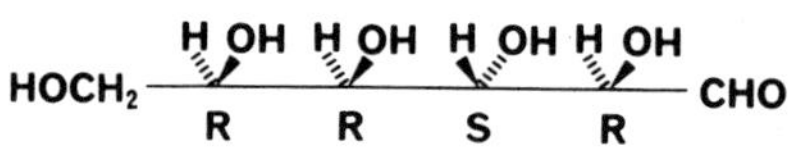

D-(+)-Glucose *****

297. 2-Deoxy-D-arabinohexose ($C_6H_{12}O_5$)

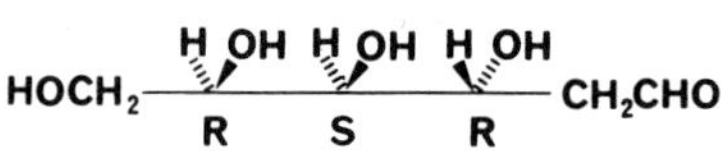

2-Deoxy-D-glucose **

298. L-Glucose ($C_6H_{12}O_6$)

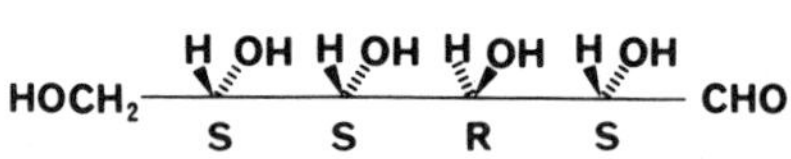

L-(−)-Glucose *

299. 6-*O*-Benzoyl-1,2-*O*-(1-methylethylidene)-α-D-glucofuranose 3-acetate 5-(methylsulfonate) ($C_{19}H_{24}O_9S$)

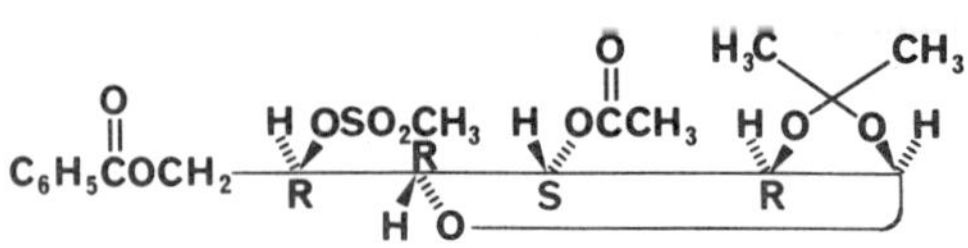

3-*O*-Acetyl-6-*O*-benzoyl-5-*O*-methylsulfonyl-1,2-isopropylidene-α-D-glucofuranose **

300. 1,6-Anhydro-β-D-glucopyranose ($C_6H_{10}O_5$)

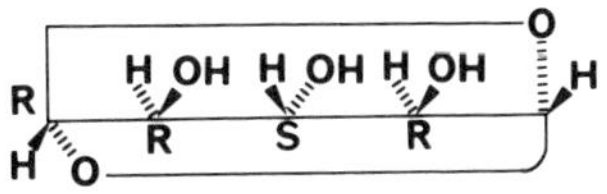

1,6-Anhydro-β-D-glucopyranose [levoglucosan] *

301. (*R*)-1,2-*O*-(2,2,2-Trichloroethylidene)-α-D-glucofuranose ($C_8H_{11}Cl_3O_6$)

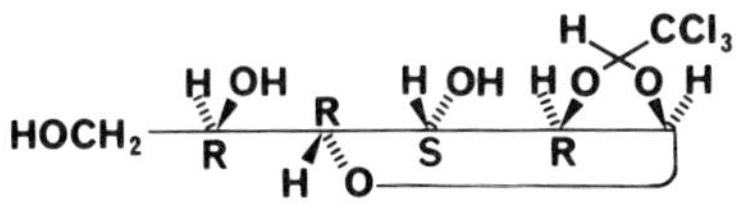

α-Chloralose *****

302. 1,2 : 5,6-Bis-*O*-(1-methylethylidene)-α-D-glucofuranose ($C_{12}H_{20}O_6$)

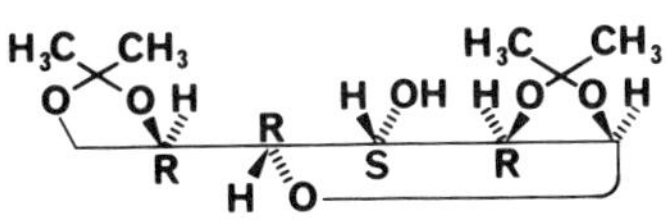

1,2 : 5,6-Di-*O*-isopropylidene-D-glucofuranose [diacetone-D-glucose] ****

303. 1,2-*O*-(1-Methylethylidene)-α-D-glucofuranose ($C_9H_{16}O_6$)

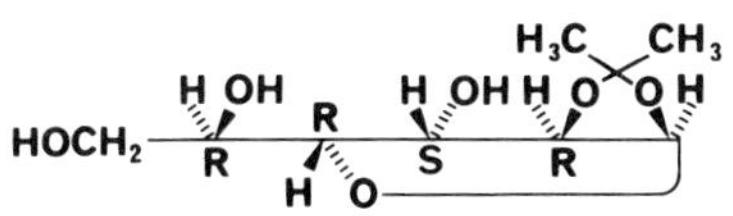

1,2-*O*-Isopropylidene-D-glucofuranose ***

304. Methyl-4,6-*O*-phenylmethylene)-α-glucopyranoside ($C_{14}H_{18}O_6$)

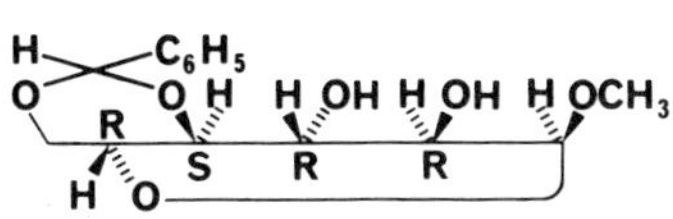

Methyl-4,6-*O*-benzylidene-α-D-glucopyranoside **

305. D-Glucuronic acid ($C_6H_{10}O_7$)

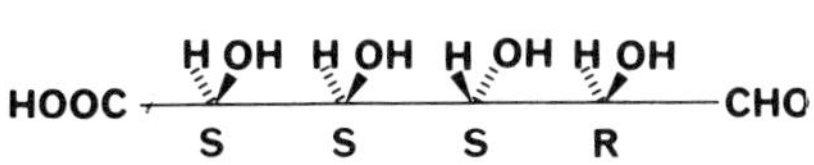

D-Glucuronic Acid ****

306. D-Glucuronic acid γ-lactone ($C_6H_8O_6$)

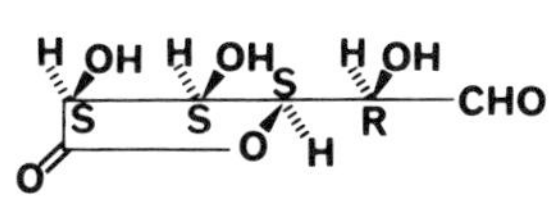

D-Glucurono-6,3-lactone [D-glucurone] *****

307. D-Gulonic acid γ-lactone ($C_6H_{10}O_6$)

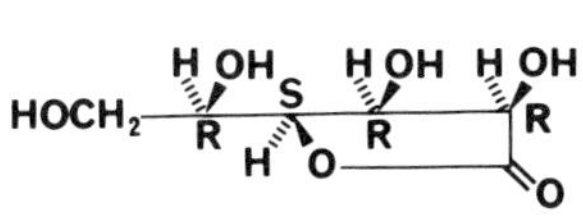

D-Gulonic acid γ-lactone [D-gulono-1,4-lactone] ****

308. 2,3 : 4,6-Bis-*O*-(1-methylethylidene)-α-L-xylo-2-hexulofuranosonic acid ($C_{12}H_{18}O_7$)

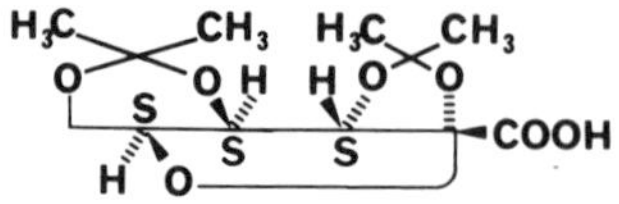

2,3 : 4,6-di-*O*-Isopropylidene-2-keto-L-gulonic acid ****

309. 1,2 : 5,6-Bis-*O*-(1-methylethylidene)-α-D-gulofuranose ($C_{12}H_{20}O_6$)

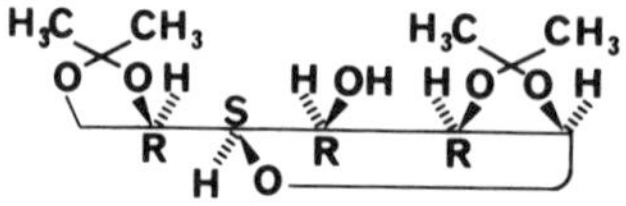

1,2 : 5,6-Di-*O*-isopropylidene-α-D-gulose *

310. 1,2 : 5,6-Bis-*O*-(1-methylethylidene)-α-D-ribohexofuranos-3-ulose ($C_{12}H_{18}O_6$)

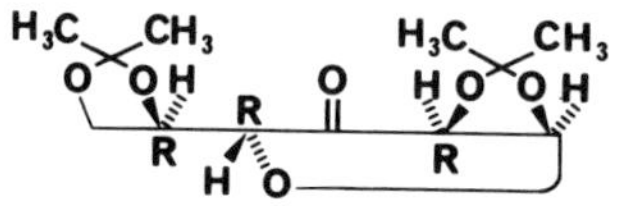

1,2 : 5,6-Di-*O*-isopropylidene-α-D-ribo-3-hexofuranulose *

311. 1,2 : 5,6-Bis-*O*-(1-methylethylidene)-α-D-allofuranose ($C_{12}H_{20}O_6$)

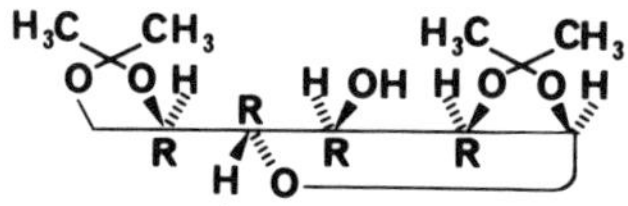

1,2 : 5,6-Di-*O*-isopropylidene-α-D-allo-furanose **

312. D-Mannitol ($C_6H_{14}O_6$)

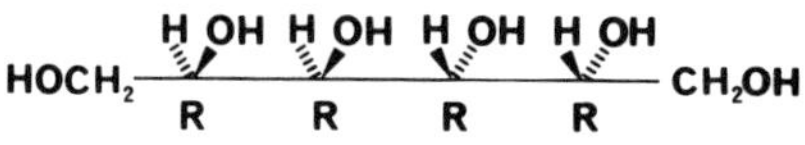

D-Mannitol *****

313. 1,2 : 5,6-Bis-*O*-(1-methylethylidene)-D-mannitol ($C_{12}H_{22}O_6$)

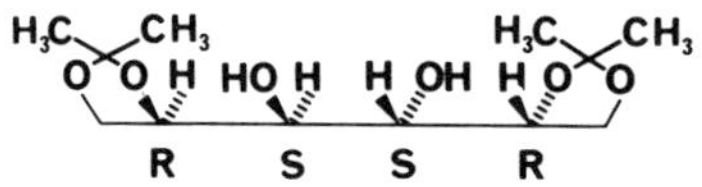

1,2 : 5,6-Di-*O*-isopropylidene-D-mannitol *

1 step (55%) ex D-mannitol (*59, 84*)

314. L-Mannonic acid γ-lactone ($C_6H_{10}O_6$)

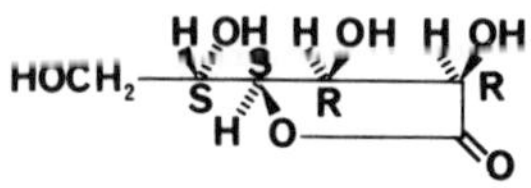

L-Mannonic acid γ-lactone [L-mannono-1,4-lactone] **

315. L-Mannose ($C_6H_{12}O_6$)

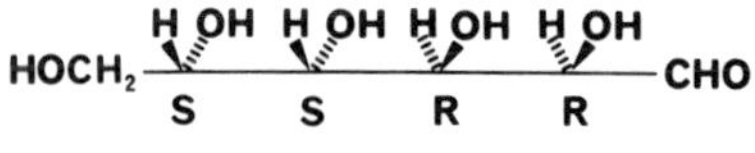

L-(−)-Mannose *

316. D-Mannose ($C_6H_{12}O_6$)

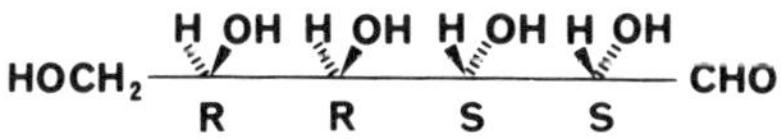

D-(+)-Mannose ****

317. 6-Deoxy-L-mannose ($C_6H_{12}O_5$)

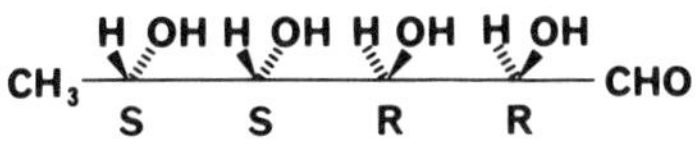

L-(+)-Rhamnose ***

318. D-Glucaric acid ($C_6H_{10}O_8$)

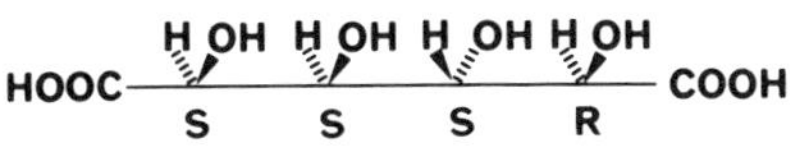

D-Saccharic acid
[glucaric acid] *****

319. D-Glucaric acid γ-lactone ($C_6H_8O_7$)

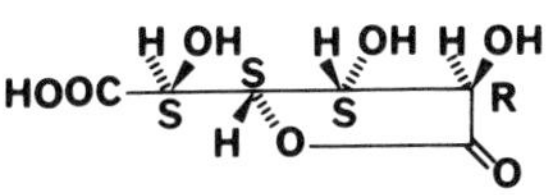

D-Saccharic acid 1,4-lactone *

320. D-Glucitol ($C_6H_{14}O_6$)

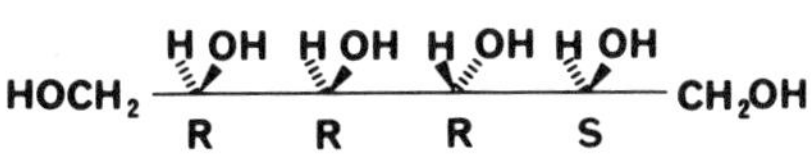

D-Sorbitol *****

Six- or More Carbon Chain: Carbon Substituted

321. (*R*)-6-(Phenylmethoxy)-3-methyl-1-hexanol ($C_{14}H_{22}O_2$)

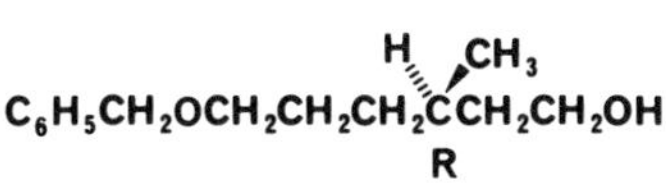

(*R*)-6-Benzyloxy-3-methylhexanol

5 steps (94%) ex (*R*)-citronellol (*111*)

322. (*R*)-3,7-Dimethyl-6-octen-1-al ($C_{10}H_{18}O$)

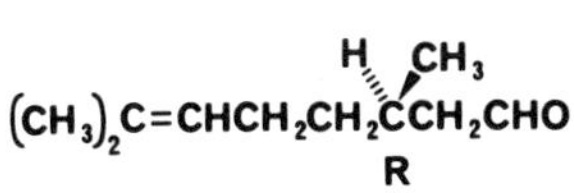

(+)-Citronellal ****

323. (*S*)-3,7-Dimethyl-1,6-octadiene ($C_{10}H_{18}$)

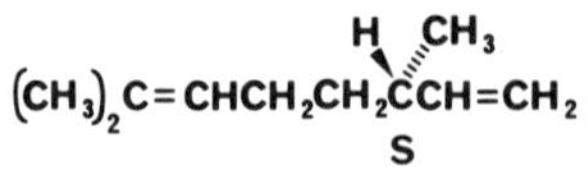

(+)-β-Citronellene **

324. (*S*)-3,7-Dimethyl-6-octen-1-ol ($C_{10}H_{20}O$)

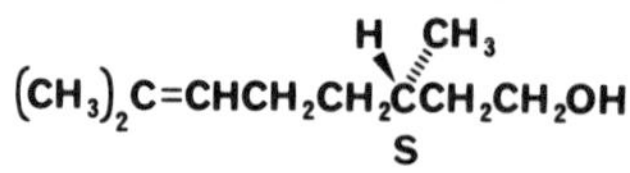

(*S*)-(−)-β-Citronellol *

325. (*R*)-5,7-Dimethyl-1,6-octadiene ($C_{10}H_{18}$)

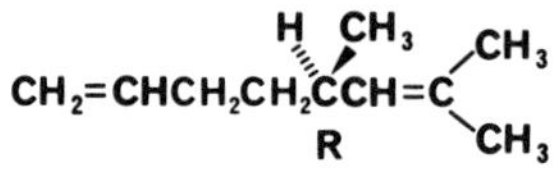

(*R*)-(+)-5,7-Dimethyl-1,6-octadiene [isocitronellene] *

326. (*R*)-3-Methylhexanedioic acid ($C_7H_{12}O_4$)

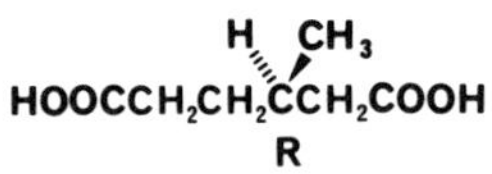

(+)-3-Methyladipic acid **

327. (*R*)-3-Methylcyclohexanone ($C_7H_{12}O$)

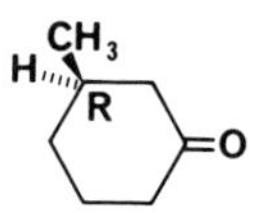

(+)-Methylcyclohexanone **

328. (*S*)-6-Methyl-2-cyclohexen-1-one ($C_7H_{10}O$)

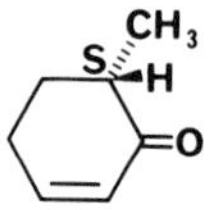

(*S*)-(+)-6-Methylcyclohex-2-enone

4 steps ex (−)-carvone (*75*)

329. (*S*)-4-Methyl-1-hexanol ($C_7H_{16}O$)

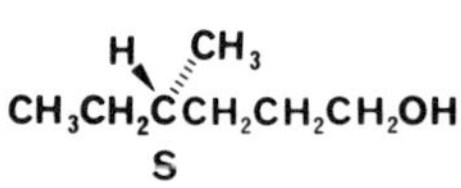

(*S*)-(+)-4-Methylhexan-1-ol

4 steps (54%) ex (*R*)-citronellol (*116*)

330. (*R*)-7-Hydroxy-4-methylheptanoic acid methyl ester ($C_9H_{18}O_3$)

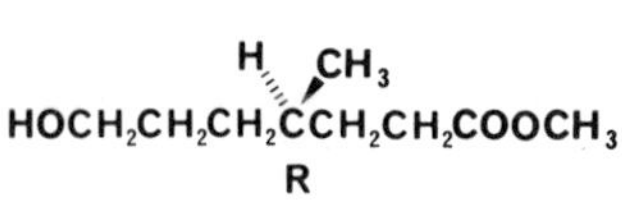

Methyl (*R*)-(−)-7-hydroxy-4-methylheptanoate

7 steps (41%) ex (*R*)-(+)-citronellol (*90*)

331. (*R*)-1-Methyl-4-(1-methylethenyl)cyclohexene ($C_{10}H_{16}$)

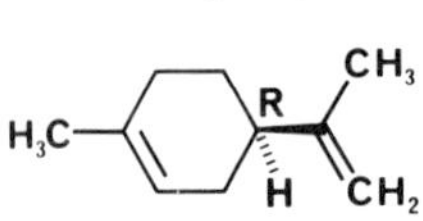

d-(+)-Limonene *****

332. (*S*)-1-Methyl-4-(1-methylethenyl)cyclohexene ($C_{10}H_{16}$)

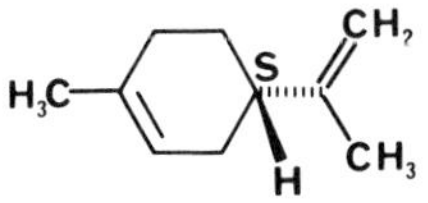

l-(−)-Limonene *****

333. (*R*)-1-Methyl-4-(1-methylethyl)cyclohexene ($C_{10}H_{18}$)

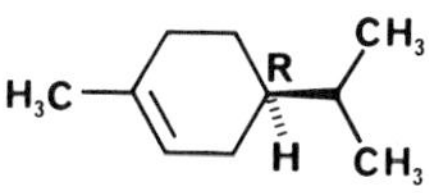

(+)-*p*-Menth-1-ene *

334. (3*R*)-*trans*-3-Methyl-6-(1-methylethenyl)cyclohexene ($C_{10}H_{16}$)

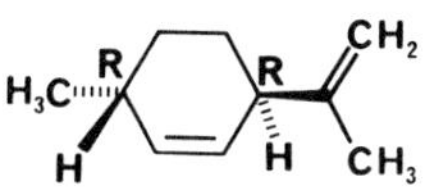

(*R*)-(+)-*trans*-Isolimonene [*p*-mentha-2,8-diene] *

335. [*R*-(*R**,*R**)]-β,4-Dimethyl-3-cyclohexene-1-ethanol ($C_{10}H_{18}O$)

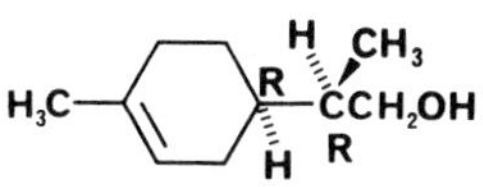

(+)-*p*-Menth-1-en-9-ol *

336. (*S*)-4-(1-Methylethenyl)-1-cyclohexene-1-carboxaldehyde ($C_{10}H_{14}O$)

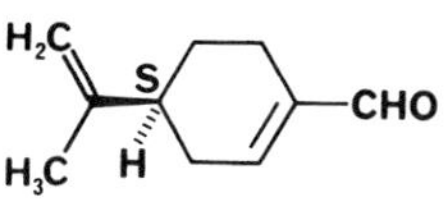

l-Perillaldehyde **

337. (*S*)-4-(1-Methylethenyl)-1-cyclohexene-1-carboxylic acid ($C_{10}H_{14}O_2$)

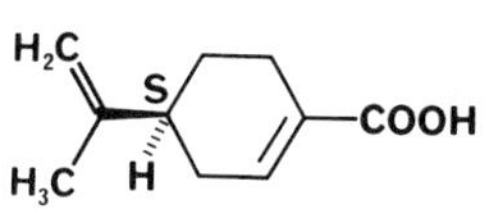

l-Perillic acid *

338. (*S*)-4-(1-Methylethenyl)-1-cyclohexene-1-methanol ($C_{10}H_{16}O$)

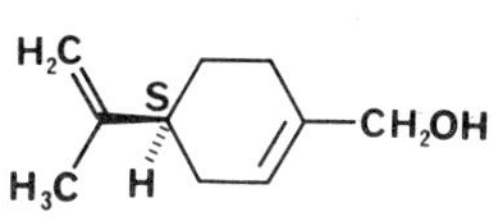

l-Perillyl alcohol ***

339. (*R*)-2-Methyl-5-(1-methylethenyl)-2-cyclohexen-1-one ($C_{10}H_{14}O$)

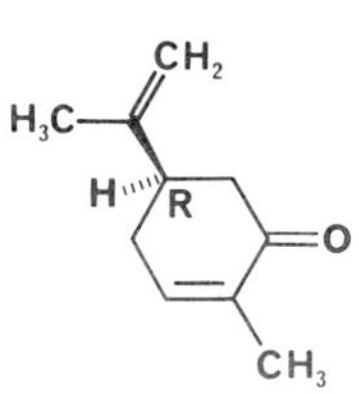

l-(−)-Carvone ****

340. (*S*)-2-Methyl-5-(1-methylethenyl)-2-cyclohexen-1-one ($C_{10}H_{14}O$)

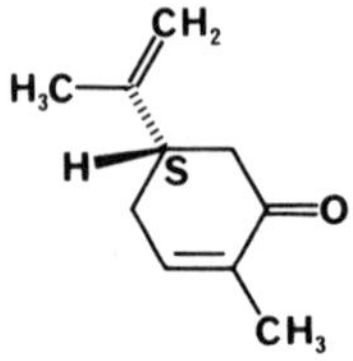

d-(+)-Carvone ***

341. (2*S*)-*trans*-5-Methyl-2-(1-methylethyl)cyclohexanone ($C_{10}H_{18}O$)

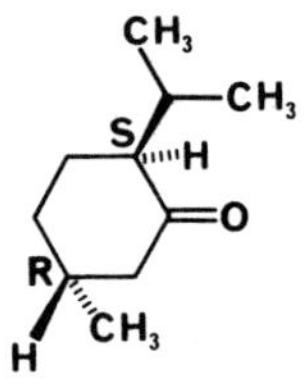

l-Menthone ****

342. [1*R*-(1α,2β,5α)]-5-Methyl-2-(1-methylethyl)cyclohexanol ($C_{10}H_{20}O$)

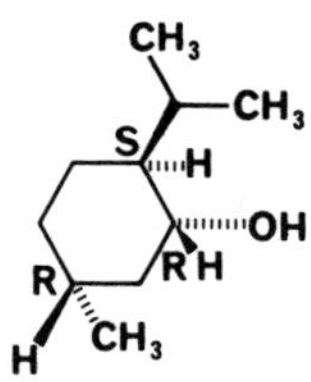

l-(−)-Menthol ****

343. [1*S*-(1β,2α,5β)]-5-Methyl-2-(1-methylethyl)cyclohexanol ($C_{10}H_{20}O$)

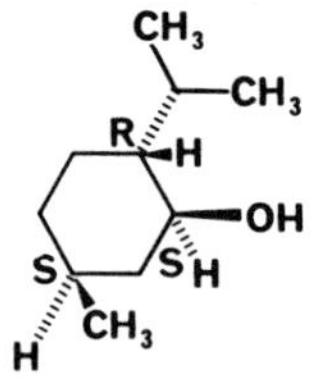

d-(+)-Menthol ****

344. [1*S*-(1β,2α,5α)-5-Methyl-2-(1-methylethyl)cyclohexanol ($C_{10}H_{20}O$)

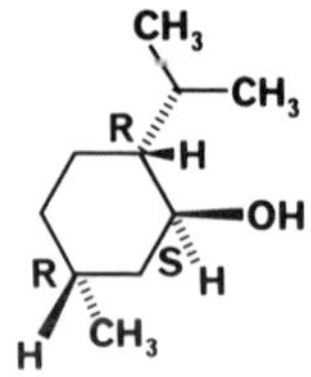

(+)-Isomenthol ****

345. [1*S*-(1β,2β,5α)]-5-Methyl-2-(1-methylethyl)cyclohexanol ($C_{10}H_{20}O$)

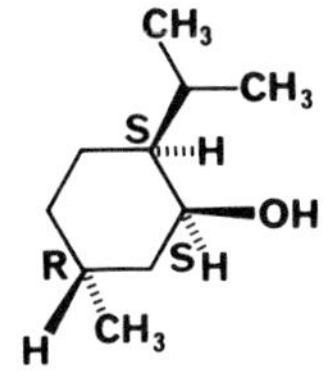

(+)-Neomenthol *

346. (*R*)-5-Methyl-2-(1-methylethylidene)cyclohexanone ($C_{10}H_{16}O$)

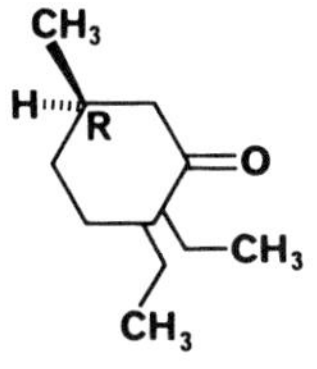

(*R*)-(+)-Pulegone ****

347. (1*R*)-4,7,7-Trimethyl-3-oxo-2-oxabicyclo[2.2.1]heptane-1-carboxylic acid ($C_{10}H_{14}O_4$)

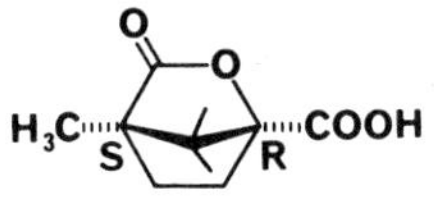

(−)-Camphanic acid *

348. (1*R*)-2,2-Dimethyl-3-methylenebicyclo[2.2.1]heptane ($C_{10}H_{16}$)

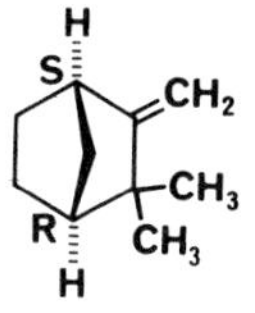

(+)-Camphene *****

349. (1*R*)-*cis*-1,2,2-Trimethyl-1,3-cyclopentanedicarboxylic acid ($C_{10}H_{16}O_4$)

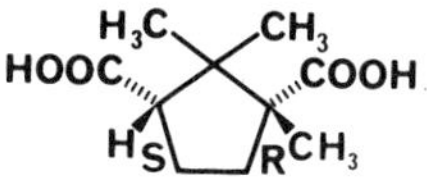

D-(+)-Camphoric acid ****

350. (1*R*)-1,7,7-Trimethylbicyclo[2.2.1]heptan-2-one ($C_{10}H_{16}O$)

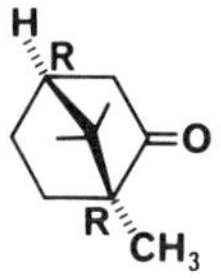

d-(+)-Camphor *****

351. (1*S*)-1,7,7-Trimethylbicyclo[2.2.1]heptan-2-one ($C_{10}H_{16}O$)

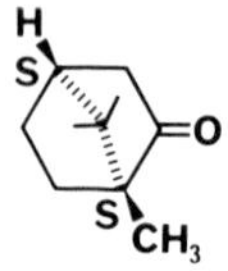

l-(−)-Camphor **

352. (1*S*)-*endo*-1,7,7-Trimethylbicyclo[2.2.1]heptan-2-ol ($C_{10}H_{18}O$)

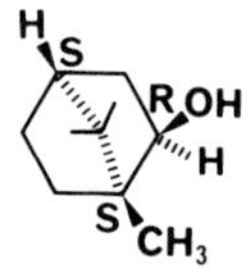

l-(−)-Borneol ****

353. (1*R*)-*endo*-3-Bromo-1,7,7-trimethylbicyclo[2.2.1]heptan-2-one ($C_{10}H_{15}BrO$)

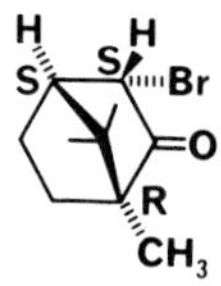

endo-3-Bromo-*d*-camphor ****

354. (1*S*)-7,7-Dimethyl-2-oxobicyclo[2.2.1]heptane-1-methanesulfonic acid ($C_{10}H_{16}O_4S$)

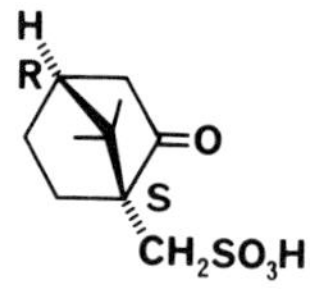

d-10-Camphorsulfonic acid ****

355. (1*R*)-7,7-Dimethyl-2-oxobicyclo[2.2.1]heptane-1-methanesulfonic acid ($C_{10}H_{16}O_4S$)

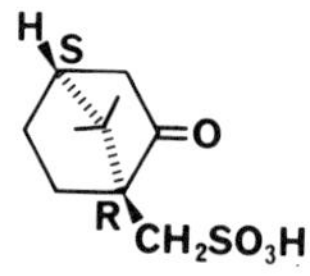

l-10-Camphorsulfonic acid ***

356. (1*S*)-3,7,7-Trimethylbicyclo[4.1.0]hept-3-ene ($C_{10}H_{16}$)

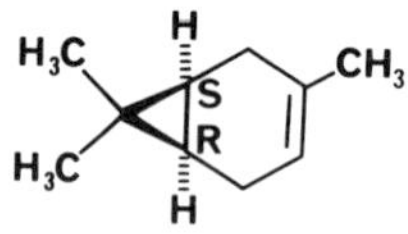

(+)-3-Carene *

357. (1*S*)-1,3,3-Trimethylbicyclo[2.2.1]heptan-2-one ($C_{10}H_{16}O$)

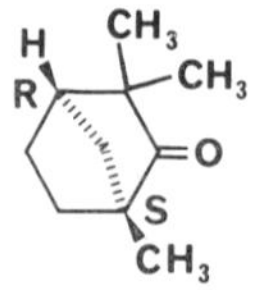

d-(+)-Fenchone ****

358. (1*R*)-1,3,3-Trimethylbicyclo[2.2.1]heptan-2-one ($C_{10}H_{16}O$)

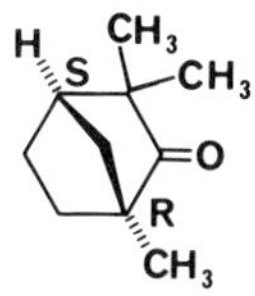

l-(−)-Fenchone ****

359. [1*S*-(1α,2α,5α)]-6,6-Dimethylbicyclo[3.1.1]heptane-2-methanol ($C_{10}H_{18}O$)

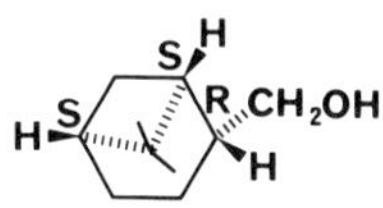

(−)-*cis*-Myrtanol *

360. [1*S*-(1α,2β,5α)]-6,6-Dimethylbicyclo[3.1.1]heptane-2-methanol ($C_{10}H_{18}O$)

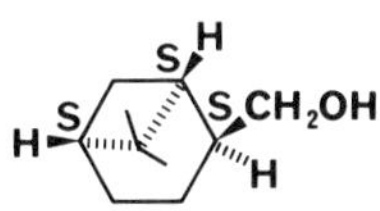

(−)-*trans*-Myrtanol *

361. (1*R*)-6,6-Dimethylbicyclo[3.1.1]hept-2-ene-2-methanol ($C_{10}H_{16}O$)

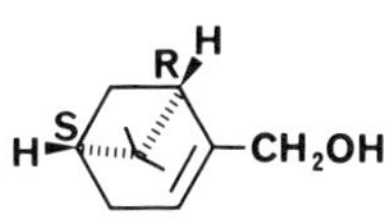

(−)-Myrtenol ***

362. (1*R*)-6,6-Dimethylbicyclo[3.1.1]hept-2-ene-2-carboxaldehyde ($C_{10}H_{14}O$)

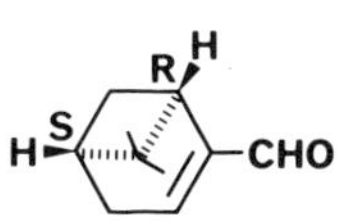

(−)-Myrtenal ***

363. (1*S*)-6,6-Dimethylbicyclo[3.1.1]hept-2-ene-2-ethanol ($C_{11}H_{18}O$)

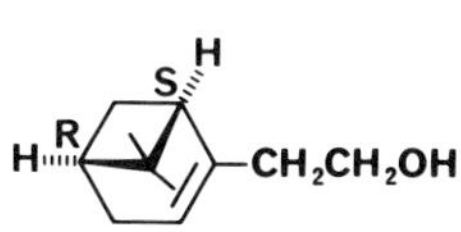

Nopol *****

364. (1*R*)-2,6,6-Trimethylbicyclo[3.1.1]hept-2-ene ($C_{10}H_{16}$)

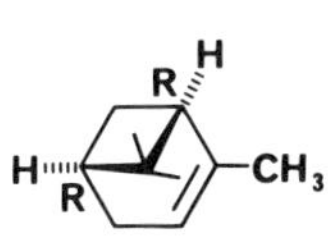

(−)-α-Pinene *****

365. (1*S*)-2,6,6-Trimethylbicyclo[3.1.1]hept-2-ene ($C_{10}H_{16}$)

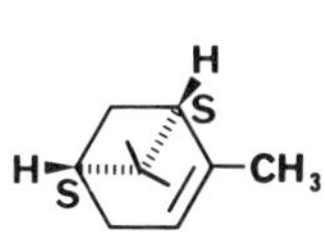

(−)-α-Pinene *****

366. (1*S*)-6,6-Dimethyl-2-methylenebicyclo[3.1.1]heptane ($C_{10}H_{16}$)

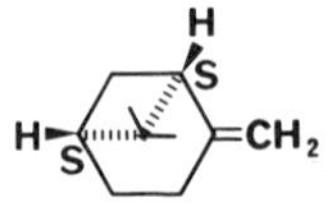

(−)-β-Pinene *****

367. [1*R*-(1β,2β,3α,5β)]-2,6,6-Trimethylbicyclo[3.1.1]heptan-3-ol ($C_{10}H_{18}O$)

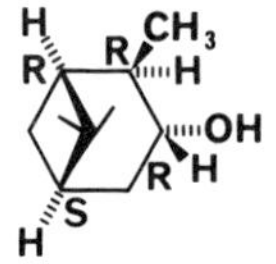

(−)-Isopinocampheol ***

368. (1*S*)-*cis*-4,6,6-Trimethylbicyclo[3.1.1]hept-3-en-2-one ($C_{10}H_{14}O$)

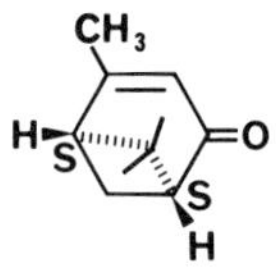

l-Verbenone **

Seven- or More Carbon Chain: Heteroatom Substituted

369. D-Glycero-D-guloheptonic acid ($C_7H_{14}O_8$)

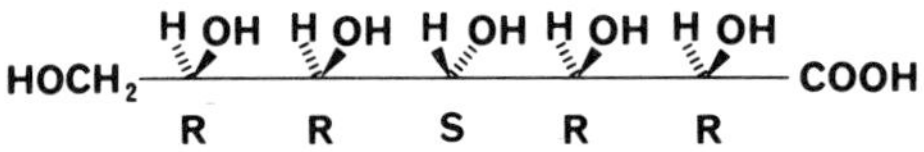

Glucoheptonic acid *****

370. D-Glycero-D-guloheptonic acid γ-lactone ($C_7H_{12}O_7$)

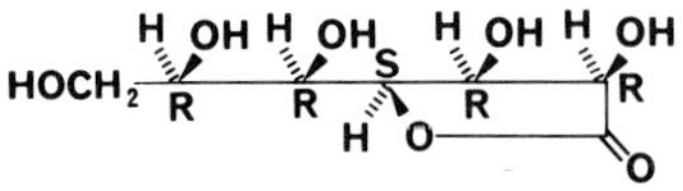

α-D-Glucoheptonic acid γ-lactone
[D-glucoheptono-1,4-lactone] *****

371. D-Glucoheptose ($C_7H_{14}O_7$)

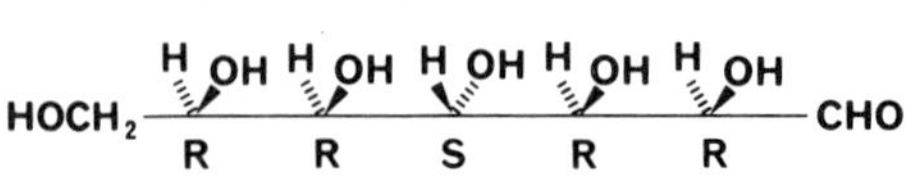

D-Glucoheptose *

372. D-Manno-2-heptulose ($C_7H_{14}O_7$)

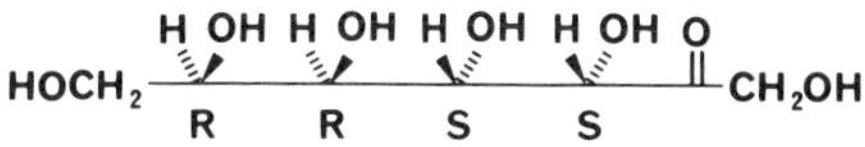

D-Mannoheptulose *

373. (*R*)-2-Octanol ($C_8H_{18}O$)

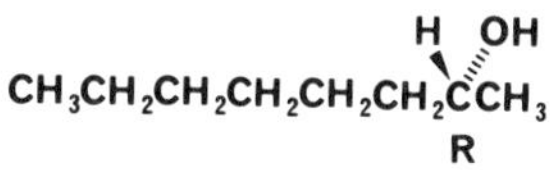

l-2-Octanol *

374. (*S*)-2-Octanol ($C_8H_{18}O$)

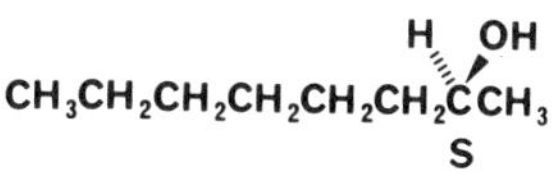

d-2-Octanol *

375. [1*R*-(1α,3α,4α,5β)]-1,3,4,5-Tetrahydroxycyclohexanecarboxylic acid ($C_7H_{12}O_6$)

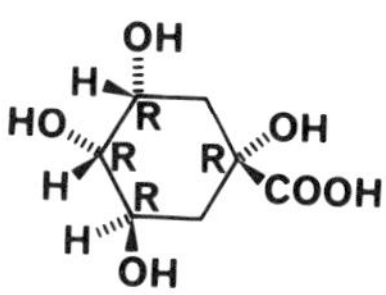

D-Quinic Acid ****

III. Approaches to Synthetic Targets Using Chiral Carbon Building Blocks

The commercial sources of chiral carbon atoms, and the chiral synthons derived therefrom, have been employed in syntheses of a wide variety of compounds. The majority of these synthetic targets have been, in themselves, naturally occurring compounds whose preparation was desired for the reasons delineated previously. These synthetic targets can be divided into the following groups: (A) alkaloids, (B) arachidonic acid metabolites, (C) compounds with chemotherapeutic activity, (D) compounds with pharmacological activity, (E) insect pheromones and related compounds, (F) sugars, (G) vitamins, and (H) miscellaneous compounds (including terpenoids). These rather arbitrary groups have been chosen because they (usually) include structurally related compounds toward which substantial synthetic effort has been invested. Other groups or subgroups of compounds for which this synthetic approach could be, but has not yet been applied should become apparent to the reader.

Our goal in the analysis of the existing syntheses is to demonstrate the conceptual and mechanical means by which the starting compounds, often of amazingly divergent structure, are converted to structurally related products. This goal has had several effects on the material presented. Virtually all of what might be termed "insertion syntheses," that is, building onto a chiral framework without creating new asymmetric centers or modifying the existing ones, have been excluded. The syntheses are presented within the narrow focus of the way in which one chiral molecule is converted to another; novel chemistry and/or reaction conditions (often employed) are not discussed. Finally, other approaches (e.g., asymmetric synthesis) do exist for many of the target molecules. These too are beyond the scope of this work.

As an aid in the conceptualization process, we have attempted to represent starting materials and products from the same visual perspective. Chiral centers carried intact into the target molecule are circled. Chiral centers that are modified by inversion (including double inversion) or by destruction of the chirality are starred.

A. Alkaloids

1. Syntheses Using Pictet–Spengler or Bischler–Napieralski Ring Closures

Virtually all of the chiral syntheses of alkaloids have employed amino acids as starting materials. The obvious objective in these syntheses was to make use of the chiral carbon–nitrogen system. Particularly elegant uses of the approach are seen in the syntheses of tetrahydroisoquinoline and tetrahydro-β-carboline alkaloids done by Yamada and co-workers (Schemes 1 and 2). These syntheses are based on the observation that 1,3-asymmetric induction occurs in the Pictet–Spengler reaction of enantiomeric amino acids with aldehydes. Thus, condensation of (*S*)-tryptophan with acetaldehyde gave the *cis*-β-carboline **378a**. Conversion to the *N*-benzylnitrile **379a** was followed by reductive decyanation with sodium borohydride and catalytic debenzylation to give (*S*)-tetrahydroharman (**380a**) (*117–119*). Similarly, (*S*)-trytophanamide (**376**) was reacted with 5-chloropentanal (**377b**) to give, ultimately, the Anacardiaceae alkaloid (*S*)-1,2,3,4,6,7,12,12b-octahydroindolo[2,3-*a*]quinolizine (**380b**) (*119*). The optical rotations of synthetic **380a** and **380b** compared well with values reported for the natural isolates, indicating that the materials were probably of high enantiomeric purity. It is known (*120–122*), however, that the Pictet–Spengler reaction produces mixtures of cis and

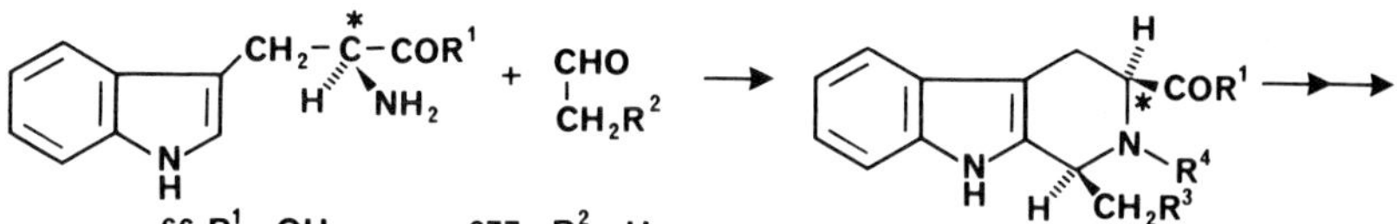

66 $R^1 = OH$ 377a $R^2 = H$ 378a $R^1 = OH$ $R^3 = R^4 = H$

376 $R^1 = NH_2$ 377b $R^2 = (CH_2)_3Cl$ 378b $R^1 = NH_2$ $R^3, R^4 = (CH_2)_3$

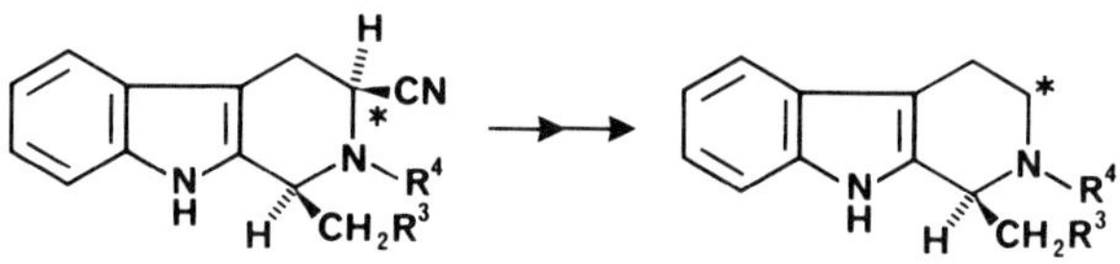

379a $R^3 = H$ $R^4 = CH_2C_6H_5$ 380a $R^3 = R^4 = H$

379b $R^3, R^4 = (CH_2)_3$ 380b $R^3, R^4 = (CH_2)_3$

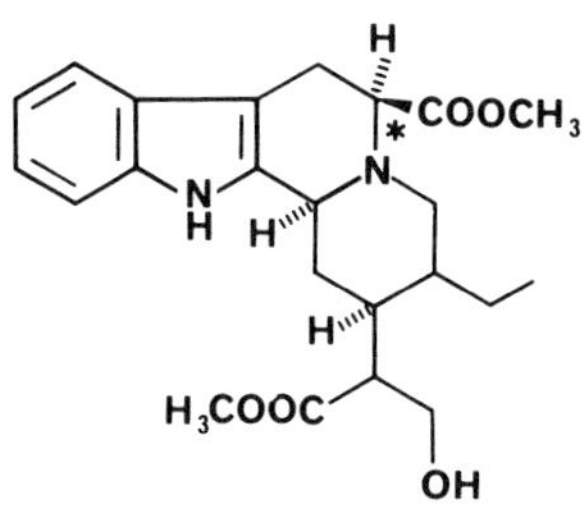

381

Scheme 1. Tetrahydro-β-carboline synthesis.

trans isomers. The enantiomeric results in the syntheses of **380a** and **380b** result from enantiomeric purification of **378a** and **378b** by crystallization, followed by nonracemizing removal of the carboxylate function. Cook (*120*) has shown that a stereospecific Pictet–Spengler reaction is observed when *N*-benzyltryptophan methyl esters are employed. Interestingly, this reaction leads exclusively to the *trans*-β-carbolines, as opposed to the *cis*-β-carbolines, which are obtained from N-unsubstituted tryptophans as the major products. To employ the Cook method in the preparation of **380a** and **380b** would require the unnatural (*R*)-*N*-benzyltryptophan.

The Pictet–Spengler reaction of tryptophans has also been employed in the synthesis of methyladirubine (**381**) (*123*). The use of a racemic aldehydic synthon led to a major diastereomer separation problem and a very low yield of the desired intermediate, precluding determination of the degree of asymmetric induction in the condensation reaction.

(*S*)-Dihydroxyphenylalanine [(*S*)-dopa] derivatives also led to 1,3 asymmetric induction. Condensation of **382a** with the glycidate **383a**, which serves as an aldehyde precursor, gave a diastereomeric mixture of *trans*-

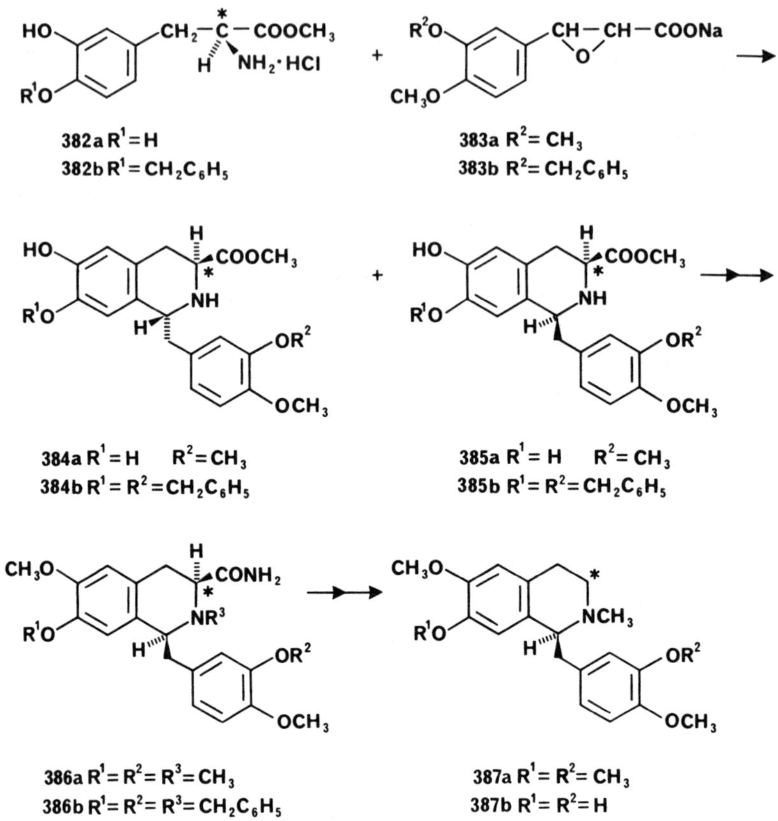

Scheme 2. Isoquinoline syntheses.

and *cis*-isoquinolines **384a** and **385a**, with the latter predominating (ratio 1 : 3.2) (*124–126*). Separation of the major diastereomer **385a** was followed by conversions comparable to those described in the carboline series to give (*S*)-laudanosine (**387a**). In an entirely similar manner, the monobenzylated (*S*)-dopa **382b** [actually prepared from (*S*)-tyrosine] was converted to (*S*)-reticuline (**387b**) (*127*). In this sequence, however, the ratio (1 : 2.1) of diastereomeric condensation products **384b** and **385b** was less favorable.

It has been shown (*126*) that the dimethyl ether of **385a** can be isomerized with sodium methylate to an equilibrium 2 : 1 trans:cis mixture. This provided sufficient intermediate to accomplish a synthesis of the unnatural (*R*)-laudanosine. A similar preparation of (*R*)-reticuline (not reported) would constitute a formal synthesis of morphine (*128*).

Pictet–Spengler reactions in which the chirality resides in the aldehyde have also been reported (Scheme 3). In a particularly illustrative example, lactone **198** (from glutamic acid) was converted to both (+)- (*129*) and (−)-quebrachamine (*130*). Thus, the trityl ether (**202**) of **198** was stereoselectively diakylated, giving **388**. Deprotection and oxidation with periodate gave aldehyde **389**, which was condensed with tryptamine to give the diastereomeric amides **390** in approximately equal amounts. That a mixture was obtained was of little consequence since both diastereomers could be converted to the desired *Aspidosperma* alkaloid **391**. To prepare (−)-quebrachamine (**394**) a formal inversion of the quaternary center of

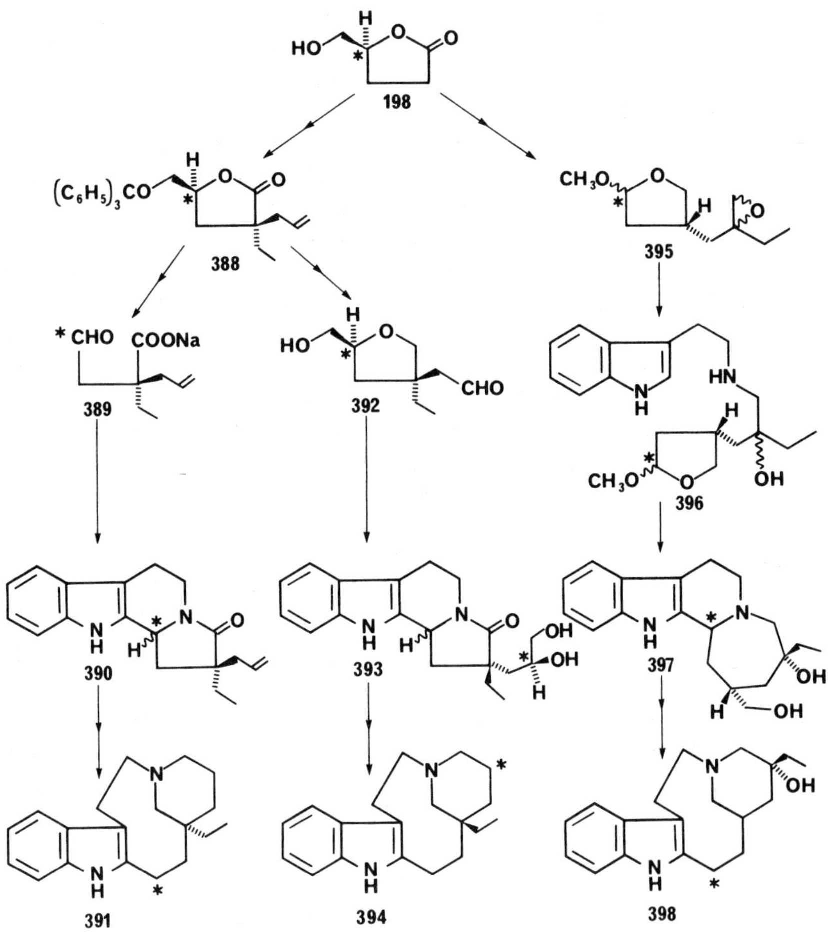

Scheme 3. (+)- or (−)-Quebrachamine and (−)-velbanamine.

388 was required. This was achieved by detritylation and ozonolysis to aldehyde **392**. Reaction of this aldehyde with tryptamine gave the diastereomer mixture **393** which, in a sequence of reactions similar to the **390** → **391** transformation, was converted to (−)-quebrachamine (**394**).

The lactone **198** was also converted, in several steps including a nonselective epoxidation, to the diasteromeric acetals **395**. Tryptamine led to epoxide opening, giving **396**. Treatment of **396** with hot aqueous acetic acid caused deacetalyzation and ring closure. Evidence of stereoselectivity in the cyclization was the finding of two major (because of the carbinol center) and two minor diastereomers. One of the major diastereomers (**397**, C-1 configuration unreported) was converted to (−)-velbanamine (**398**) (*131*), a degradation product of the important antitumor agents vincristine and vinblastine (*132*). The second major diastereomer gave (+)-isovelbanamine. The stereospecific dialkylation of the trityl ether of **198** also formed the basis of preparations (*133*) of two compounds having the gibbane framework.

A further use of the lactone **198** in the synthesis of chiral alkaloids is found in the preparation of (−)-antirhine (**409**, Scheme 4) (*134*). In this case, however, a Bischler–Napieralski ring closure was employed. Careful allylation of the trityl ether (**202**) of **198** provided **399**, which was reduced and deprotected to **400**. Periodate cleavage and oxidation yielded lactone **401**, which was allylated a second time giving **402**. Ring opening with cyanide produced the nitrilo acid **403**. Iodolactonization followed by a four-step removal of the resulting iodomethyl group gave lactone **404**. Ring opening, this time with sodiumphenylselenide, followed by acid reduction gave **405** which, on treatment with acid, yielded the δ-lactone **406**. Ozonolysis and treatment with base gave the unsaturated aldehyde **407**. Reductive condensation of this compound with tryptamine occurred with simultaneous lactone → lactam rearrangement. The resultant **408** was then converted in a straightforward manner to (−)-antirhine.

1,3-Transfer of a symmetry similar to that in Pictet–Spengler reactions has also been reported in phenolic oxidative couplings of *N*-benzyltyrosine derivatives (Scheme 5). Treatment of the *N*-trifluoroacetyl ester **410** with thallium trifluoroacetate was followed by ester → amide conversion yielding **411**. This compound was apparently a diastereomeric mixture. Removal of the trifluoroacetyl group, however, accompanied by spontaneous Michael reaction, gave only the diastereomerically homogeneous **412**. The second diastereomer of **411** did not undergo the Michael addition, presumably because of unfavorable steric interactions. Compound **412** was then converted to (+)-maritidine (**413**) (*135, 136*). Use of a more complex starting material resulted in an analogous synthesis of (−)-galanthamine (**414**), but involving inversion of a key intermediate (*137, 138*).

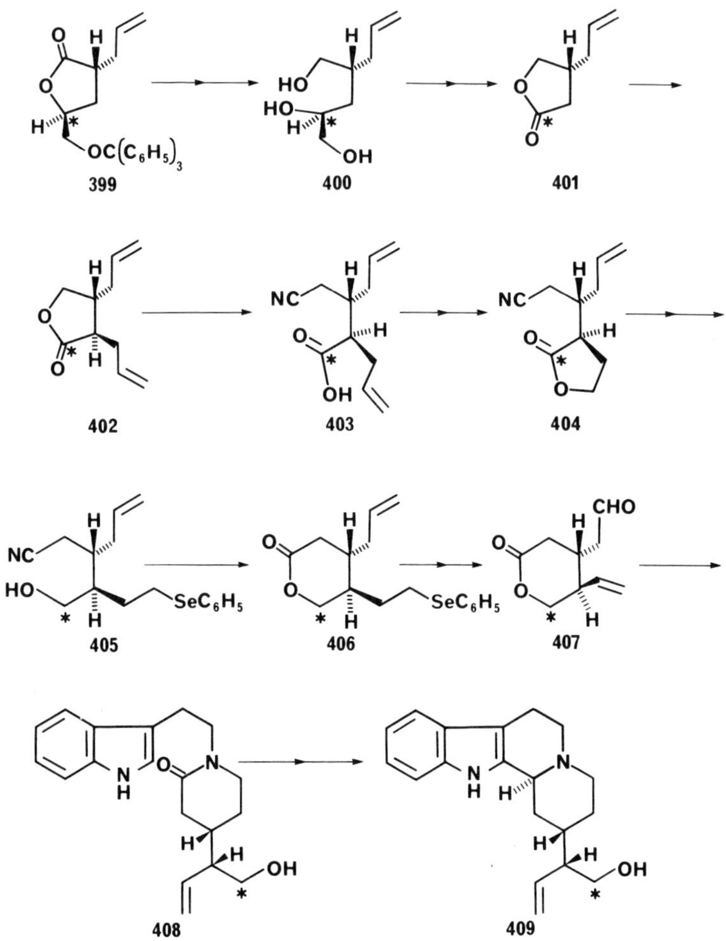

Scheme 4. (−)-Antirhine.

Other preparations of chiral isoquinoline alkaloids in which a Bischler–Napieralski reaction is the key step have been reported (Scheme 6). Thus, the Bischler–Napieralski product **415** from an *N*-acetyl (*S*)-dopa derivative was acylated to **416**. Photocyclization to **417** was followed by conversions, including a reductive decyanation by the method of Yamada (described above), to give (−)-xylopinone (**418**) (*139*), whose optical rotation was ~95% of the value of the natural product. Acid-catalyzed ring annulation of **419** [from (*S*)-tryptophan] with 3-methylenepentan-2-one was a key step in syntheses of dregamine (**420a**) and epidregamine (**420b**) (*140*).

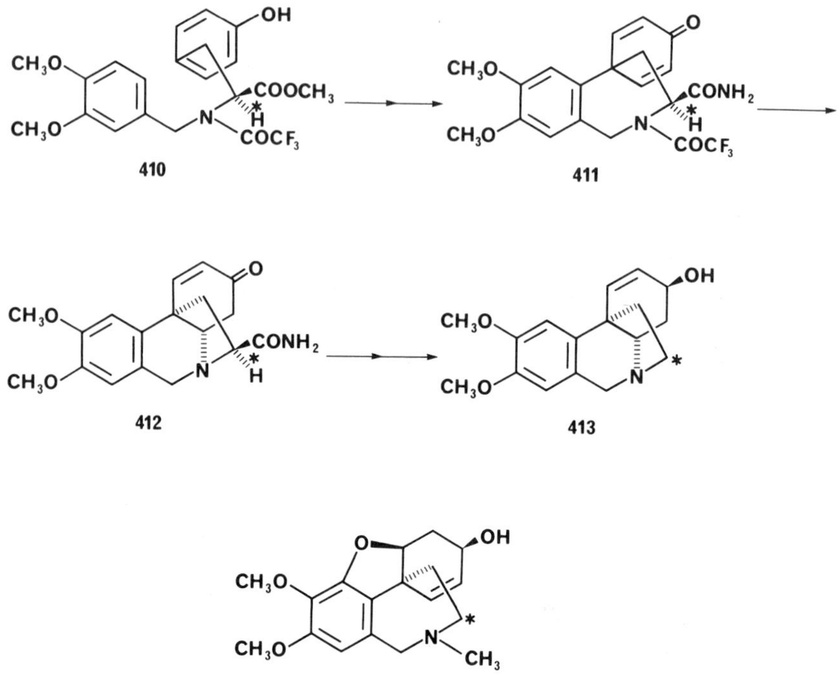

Scheme 5. (+)-Maritidine.

The preparation of water-soluble isoquinolines from tryptamine and a tartaric acid derivative has been described (*141*).

2. Pyrrolidine-Containing Alkaloids

Alkaloids containing a pyrrolidine ring are widely distributed in nature. Obvious chiral synthons for these alkaloids are proline and hydroxyproline (Scheme 7). Dipolar cycloaddition of ethylpropiolate in acetic anhydride to the bisformylated hydroxyproline **421** gave the dihydropyrrolizine **422**. The C-2 chirality, destroyed in this reaction, was reestablished in diastereomeric form by catalytic hydrogenation. Pyrrolizidine **423** was then converted to (+)-isoretronecanol [**424**, 89% optical purity (op)], (+)-laburine (**425**, 95% op) and (+)-supidine (**426**, 83% op). The chiral center of **422** could be inverted, thus providing access to the enantiomers of **424** and **425** [known as (−)-trachelanthamidine], and **426** in 90, 98, and 81% op, respectively (*142*). The variations in estimated enantiomeric

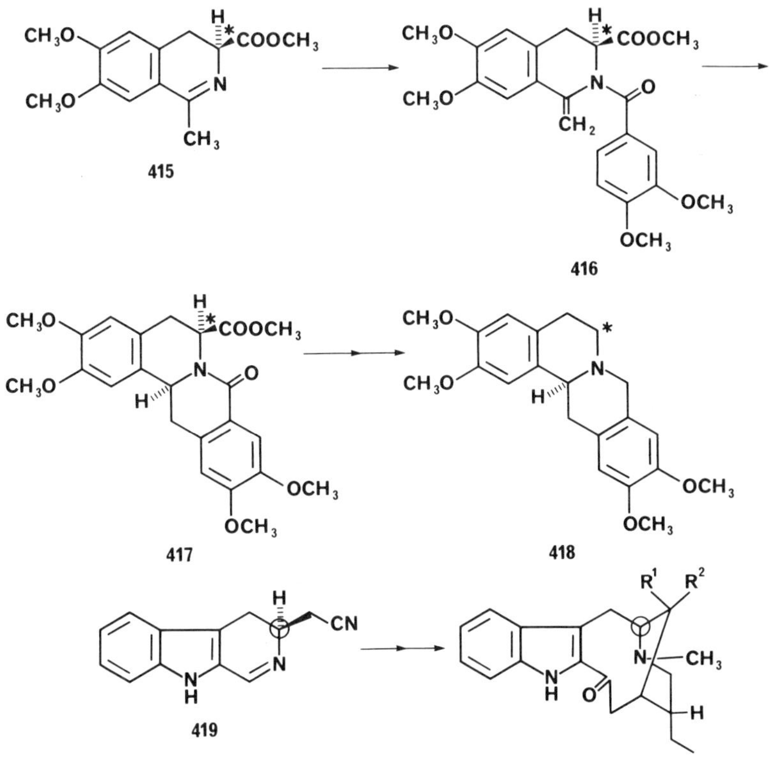

420a R^1 = $COOCH_3$ R^2 = H
420b R^1 = H R^2 = $COOCH_3$

Scheme 6. (−)-Xylopinine and dregamine.

purities of these compounds as based on optical rotations probably reflect inaccurate standards more than nonselective reactions. (*S*)-Proline has served as a common chiral precursor of the dendrobatid toxin 251D (**427**) (*143*), the enantiomer **428** of (+)-odorine (*144*), and of septicine (**429**) (*145*). However, the pyrrolidine ring of the enantiomer **430** of (−)-antofine was derived from (*S*)-glutamic acid via pyroglutamic acid. Partial racemization during reduction to 5-oxoprolinol gave final product of only ~50% op (*146*). (*S*)-Pyroglutamic acid also served as starting material in a synthesis of the antipode **431** of gephyrotoxin (*147*). This work established that the original assignment of configuration made on the basis of an X-ray crystal analysis was, in fact, incorrect. (*S*)-Proline was the starting material for two syntheses of the fungal metabolite deoxybrevianamide E (**432**) (*148–150*) and a synthesis of brevianamide E (**433**)

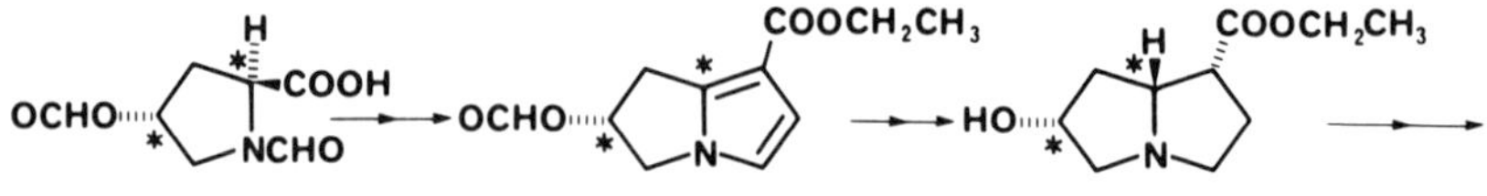

421 422 423

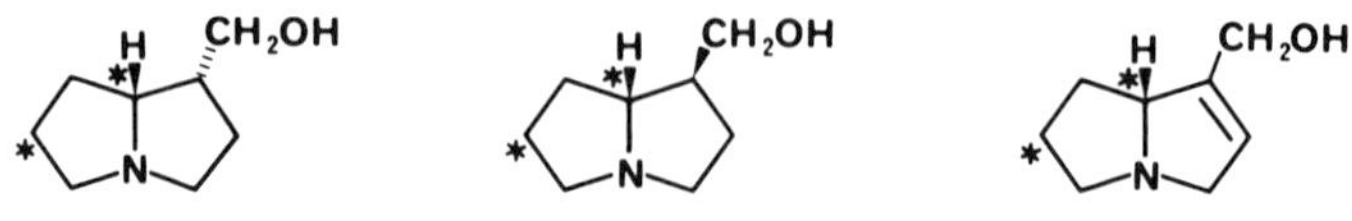

424 425 426

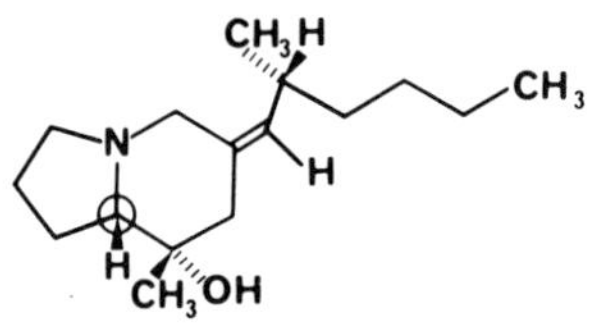

427

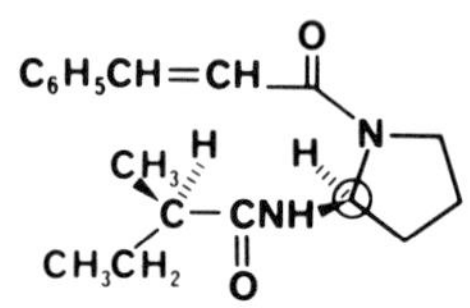

428

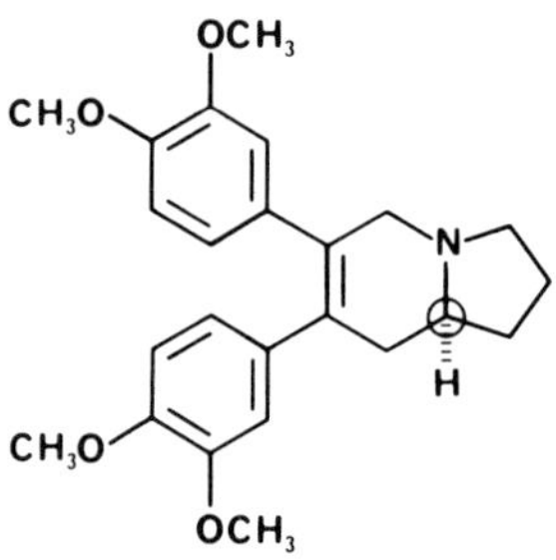

429

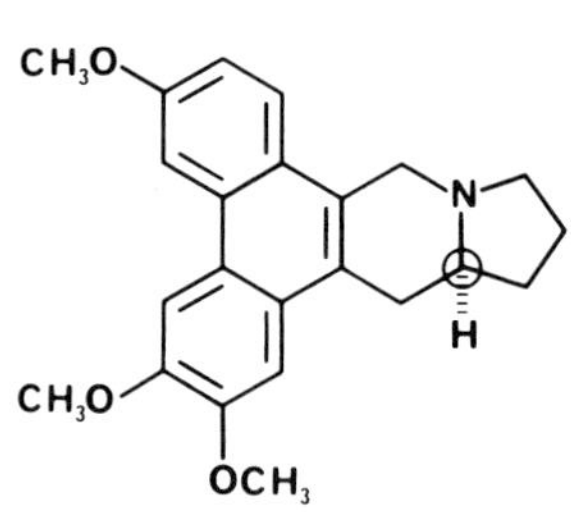

430

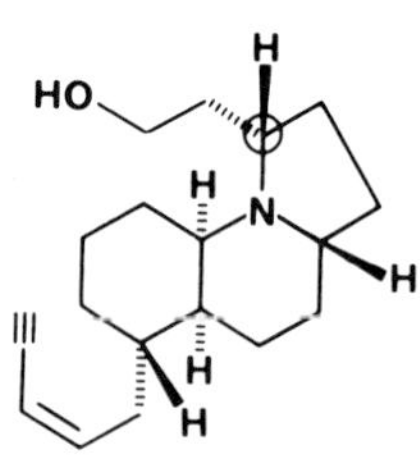

431

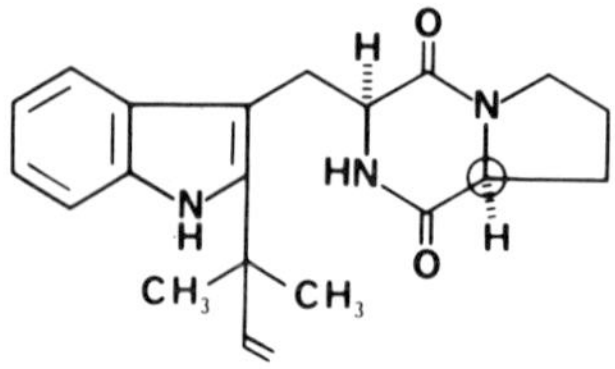

432

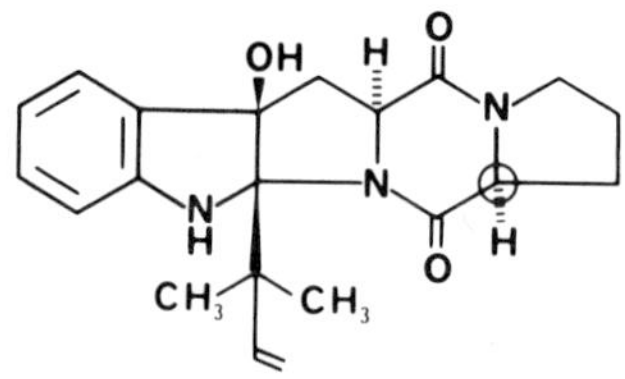

433

Scheme 7. Pyrrolidine alkaloids.

(*149, 151*). The value of these syntheses is mitigated, however, by the necessity for diastereomer separations and, in one case (*149, 151*), by partial racemization. A similar diasteromer separation problem was observed in the preparation of echinulin (*152, 153*) in condensations involving a diketopiperazine from (*S*)-alanine.

A synthesis (*154*) of the *Sceletium* alkaloid (−)-mesembrine (**438**, Scheme 8) employed an entirely different principle for formation of the pyrrolidine ring. Rather than using a preformed synthon, the ring nitrogen was introduced in the last step. Ring opening of the epoxide **22** (derived from D-mannitol) with 3,4-dimethoxybenzyl cyanide gave the diastereomer mixture **434**. Hydrolytic conversion to the lactone and alkylation

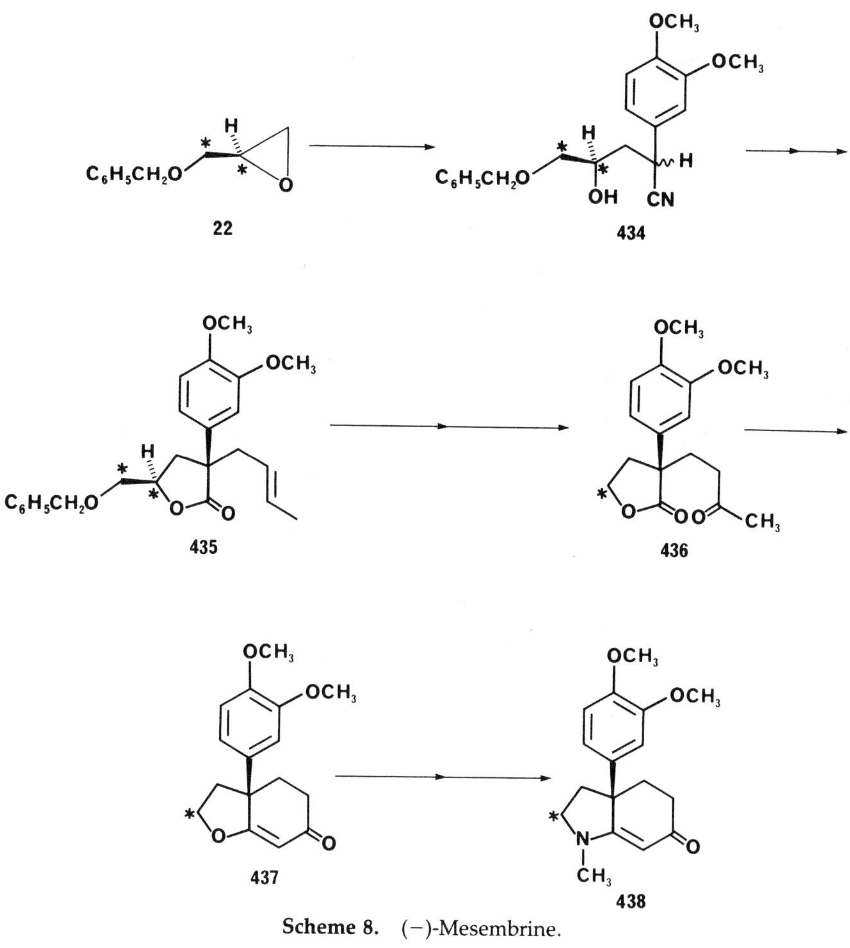

Scheme 8. (−)-Mesembrine.

with crotyl bromide gave **435** along with 15% of its diastereomer. Removal of the benzyloxymethyl group in four steps, followed by oxidative hydration of the olefin, produced ketone **436**, which was cyclized with base to **437**. At this point racemate, resulting from the nonstereospecific formation of **435**, was removed by crystallization. A two-step sequence then gave (−)-mesembrine.

3. Piperidine-Containing Alkaloids

Piperidine, as well as pyrrolidine alkaloids, have been prepared with amino acids as chiral sources (Scheme 9). Methods have been developed for conversion of (*S*)-lysine into either (*R*)-pipecolic acid (≥90% op) or (*S*)-pipecolic acid (78% op). The former, after enantiomeric enrichment, was converted to (*S*)-coniine (**439**) (*155*). (*S*)-Serine was the precursor of

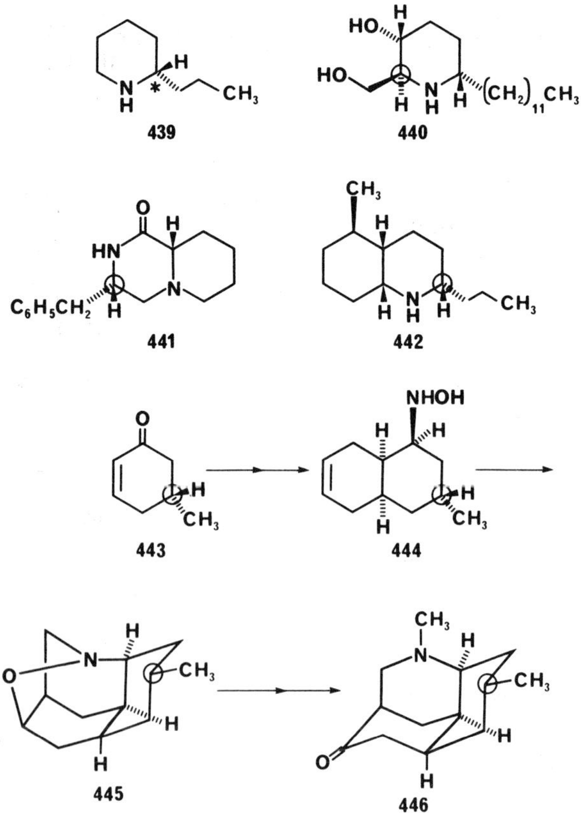

Scheme 9. Piperidine alkaloids.

the enantiomer **440** of deoxyprosopinine (*156*), while (*S*)-phenylalanine was converted to verruculotoxin (**441**) (*157*). The preparations of both **440** and **441** involved diastereomer separations. Two syntheses (*158, 159*) of pumiliotoxin-C (**442**) have been reported. The second of these (*159*), which corrects the structure, proceeded from (*S*)-norvaline. The other three chiral centers were introduced by a highly selective (≥97%) intramolecular Diels–Alder reaction. A particularly elegant synthesis of a piperidine-ring-containing alkaloid, (+)-luciduline (**446**) (*160*), began not with an amino acid but with (*R*)-5-methylcyclohex-2-enone (**443**) (*161*), which was prepared from (+)-pulegone. Diels–Alder reaction of this compound with butadiene, followed by oximation and reduction, gave **444**. Heating this material with formaldehyde gave a nitrone that spontaneously cyclized to **445**. Methylation, N,O-fragmentation, and oxidation completed the synthesis.

A synthesis of *N*-benzoylrugulovasine-A from (*S*)-tryptophan has been reported (*162*).

B. Arachidonic Acid Metabolites

In recent years it has been shown that the oxidative metabolism of arachidonic acid (**447**) leads to a variety of substances (*163*). These compounds, now designated the "eicosanoids," (*14*) include the prostaglandins and thromboxanes, occurring as a result of a cyclooxygenase pathway, and the leukotrienes, resulting from the alternative lipoxygenase pathway. (For reviews of syntheses of prostaglandins and thromboxanes, see refs. *19* and *164*.) A summary of some of the relevant structures and an indication of their biosynthesis is given in Scheme 10. These compounds are extremely bioactive (*165, 166*) and present in natural tissues in only very small amounts. A desire to confirm the structures, to have additional substance for biological testing, and to prepare analogs has led to intense synthetic efforts. The unusually sensitive nature of these materials and the often complex array of chiral centers represented has led to the frequent choice of synthesizing these compounds by manipulation of existing chiral molecules.

1. Prostaglandins

The first practical synthesis of prostaglandins, done by Corey and coworkers, proceeded via the bicyclic alcohol **466** (*167*) (Scheme 11), the preparation of which in enantiomerically pure form involved a resolution. A chiral synthesis of this compound from (*S*)-malic acid (**127**) has now been reported (*168*). Homologation to **463** was followed by cyclization to

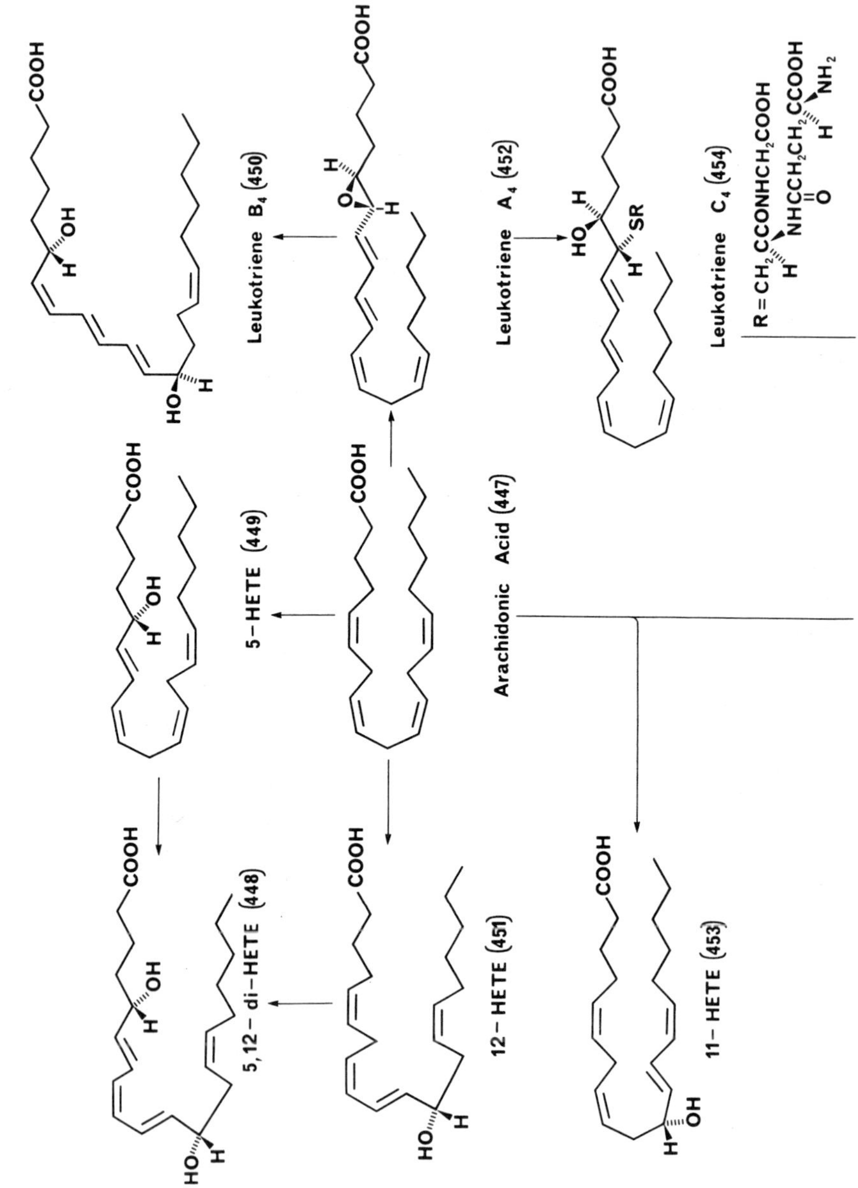

Leukotriene B4 (450)
5-HETE (449)
5,12-di-HETE (448)
Arachidonic Acid (447)
Leukotriene A4 (452)
12-HETE (451)
Leukotriene C4 (454)
R = CH2CCONHCH2COOH
NHCCH2CH2CCOOH
11-HETE (453)

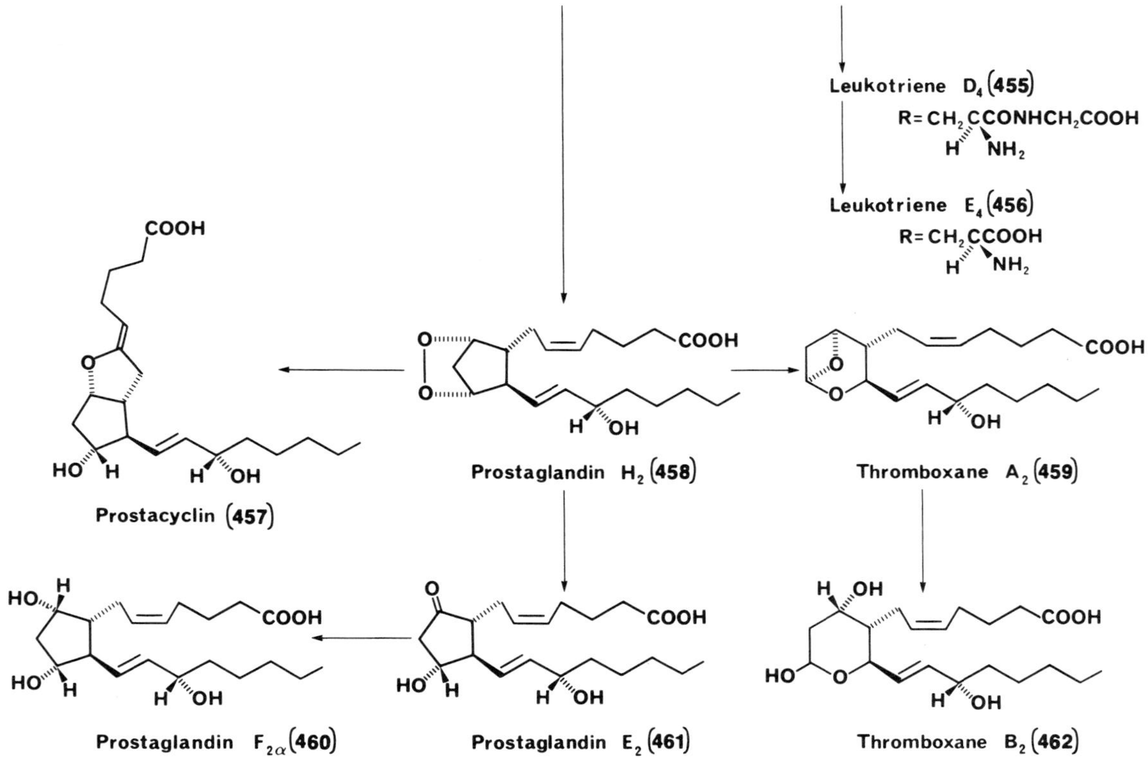

Scheme 10. Arachidonic acid metabolites.

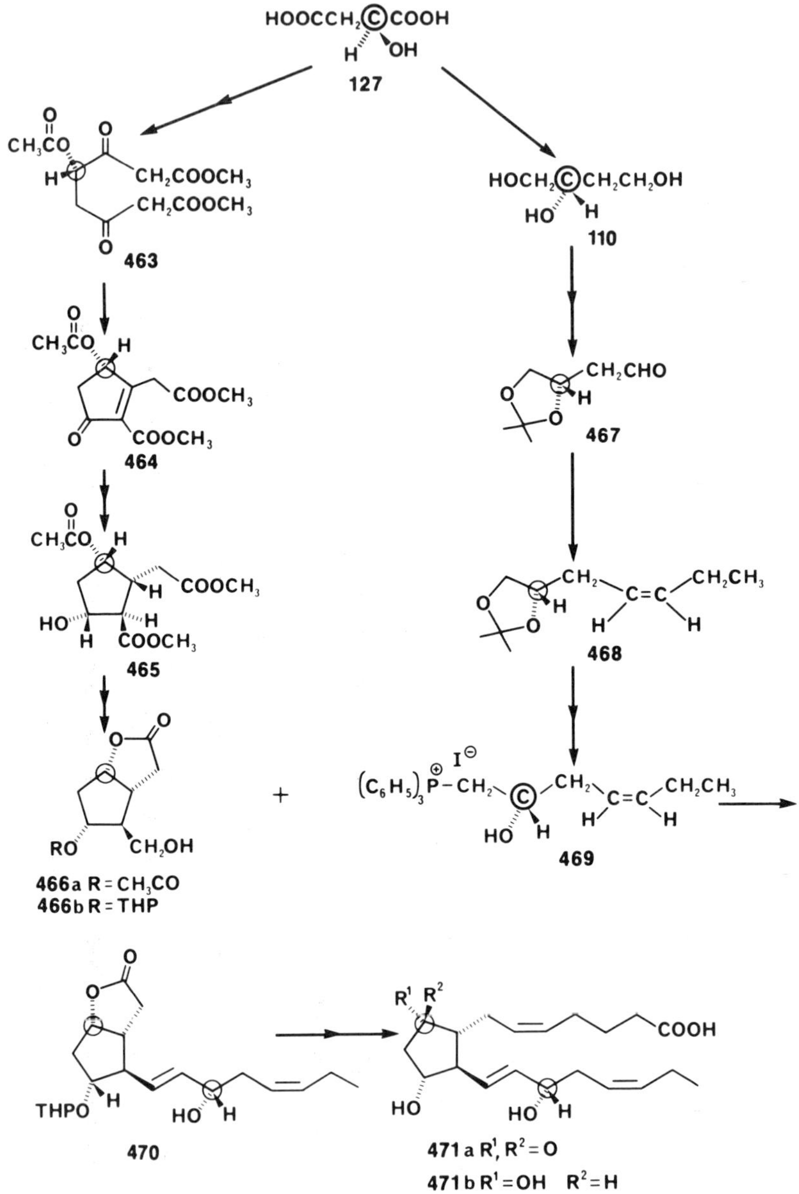

Scheme 11. Prostaglandins E_3 and $F_{3\alpha}$.

give primarily regioisomer **464**. The remaining chiral centers were introduced by hydrogenation (with epimerization) and sodium borohydride reduction to **465**, which was then transformed to the desired **466a**. The related **466b** (*169*) has been converted to the less common prostaglandins E_3 (**471a**) and $F_{3\alpha}$ (**471b**) (*81*). The side-chain synthon **469** required for these syntheses was also prepared from (*S*)-malic acid. Reduction to triol **110** was followed by acetalization and oxidation to **467**. Introduction of the remaining three carbon atoms by Wittig reaction was followed by a straightforward series of conversions of **468** to **469**. Oxidation of **466b** to the aldehyde level and Wittig reaction with **469** gave adduct **470**. Addition of the second chain and functional-group manipulation then yielded the desired prostaglandins.

A synthon functionally equivalent to **466a** (*170*) is the epoxy lactone **478** (Scheme 12), which has been synthesized starting from D-glucose

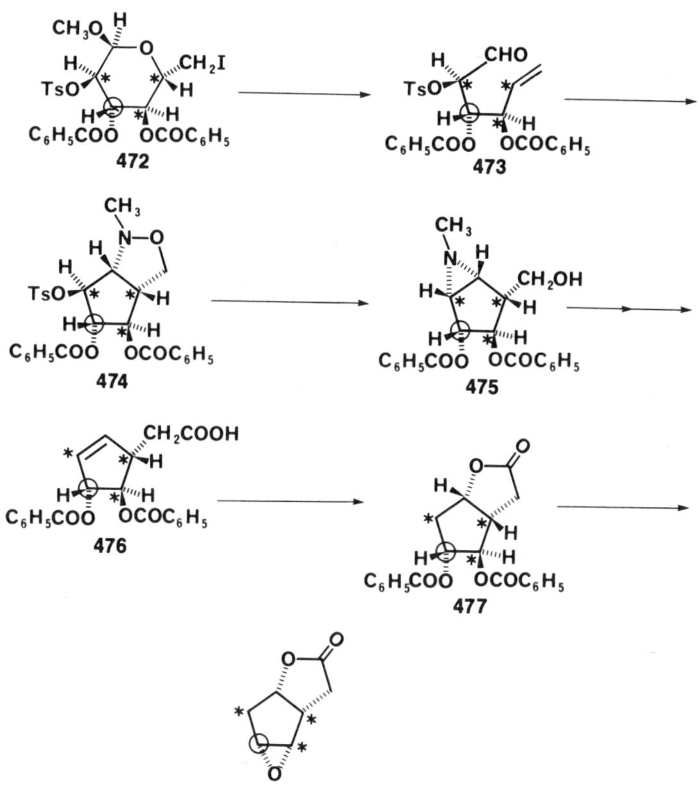

Scheme 12. Corey epoxylactone.

(*171*). Reductive ring opening of **472**, readily derived in four steps from D-glucose, gave aldehyde **473**. Reaction of this material with *N*-methylhydroxylamine was followed by a spontaneous and apparently stereospecific nitrone cycloaddition to give isoxazolidine **474**. Reductive ring opening was accompanied by an unexpected aziridine formation. Elimination of the methylamine unit from **475** with *m*-chloroperbenzoic acid and a standard four-step chain lengthening gave acid **476**. Iodolactonization and reductive iodine removal then gave lactone **477** which, on debenzoylation and treatment with *N*-tosylimidazole and sodium hydride, yielded **478**. Interestingly, the diastereomeric epoxylactone was obtained on reaction of the intermediate diol with diethyl azodicarboxylate and triphenylphosphine.

Stork and co-workers developed extremely elegant syntheses of PGA_2 (**484**, Scheme 13) (*172*) and PGF_2 (**460**, Scheme 14) (*173*) in which sugars serve as the chiral sources. Key to the success of these syntheses was the use of orthoester Claisen rearrangements to effect 1,3-transposition of chirality and secure proper location and geometry of the 13,14*E* olefin. The hemiacetal **479**, prepared from L-rhamnose, was reacted with vinylmagnesium chloride and then subjected to orthoester Claisen rearrangement to give **480**. Hydrolysis of the acetonide and recyclization gave

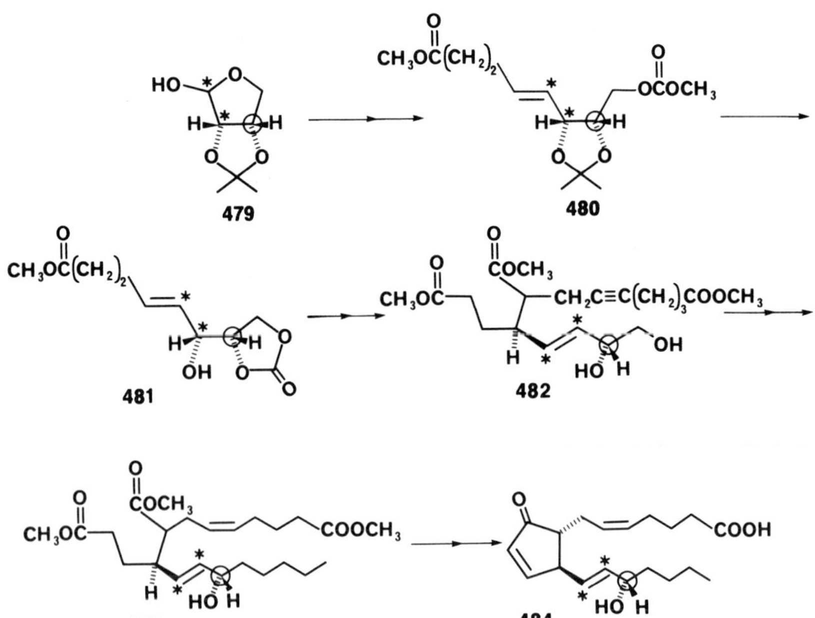

Scheme 13. Prostaglandin A_2.

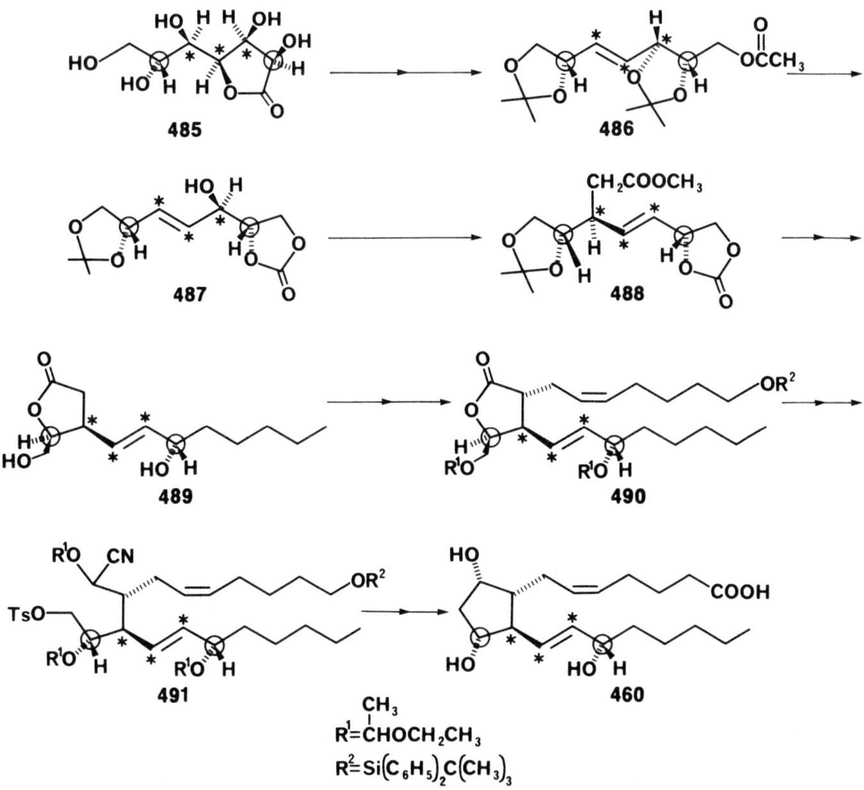

Scheme 14. Prostaglandin $F_{2\alpha}$ synthesis I.

the carbonate **481** possessing a second allylic alcohol capable of undergoing the Claisen reaction. Indeed, reaction of **481** with the trimethyl orthoester of 8-carbomethoxy-4-octynoic acid gave **482**, in which the second new asymmetric center, next to the ester, was a mixture. Semihydrogenation of **482** and conversion of the primary alcohol to a butyl group gave **483**. Ring closure and introduction of the olefinic bond occurred with equilibration to give the desired PGA_2. Thus, the nonstereospecificity in the **481** → **482** conversion was of no consequence.

For the synthesis of $PGF_{2\alpha}$, the commercially available lactone **485** (prepared from D-glucose) was transformed in several steps to the bisacetonide **486**. In a sequence similar to that used in the PGA_2 synthesis, **486** was converted to cyclic carbonate **487**, the Claisen rearrangement of which with triethyl orthoacetate gave **488**. Completion of the allylic alcohol side chain, again in analogy to the PGA_2 synthesis, was followed by acidic deacetalization and lactonization, yielding **489**. After protection of the alcohol groups, stereospecific alkylation introduced the second chain. The lactone **490** was then reduced to the corresponding hemiacetal. Cyanohydrin formation, removal of the alcohol protecting groups, selective tosylation of the primary alcohol, and reprotection gave **491**. Basic cyclization of **491** was followed by a straightforward set of manipulations that completed the $PGF_{2\alpha}$ (**460**) synthesis. The series of steps for the hydroxymethyllactone → hydroxycyclopentanone conversion using a

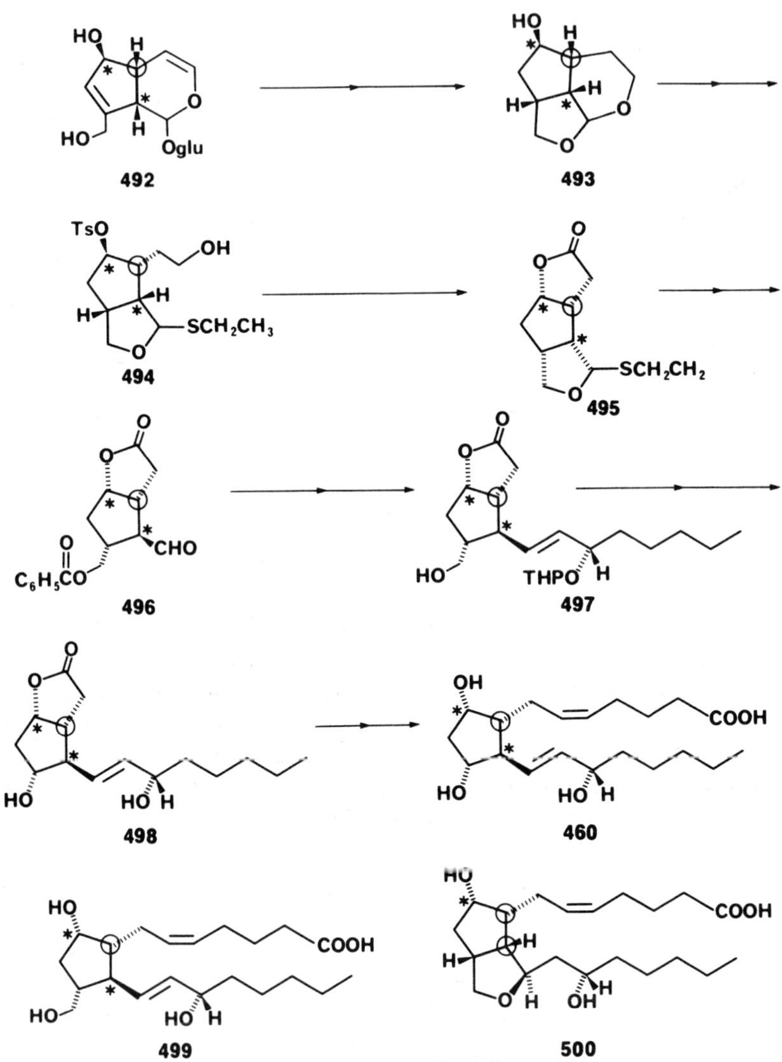

Scheme 15. Prostaglandin $F_{2\alpha}$ synthesis II.

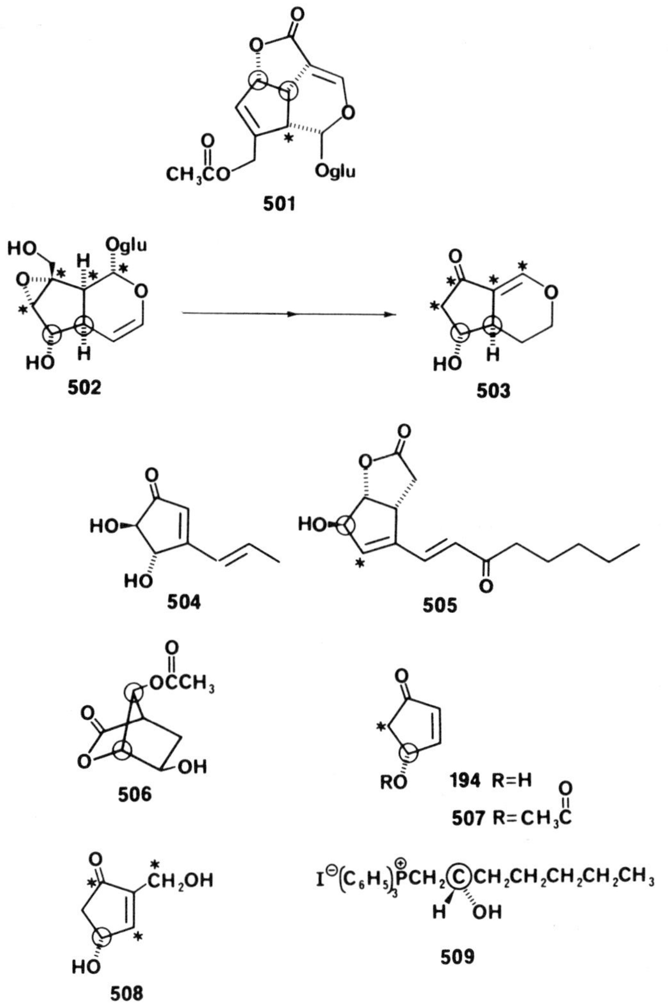

Scheme 15 (*Continued*)

protected cyanohydrin had been worked out in an earlier synthesis (*174*), in which isopropylidene D-glyceraldehyde had been converted to prostaglandin E_1.

There are a number of naturally occurring materials with cyclopentane rings possessing substitution patterns recognizably similar to those of the prostaglandins. Perhaps the most significant work to date in using these compounds for prostaglandin synthesis is the conversion of the glycoside aucubin **492** to $PGF_{2\alpha}$ (**460**, Scheme 15) (*175*). Thus, hydrogenation of aucubin followed by treatment with polyphosphoric acid gave the tricyclic alcohol **493**. Tosylation and treatment with ethanethiol in the presence of acid led to hemithioketal **494**. Oxidation of **494** was accompanied by tosylate displacement to yield tricyclic lactone **495**. Further reaction of **495** with ethanethiol, benzoylation, dithioacetal hydrolysis, and aldehyde isomerization gave aldehyde **496**. A standard set of conversions introduced the allylic alcohol chain yielding **497**. The extraneous carbon atom was excised by oxidation to the acid, conversion to the mixed anhydride with *m*-chloroperbenzoic acid, decarboxylative rearrangement, and ester hydrolysis. The resulting diol **498** is a known (*167*) precursor of $PGF_{2\alpha}$ (**460**). An intermediate similar to **497** was converted to 11-homolog **499** of $PGF_{2\alpha}$ (*176*), while a different modification in the reaction sequence gave the cyclic analog **500** (*177*). A second approach to the use of aucubin as a prostaglandin starting material has been reported (*178*). This same group (*179*) has devised a conversion of asperuloside (**501**) to intermediates in the Ohno synthesis (*175*) of **460**, while catapol (**502**) has been transformed (*180*) into the potential prostanoid synthon **503**. The mold metabolite terrein (**504**) has been converted to prostaglandin and/or prostaglandin analog synthons **505** (*181*), **506** (*182*), and **507** (*183*). The cumyl ester of the racemate of **194** has been employed by the Stork group (*184*) in another $PGF_{2\alpha}$ synthesis. As a result, a synthesis of **194** and its enantiomer from (*S*,*S*)- and (*R*,*R*)-tartaric acid, respectively, has been developed (*98*). The conversion of (−)-quinic acid to the more highly substituted **508** has been reported (*185*). Finally, we note the conversion of L-arabinose to a 9,11-bis(hydroxymethyl)prostaglandin intermediate (*186*), of D-glucose to an 11-hydroxymethylprostaglandin synthon (*187*), and of camphor-10-sulfonic acid to a trimethylcyclopentanone derivative (*188*) of potential use in analog synthesis.

Two preparations (*189*) of the phosphonium salt **509** (of potential use in condensations with, e.g., aldehyde **496**) have been described. Use of this material would avoid the separation of diastereomers required when the prostaglandin lower chain is introduced at the ketone level and is then subjected to hydride reduction. The chiral sources in these syntheses of **509** were D-mannitol and (*S*)-malic acid.

The desire to prepare unique analogs of the prostaglandins has led to two syntheses (*190–192*) of 11-oxaprostaglandin $F_{2\alpha}$ (**515**) from D-glucose. While similar in conception, these syntheses vary somewhat in execution. In the second, higher-yield synthesis (*190, 191*) (Scheme 16), ketone **510** was subjected to Horner reaction with the anion of triethyl phosphonoacetate, followed by catalytic hydrogenation to give ester **511**. Selective deacetalization followed by acetylation gave **512**. Heating this compound with aqueous acetic acid cleaved the second acetal and induced

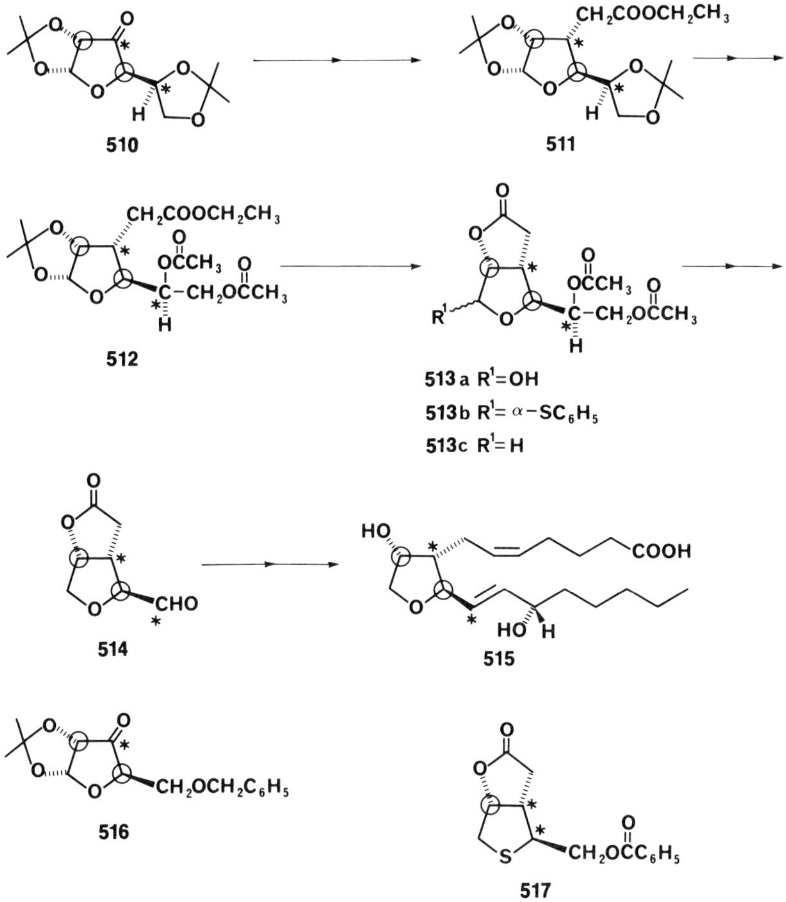

Scheme 16. 11-Oxaprostaglandin $F_{2\alpha}$.

lactone formation. The anomeric oxygen of the resulting **513a** was removed in several steps, the last of which was desulfurization of **513b**. Deacylation of the resultant **513c** and periodate oxidation gave **514**, which was converted in the manner developed for the synthesis of $PGF_{2\alpha}$ to the 11-oxa analog **515**. A very similar sequence of reactions for the preparation of **514** from the D-xylose derivative **516** was also reported (*190*). With this starting material, the need for the periodate cleavage to remove the extra carbon atom was obviated. A xylose derivative similar to **516** has also been converted, by the same group, to a 10-hydroxy-11-oxaprostaglandin (*193*).

The 11-thiaprostaglandin intermediate **517** has been obtained either from L-arabinose or D-xylose (*194*). The synthesis of 11-α-hydroxy-13-

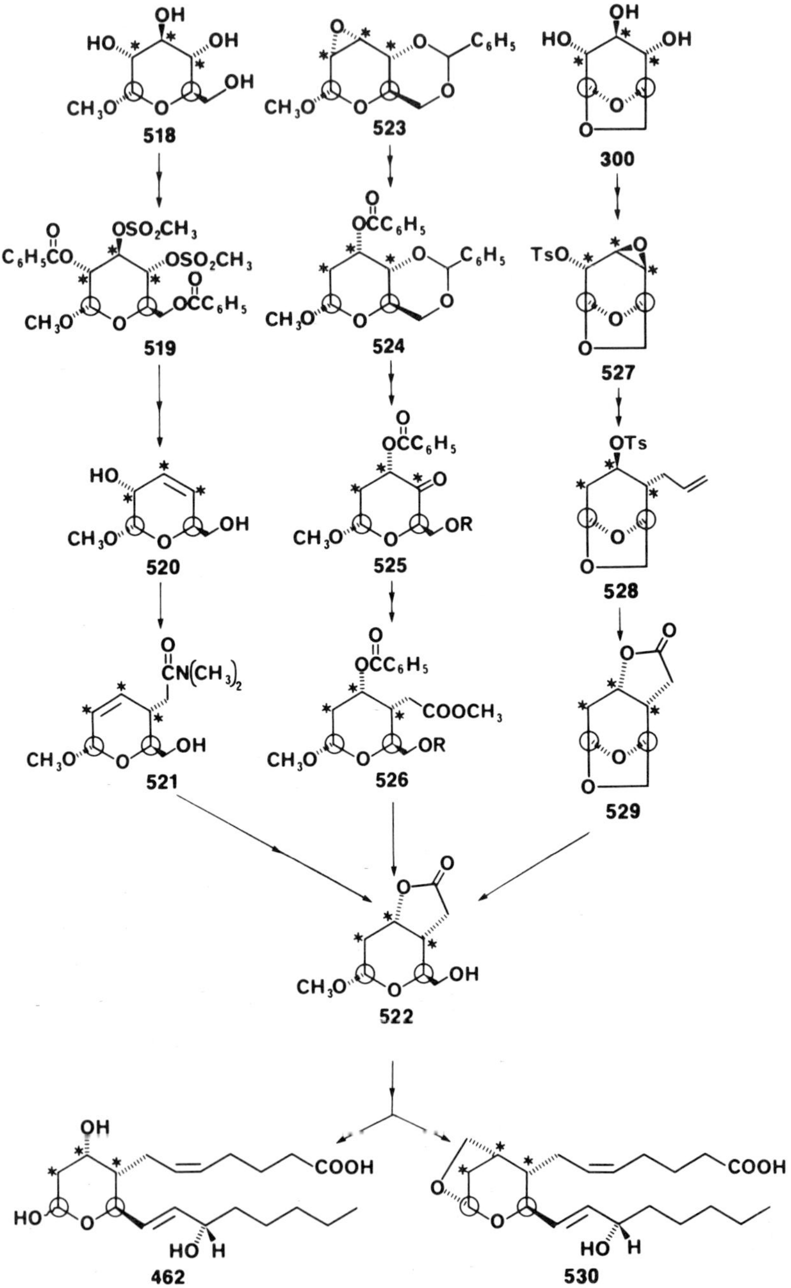

Scheme 17. Thromboxane B_2. R = $Si(C_6H_5)_2C(CH_3)_3$.

oxaprostanoic acid from (−)-quinic acid (*195*), of 9,13- and 7,11-dioxaprostanoic acids from L-xylose (*196*), and of 11-deoxy-8-azaprostaglandin E_1 from (*S*)-glutamic acid (*197*) have also been reported. The preparation of 15-deoxy-(16*S*)-methylprostaglandins from (*S*)-citramalic acid has been described (*198*).

2. Thromboxanes

The presence in the thromboxanes of the tetrahydropyranyl ketal (TXA_2) or hemiketal (TXB_2) unit makes the use of sugars for the synthesis of this class of compounds an obvious choice. Although this approach has not yet yielded a synthesis of TXA_2, there are reports from five separate groups of the preparation of TXB_2 (or intermediates thereof) from D-glucose (Scheme 17). All of these syntheses proceed via the lactone **522**, which had previously (*199*) been converted, using standard prostaglandin technology, to TXB_2 (**462**). The approaches to **522** have taken three forms. In the first of these (*200, 201*), the acetic acid unit was introduced by Claisen rearrangement. As exemplified by the work of Corey (*200*), the α-methylglucoside **518** was converted to the dibenzoate dimesylate **519**. Reductive elimination and debenzoylation gave **520**, which was subjected

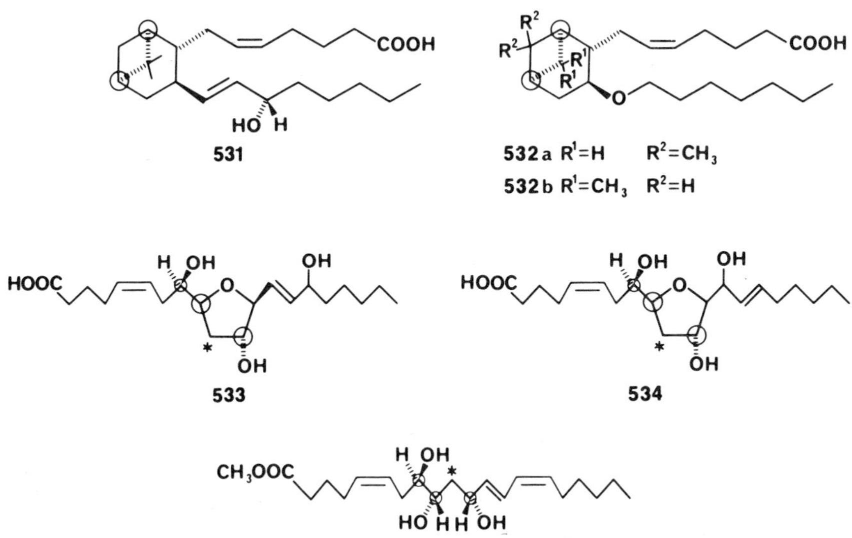

531

532a R¹=H R²=CH₃
532b R¹=CH₃ R²=H

533

534

535

Scheme 17 (*Continued*)

to Claisen rearrangement by heating with the dimethylaminal of *N,N*-dimethylacetamide, yielding **521**. Treatment of the amide **521** with iodine led to iodolactonization, without need for prior hydrolysis to the acid. Deiodination with tributyltin hydride then gave the desired lactone **522**.

In the second approach (*202–204*) to **522**, the extraneous oxygen at C-2 (sugar numbering) was removed in a process that also effected the necessary inversion of configuration at C-3. Thus, hydride reduction of the epoxide **523** and benzoylation gave **524**. Hydrogenolysis (*203, 204*) over palladium hydroxide on carbon, silylation, and oxidation gave ketone **525**. At this point the acetic acid unit was introduced by Horner reaction followed by hydrogenation, yielding **526**. Treatment of this compound with potassium carbonate in methanol completed the synthesis of the silyl ether of **522**. A variation of the normal sequence was used to convert this silyl ether to TXB_2. Hanessian (*203, 204*) first introduced the remaining five atoms of the heptenoic acid chain, then liberated the alcohol and constructed the allylic alcohol chain, in a reversal of the usual route.

The third route (*205*) to **522** began with the "anhydro" form **300** of glucose, L-glucosan, which was converted to the epoxy tosylate **527** by ditosylation and regiospecific epoxide formation with base. Nucleophilic opening of the epoxide with allylmagnesium chloride in the presence of cuprous iodide was followed by reductive removal of the tosylate with lithium triethylborohydride and tosylation of the remaining alcohol, yielding **528**. Ruthenium tetroxide–sodium metaperiodate oxidation of **528** directly gave the tricyclic lactone **529** with the necessary inverted configuration at C-3. Acid catalyzed acetal exchange then gave **522**.

Although designed as an intermediate for TXB_2 (**462**), the acetal lactone **522** has also been converted to the TXA_2 homolog **530** (*206*). This compound is of substantial interest since it appears to be a TXA_2 agonist without thromboxane synthetase-inhibiting activity. Syntheses of the carbocyclic TXA_2 analog **531** from myrtenol (*207*) and nopol (*208*) have been reported. The more distantly related **532a,b** have been prepared (*209*) from (+)-α-pinene and (−)-β-pinene, respectively. These compounds were prepared as potential thromboxane synthesis inhibitors. Among the minor oxidation products of arachidonic acid are the tetrahydrofurans **533** and **534** of unknown configuration at the allylic alcohol centers. Syntheses of these compounds from diacetone glucose have now been reported (*210, 212*). A plausible biosynthesis (*213*) of **533** might also result in synthesis of the triol **535**. Methods similar to those employed in the preparation of **533** have been used to convert diacetone glucose to **535** (*213*) and three of its diastereomers (*214*). It has not yet been reported whether any of these compounds are, in fact, arachidonic acid metabolites.

3. Leukotrienes

A key compound in the biosynthesis of the leukotrienes is leukotriene A_4 (LTA_4) (**452**) since this compound leads to leukotrienes B_4, C_4, D_4, and E_4 (Scheme 10). Similarly, LTA_4 is a logical chemical precursor for the other leukotrienes, and a substantial effort has been invested in the synthesis of this compound. The initial synthesis of LTA_4 methyl ester (**542**) by Corey (*215*) proceeded via the epoxy aldehyde (**541**) (Scheme 18). Since that report, three additional syntheses (*216–219*) of **541** have appeared. Three of the four syntheses are similar in concept; that is, a 1*S*,2*R* configurated sugar was converted to the (1*S*,2*S*)-epoxide by conversion of an alcohol to a leaving group and by internal displacement with inversion. The syntheses differ in the source of the (*S*,*R*)-diol unit. In Corey's synthesis (*215*), these carbons were obtained from C-3 and C-4 of the D-ribose derivative **536**. Wittig reaction of **536** with ethoxycarbonylmethylenetriphenylphosphorane and acetylation gave **537**. The allylic benzoate was reductively removed with zinc amalgam in hydrogen chloride saturated ether. This reduction occurred with double bond migration, yielding **538**, which was hydrogenated, selectively deacylated, and tosylated to give **539**. Treatment of this compound with potassium carbonate in methanol led to deacylation and epoxide formation. The resultant alcohol **540** was then oxidized to **541**.

In an alternative synthesis of **540** (and thus of **541**), Cohen (*216*) used C-4 and C-5 of D-araboascorbic acid (**268**, erythorbic acid) as the epoxide synthon. This starting material was converted, in four steps, to the hemiketal **543** [the enantiomer of **479** (Scheme 13)]. Wittig reaction with the appropriate phosphorane and benzoylation gave **544**, which was hydrogenated and ozonized to ester **545**. Treatment of this compound with aqueous trifluoroacetic acid caused deacetalization and lactone formation. Mesylation of **546** and reaction with potassium carbonate in methanol then yielded the desired **540**. These same authors also prepared diastereomers of **544**, **545**, and **546** from (*R*,*R*)-tartaric acid en route to the cis 5*R*,6*S* analog of **540**.

The third synthesis (*217*) of epoxyaldehyde **541** was entirely different in concept. The less readily available (*S*)-isopropylideneglyceraldehyde (**547**, from L-arabinose) was chain lengthened by Wittig reaction to the acid **548**. Photoisomerization and esterification gave the ester **549**. Epoxidation of this compound was nonspecific, yielding a 1 : 2 mixture of **550** and the desired **551**. Chromatographic isolation of **551** and subsequent periodate oxidation in aqueous acetic acid led directly to epoxyaldehyde **541**. The diastereomeric epoxide **550** and the epoxides derived from the *Z* olefin **548** were of use in analog syntheses.

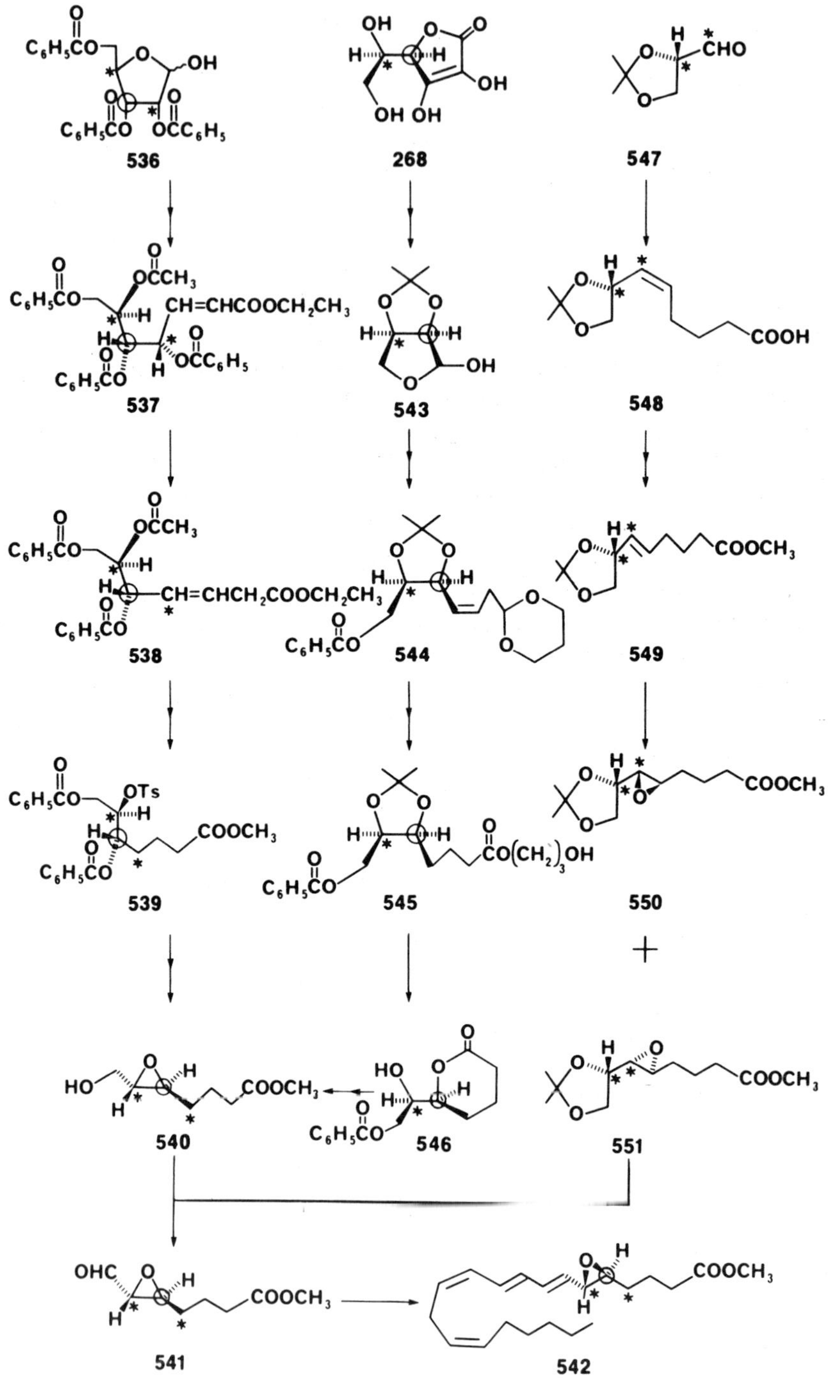

Scheme 18. Leukotriene A_4 methyl ester.

In addition to its use in the synthesis of the natural leukotrienes, **541** has also been converted to 11-(*E*)-LTC_4 and LTD_4 (*220*).

The Merck Frosst group, which devised the third systhesis (*217*) of LTA_4 methyl ester (**542**), was also responsible for the fourth preparation (*218, 219*) of this compound. The approach employed in this latter work (Scheme 19) is particularly instructive since one starting material (2-deoxy-D-ribose, **234**) was converted, by an appropriate series of stereocontrolled reactions, not only to **542**, but also to its 5-epi- and 6-epidiastereomers (**562** and **559**) and its 5-epienantiomer (**556**). Thus, inversion at C-4 during epoxide formation gave **542**, inversion at C-3 led to **556**, double inversion at C-3 gave **559**, and inversion at C-3 and C-4 produced **562**.

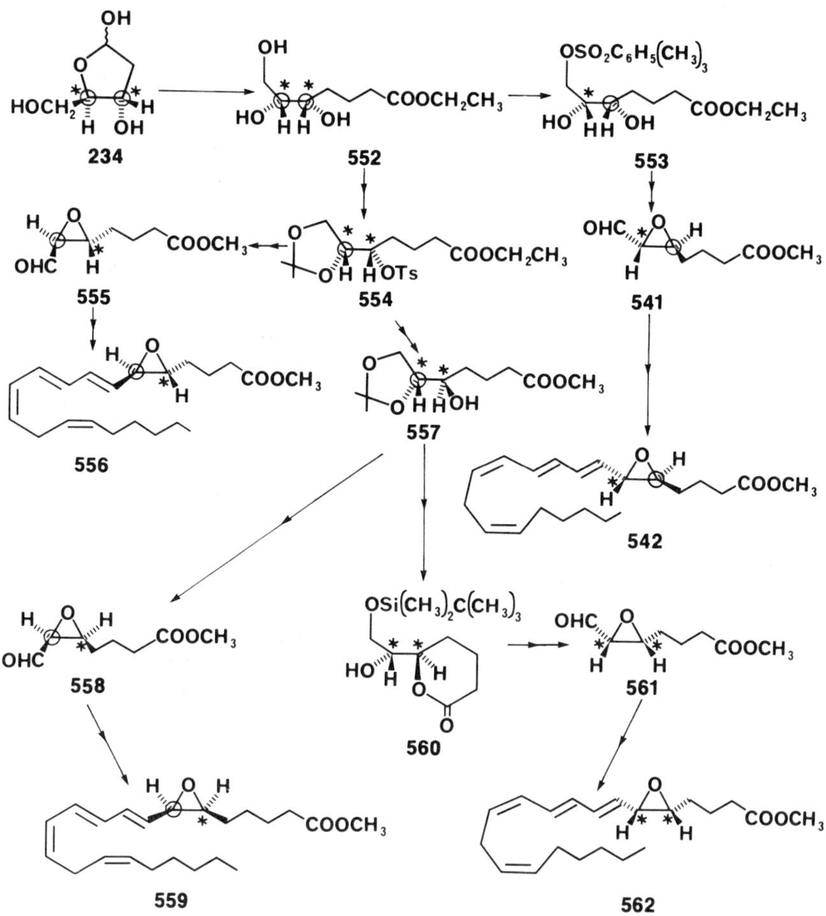

Scheme 19. Leukotriene A_4 methyl ester diastereomers.

In more detail, the synthesis began with Wittig two-carbon homologation of **234**. Hydrogenation of the product gave the triol **552**, which was converted by selective reaction with mesitylenesulfonyl chloride to **553**. Treatment of this compound with sodium methylate gave, via the epoxy secondary alcohol, an epoxy primary alcohol (formed with inversion at C-4), which was oxidized to **541** and subsequently converted to LTA_4 methyl ester (**542**). An alternate use of triol **552** involved acetonide and tosylate formation, yielding **554**. With C-3 thus activated for inversion, freeing of the diol, epoxidation with methanolic potassium carbonate, and

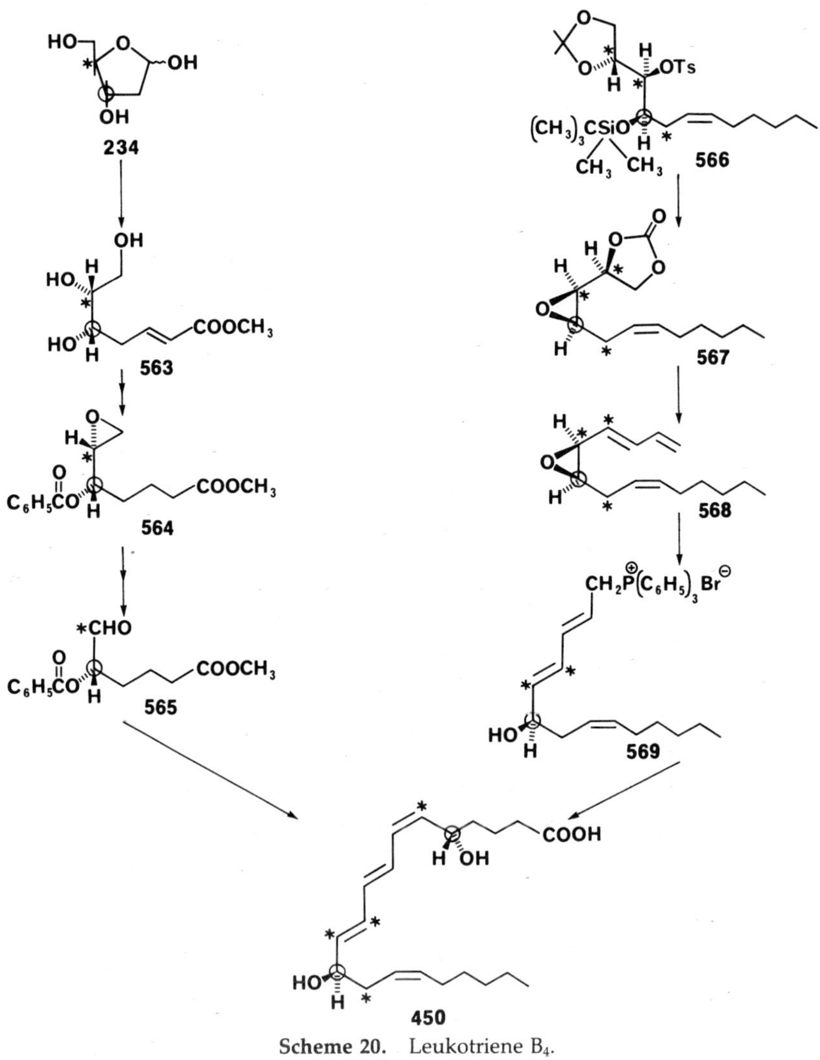

Scheme 20. Leukotriene B_4.

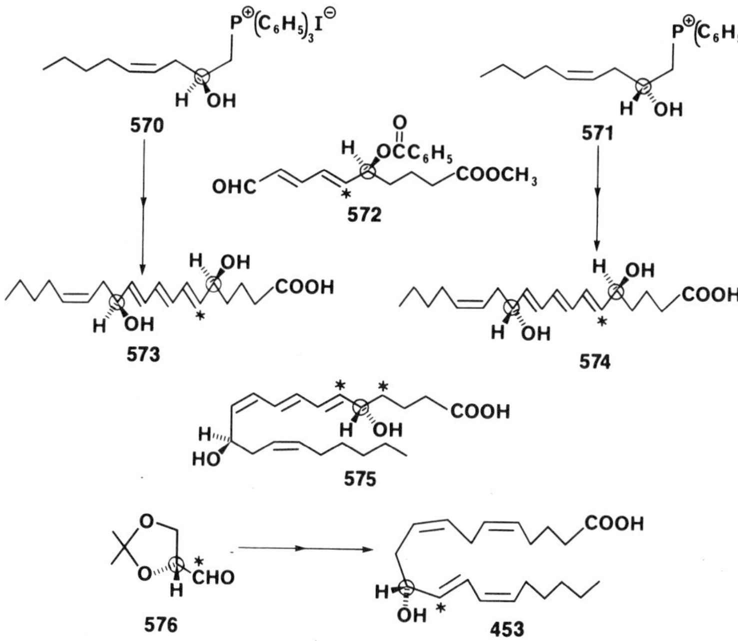

Scheme 20 (*Continued*)

oxidation gave **555**, a precursor of 5-epi,6-epi-LTA$_4$ methyl ester (**556**). Inversion of C-3 of **554** prior to acetonide hydrolysis by treatment with sodium benzoate followed by methanolic sodium methoxide transesterification gave alcohol **557**. Tosylation, acetonide hydrolysis, epoxide formation (involving a second inversion at C-3), and oxidation gave **558**, and subsequently 6-epi-LTA$_4$ methyl ester (**559**). The fourth diastereomer of LTA$_4$ methyl ester was prepared by initial hydrolysis of the C-3 inverted **557**, silylation, saponification, and lactonization to **560**. Tosylation of the remaining alcohol group, lactone opening, and epoxide formation (with C-4 inversion) effected with sodium methoxide, deprotection, and oxidation gave aldehyde **561**, which then was converted to 5-epi-LTA$_4$ methyl ester (**562**).

A second report (*221*) of the preparation of **541** from 2-deoxy-D-ribose has been made. Corey (*222*) employed synthetic technology very similar to that used for **542** in the preparation of 6-epi analog **559** from D-mannitol.

The diol, LTB$_4$ (**450**), presents an even more exciting synthetic challenge than does LTA$_4$ because of the two widely separated chiral allylic alcohol functions. The route chosen by Corey (*223, 224*) for the synthesis of this compound (Scheme 20) was convergent, having the two chiral centers derived from deoxyribose and mannose.The 5*S* carbon was obtained from deoxyribose (**234**) by initial Wittig chain extension to **563**. A sequence of hydrogenation, selective tosylation of the primary alcohol, epoxide formation, and benzoylation gave **564**. Hydrolytic epoxide opening to the diol and oxidative cleavage then provided the aldehyde **565**. The 12*R* carbon was obtained from mannose via its 2,3:5,6 bisacetonide, which was converted in several steps to tosylate **566**. Deprotection of this compound to the tosylate triol was followed by cyclic carbonate and epoxide formation, yielding **567**. Hydrolysis of the carbonate, oxidative cleavage to the aldehyde, and Wittig reaction gave **568**, which on treatment with dry hydrogen bromide in methylene chloride followed by triphenylphosphine yielded the salt **569**. Wittig condensation of **569** with **565** and basic hydrolysis then gave LTB_4 (**450**).

The intermediates prepared for the LTA_4 and LTB_4 syntheses have also been employed in preparations of analogs of LTB_4. Thus, reaction of the (*S*)- and (*R*)-phosphonium salts **570** and **571** with the homologated form **572** of aldehyde **565** gave the diastereomeric 6*E*-LTB_4's **573** and **574** (*87*). The salts **570** and **571** were derived from (*S*)- and (*R*)-malic acids, respectively. The preparation of **571**, however, required the design of a novel preparation of the unnatural (*R*)-dimethylmalate from (*R*,*R*)-tartaric acid (*87*). Compound **570** was initially synthesized for, and used in, a preparation of 12-HETE (**451**) (*82*). The LTA_4 intermediate **541** was converted to 6*E*,10*Z*-LTB_4 (**575**) (*225*). The second chiral center (C-12) in this molecule was obtained by asymmetric epoxidation. Reaction of this latter synthon with a two-carbon homolog of **565** gave (5*S*,12*S*)-di-HETE (**448**), another naturally occurring eicosanoid (*226*). The mannitol-derived aldehyde **576** was converted to (11*R*)-HETE (**453**) (*227*).

C. Compounds with Chemotherapeutic Activity

The fight against infectious disease has been a major concern of mankind since the dawn of civilization. It is only during this century, however, that truly useful chemotherapeutic agents have been identified and applied in clinical medicine. Many of the best of such agents are natural products (or modifications thereof). The compounds, of widely diverse nature, have structures that are usually quite complex and often possess many asymmetric centers. Aside from their industrial significance, these compounds present tempting challenges to the synthetic chemist. As will be seen, there are many examples in which the challenge has been successfully met. Critical to many of these preparations was the use of chiral starting materials and stereoselective introduction of additional chiral centers as required.

1. β-LACTAM ANTIBIOTICS

The first naturally occurring antimicrobial substances isolated, identified, and applied in clinical medicine were the penicillins. These compounds, along with their congeners, the cephalosporins, clavulanic acid, the nocardicins, and thienamycin, are today still of tremendous medicinal and economic significance in the fight against infectious diseases. It has long been recognized that the β-lactam moieties of these antibiotics are required for biological activity. This observation, along with the difficulties in synthesizing this β-lactam ring unit, has led to a major thrust in antibiotic research: the modification of microbiologically produced penam or cephem ring units. Those partial synthetic efforts will not be considered here. Rather, we shall concentrate on total syntheses of β-lactam antibiotics. These syntheses, all of which are based on amino acid starting materials (the biosynthetic precursors), vary in the manner in which the amino acid is incorporated into the ring system and in the bond that is formed in the ring-closure reaction.

The initial syntheses (*228*) of β-lactam antibiotics were based on formation of the C-2—N-1 amide bond during ring closure. The premier example of this type of synthesis, and the one against which all others must be judged, is Woodward's 1966 preparation of cephalosporin C (*229, 230*) (**584**, Scheme 21), which was announced during the acceptance of the Nobel Prize in chemistry. Treatment of the blocked ester **577**, derived from (*S*)-cysteine, with dimethyl azodicarboxylate gave the trans hydrazo diester **578**. Oxidation of this material with lead tetraacetate followed by heating with sodium acetate gave alcohol **579**. Mesylation, displacement with azide, and reduction by aluminum amalgam gave the necessary nitrogen introduction with inversion, yielding cis amino ester **580**, which was ring closed with triisobutylaluminum to the bicyclic β-lactam **581**. Condensation of **581** with the trichloroethyl ester of (carbomethoxymethylene)malondialdehyde gave the adduct **582**. Treatment of this material with trifluoroacetic acid liberated the cysteine N and S groups and caused subsequent formation of the thiazine ring, giving **583**. Introduction of the side chain, double bond migration, and modification of the ring substituents completed the synthesis.

The strategy of amide ring closure to form β-lactams has been employed by two groups (*231, 232*) in syntheses of thienamycin (**592**). In these syntheses, however, the original amino acid amine function now appears in the β-lactam ring by virtue of reaction with the β-carboxylate of aspartic acid, and not as a ring substituent. As carried out by the Merck group (*231*), the synthesis (Scheme 22) involves initial Grignard reagent–induced cyclization of dibenzyl (*S*)-aspartate (**585**) to the β-lactam **586**. Conversion in several steps of the carbobenzoxy group to iodomethyl and

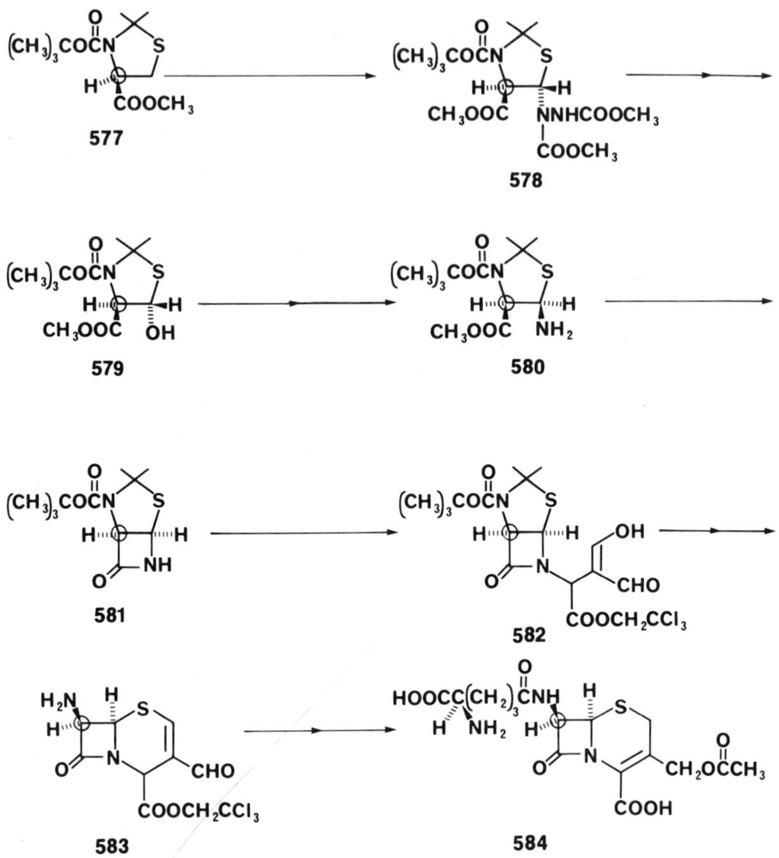

Scheme 21. Cephalosporin C.

N-protection yielded **587**, which was homologated by the dithiane method to **588**. The (*R*)-hydroxyethyl unit was introduced by stereospecific acylation followed by a 90% stereoselective reduction with potassium *sec*-butylborohydride. The resultant alcohol **589** was deprotected and chain lengthened to give β-keto ester **590**. Diazoketone formation and rhodium(II) acetate–induced decomposition gave the bicyclic lactam ester **591**. Introduction of the β-(aminoethyl)thio unit then led, after deprotection, to thienamycin (**592**). The unprotected analog of **587** was also converted (*233*) to hemothienamycin (**593**) using similar chemistry. Although this compound was much more stable than thienamycin, it was also virtually devoid of antibacterial activity. Merck chemists have also devised another thienamycin synthesis, in this case starting from 6-aminopenicil-

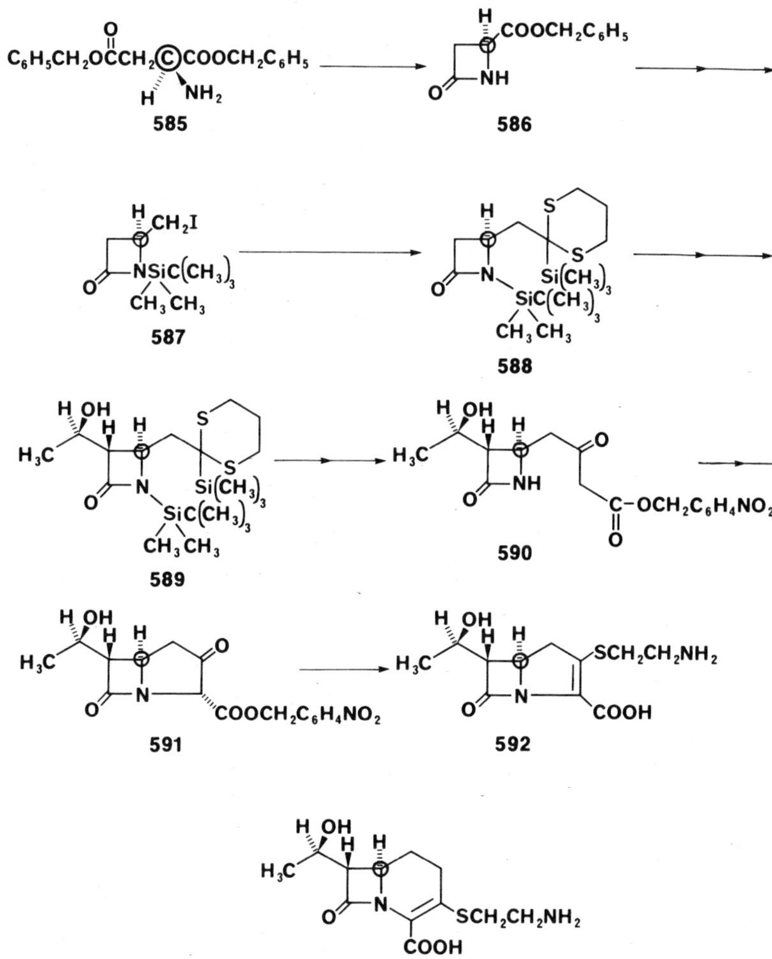

Scheme 22. Thienamycin synthesis I.

lanic acid (*234*). In this synthesis, the β-lactam unit is carried intact from 6-APA to thienamycin.

A second synthesis of thienamycin (Scheme 23) involves C-3—C-4 ring closure to form the β-lactam ring (*235*). This work is also noteworthy in its use of the terminal portion of L-threonine to give the thienamycin side chain without the chemical manipulations required in the Merck synthesis. Thus, L-threonine (**168**) was converted with retention of configuration, by diazotization, to bromoacid **594**. Amide formation with the *tert*-butyl ester of 2,4-dimethoxybenzylglycine gave **595**. Treatment of this

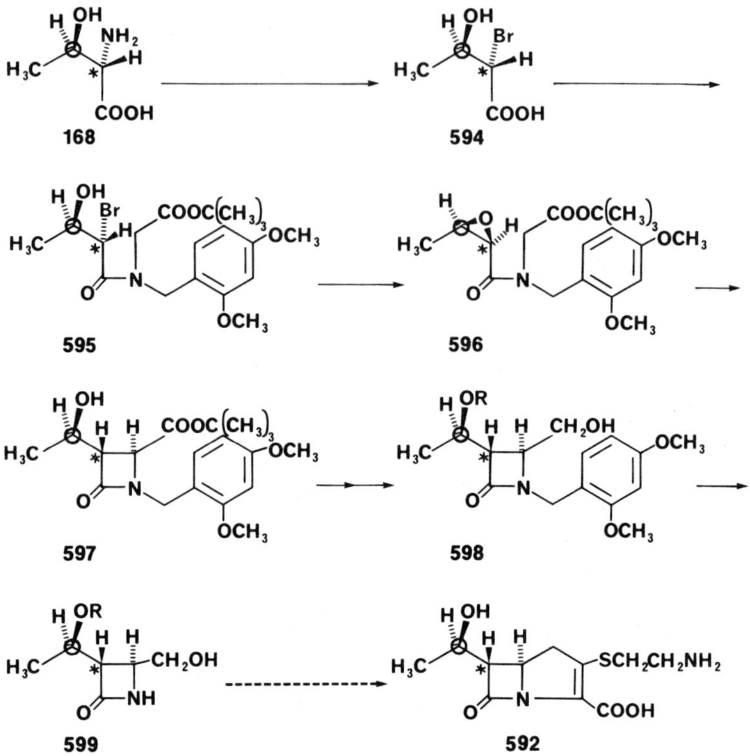

Scheme 23. Thienamycin synthesis II. R = $Si(CH)_2C(CH_3)_3$.

material with one equivalent of base gave the epoxide **596**. Rather than isolate this material, it was treated with a second equivalent of base to cause β-lactam formation with concomitant epoxide opening. Removal of the ester from the resultant **597**, alcohol protection, and acid reduction gave alcohol **598**. Oxidative removal of the dimethoxybenzyl unit with potassium persulfate gave **599**, considered by the authors as a synthetic equivalent of, for example, compound **589** (Scheme 22) in the Merck synthesis.

Although the details of the biosynthesis of the penicillins remain a mystery, it is now generally assumed that a tripeptide is involved in this process. If this is so, the β-lactam ring would be formed by an N-1—C-4 union. Two groups (*236–240*) have reported the synthesis of penam and cephem ring systems from peptide derivatives in analogy with the biosynthesis. Both approaches involved the N-1—C-4 β-lactam ring closure. In Baldwin's (*236*) synthesis of *N*-benzoyl-6-aminopenicillanic acid methyl ester (**607**, Scheme 24), the protected cysteine **600** was reacted with (*R*)-

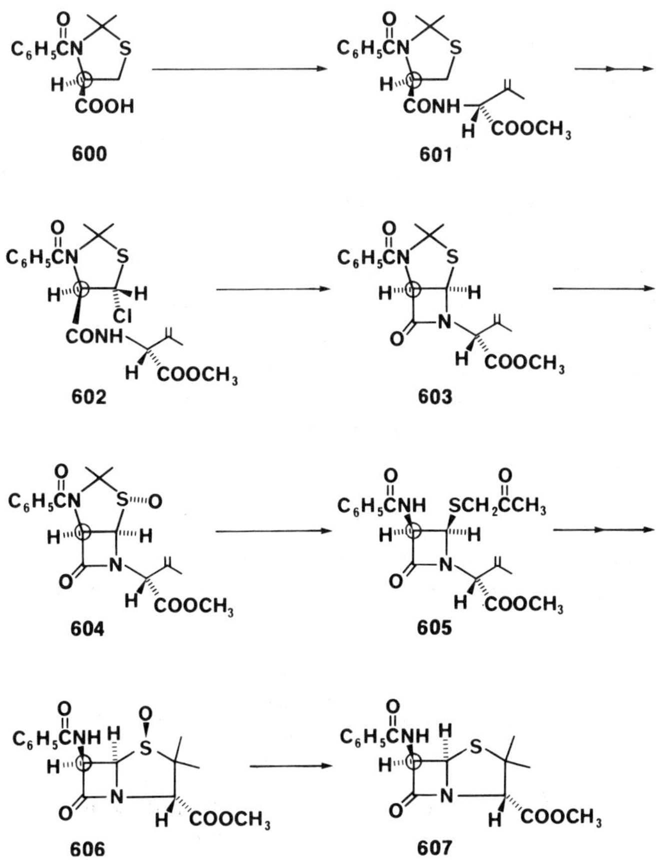

Scheme 24. 6-Aminopenicillanic acid.

isodehydrovaline methyl ester to give the amide **601**.[2] Activation of C-4 was obtained in a stereospecific manner by reaction with benzoyl peroxide. Treatment of the intermediate benzoate with hydrogen chloride led to **602**, which was cyclized with inversion by reaction with sodium hydride, yielding **603**. Oxidation of **603** to **604** was followed by acidic ring opening to **605**. A several-step sequence, involving thioacetone → epoxide → thioisobutyraldehyde conversion, sulfoxide formation, and thermolysis generated the penam **606**. Deoxygenation then gave the desired 6-APA derivative **607**.

The synthesis outlined in Scheme 24 has been used in modified form to prepare the interesting monocyclic β-lactam antibiotic nocardicin A (**613**,

[2] Note the similarity between the starting material **600** and **577** in Scheme 21. The difference in approach to the β-lactam ring from these two materials is illustrative.

Scheme 25) (*241*). Thus, reaction of the protected cysteine **600** with a blocked form of (*R*)-*p*-hydroxyphenylglycine gave the amide **608**, which was converted in three steps (as outlined above) to the β-lactam **609**. Oxidation of this compound with mercuric acetate gave the oxazoline **610**, which upon treatment with phosphorous pentachloride underwent ring opening to **611**. Reduction and hydrolysis of the Schiff base gave the protected 3-aminonocardicinic acid **612**. Coupling of this material with a derivative of (*R*)-homoserine, deprotection, and oximation yielded nocardicin A (**613**). The (*R*)-homoserine required for this final sequence of the synthesis was prepared in a novel manner from (*R*)-methionine (*242*).

N-1—C-4 β-Lactam ring formations using serine, rather than cysteine, starting materials were reported (*243–246*). This efficient synthesis of monocyclic β-lactams, devised by Miller, has now been applied to the synthesis of nocardicin A (Scheme 26) (*247*). Treatment of the (*S*)-serine hydroxamate **614** with triphenylphosphine and carbon tetrachloride in triethylamine gave the lactam **615**. Reductive removal in two steps of the *N*-benzyloxy group gave the unsubstituted compound **616**. Condensation of this compound with the appropriate carbene led to N—H bond insertion to give a mixture of diastereomers. These were separable, and in

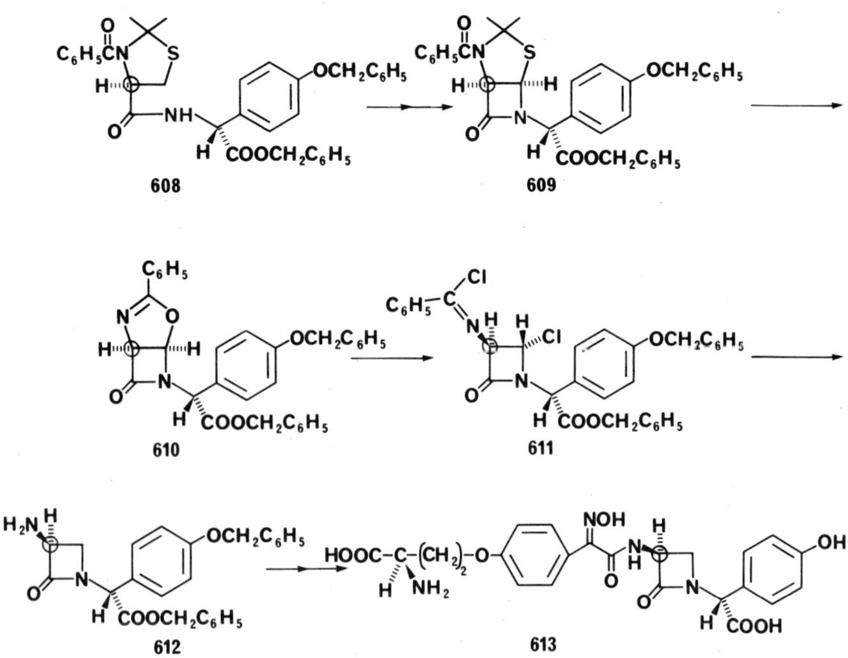

Scheme 25. Nocardicin A synthesis I.

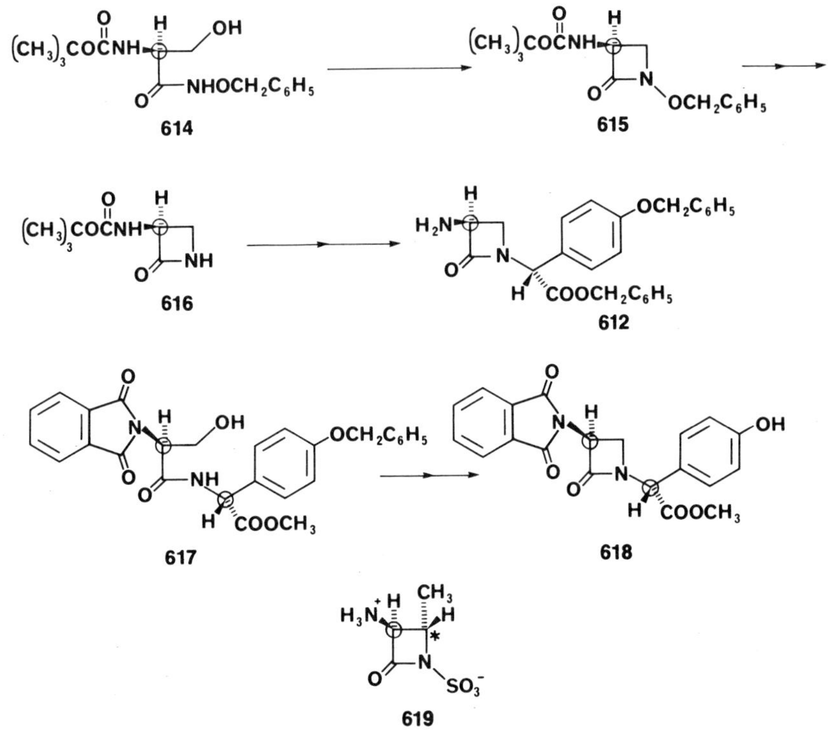

Scheme 26. Nocardicin A syntheses II and III.

addition, the equilibrium mixture could be reestablished with base. Thus, the nonspecificity in this reaction was not of serious consequence. After removal of the *tert*-butyloxycarbonyl group, the dibenzyl-3-aminonocardicinic acid **612**, a previously established (Scheme 25) nocardicin A intermediate, was obtained.

More recently, it was shown (*248, 249*) that the hydroxamate unit is not required for this type of β-lactam formation. Thus, treatment of the protected dipeptide **617** obtained from (*S*)-serine and (*R*)-*p*-hydroxyphenylglycine with triphenylphosphine and diethyl azodicarboxylate gave cyclized material as a 2 : 1 diastereomer mixture (at the phenylglycine center). After debenzylation, the stereochemically homogeneous **618** was obtained. This material (Scheme 27, and see below) had previously been converted (*250, 251*) to nocardicin A.

The basic Miller strategy has also been employed (*252*) in the conversion of threonine to the β-lactam **619**, a basic structural unit of several of the so-called monobactams.

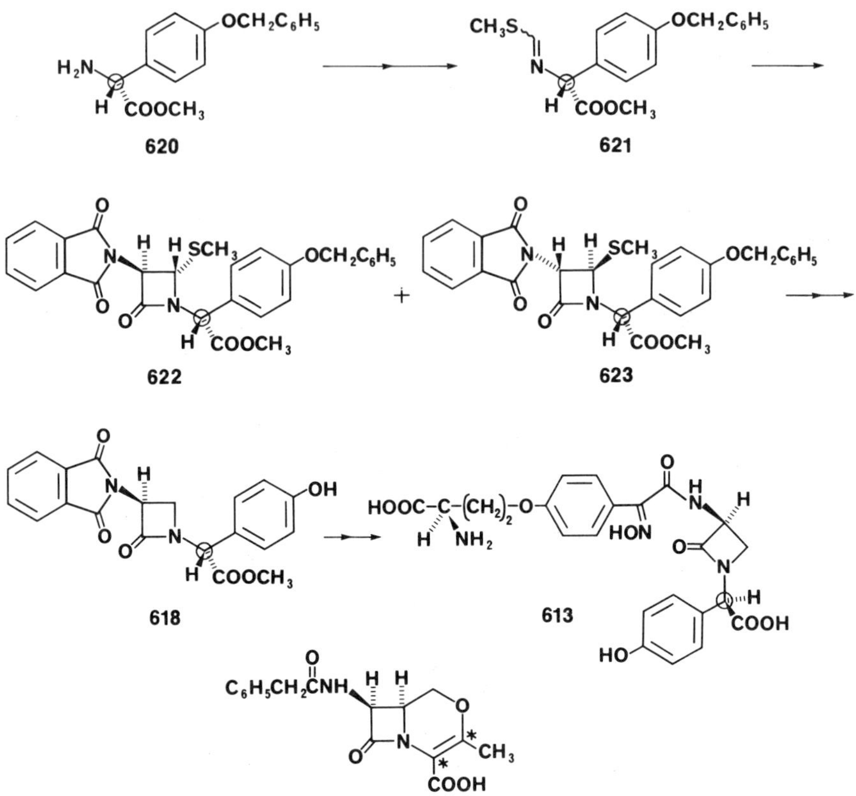

624

Scheme 27. Nocardicin A synthesis IV.

A third approach to β-lactam antibiotics involved cycloaddition reactions with simultaneous formation of the N-1—C-2 and C-3—C-4 bonds. This synthetic methodology was employed in what was the first synthesis of nocardicin A (Scheme 27) (*250, 251*). The protected (*R*)-*p*-hydroxyphenylglycine **620** was thioformylated and then S-methylated to give **621**. The critical β-lactam formation was effected by condensation of this material with the ketene generated *in situ* from phthalimidoacetyl chloride. The resultant diasteriomeric mixture (**622**:**623**, 3:2 mixture) could not be conveniently separated. A two-step desulfurization and benzylation yielded **618**, which could, however, be isolated in diastereomerically pure form by crystallization. Deprotection of **618** was followed by chain addition to give nocardicin A (**613**). In this synthesis, the (*R*)-homoserine unit was obtained by resolution.

The cycloaddition type β-lactam synthesis has also been used to convert (*R*)-threonine to the biologically active *O*-2-isocephem antibiotic **624** (*253*). In this case, the cycloaddition was more highly selective (9:1 ratio).

2. Macrolide Antibiotics

Since the mid-1950s, over 100 antibiotic substances characterized by the presence of a macrocyclic lactone have been reported (*11*). In many cases, these compounds contain several contiguous chiral centers, making their synthesis an imposing task. Due to the flexible nature of the large rings, the preparation of these chiral centers in the correct relative and absolute stereochemistry has amounted to a problem in acyclic stereoselection. As we have previously commented, this type of stereochemical problem is one of the most difficult to solve, and it should therefore come as no surprise that attempts to prepare the macrolides from smaller molecules possessing the desired chirality have been more fruitful, on the whole, than alternative syntheses.

Some of the macrolides present the conceptually simpler synthetic problem of only one type of chiral center. Such a compound is the diolide (*R*,*R*)-pyrenophorin (**630**), which possesses C-2 symmetry. The synthesis of this compound by Seebach (*254, 255*) (Scheme 28) was initiated by alkylation of dithiane with the iodide **625** derived from (*S*)-β-hydroxybutyric acid. The resultant **626** was lithiated and reacted with dimethylformamide to give aldehyde **627**. Wittig chain extension was followed by basic and acidic protecting group removal to give the hydroxy acid **628**. Dimerization with inversion of **628** to **629** was then followed by hydrolysis of the dithioacetal groups to yield pyrenophorin. The same general synthetic route was also used to convert epoxide **104** [from (*S*)-malic acid] to the *R*,*R*-enantiomer **631** of vermiculin (*254*). These syntheses established the absolute configurations of pyrenophorin and vermiculine. The Seebach group reported a second synthesis of pyrenophorin (*256*), as well as of a dideoxy analog thereof (*256*), using (*S*)-methyloxirane as the source of chirality. (*S*)-Methyloxirane and (*S*)-β-hydroxybutyric acid were also converted, by the same group (*257*), to the enantiomer **632** of the unsymmetrical diolide (*R*,*R*)-grahamimycin A_1. The use of (*R*)-methyloxirane has led, in a straightforward synthesis (*258*), to (*R*)-recifeiolide.

The tetrolide antibiotic nonactin (**638**) is composed of alternating units of (*R*,*R*,*S*,*S*)-(−)- and (*S*,*S*,*R*,*R*)-(+)-nonactic acids. Because of the four chiral centers in these molecules, a stereorandom synthesis would yield

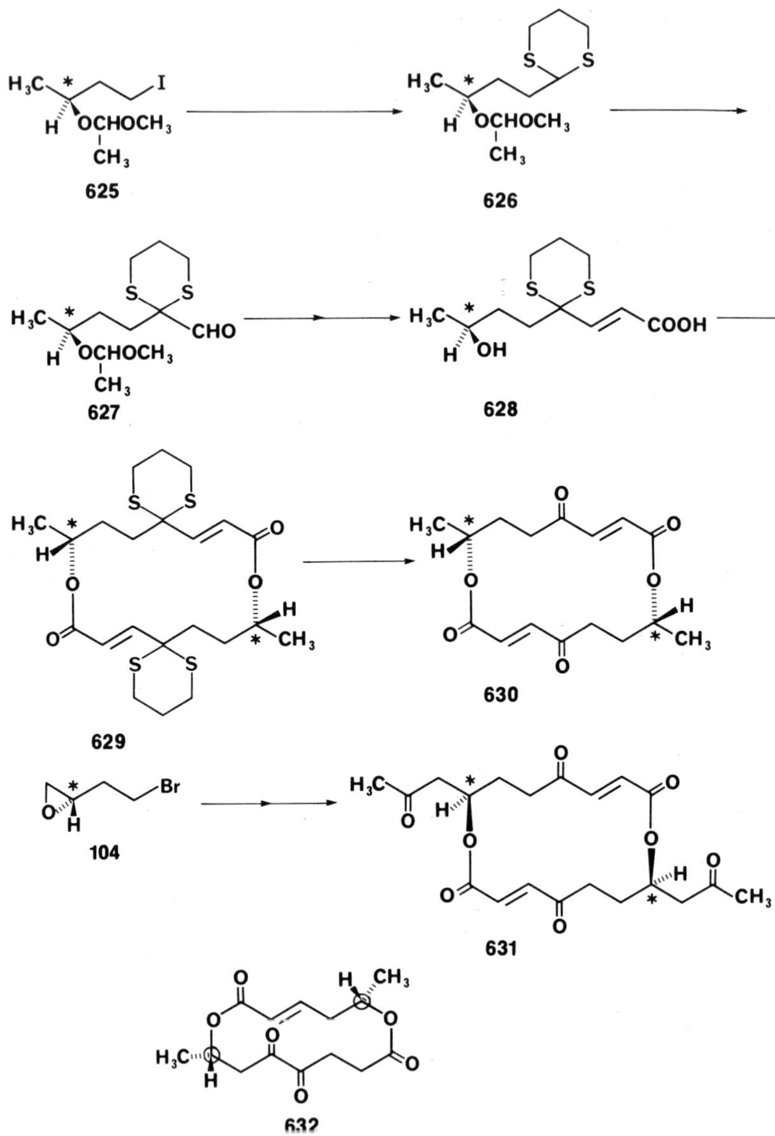

Scheme 28. (*R,R*)-Pyrenophorin.

eight diastereomer pairs or sixteen compounds. The synthesis of nonactin by Schmidt (*259*) used (*S*)-2-methyloxirane (**61**) [from lactic acid (*260*)] as starting material and a specific cis hydrogenation to reduce the number of compounds obtained to four (Scheme 29). Alkylation of **61** with furyl-

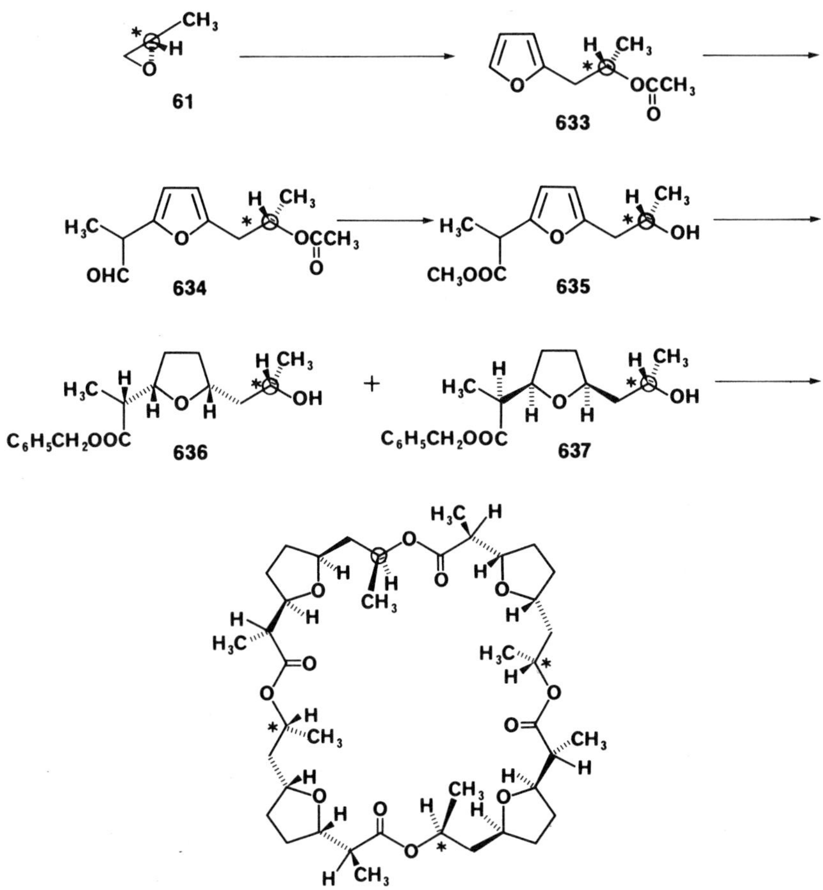

638

Scheme 29. Nonactin.

lithium and acetylation gave **633**. A three-step preparation of the second furan substituent, giving **634**, was followed by silver carbonate oxidation and esterification to yield **635**. Catalytic reduction of this compound gave four diastereomeric compounds. After transesterification to the more readily separated benzyl esters, these materials were separated by chromatography. Two moles of each of two of the isolated diastereomeric esters (**636** and **637**) were then combined to complete the nonactin synthesis. Three of the esterification reactions involved inversion of configuration at the alcohol center, while the fourth proceeded with retention.

The presence in the nonactin molecule of four tetrahydrofuran rings has led two groups (*261, 262*) to employ sugars as starting materials in the preparation of the (*R,R,S,S*)- and (*S,S,R,R*)-nonactic acids. The synthesis developed by Fraser-Reid (Scheme 30) (*261*) used D-ribose as a common point of origin and depended on epimerization of the appropriate centers. The aldehyde **639**, readily prepared from D-ribose, was subjected to two-carbon homologation to **640**. Hydrogenation of this ketone over Raney nickel gave the alcohol **641** along with 10% of the wrong diastereomer. The wrong diasteremeric alcohol could be isolated and converted

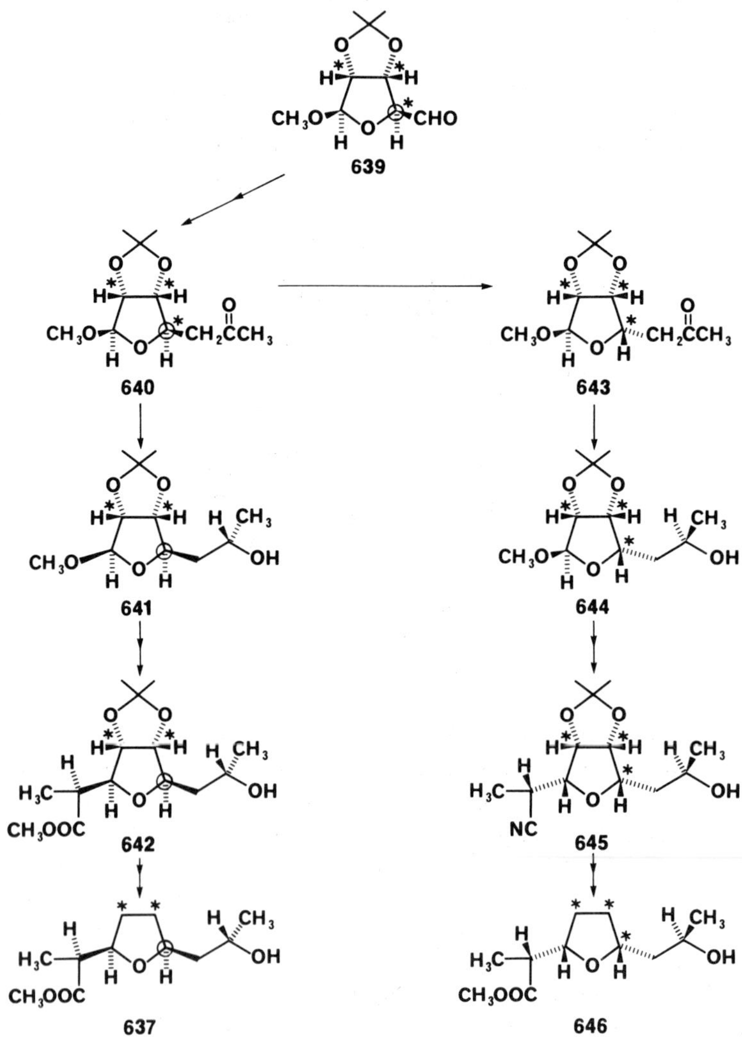

Scheme 30. Nonactic acid.

to **641** by tosylation and displacement. Freeing of the masked aldehyde of **641**, followed by Wittig reaction under conditions of kinetic control gave exclusively the *cis*-tetrahydrofuran. The product was, however, a diastereomeric mixture (α to the carboxyl ester). After separation of the desired **642**, the diastereomeric ester could be equilibrated with base to a mixture again containing **642**. With three cycles of equilibration and separation, the yield of **642** reached 90%. A five-step deoxygenation of this compound then gave the (−) ester **637**. For preparation of the (+) ester **646**, ketone **640** was equilibrated with base, giving a 1:9 mixture of **640** : **643**. Hydrogenation of the latter gave a 3 : 1 mixture of **644** and its diastereomer. As before, the unwanted material could be epimerized to the desired compound. Subjecting the deacetalized product from **644** to Wittig reaction and base treatment, after recycling of the wrong diastereomer, gave an 84% yield of the nitrile **645**. The choice of nitrile over carbomethoxyl at this point was dictated by the fact that basic equilibration of only the former led to the desired *cis*-tetrahydrofuran. Conversion of the nitrile to carbomethoxy and deoxygenation completed the synthesis of (+)-ester **646**.

The second synthesis of the nonactic acids from sugars, done by Ireland (*262*), involved Claisen rearrangement of carbohydrate-derived furanoid glycals. This synthetic methodology, by which D-mannose and D-gulono-γ-lactone were converted to the (−)- and (+)-nonactic acids, respectively, is discussed in detail in connection with the description of the preparation of lasalocid (see Section III,C,3 and Scheme 41).

Another group of oligomeric macrolide antibiotics is represented by azimine and carpaine, which are dimers of azimic acid (**654a**) and carpamic acid (**654b**), respectively. The acids are unusual in containing the cis 2,3,6-trisubstituted piperidine nucleus. Hanessian's synthesis (*263*) of these compounds (Scheme 31) was based on removal of the C-6 oxygen of D-glucose and conversion of the C-5 pyranose oxygen to nitrogen with inversion. The readily available glucose derivative **647** was hydrogenated and then treated with *N*-bromosuccinimide to give ester **648**. Saponification, reductive debromination, and benzylation yielded **649**, which was converted with ethanethiol in the presence of acid to the thioacetal **650**. Tosylation, reaction with sodium azide, and deacetalization then gave the azidoaldehyde **651**. Introduction of the remaining carbon atoms by Grignard reaction was followed by oxidation to yield ketones **652a,b**. Conversion of these compounds to the *cis*-piperidines **653a,b** was accomplished in one step by catalytic hydrogenation over palladium on carbon. A sequence involving N-protection, cleavage of the tetrahydropyranyl ether, oxidation, and removal of the remaining protecting groups then provided the desired acids **654a,b**.

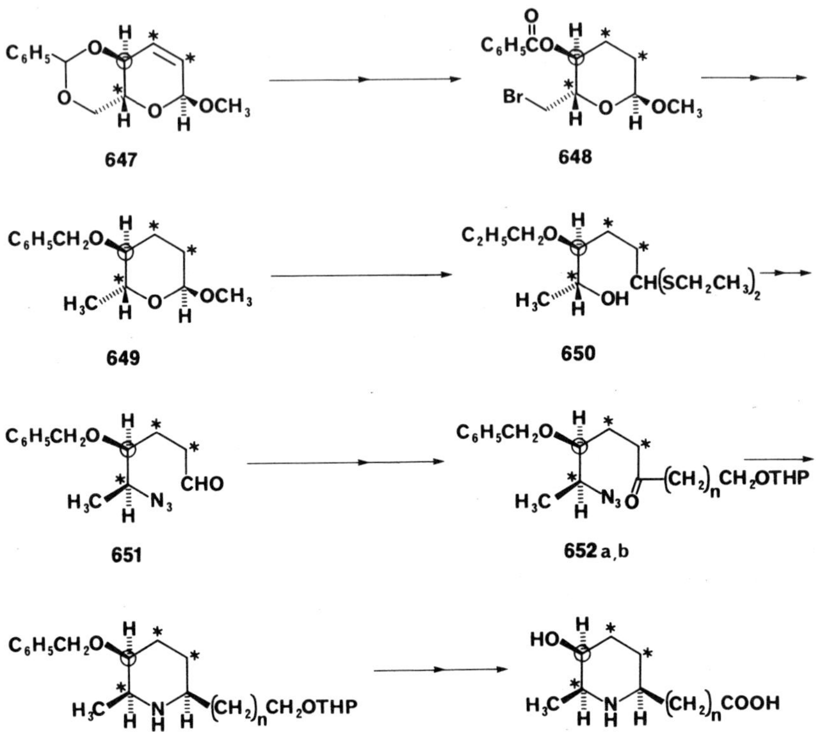

647 648 649 650 651 652 a,b 653 a,b 654 a,b

Scheme 31. Azimic and carpamic acids: **a** series (n = 5), azimic acid; **b** series (n = 7), carpamic acid.

The macrolide syntheses discussed up to this point have been simplified in most cases by the symmetry elements present in the molecules. Many of the macrolides possess no such symmetry elements, and thus their preparation is often more of a challenge. With some of the compounds, however, the fusion of the macrolide to a carbocyclic ring reduces the stereochemical problems to some extent. One such compound is brefeldin A (**666**). Despite the apparent simplification resulting from the ring fusion, however, the synthesis of this compound (Scheme 32) nevertheless required over 40 steps from D-mannitol and (R)-glutamic acid (*109*). Thus, the iodoacetal **655** from D-mannitol was used to alkylate diethyl malonate. Saponification of the resulting diester **656** was followed by acidic deacetalization, lactone formation and a three-step introduction of the cyanomethyl unit, giving **657**. This compound was approximately a 2 : 1 mixture with its diastereomer; interestingly, however, both of these materials could be converted to cyano tosylate **658**. The **657** → **658** transformation

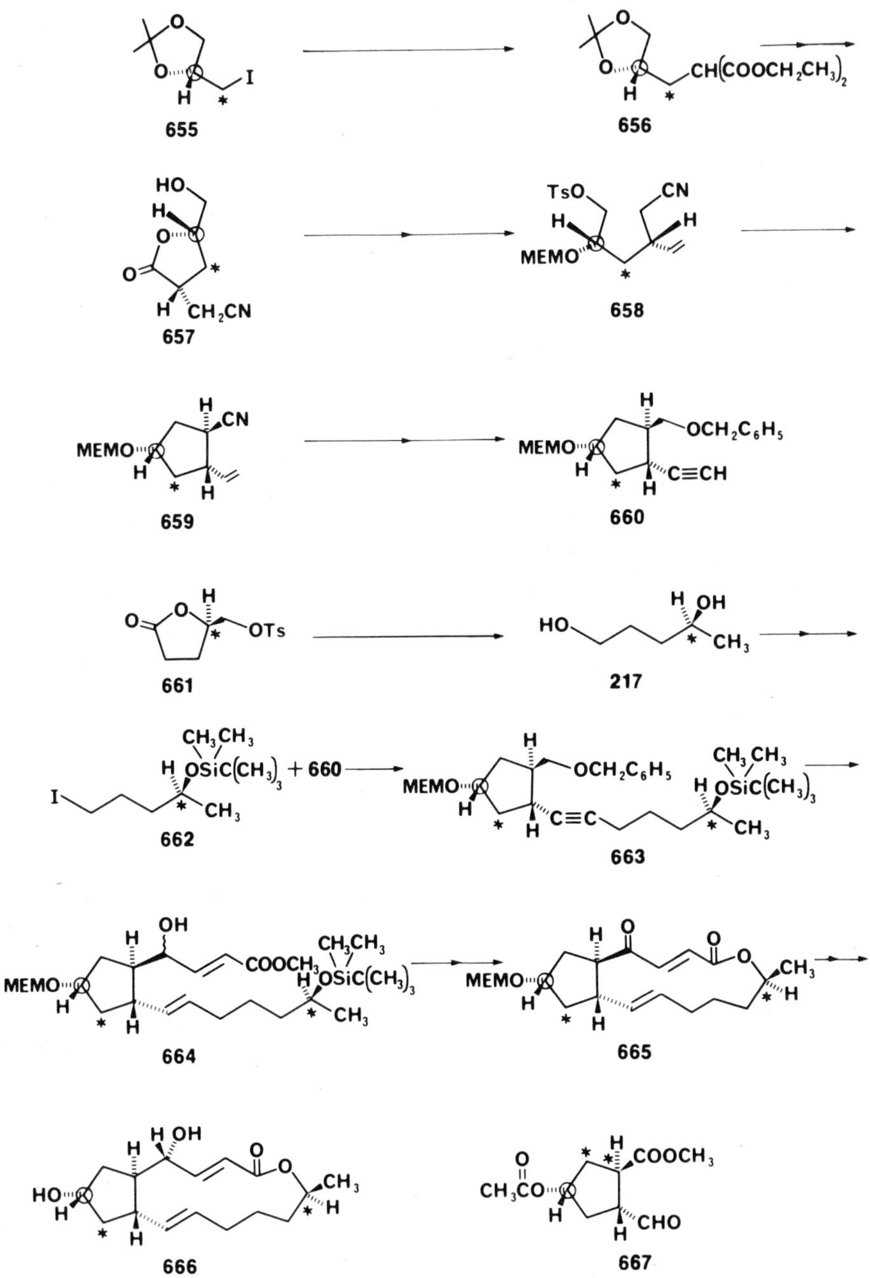

Scheme 32. Brefeldin A. MEM = Methoxyethoxymethyl.

was effected by protection of the primary alcohol, DiBAlH reduction, Wittig reaction (epimerization probably occurring at this point), protection of the secondary alcohol, selective deprotection of the primary alcohol, and tosylation. Cyclization of **658** with base gave **659** along with a small amount of its cis isomer. Separation was unnecessary, however, since epimerization of the unwanted diastereomer occurred during the subsequent straightforward conversion to **660**.

The synthesis of the chiral alcohol portion of brefeldin A began with tosylate **661**, derived from (*R*)-glutamic acid. Hydride reduction of this compound yielded the diol **217**, which was converted in a series of unexceptional reactions to the protected iodide **662**. Alkylation of **660** with **662** proceeded smoothly, yielding **663**. Sodium in ammonia simultaneously reduced the acetylene to the *E* olefin and cleaved the benzyl ether. After oxidation of the alcohol to the aldehyde level, the remaining carbon atoms were introduced using methyl-β-nitropropionate. The diastereomeric mixture **664** was then converted in several steps to **665**. Borohydride reduction and protecting group removal yielded brefeldin A.

The conversion of D-glucose to **667**, of potential use in the synthesis of brefeldin A, has been recorded (*187*).

One of the most spectacular total syntheses of a chiral molecule, independent of the source of the starting materials, is Stork's preparation (*264*) of cytochalasin B (**677**, Scheme 33). This cytotoxic compound contains eight asymmetric centers, three of which were derived from citronellol, malic acid, and phenylalanine, respectively. Four of the remaining centers were generated in a Diels–Alder reaction, while the last resulted from a stereospecific epoxidation. Kolbe coupling of the acid **668**, which was obtained by ozonolysis of citronellyl acetate, with the acetate of the 1-ethyl ester of malic acid gave triester **669**. Hydride reduction to the triol, acetonide formation, and oxidation gave the protected aldehyde **670**. Horner coupling of this compound with the appropriate phosphonate, derived in seven steps from glycidaldehyde, yielded the trienic aldehyde **671**, which was to serve as diene in the Diels–Alder reaction. The dienophilic component for this reaction was prepared from ester **672**, itself prepared by Arndt–Eistert homologation of *N*-carbobenzyloxy-(*S*)-phenylalanine. Hydroxypyrrolone ester formation from **672** was followed by acetylation and decarboethoxylation to give **673**. The Diels–Alder reaction of **673** with **671** gave, after desilylation, adduct **674** along with some regioisomeric (but apparently no stereoisomeric) product. With seven of the eight chiral centers of cytochalasin B installed in the correct orientation, the eighth was then constructed by epoxidation, alcohol to bromide conversion, and reductive β-elimination with zinc. Conversion of the resultant allylic alcohol **675** to cytochalasin B, via the acid **676**, was then straightforward.

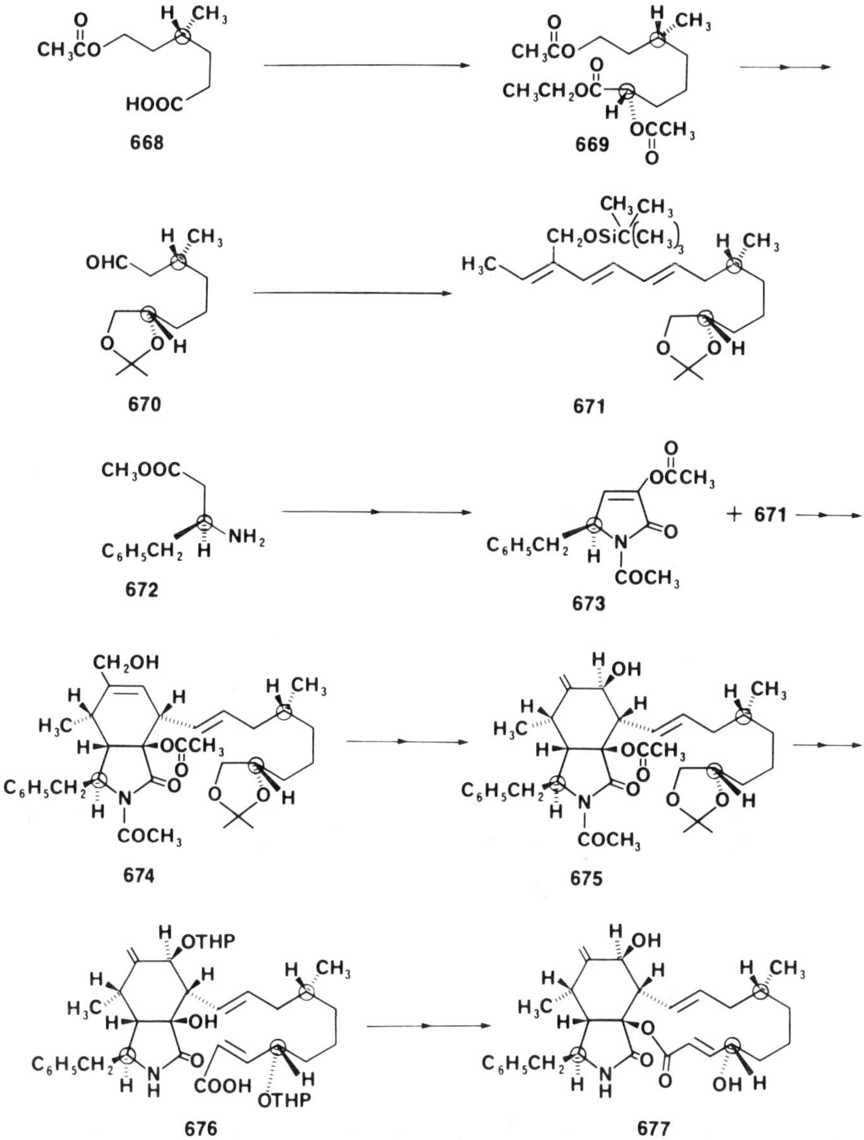

Scheme 33. Cytochalasin B.

From a conceptual viewpoint, the macrolides possessing neither ring fusion nor symmetry elements present a particularly demanding synthetic challenge. One molecule that has been successfully synthesized is the antibiotic A26771B (**688**, Scheme 34) (*265*). In the preparation of this compound, the two remote chiral centers were both derived from

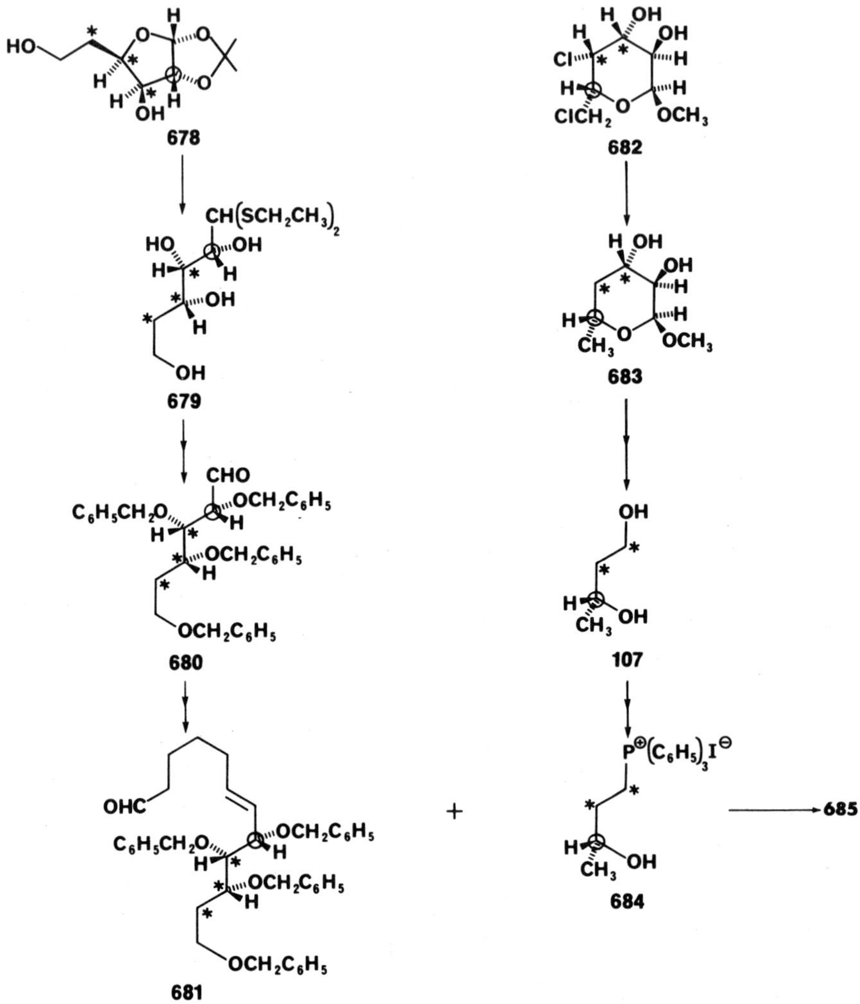

Scheme 34. Antibiotic A26771B.

D-glucose, although in entirely different fashions. For the first segment of **688**, glucose was converted to the deoxygenated acetal **678**. Treatment of this compound with ethanethiol under acid catalysis gave the dithioacetal **679** which, on benzylation and hydrolysis, gave aldehyde **680**. Wittig chain lengthening by five carbons gave, after aldehyde deprotection, the *E* olefin **681**. The remaining carbon atoms of A26771B were obtained from glucose by initial conversion via **682** to the dideoxy derivative **683**. Peri-

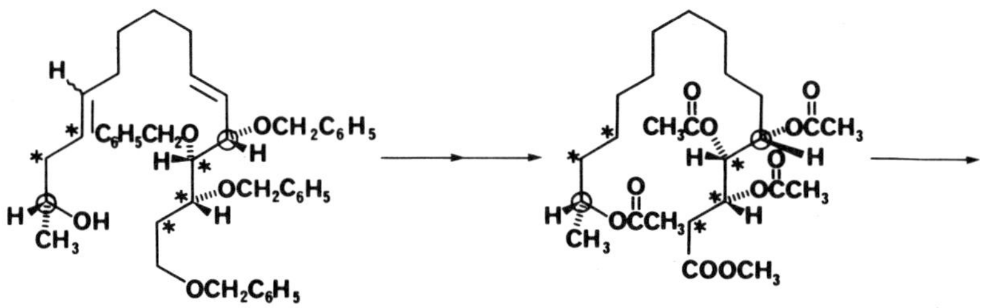

685 686

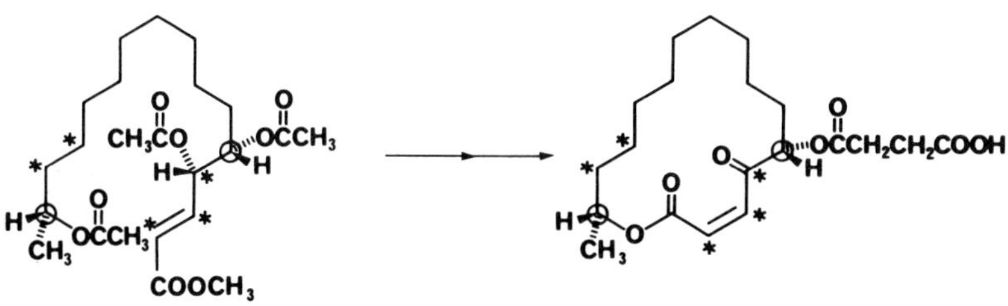

687 688

Scheme 34 (*Continued*)

odic acid oxidation and reduction gave the diol **107** (for entirely different preparations of this compound, see refs. *37* and *77*), which was converted in the usual manner to **684**. Combination of **681** and **684** gave the *E*/*Z* olefin mixture **685**, in which all the ring carbon atoms of the final synthetic target were present. After hydrogenation to remove the benzyl groups and to saturate the olefinic bonds, oxidation of a suitably protected intermediate led to ester **686**. Elimination of the elements of acetic acid with base then gave the unsaturated ester **687**. Protective group hydrolysis, lactone ring formation, selective succinoylation and oxidation completed the synthesis of A26771B(**688**).

After completion of the synthesis of A26771B, Tatsuta's group was able to use a similar synthetic approach to achieve preparations of the structurally similar, but more complex, antibiotics carbomycin B (magnamycin B, **702**) and leucomycin A_3 (josamycin, **703**) (*266*) (see Scheme 35). Thus, acidic hydrolysis of the methyl ether of **678** (Scheme 34) followed by reaction with carbomethoxymethylenetriphenylphosphorane gave the unsaturated ester **689**. Conjugate addition of methyl methylthiomethyl sulfoxide to this material was sterospecific. Acid-catalyzed exchange with ethanethiol then occurred with simultaneous lactonization to give **690**. Lactone to lactol reduction, protection as the mixed acetal (stereochemically homogeneous, but of unknown configuration), benzylation, and

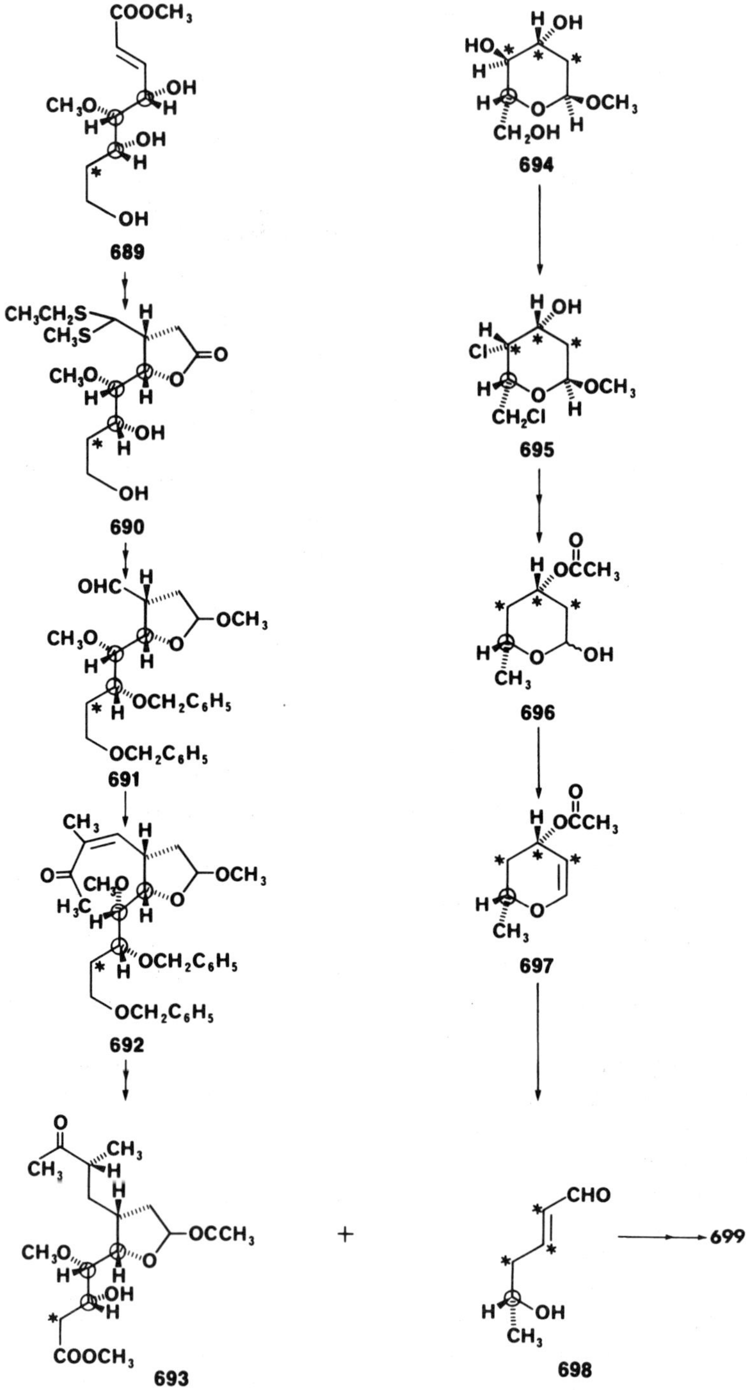

Scheme 35. Carbomycin and leucomycin A_3 synthesis I.

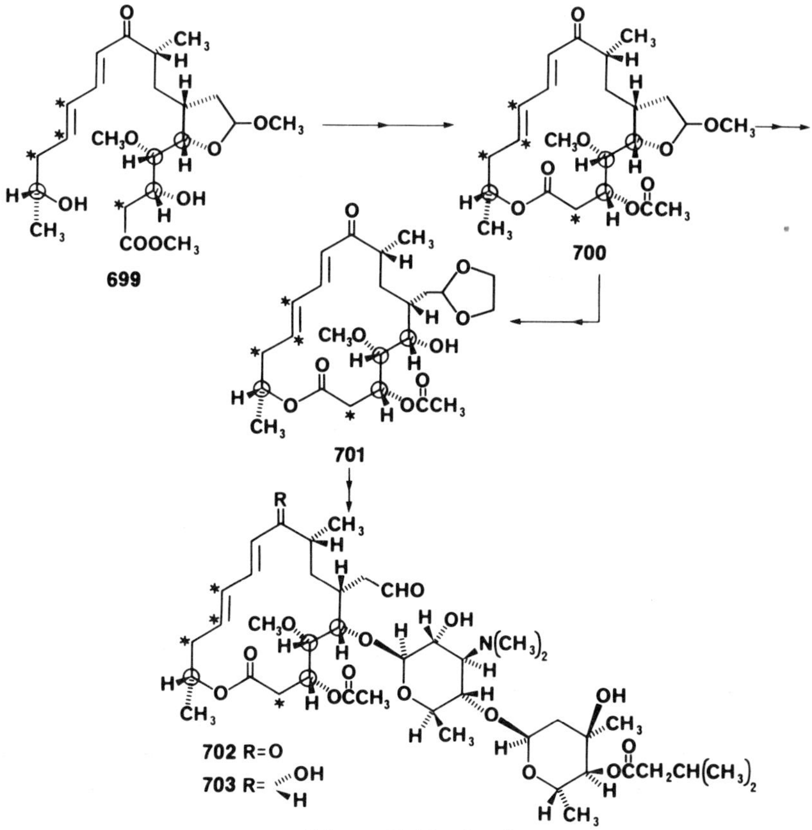

Scheme 35 (*Continued*)

thioacetal hydrolysis then gave aldehyde **691**. Chain lengthening by four carbon atoms was achieved by Wittig reaction. Hydrogenation of the resulting unsaturated ketone **692** was accompanied by hydrogenolysis. After catalytic oxidation of the primary alcohol and esterification, keto ester **693** was obtained along with a minor diastereomeric compound. After separation, the undesired compound was equilibrated with acid to a mixture containing additional **693**, thus maximizing the yield in this reaction.

The remaining chiral carbon atom was introduced as the unsaturated hydroxyaldehyde **698**. This compound was prepared (*267*) from the 2-deoxyglucose **694** by initial dechlorination to **695**. As for **682**, reductive chlorine removal gave, after acetylation and acetal hydrolysis, acetate **696**. Tosylation of this compound in the presence of triethylamine gave the unstable glycal **697**, which on treatment with mercurous acetate followed by sulfuric acid was converted to the desired **698**. Reaction of this latter compound with the anion of **693** yielded **699**, thereby completing the synthesis of the basic carbon skeleton. Reduction of the unsaturated ketone, ester hydrolysis, lactonization, selective reoxidation, and reacetalization with ethylene glycol gave, via **700**, the protected aglycone **701**. This compound was subsequently converted to carbomycin B (**702**) (*268*) and leucomycin A_3 (**703**) (*269*).

In an independent investigation, Nicolaou (*270, 271*) used a synthetic strategy very similar to that of Tatsuta to also prepare the macrocyclic acetal **701** (Scheme 36). Thus, the 3-methyl ether of acetal **678** (Scheme 34) was benzylated at C-6 and then converted to the benzyl ether of **689** (Scheme 35). Protection of this material as its acetonide and conjugate addition of methallyllithium gave, with a high degree of stereoselectivity, ester **704**. Coversion in several steps to the fully protected pentaol **705** was followed by hydroboration and oxidation to give aldehyde **706** as a mixture of diastereomers. Reaction of **706** with the anion from dimethyl methylphosphonate and then oxidation gave ketophosphonate **707** which, when subjected to hydrogenolysis and further oxidation, gave acid **708**, still as a mixture of diastereomers. Esterification with hydroxyaldehyde **698**, derived in this case from the resolution-produced (*R*)-β-hydroxybutyric acid, gave **709**. Intramolecular cyclization gave the stereochemically homogenous macrocycle **710**, but only in 20% yield. This low yield, due in part to the required diastereomer separation, detracts from the overall value of this synthesis. With lactone **710** in hand, functional group and oxidation level manipulation in seven steps gave the desired carbomycin B and leucomycin A_3 precursor **701** (*272*).

In addition to the syntheses of Tatsuta and Nicolaou, reports (*273, 274*) of the early stages of the third synthesis of carbomycins from glucose have been made.

The same basic synthetic concepts that led to the carbomycin preparation have now been employed to synthesize (*275–277*) the protected form **711** of tylonolide, the aglycone of the medicinally important antibiotic tylosin, and then the antibiotic itself (*278*).

Among the most structurally complex of the macrolide antibiotics is erythromycin. The aglycone, erythronolide A (**725**), contains ten asymmetric centers in which there are two groups (one of four and one of five) of contiguous chiral carbon atoms. Although the completion of the synthesis of erythronolide A has yet to be reported, Hanessian (*279*) achieved preparation of the intermediate **724** (Scheme 37), which contains eight of the ten chiral centers in the proper orientation. Even though the synthetic target has yet to be reached, this synthesis is of significance because of its convergent nature and the manner in which glucose is converted in a

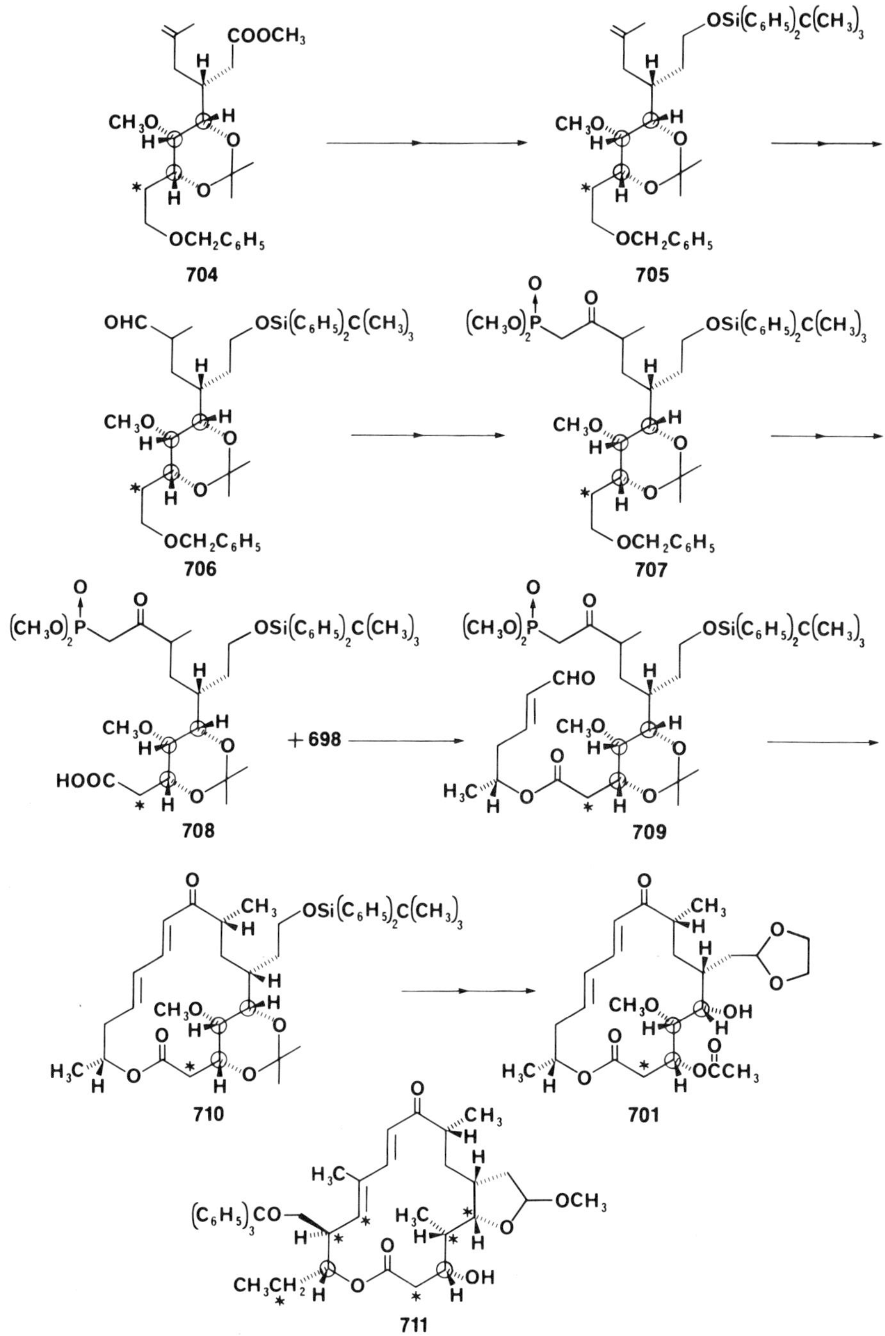

Scheme 36. Carbomycin B and leucomycin A_3 synthesis II.

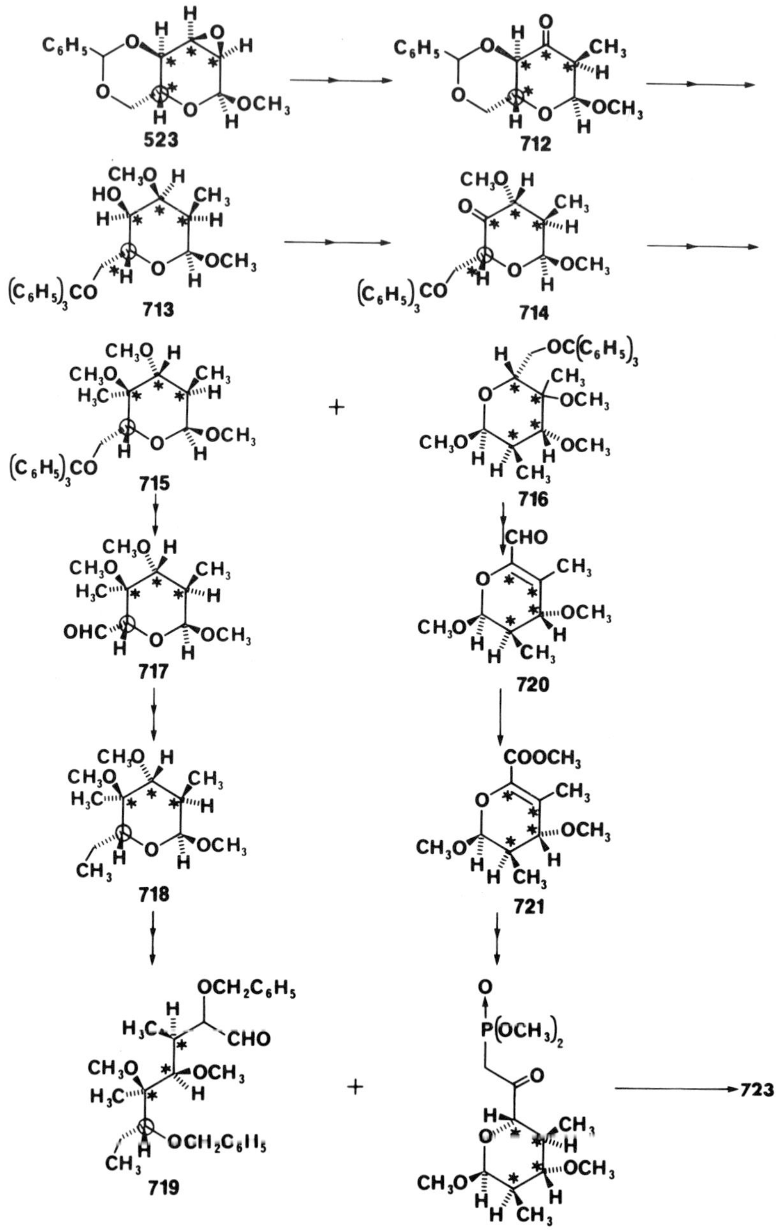

Scheme 37. Erythronolide A (partial synthesis).

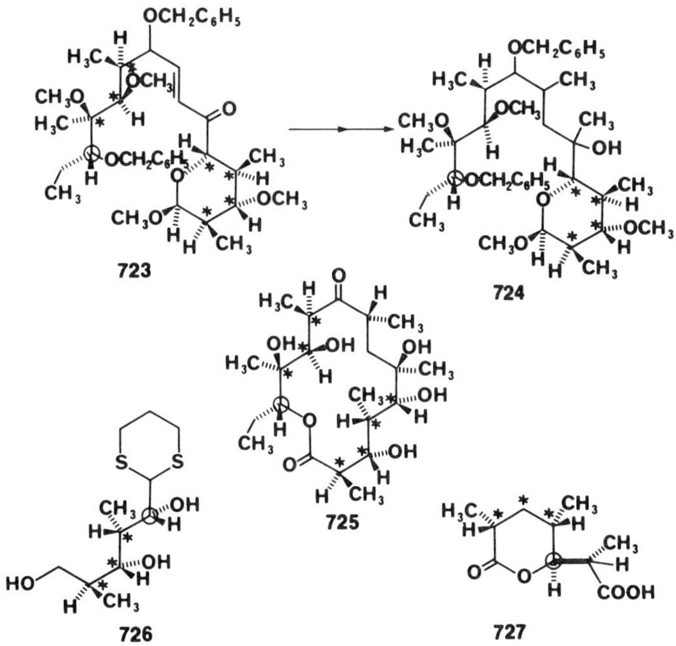

Scheme 37 (*Continued*)

number of steps to a central intermediate (**714**), which then is transformed to the two "wings" of **724**. Of additional import is the manner in which the synthetic design was conceived and executed: of the eight centers of asymmetry present in two molecules of glucose, only one appears in **724** in unchanged form.

The epoxide **523**, prepared in several steps from glucose, was treated with lithium dimethylcuprate. After oxidation and methyl group epimerization, the ketone **712** was obtained. Borohydride reduction, methyl ether formation, hydrogenolysis of the acetal, and selective tritylation of the primary alcohol gave alcohol **713**, which on oxidation and methoxyl epimerization gave ketone **714**. Reaction of this ketone with methyllithium gave, after etherification, the crystalline-protected triol **715**. The mother liquors contained the mixture **716** (drawn in 120° rotated form) of **715** with its diastereomer. It was at this point that the synthesis diverged. The crystalline **715** was detritylated and oxidized to **717**, the methylenation and hydrogenation of which yielded **718** (*280*). Acetal hydrolysis, reaction with vinylmagnesium bromide, benzylation, and ozonolysis then gave the C-8—C-15 left-wing fragment **719** as a diastereomeric mixture (*281*). The noncrystalline **716** was oxidized after detritylation. Treatment of the resultant aldehyde with calcium hydroxide gave the unsaturated **720**. The aldehyde **720** was then oxidized by the Corey manganese dioxide–cyanide method to **721**. Catalytic hydrogenation, which occurred with high stereoselectivity, was followed by reaction with lithium dimethyl methylphosphonate to give the β-ketophosphonate **722**, representing the C-1—C-7 right wing of erythronolide A.

Combination of the segments **719** and **722** gave the diastereomer mixture **723**. The conjugate addition of lithium dimethylcuprate and subsequent treatment with methyllithium gave **724**. Although this material was a mixture of eight diastereomers, the stereochemical problem was not as bad as it might have appeared, since the protected secondary alcohol was destined for oxidation to the carbonyl level. In addition, it was known that the adjacent methyl center was epimerizable in **725** to the requisite configuration.

The preparation of **726**, a synthetic equivalent of C-1—C-9 segment **722**, from D-glucose has been reported (*282*) by another group.

The Prelog–Djerassi lactone (**727**) is a degradation product of the antibiotics narbomycin and methimycin, as well as a synthon (*283, 284*) for the latter. Three syntheses (*285–287*) of this compound from D-glucose have been disclosed. The preparation of eight diastereomeric pyranosides containing the chiral functional unit R^1—$CH(CH_3)CH(OR^2)CH(CH_3)CH$-$(OR^2)$—$R^4$ has been effected (*288*). These materials are of potential usefulness in macrolide and polyether antibiotic syntheses.

Because of its nine-membered diolide structure, antimycin A_3 (**737**, Scheme 38) is not generally included among the macrolide antibiotics. A discussion of the synthesis of this molecule at this point, however, is appropriate because the approach employed (*289–291*) is similar to that used in the preparation of the macrolides. In fact, this synthesis uses the same epoxide **523** used for erythronolide A (Scheme 37). Thus, epoxide opening with butylmagnesium chloride, acetylation, and acetal rearrangement with *N*-bromosuccinimide gave **728**. Olefin formation with silver fluoride in pyridine and hydrogenation gave the *C*-methyl compound **729** of inverted configuration. Reaction of this compound with acetic anhydride and sulfuric acid, followed by sodium methoxide, gave the deprotected anomeric mixture **730**. Removal of one carbon atom by periodate oxidation and subsequent deformylation and acetalization then yielded furanose **731**. Inversion of the secondary alcohol was accomplished by mesylation, displacement with benzoate, and saponification. The resultant **732** was benzylated, deacetalized, and reduced to diol **733**. Selective tritylation, condensation with *N*-(benzyloxycarbonyl)-(*S*)-threonine-*tert*-butyl ether, and detritylation gave ester **734**. Functional group manipulation yielded **735**, which was lactonized and subjected to hydrogenolytic

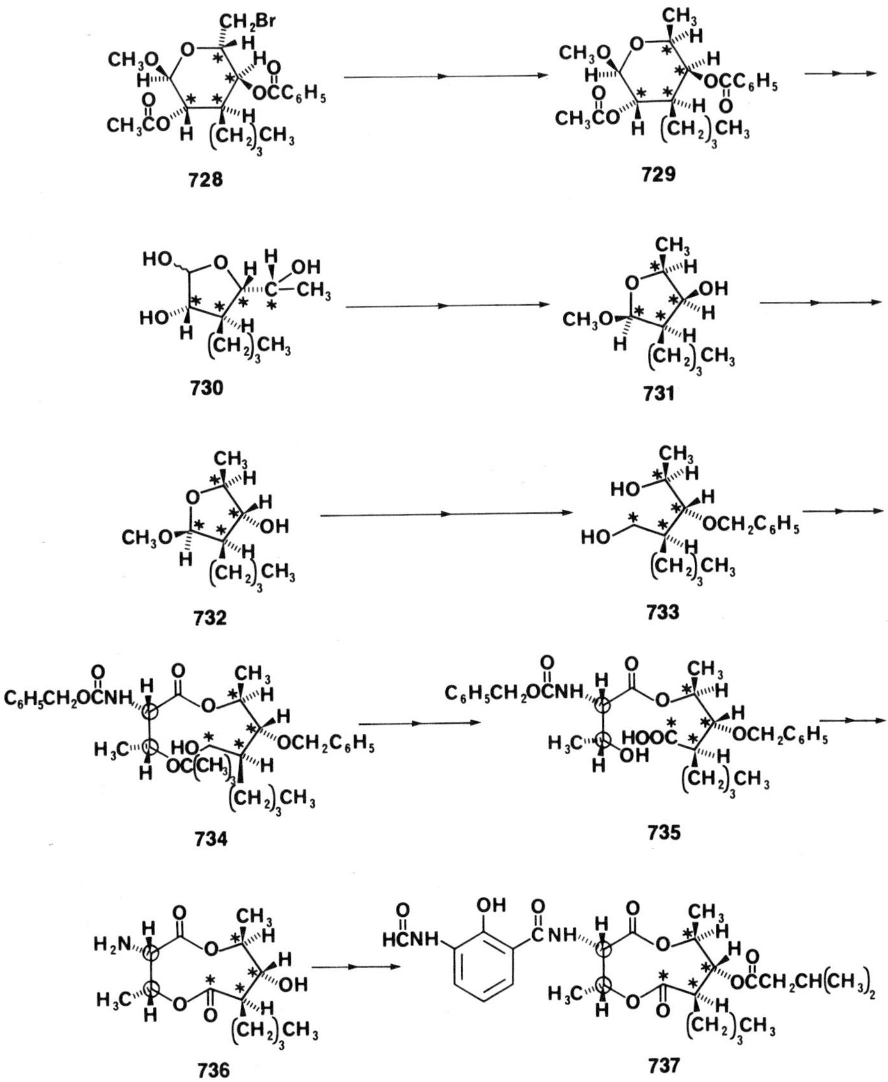

Scheme 38. Antimycin A_3.

removal of the aromatic protecting groups to give **736**. Addition of the requisite appendages then yielded antimycin A_3 (**737**).

The lactam-containing, as opposed to lactone-containing, macrocyclic compounds are not macrolides. Because the synthetic approaches to these "azamacrolides" are similar to those of the macrolides, their preparation is considered in this section.

Because of its promising antitumor activity, maytansine (**748**) has generated intense synthetic activity in recent years. Despite this substantial effort, only one synthesis of maytansine in enantiomerically pure form (Scheme 39) has been reported (*292, 293*). In this synthesis, triacetyl-D-glucal (**270**) was methanolyzed to the triol, which was acetoxymercurated and reduced to acetal **738**. After tritylation, dianion formation followed by reaction with triisopropylbenzenesulfonylimidazole gave the epoxide **739**, the ring opening of which with methyllithium gave **740**. Reaction of this

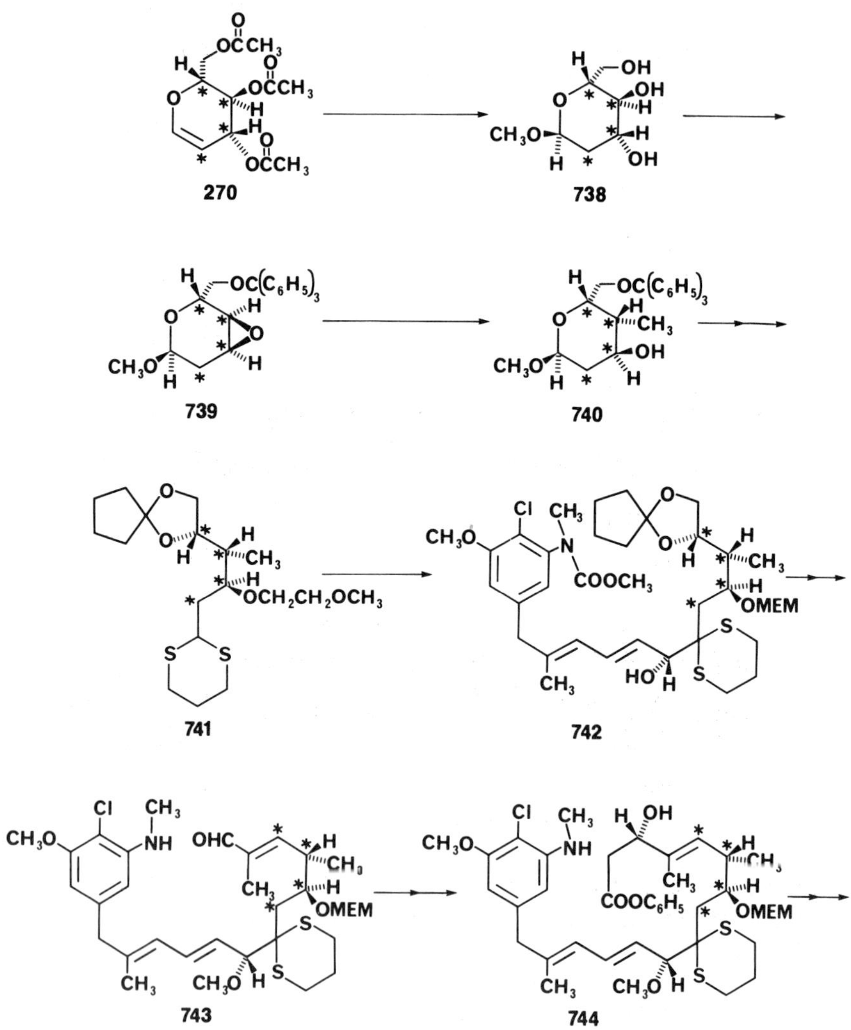

Scheme 39. Maytansine. MEM = Methoxyethoxymethyl.

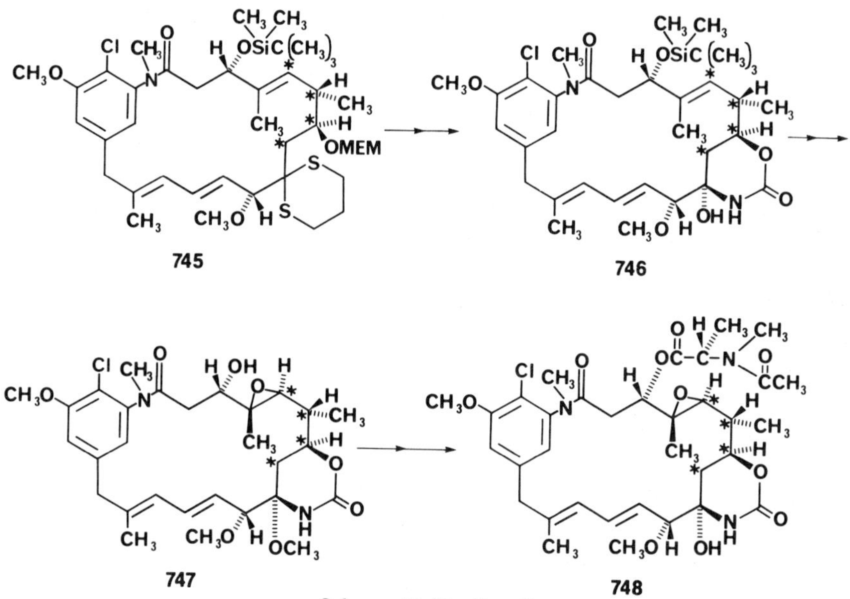

Scheme 39 (*Continued*)

compound with propane-1,3-dithiol and protection of the resulting 1,2-diol unit yielded, after conversion of the remaining alcohol to its methoxyethoxymethyl ether, the fully protected **741**. The dithiane anion of **741** was condensed with the appropriate achiral aldehyde representing the left-hand side of the molecule to give a 1 : 1 mixture of alcohol diastereomers. After isolation of **742**, the epimeric alcohol was oxidized and then reduced with lithium *n*-butylborohydride to give mainly additional **742**. The nonselectivity in formation of the allylic alcohol thus did not result in any substantial loss of material. A series of reactions involving ether formation, urethane cleavage, acetal hydrolysis, lead tetraacetate oxidation of the resulting diol, and chain extension with the anion of α-trimethylsilylpropionaldehyde-*N*-*tert*-butylimine produced the unsaturated aldehyde **743**. The remaining carbons of the lactam ring were introduced in a highly stereoselective manner by reaction of **743** with (*R*)-*p*-tolyl(carbophenyloxy)methyl sulfoxide. After reductive cleavage with aluminum amalgam, hydroxyester **744** was obtained. Silylation, saponification, and ring closure produced lactam **745**. The cyclic urethane portion of the molecule was established by cleavage of the methoxyethoxymethyl ether, conversion of the resultant alcohol to its carbamate ester, and hydrolysis of the dithioacetal, which was accompanied by spontaneous cyclization to **746**. After desilylation of **746** and stabilization of the hemiaminal unit as the *O*-methyl derivative, epoxidation with a selectivity of >200:1 was achieved with *tert*-butyl hydroperoxide and oxyvanadium(IV) bis(acetylacetonate) as catalyst. Conversion of the resultant **747** to maytansine was then straightforward.

In addition to the total synthesis of maytansine, there have been several reports of the preparations of chiral fragments of possible use in the preparation of this molecule. The structures of these fragments are shown in Scheme 40. Thus, mannose served as a precursor of **749** and **750** (*294*), quinic acid was converted to **751** (*295*), and glucose was transformed into **752** (*296*). The C-1—C-8 segment **753** was prepared from (*S*)-malic acid and (*R*,*R*)-2-hydroxy-3-methylsuccinic acid (*297*). Although the source of this latter molecule is not indicated in the abstract, it can be prepared (*4*) from either malic or tartaric acids. Preliminary work on the preparation of methyl-branched sugars as maytansine synthons has also been reported (*298*).

Preparations of **754** from (*R*)-cystine (*299*), of **755** from glucose (*300*), and of **756** also from glucose (*301*) have been described. The first is projected as a synthon for the ansa macrolide griseovirdin, and the second and third compounds represent segments of pyridomycin.

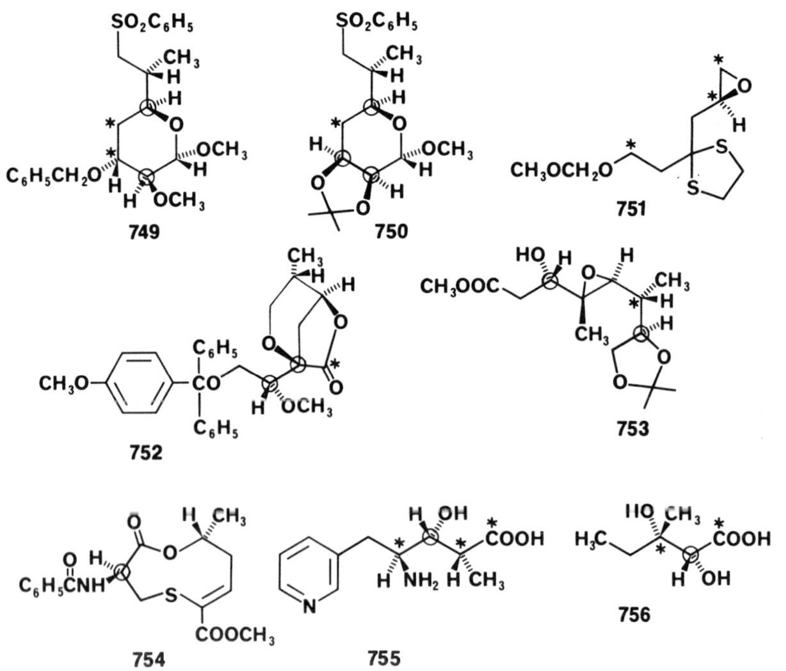

Scheme 40. Maytansine, griseovirdin, and pyridomycin synthons.

3. Polyether Antibiotics

The polyether antibiotics have many structural features in common with the macrolide antibiotics, and thus it should come as no surprise that similar synthetic strategies have been employed with both classes of compounds (for a review of the synthesis of this class of compounds, see ref. *18*). Two of the polyether antibiotics, lasalocid A and monensin, are of substantial economic importance in animal husbandry, not so much for their antibiotic activity as for their growth promotant properties. Reports of the syntheses of both of these compounds have appeared.

Ireland's preparation (*302*) of lasalocid A (*775*, Scheme 41) is highly convergent, employing (−)-citronellene, D-glucorono-γ-lactone, and D-fructose as starting materials for the three chiral fragments **762**, **766**, and **770**, respectively, and relies on the Claisen rearrangement of ester enolates (*303*) derived from furanoid and pyranoid glycals (*304*) for control of chirality. The aromatic portion of the molecule and the adjacent chiral center, in the form of aldehyde **762**, was synthesized from (−)-citronellene (**757**) (*112*). Selective oxidation of this compound, in several steps, gave acid **758**, which was converted in a straightforward manner to bromide **247**. The Grignard reagent from **247** underwent conjugate addition to 2-carbomethoxy-α-pyrone to give, after oxidation with manganese dioxide, compound **759**. Diels–Alder reaction with *N*,*N*-dibenzyl-1-aminopropyne then gave the aniline **760**, which was converted via **761** to the desired **762**.

The ethyl ketone fragment **774** required for ulitmate condensation with **762** to give lasalocid A (*305*) was prepared from D-glucosaccharino-1,4-lactone (**763**), which itself was derived from D-fructose (**272**).

Conversion of **763** to the furanoid glycal **764** was followed by esterification, enolate formation and rearrangement, reesterification, and hydrogenation. Of the three new chiral centers formed in these reactions (two in the Claisen rearrangement and one in the hydrogenation), two were expected to be homogeneous while the third (α to the ester) was not. In fact, the product **765** was a 3:1 mixture with its diastereomer. After purification, manipulation of oxidation states gave acid **766**, which was designed to serve as the acid component in a second ester enolate Claisen rearrangement. The pyranoid glycal **770** required for this second rearrangement was obtained from D-glucorono-γ-lactone (**306**) by initial thioacetal and acetonide formation. Desulfurization of the resultant **767** was followed by alcohol protection, lactone to hemiacetal reduction, and protecting-group removal, yielding 6-deoxy-L-gulose (**768**) (*306*). Addition of the appropriate protecting groups gave **769**, which was then transformed into glycal **770**. The esterification of **766** with **770** was followed by ester

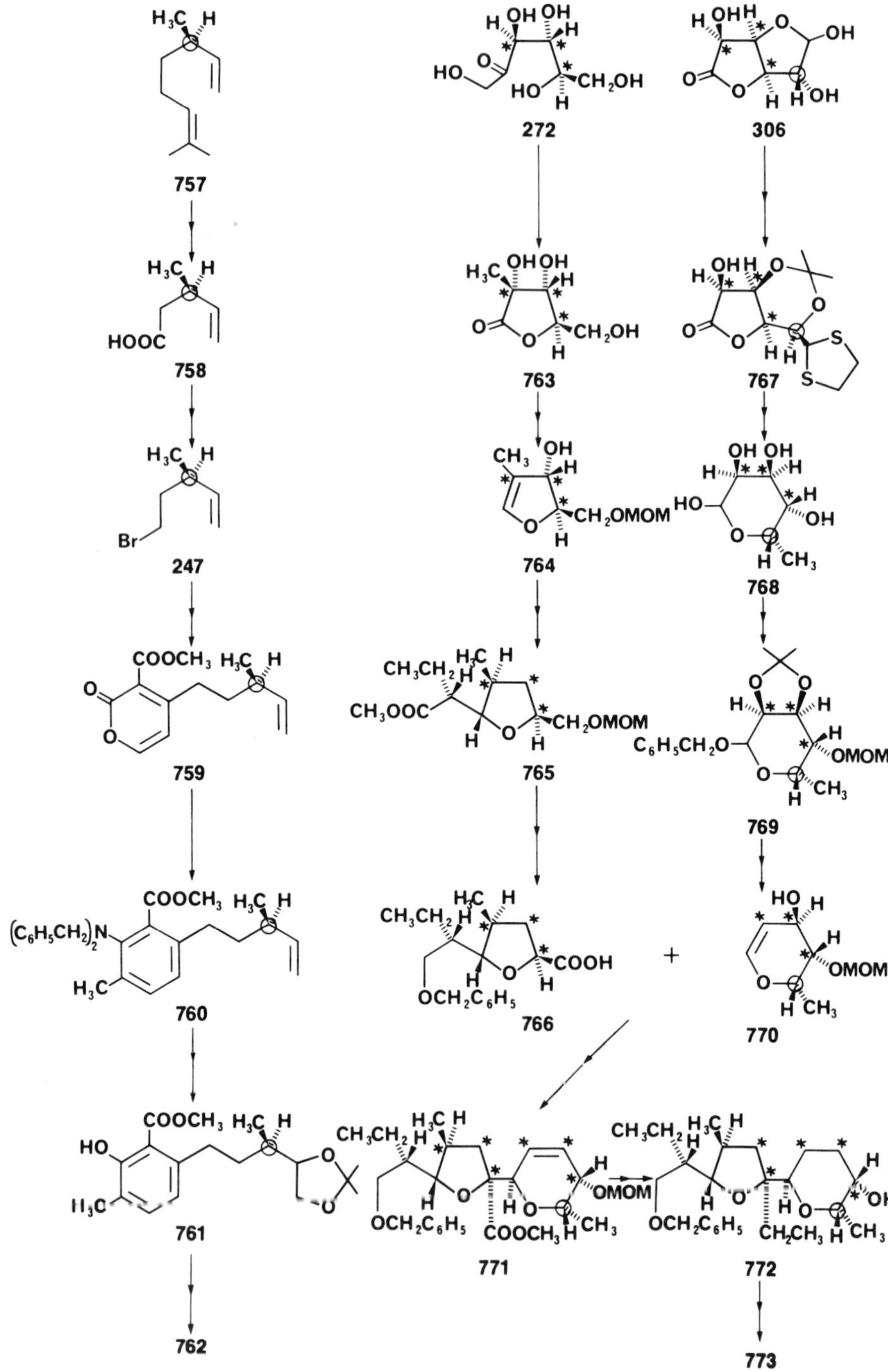

Scheme 41. Lasalocid A. MOM = Methoxymethyl.

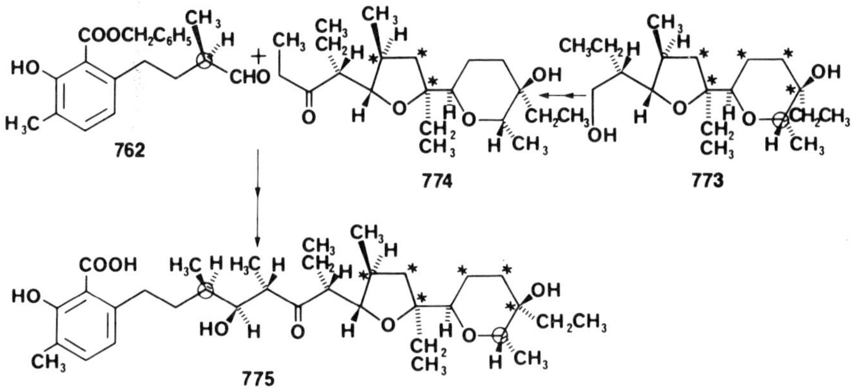

Scheme 41 *(Continued)*

enolate formation and rearrangement and reesterification to give **771**, once again as an approximate 3 : 1 diastereomer mixture. After isolation of the pure **771**, reduction of the olefin and conversion of the carbomethoxyl to an ethyl unit yielded **772**. Introduction of the terminal ethyl group of **773** in a stereoselective manner was followed by two-carbon chain lengthening to give the desired **774**. The union of **762** and **774** then gave lasalocid A.

As noted in previous sections, the general synthetic approach involving Claisen rearrangement of ester enolates derived from furanoid or pyranoid glycals has also been applied to syntheses of the nonactic acids (*262*) and the Prelog–Djerassi lactone (*286*). In addition, the method has been used as the cornerstone of a preparation (*307*) of tirandamycic acid, which includes all the chiral centers of the antibiotic tirandimycin.

The ionophore monensin (**791**, Scheme 42) contains no less than 17 centers of asymmetry and thus represents a monumental synthetic challenge. As in the case of Ireland's preparation of lasalocid A, Still's synthesis (*86, 308, 309*) of monensin is highly convergent. The syntheses are quite different, however, in that Ireland used the Claisen rearrangement to introduce new asymmetric centers while Still relied on chelation-controlled aldol and Grignard reactions to achieve the same goal. Interestingly, despite the presence of two pyran and three furan rings in monensin, sugars were not employed as chiral starting materials. Rather, (*S*)-malic acid, (*R*)-citronellic acid, and the microbiologically derived (*S*)-β-hydroxyisobutyric acid, each possessing only a single chiral carbon atom, were the building blocks in this synthesis. Thus, the lactone acetonide **776** (derived from malic acid), after reduction of the free carboxylic acid with borane, acidification (which led to rearrangement), and alcohol

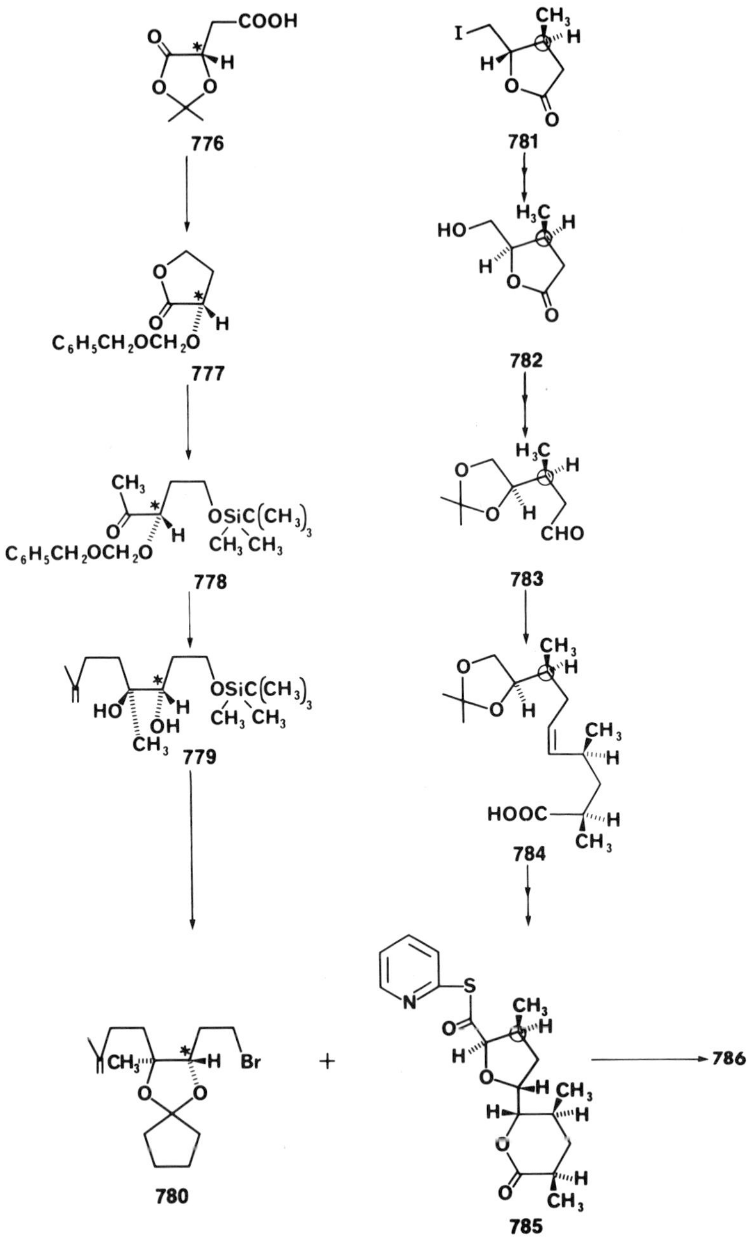

Scheme 42. Monensin.

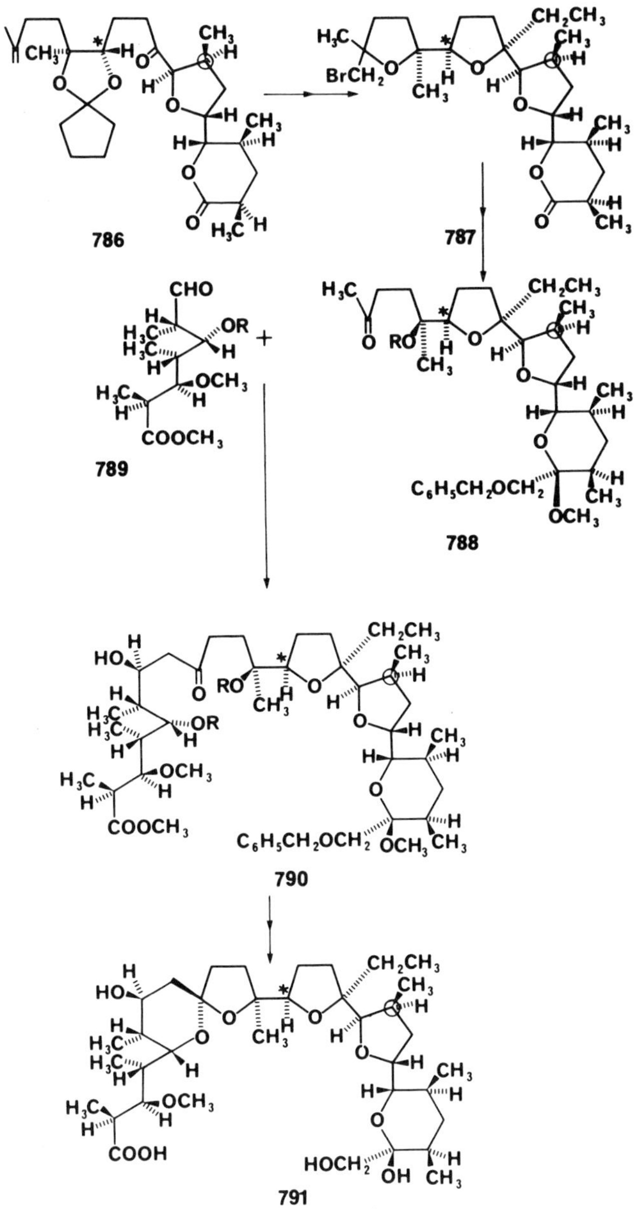

Scheme 42 (*Continued*) **789, 790** R = $Si(CH_2CH_3)_3$.

protection, gave ketone **777**. Selective Grignard reaction, followed by alcohol protection, then gave methyl ketone **778**. This ketone underwent a highly selective (50 : 1 diastereomer mixture) Grignard reaction with 3-methyl-3-butenylmagnesium bromide. The diol **779** obtained after debenzylation was then transformed in routine fashion to bromide **780**.

The synthesis of the second major fragment **785** of monensin began with (*R*)-citronellic acid, which was converted in five steps to iodolactone **781** with a high degree of selectivity in formation of the second chiral center. Inversion of this center was achieved by reaction with the potassium salt of benzyl alcohol followed by hydrogenolysis of the resulting epoxyester. Reduction of lactone **782** to the triol, acetonide formation, and oxidation gave aldehyde **783**, which was condensed with the appropriate ylid derived in several steps from (*S*)-β-hydroxyisobutyric acid. The thus prepared acid **784** was subjected to iodolactonization, treatment with silver trifluoroacetate (causing formation of the tetrahydrofuran ring), oxidation, and esterification to yield thioester **785**.

Reaction of the Grignard reagent derived from **780** with **785** gave ketone **786**. Because of the adjacent oxygen atom, reaction of the ketone with ethylmagnesium bromide was stereospecific. The resultant alcohol was deketalized with concomitant formation of the terminal tetrahydrofuran ring, mesylated, and solvolyzed to give **787**. Modification of the δ-lactone and bromomethyltetrahydrofuran moieties then gave methyl ketone **788**. The aldol reaction of **788** with aldehyde **789** (prepared in a number of steps, including a chelation-controlled aldol reaction, from β-hydroxyisobutyric acid), occurred with a fair degree of selectivity, giving a 3 : 1 mixture of **790** and its diastereomer. A relatively unexceptional series of transformations (*310*) then gave monensin (**791**) as its sodium salt.

A third ionophore antibiotic for which a total synthesis has been described is X-14547A (**800**, Scheme 43). The synthesis (**89, 311**) is characterized by the creative manner in which two molecules of (*S,S*)-tartaric acid are transformed into the chiral left-hand portion of the molecule. Thus, the epoxide **148**, which was obtained in five steps from diethyl tartrate, was treated with lithium dimethylcuprate and then subjected to hydrogenolysis, giving triol **185**. Selective acetonization gave **792**. At this point, the synthesis diverged. One portion of the material was benzylated and deprotected to **793**. Selective protection of the primary and secondary alcohols followed by deprotection and oxidation of the primary alcohol provided **794**. The other portion of **792** was transformed into phosphonium salt **795**. Wittig condensation of **794** and **795** gave adduct **796** as a mixture of olefin isomers. A series of five standard operations converted this compound to the epoxide **797**. A critical step in the preparation of X-14547A was the stereospecific intramolecular cyclization of **797**, which

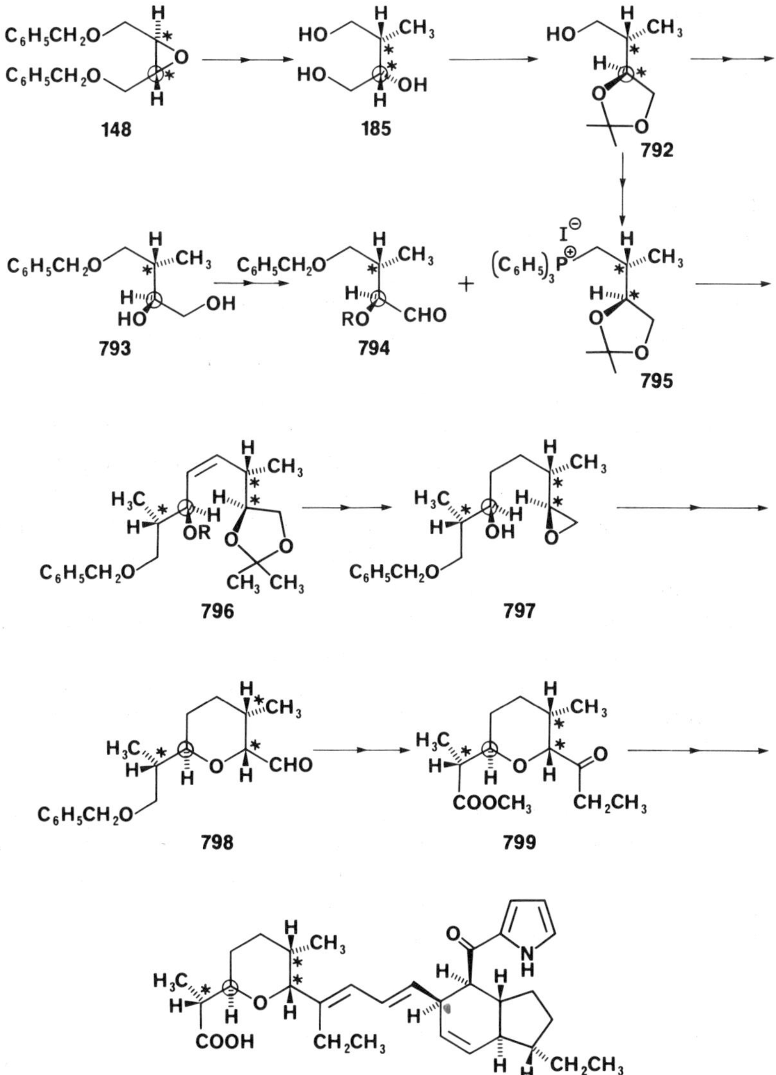

Scheme 43. Antibiotic X-14547A. R = $Si(C_6H_5)_2C(CH_3)_3$.

was accomplished with camphorsulfonic acid as catalyst. Oxidation of the cyclization product gave aldehyde **798**, which was transformed in several steps to **799**. The chiral right-hand portion of X-14547A was obtained by an elegant asymmetric synthesis. Combination of the units was then successfully effected to give **800**.

The synthesis of an 18-carbon degradation product of the antibiotic boromycin has been described (*312*). In addition, two groups (*313, 314*) have described carbohydrate-based approaches to the chiral portions of the polyether ionophore A-23187.

4. γ-Butyrolactones and α-Methylene-γ-butyrolactones

The butyrolactone structural unit is common to many compounds possessing, in particular, cytotoxic properties. One such group of compounds is the lignan lactones. The syntheses of these compounds, done by Koga, depend on the chiral lactone **202** [from (*S*)-glutamic acid] (Scheme 44), to which two groups are stereoselectively appended. Subsequent removal of the original chiral center, with 1,3-carbonyl transposition, then gives the desired products. This approach, in its most advanced form, is illustrated by the synthesis of (+)-steganacin (**806**) (*105*). By selenation, oxidation, and deselenation, **202** was converted to the unsaturated lactone **205**. Dithiane conjugate addition to **205** was stereospecific, giving **801**. After desulfurization and detritylation, alkylation with piperonyl bromide gave exclusively stereoisomer **802**. The lactone ring was then restructured by hydride reduction to the triol, oxidative glycol cleavage and Collins oxidation. Nonphenolic oxidative coupling with vanadium oxyfluoride of the resulting **803** led to (−)-isostegane (**804**). Thermal biphenyl isomerization gave the diastereomeric **805**, which was cleanly and stereospecifically oxidized to (+)-steganacin (**806**). Surprisingly, this material was enantiomeric with the natural material whose structure had been assigned as **806** on the basis of an X-ray crystallographic analysis. The value of total synthesis in determination of natural product structures, despite the tremendous advances in structure analytical methods, is thus once again verified.

The approach described in Scheme 44 has also been applied to a synthesis of (+)-*trans*-burseran (*104*). An alternative approach to the use of **202** in lignan lactone synthesis has been described. In this approach, **202** is either alkylated or alkenylated and then hydrogenated. In both cases, however, mixtures of diastereomers have been obtained. Enantiomeric enrichment was possible, and syntheses of unnatural (−)-podorhizin (*315*), (−)-hinokinin (*315*), (−)-*trans*- and (+)-*cis*-burseran (*317*), and (−)-steganone (*318*), among others, were achieved.

One of the most significant structure–activity relationships to be found in recent years is the contribution of an α-methylane-γ-butyrolactone unit to the anticancer activity of many sesquiterpenes. Despite the fact that

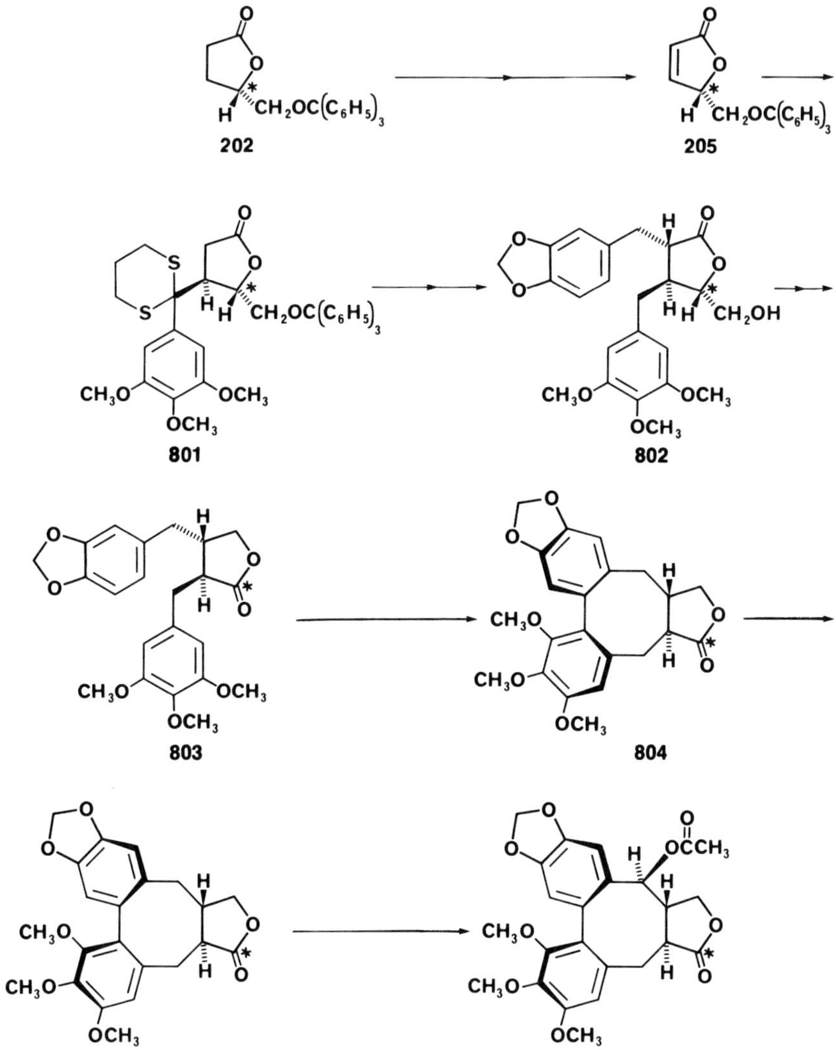

Scheme 44. (+)-Steganacin.

the lactone rings of these compounds are often chirally substituted, there has been surprisingly little work done to employ, for example, furanoid sugars as starting materials in total syntheses of these biologically interesting materials. A noteable exception is the preparation of the antifungal microbial metabolite avenaciolide (**811**), the synthesis of which (Scheme 45) has been achieved in a virtually identical manner by two groups (*319*,

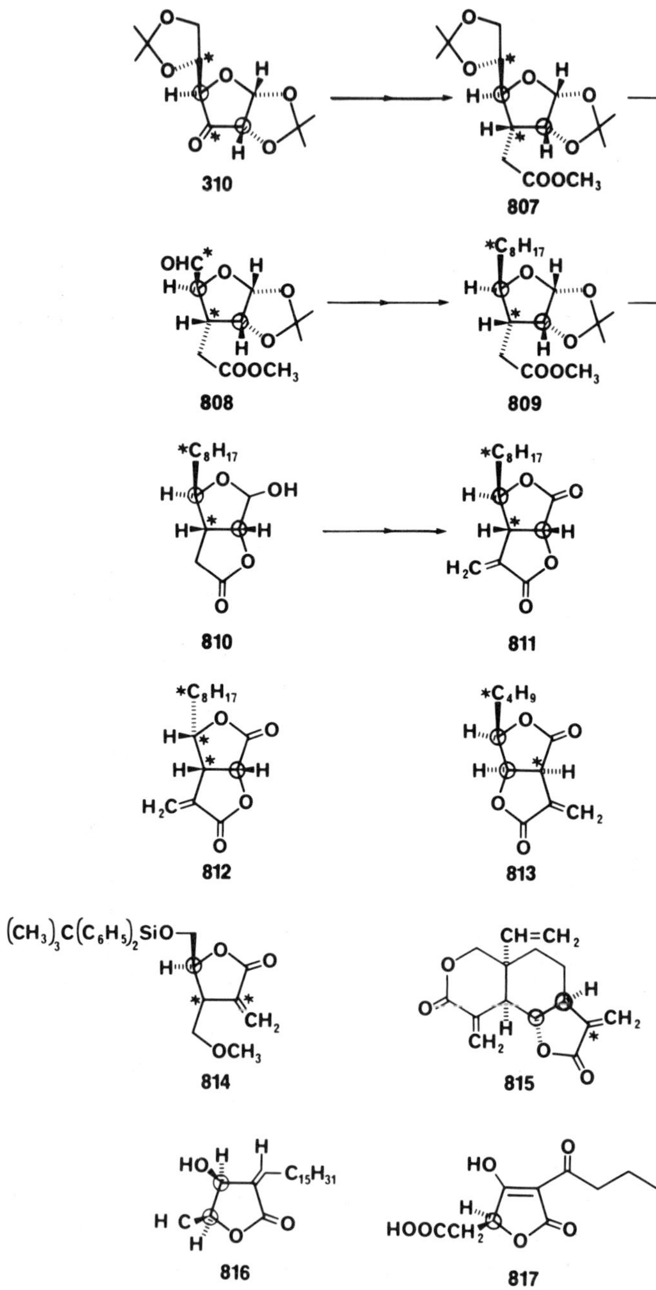

Scheme 45. Avenaciolide.

320). Thus, ketone **310**, obtained by oxidation of diacetone glucose, was subjected to Horner reaction and then to a stereospecific hydrogenation to yield ester **807** (*321*). Selective acetonide hydrolysis followed by oxidative glycol cleavage gave aldehyde **808**, which by Wittig reaction and hydrogenation was smoothly converted to **809**. Treatment of this compound with acid caused acetonide cleavage and lactonization. Oxidation of the resultant **810** and then methylenation gave the desired **811**. These syntheses corrected the absolute configuration of avenaciolide, which had been misassigned on the basis of degradation studies. The basic synthetic approach employed to prepare avenaciolide also yielded (−)-isoavenaciolide (**812**) (*322*) and two separate preparations of canadensolide (**813**) (*323, 324*).

Fraser-Reid has reported (*325*) the preparation from xylose of the diastereomeric mixture of methylene lactones **814**, which are potential sesquiterpene intermediates. Two groups have independently described (*326, 327*) the conversion of α-santonin to deoxyvernolepin (**815**), which in racemic form (*326, 327*) appears to be substantially more effective against leukemia than the naturally occuring vernolepin. Preparations of the potentially cytotoxic (−)-dihydromahubanolide B (**816**) from (*R*,*R*)-tartaric acid (*328*) and of (*S*)-carlosic acid (**817**) from (*S*)-malic acid (*329*) have been published.

5. Vicinal Amino Alcohols

A surprisingly large number of naturally occurring compounds with chemotherapeutic activity possess a vicinal amino alcohol unit (e.g., $R^1CHOHCHNR^2R^3$). In most cases, the carbon atoms bearing the oxygen and nitrogen functions are chiral. Prime examples of this type of compound are the amino sugar–containing antitumor agents, such as daunomycin, aclacinomycin A, and adriamycin. The syntheses of these particular compounds will not be discussed here. As will be seen in this section, however, there are a number of other compounds containing the amino alcohol unit in a five- or six-carbon atom chain (ring). The synthesis of these materials has often been the result of a rather straightforward sequence involving, at a minimum, conversion of the appropriate sugar to the amino sugar. One of the less obvious of such syntheses is that of the aglycon **834** of bleomycin A_2 (Scheme 46).

Bleomycin A_2 is a pentapeptide derived from pyrimidoblamic acid (**820b**), (2*S*,3*R*)-3-hydroxyhistidine (**824**), (2*S*,3*S*,4*R*)-4-amino-3-hydroxy-2-methylpentanoic acid (**831**), (*S*)-threonine (**168**), 2′-(2-aminoethyl)-2,4′-bithiazole-4-carboxylic acid (**832**), and a terminal (3-aminopropyl)di-

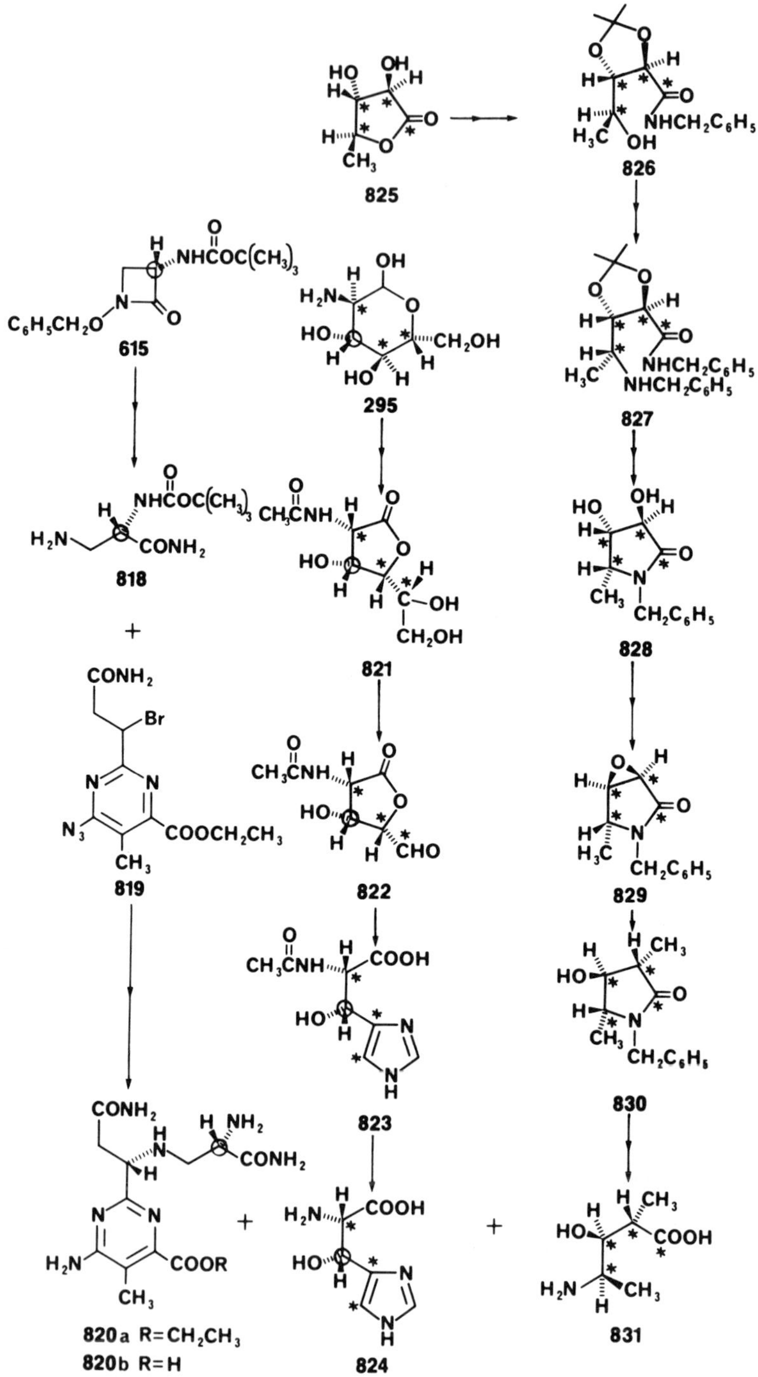

Scheme 46. Bleomycin A_2 aglycon.

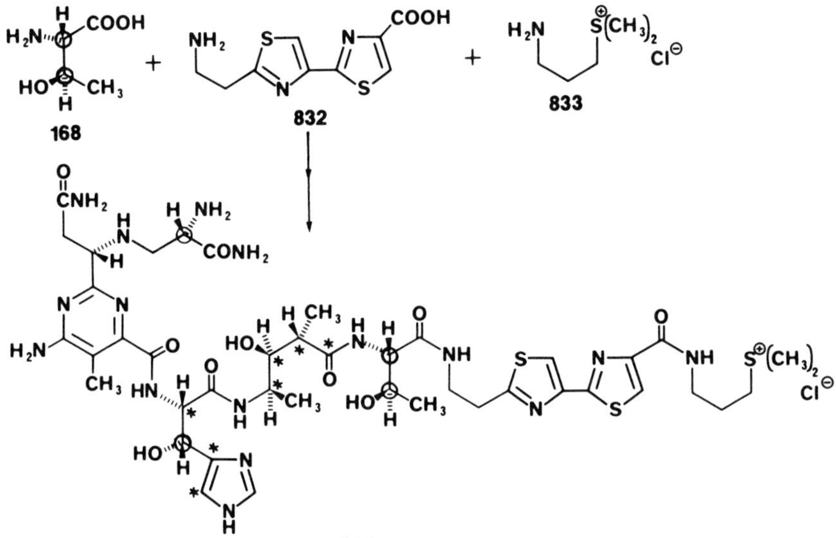

Scheme 46 (*Continued*)

methylsulfonium chloride (**833**) unit. The synthesis of these fragments has been the subject of intense efforts by both American (*330–332*) and Japanese (*333–335*) groups. Both groups have announced (*336–339*) syntheses of the compound, in one case (*336, 337*) via aglycon **834**.

The syntheses of fragments **820b** (*331*), **824** (*330*), and **831** (*332*) by Hecht are of particular interest, using (*S*)-serine, D-glucosamine, and L-rhamnose as chiral starting materials. As previously described in Section III,C,1, the β-lactam **615** (Scheme 26) is readily obtained from (*S*)-serine. Aminolysis of this compound followed by hydrogenolytic N—O bond cleavage led to amide **818**. Reaction of this compound with the racemic bromide **819** gave the expected 1 : 1 mixture of diastereomeric displacement products. After diastereomer separation, reduction of the azide group and acidic *tert*-butyloxycarbonyl removal gave the ethyl ester **820a** of pyrimidoblamic acid (**820b**) (*331*).

The second major amino acid fragment, (2*S*,3*R*)-3-hydroxyhistidine (**824**) was obtained from lactone **821**. The lactone was itself obtained either from D-glucosamine (**295**) in a process involving inversion of C-2 or from the much more expensive *N*-acetyl-D-mannosamine. Oxidative cleavage of **821** gave the aldehyde **822**, which on treatment with ammonium acetate and formaldehyde in the presence of cupric acetate was converted to the protected histidine **823**. Amide hydrolysis then gave **824** (*330*).

The third chiral fragment of **834** for which a synthesis was required, namely, (2*S*,3*S*,4*R*)-4-amino-3-hydroxy-2-methylpentanoic acid (**831**), was prepared from the L-rhamnose oxidation product **825**. Ring opening with benzyl amine and acetonide formation gave **826**, which was activated for inversion at C-4 by mesylate formation. The inversion was effected with sodium azide to give, after hydrogenation and reductive amination, aminoamide **827**. Ring closure and acetonide cleavage gave lactam **828**. To form the stereochemically required C-2—C-3 β-epoxide **829**, selective activation of the C-3 hydroxyl was required. This was done by first masking the more acidic C-2 hydroxyl by treatment with isopropylmagnesium bromide and then formation of the C-3 mesylate. Reaction with base then gave **829**, which underwent regiospecific epoxide opening with lithium dimethylcuprate to **830**. Debenzylation and lactam hydrolysis completed the preparation of **831** (*332*). Combination of the six fragments, using standard peptide technology, yielded aglycon **834** (*336, 337*). Syntheses of the sugar portion of bleomycin A_2 have been reported (*340, 341*). Coupling to give bleomycin A_2 itself was then effected (*338*).

Synthetic approaches similar to that described above for the bleomycin A_2 preparation have also been employed to obtain negamycin (**835,** from D-galacturonic acid) (*342*), a synthon of negamycin (from L-arabinose) (*343*), sorbistin A (**836**, from two molecules of D-galactose) (*344*), polyoxin J (**837**, from D-glucose and L-sorbose) (*345–348*), and spectinomycin (**838**, from D-glucose) (*349, 350*) (Scheme 47).

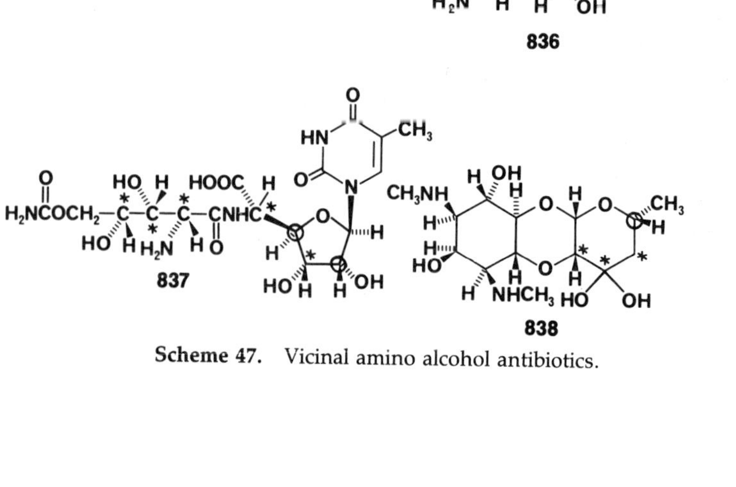

Scheme 47. Vicinal amino alcohol antibiotics.

In some cases the nitrogen atom of the amino alcohol is found in a pyrrolidine ring. An example of such an antibiotic is anisomycin (**842**). Three syntheses of this compound in chiral form have been reported (*351–354*). The first two of these syntheses, in which (2*R*,3*R*)-tartaric acid provided the vicinal glycol unit, were marred by the necessity for a nonspecific monoacetylation at the final stage (*351–353*). The synthesis (Scheme 48) devised by the Syntex group (*354*) used glucose to provide all three chiral centers and was designed to avoid the selective acetylation problem. Thus, oxidation of diacetoneglucose followed by borohydride reduction gave the allofuranose **311**. Tosylation of the free alcohol and a careful selective acetonide hydrolysis provided diol **839**, which was sequentially benzoylated and mesylated to **839a**. Treatment of this material with sodium methoxide gave an epoxide that was opened with sodium azide. After catalytic hydrogenation, the amine **839b** was obtained. Heating this compound in ethanol in the presence of sodium acetate led to internal displacement giving, after amine protection, the tricyclic compound **839c**. With the pyrrolidine ring established, the furanose portion of the molecule could be converted to the remaining functional groups of anisomycin. After initial protection of the free secondary alcohol as its benzyl ether to allow distinction from the second such alcohol (to be liberated from the furanose), acidic hydrolysis and periodate cleavage gave aldehyde **839d**. Treatment of this epimerically labile compound with *p*-methoxyphenylmagnesium bromide gave the diastereomeric mixture of diols **840**. Ionic hydrogenation (triethylsilane and trifluoroacetic acid in methylene chloride) selectively removed the benzylic alcohol to give, after acetylation, compound **841**. The normal catalytic hydrogenation then removed the benzylic protecting groups to give anisomycin (**842**). A cyclization process similar to that employed in the anisomycin synthesis was a key feature in the preparation of **843**, the synthesis of which from D-glucose established the absolute configuration of detoxinine (**844**) (*355*).

An entirely different tack was taken in the total synthesis of anthramycin methyl ether (**849**, Scheme 49) (*356*). Acylation of hydroxyproline methyl ester with the appropriate benzoyl chloride gave amide **845**. Reduction of the nitro group, cyclization, and oxidation of the secondary alcohol produced ketone **846**. Elaboration of the unsaturated chain in several steps gave **847**, which was converted by debenzylation, treatment with benzaldehyde dimethylacetal, and nitrile-to-amide conversion to **848**. Selective amide reduction with sodium borohydride followed by treatment with dilute methanolic hydrochloric acid produced anthramycin methyl ether.

The use of hydroxyproline in preliminary model experiments designed to provide an approach to the mitosenes should be noted (*357, 358*).

The vicinal alcohol unit can also be part of a tetrahydrofuran ring. One

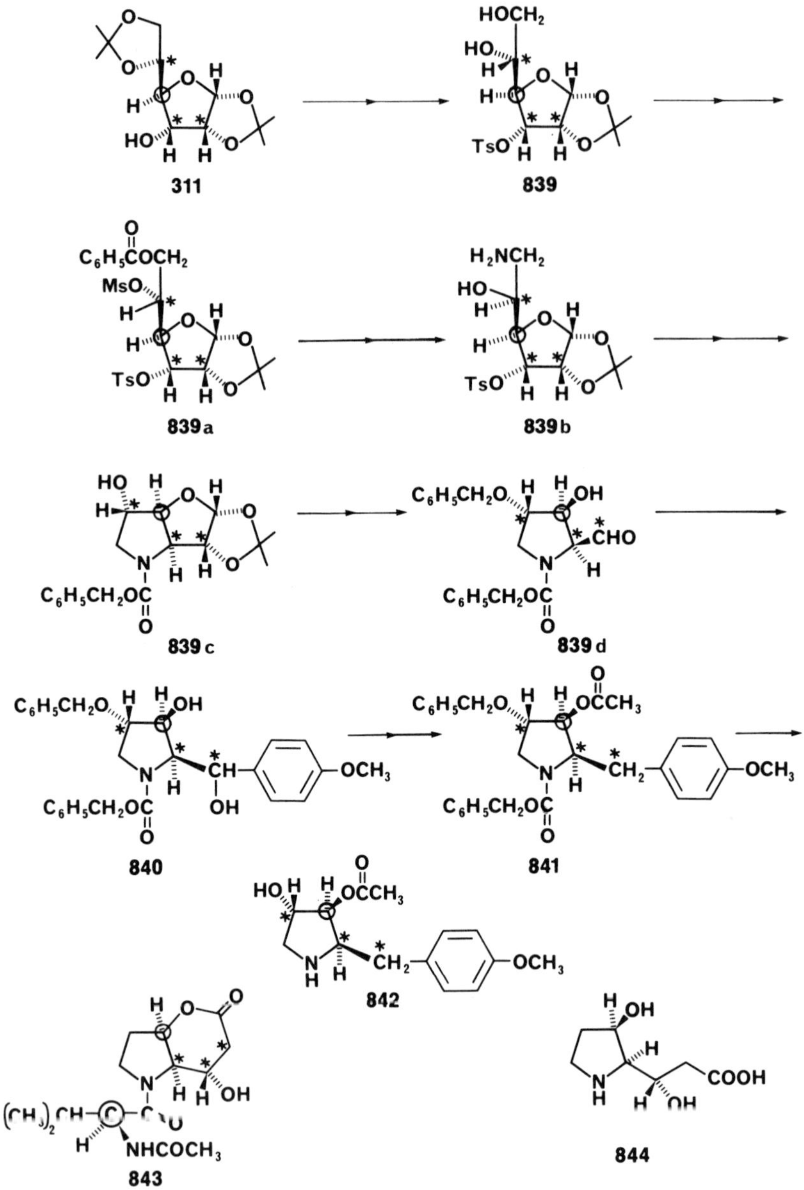

Scheme 48. Anisomycin.

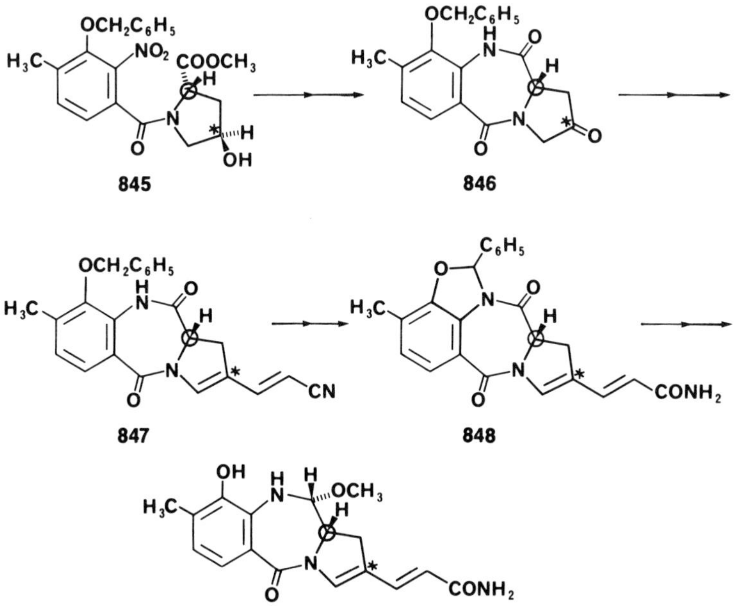

Scheme 49. Anthramycin methyl ether.

such example is the antibiotic (+)-furanomycin (**856**), the synthesis of which from D-glucose (*359*) has been achieved (Scheme 50). Treatment of the tritosylate **850** with methanolic hydrochloric acid caused acetonide cleavage, ring opening, and reformation and acetal formation. Treatment of the resultant ditosylate **851** with sodium phenylselenide gave **852**. The chemical and stereochemical outcome of this reaction was a result of initial epoxide formation followed by stereospecific ring opening. Reductive removal of the selenide groups, tosylation, and base-catalyzed elimination produced the unsaturated acetal **853**. After acetal hydrolysis, four-component condensation generated the requisite amino acid functionality. Despite the use of (*R*)-α-methylbenzylamine in this reaction, the diastereomers **854** and **855** were obtained in equimolar amounts. Hydrolysis of the latter compound gave furanomycin (**856**), thus correcting an erroneously assigned stereochemistry.

A second example of a tetrahydrofuran-containing antibiotic is anhydromyriocin. The synthesis of the enantiomer **857** of this compound from L-arabinose has been described (*360*). The rotation of the synthetic **857** was only ~75% of that of the naturally derived material, however. A nonstereospecific synthesis of antitumor agent AT-125 (**858**) from (*S*)-glutamic acid has been described (*361*).

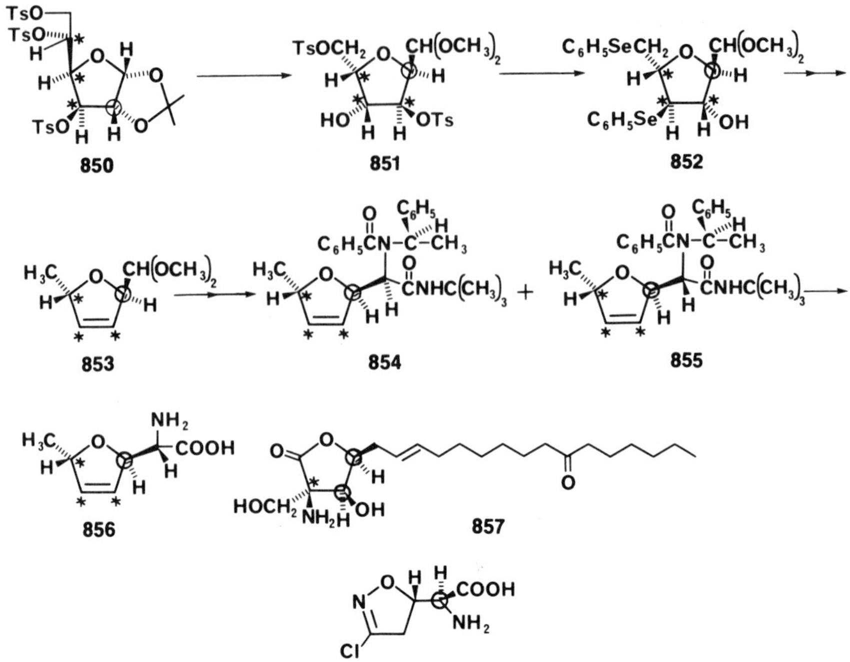

Scheme 50. Furanomycin.

In the case of roseonine (**863**), the amino alcohol is attached to (and is also part of) a guanidine-containing ring. Three syntheses of this compound have been reported (*362–364*). All employed the same principle used in the anisomycin synthesis (Scheme 48); that is, a furanose was used as a template for construction of the nitrogen-containing ring and was then destroyed with generation of the necessary ring substituents. In the best of these three syntheses (*363*) (Scheme 51), D-ribose was initially converted to the diazide **859**, which after hydride reduction and acetylation yielded aziridine **860**. Ring opening of this compound with azide was stereoselective, giving **861** as the major product. A series of routine conversions produced the triaminolactone **862**, which on treatment with cyanogen bromide was converted to roseonine.

The syntheses of two alicyclic amino alcohols, chloramphenicol base from (*S*)-phenylalanine (*365*) and sparsomycin from (*S*)-cystine (*366*), have been reported. Four diastereomeric adenosine deaminase inhibitors have been prepared (*367*) from L-rhamnose.

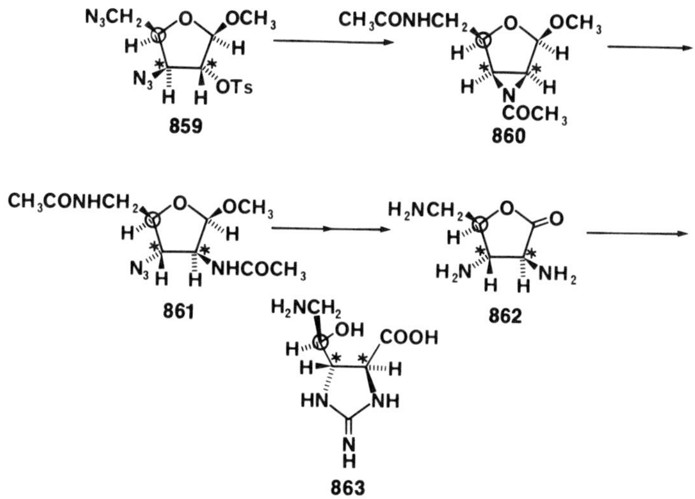

Scheme 51. Roseonine.

6. Miscellaneous Compounds with Chemotherapeutic Actvity

There are a number of other compounds with chemotherapeutic activity that have been partially or totally synthesized from available chiral molecules. These molecules, which fall outside the arbitrary classifications used above, are listed here (structures shown in Scheme 52):

1. Chromomycinone side-chain synthon **864** from D-galactose (*368*)
2. Olivin framework **865** (mixture of diastereomers) presumably from (*S*,*S*)-tartaric acid (*369*); side-chain synthon **866** from D-arabinose (*370*)
3. Anthracyclinone **867** (one isomer) from D-arabinose (*371*)
4. Enantiomer **868** of methyl 8-*O*-benzoyl-α-D-pillaroside, a degradation product of antibiotic pillaromycin A, from D-glucose (*372*)
5. Antileukemic triptolide **869** from dehydroabietic acid (*373*)
6. Antibiotic anticapsin (**870**) from (*S*)-tyrosine (*374*)
7. Potential antitumor nucleoside analogs **871** and **872** from (*S*)-cysteine (*375*)
8. Potential cytotoxic bismethanesulfonate **873** from (*R*,*R*)-tartaric acid (*376*)
9. Dictyolene (**874**), a seaweed metabolite possessing a variety of potentially useful biological activities, from α-santonine (*377*)

10. Protected hikosamine derivative **875**, a portion of the antihelmintic hikizimycin, from D-arabinose and D-galactose (*378*)
11. Methyl α-L-mycaminoside (**876**), a sugar constituent of antitumor antibiotics, from (*S*)-alanine (*379*)
12. Tetrahydropyran **877**, a degradation product of the antifungal agent ambruticin, from L-arabinose (*380*) and the protected cyclopropanecarboxylic acid **878**, a proposed synthetic intermediate for ambruticin, from (*R*)-citronellic acid (*381*)
13. Cytotoxic alkaloid sesbanine (**879**) from (*S*)-malic acid (*382*)
14. Octadienoic ester **880**, a constituent of the biologically active trichothecene family of natural products, from D-glucose (*383*)

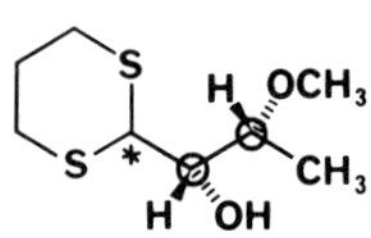
864

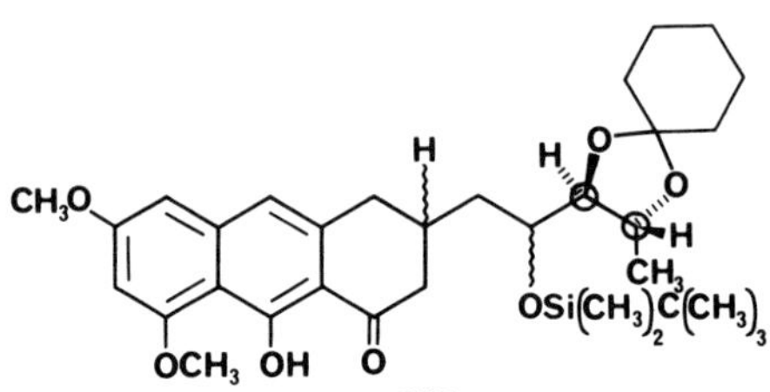
865

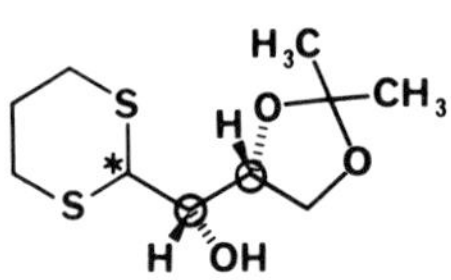
866

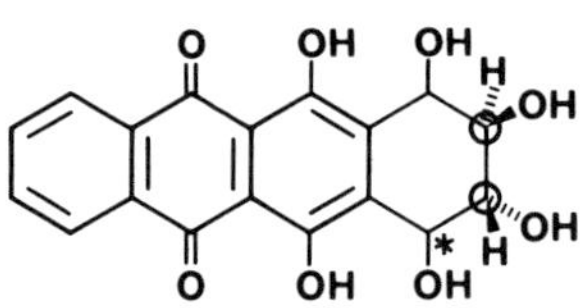
867

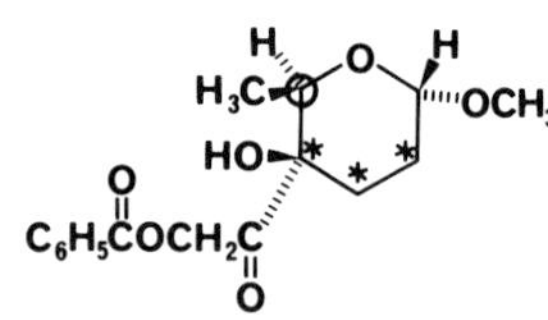
868

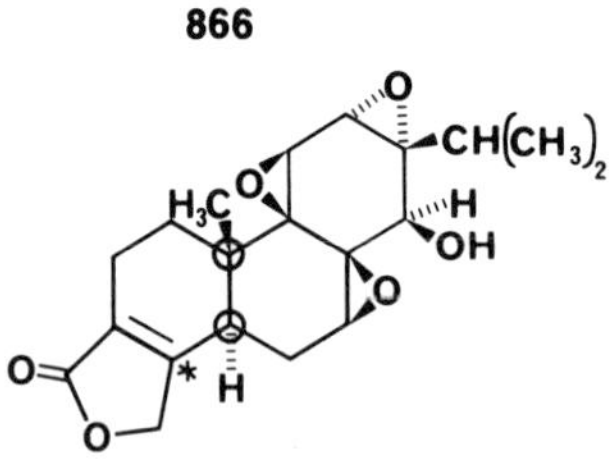
869

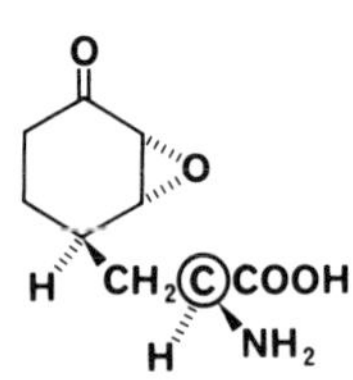
870

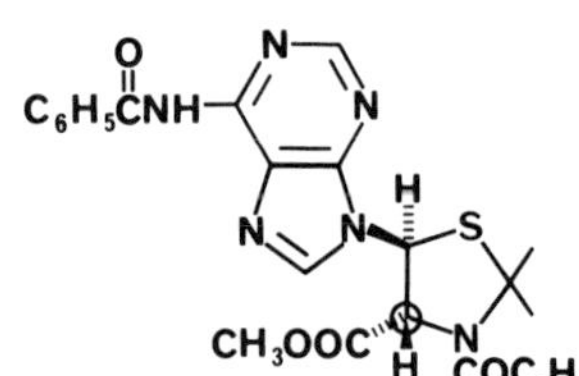
871

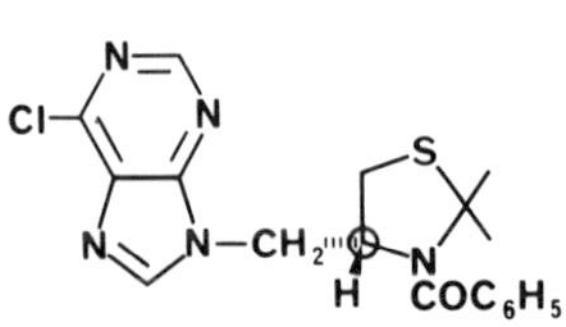
872

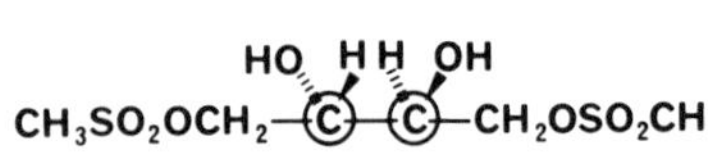
873

Scheme 52. Compounds with chemotherapeutic activity.

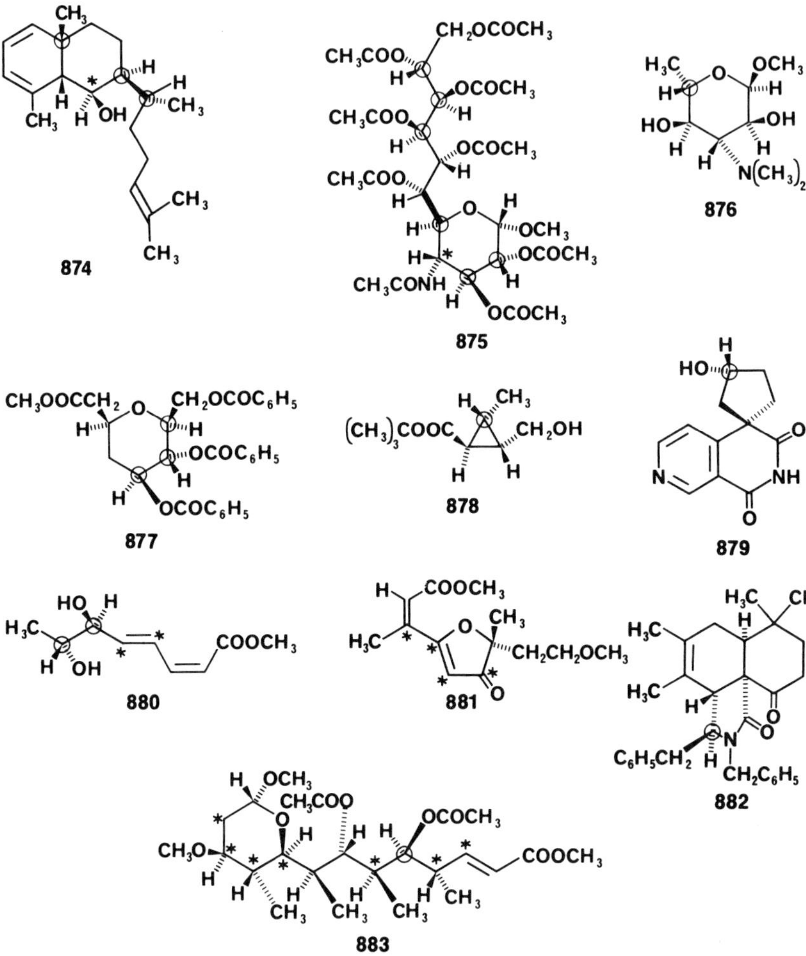

Scheme 52 (*Continued*)

15. Furenone **881** and its double-bond isomer, models for germacranolide antitumor agents, from D-mannose (*384*)
16. Tricyclic keto amide **882**, a model for cytochalasin C, from (*S*)-phenylalanine (*385*)
17. Ansa-chain segment **883** of rifamycin, from D-glucose (two molecules) (*386*)

The keto epoxy amide cerulenin (**888**) is of interest because of its antibacterial and antifungal properties. In addition, it appears to be an inhibitor of fatty acid and sterol biosynthesis. After successful preparations of tetrahydrocerulenin (*387, 388*), two groups were able to report syntheses

(*389, 390*) of the natural product. These syntheses are very similar and can be compared with those of the epoxy containing leukotrienes (Section III,B,3). In this case, the appropriate epoxide configuration could be generated from C-2 and C-3 of D-glucose (Scheme 53) by inversion at C-3. The 1,2:5,6 di-*O*-isopropylidene-α-D-glucofuranose (**302**) starting material was converted in several straightforward steps to the aldehyde **884**. A multistep elaboration of the side chain and conversion of the benzyl to a mesyl group produced **885**, setting the stage for epoxide formation. This was achieved by removal of the acetonide group with acid followed by ring closure induced by sodium methoxide. Oxidation of the resultant **886** to lactone **887** was followed by aminolysis and further oxidation to cerulinen (**888**) (*389*).

Pentenomycin (**897**, Scheme 54) has a deceptively simple structure, but

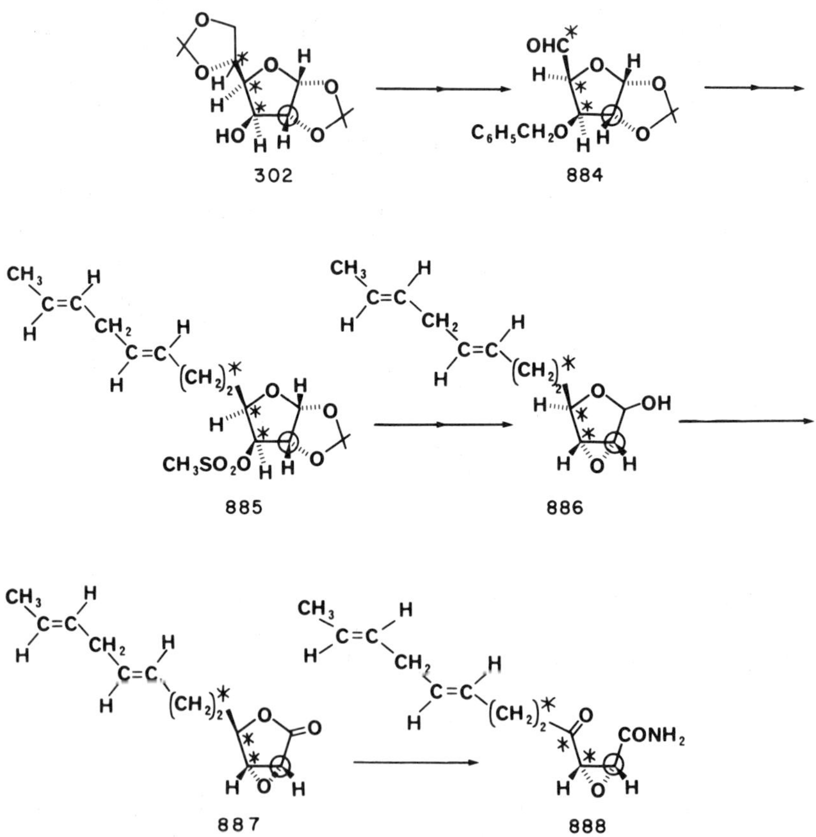

Scheme 53. Cerulenin.

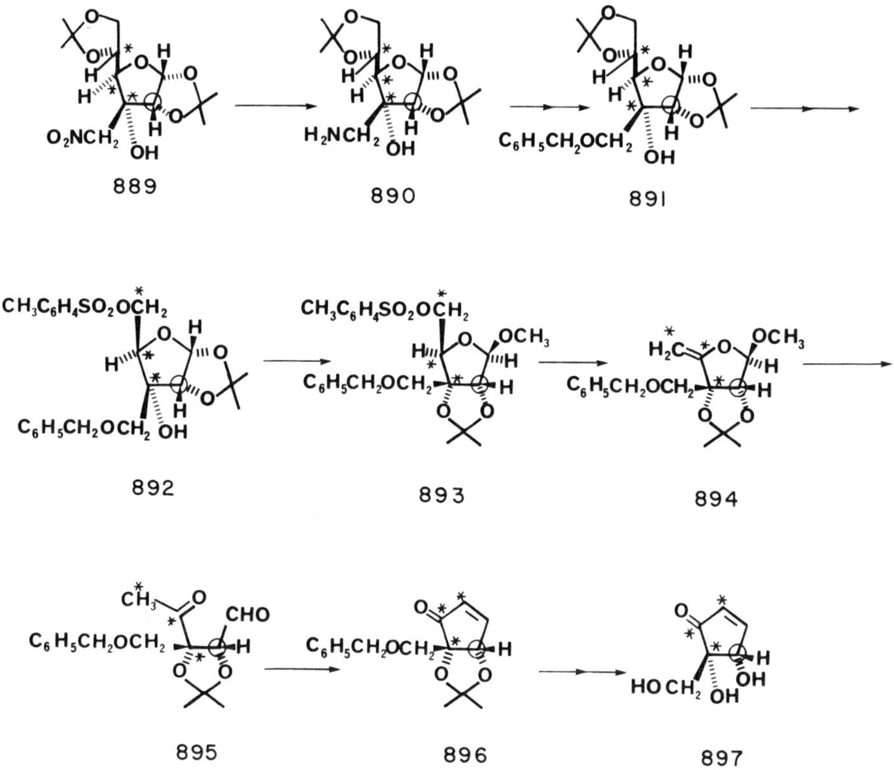

Scheme 54. Pentenomycin.

its preparation turned out to be anything but trivial. The synthesis (*354*) of this compound began with the glucose-derived ketone **310** (Scheme 45). Reaction with nitromethane gave, with a high degree of selectivity, the β adduct **889**. Catalytic hydrogenation converted this material to the amine **890**, which was quaternized and reacted with sodium benzyloxide to give **891**. A standard C-5—C-6 degradation led to tosylate **892**, which was isomerized to **893** with perchloric acid in 2,2-dimethoxypropane and methanol. This conversion set the stage for construction of the cyclopentenone ring. Elimination of the elements of *p*-toluenesulfonic acid from **893** with base gave enol ether **894**. Mild hydrolysis of **894** with acid gave keto aldehyde **895** which was cyclized in modest yield by heating in the presence of alumina at 100–180°C *in vacuo*. Removal of the protecting groups from **896** was a difficult task which was accomplished, nonetheless, to give pentenomycin.

D. Compounds with Pharmacological Activity

Medicinal agents can be divided into two broad classes, those having chemotherapeutic activity and those having pharmacological activity. As opposed to chemotherapeutic agents, which are employed to destroy foreign invaders, pharmacological agents are used to modify and/or control the physiological state. These agents have been classified by one authority (*391*) as drugs acting on somatic systems (cardiovascular and smooth muscle agents; blood; alimentary, diagnostic, and immunologic agents; hormones; and vitamins and nutritional agents) and drugs acting on the nervous system (autonomic nervous system; neuromuscular system; anesthetics and therapeutic gases; analgesics and antitussives; and sedatives and stimulants). I have chosen to discuss two groups of compounds fitting into these classifications, the arachidonic acid metabolites and the vitamins, in separate sections. In addition, some of the alkaloids and terpenes have pharmacological effects. The remaining work on the preparation of pharmacologically active compounds from chiral starting materials is covered in this section.

Comparison of chemotherapeutic and pharmacological agents indicates that the former, as discussed in the previous section, are in many cases derived from natural products. On the other hand, pharmacology has been the domain of empirically derived, synthetic organic molecules. Many of the compounds are achiral, and of those that are chiral, many are still produced only in racemic form. Despite this, there is a growing appreciation (*1*) of the often different biological effects of enantiomeric molecules. This has not yet translated, however, into a general use of chiral building blocks to prepare pharmacological agents. Most such compounds, if desired in chiral form, are still prepared by resolution.

The pharmacological agents prepared using chiral building blocks have been limited to a few structural types (asterisks indicate the centers of asymmetry):

1. Vicinal amino chiral alcohols 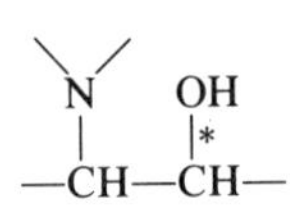

2. Vicinal hydroxy chiral amines 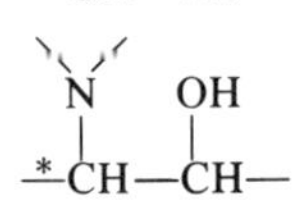

3. Vicinal chiral hydroxy amines 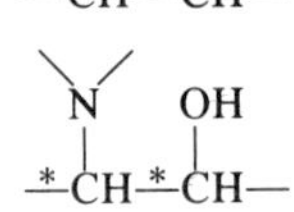

4. Vicinal hydroxy chiral alcohols 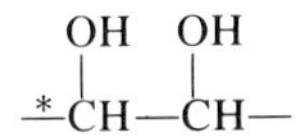

5. Chiral amines and chiral alcohols 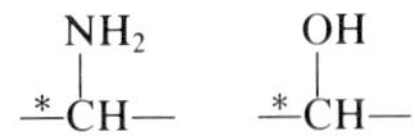

As will be seen, there is a substantial overlap in the synthetic approaches to these functional units.

1. Vicinal Amino Chiral Alcohols

A number of compounds having either α- or β-adrenergic receptor blocking activity contain a chiral alcohol flanked by an amine function. One of the better approaches to such compounds is the opening of a chiral epoxide with an amine. Several compounds (Scheme 55) have been prepared in this manner from D-mannitol via (2*R*)-glyceraldehyde-2,3-acetonide (**898**). These include practolol (**899**) (*392*) and **900** (*393*). Alternatively, conversion of the aldehyde to an aminomethylene function gave timolol (**901**) (*394*), (2*S*)-prosympal (**902**) (*395*), and (*S*,*S*)-dibozane (**903**) (*396*). Proper manipulation of functionality in intermediates allowed the conversion of **898** additionally to the enantiomers of **900**, **902**, and **903**. Using similar techniques, acetonide **898** has also been converted to the analgesic oxapadol (**904**) (*397*).

Sugars other than mannitol have been employed in syntheses of molecules having a chiral amine center. Thus, L-rhamnose was converted via 5-*O*-benzyl-L-rhamnitol to (*S*)-acetyl-β-methylcholine (**905**) (*68*). These same authors have prepared the *R* enantiomer of **905** by first synthesizing 5-*O*-benzyl-D-rhamnitol from D-mannitol.

As mentioned in our discussion of chiral synthons and noted above with regard to Scheme 55, the (*R*)-acetonide **898** can, by proper manipulation of the substituent oxidation states, serve as a precursor of either *R* or *S* product alcohols (*398*). For some compounds, however, these processes may be difficult and/or inefficient to carry out. Two groups (*57, 58*) have now shown that it is possible to obtain the *S* enantiomer of **898** in an efficient manner from ascorbic acid. This material was used in a synthesis of the hypotensive and antiepileptic drug (*R*)-γ-amino-β-hydroxybutyric acid (**906**) (*57*).

The muscarines, a series of compounds isolated from fly agaric mushroom, have been of interest as pharmacological tools for many years because of their ability to induce cholinergic effects. The gross structure

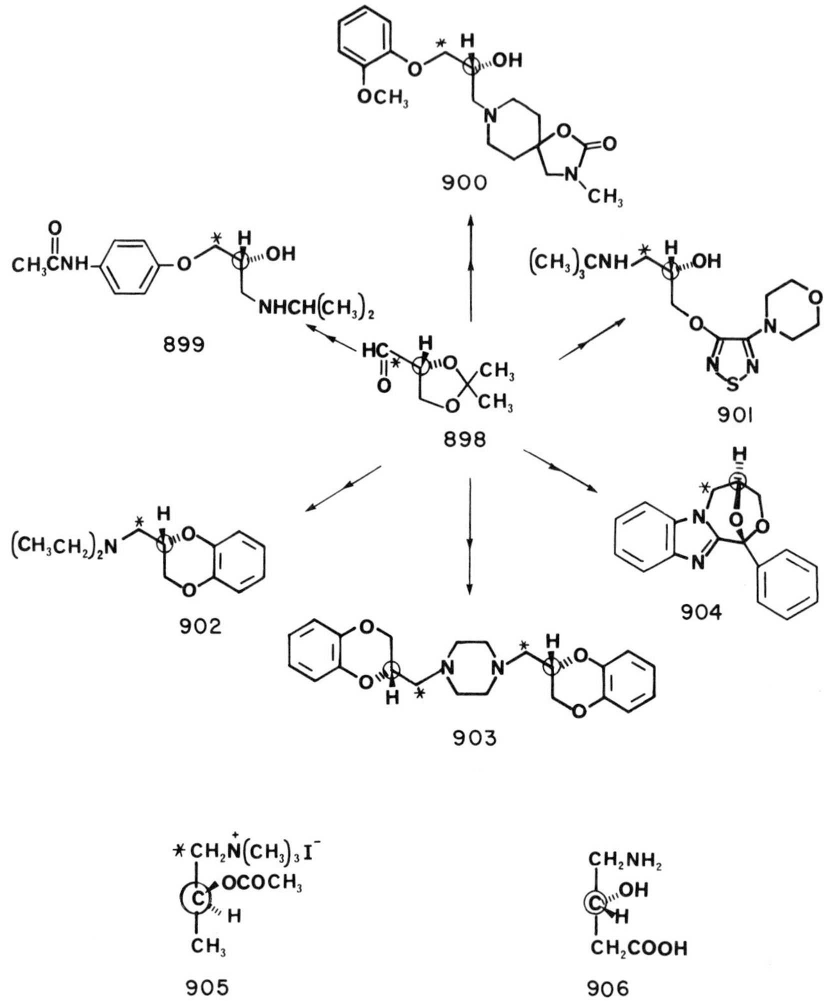

Scheme 55. Vicinal amino chiral alcohols.

(Scheme 56), with three centers of asymmetry, allows eight diasteriomers. Of these, three have been identified as natural products: L-(+)-(2*S*,4*R*,5*S*)-muscarine (**911**), D-(−)-(2*R*,4*R*,5*S*)-allomuscarine (**923**), and L-(−)-(2*S*,4*S*,5*S*)-epimuscarine (**926**).

The synthetic concepts used to synthesize the muscarines are all based in part on the work of E. Hardegger at the ETH, Zurich, in the 1940s and 1950s. Unfortunately, all these syntheses are deficient in that they require a diastereomer separation (due to a nonspecific reaction) or that they

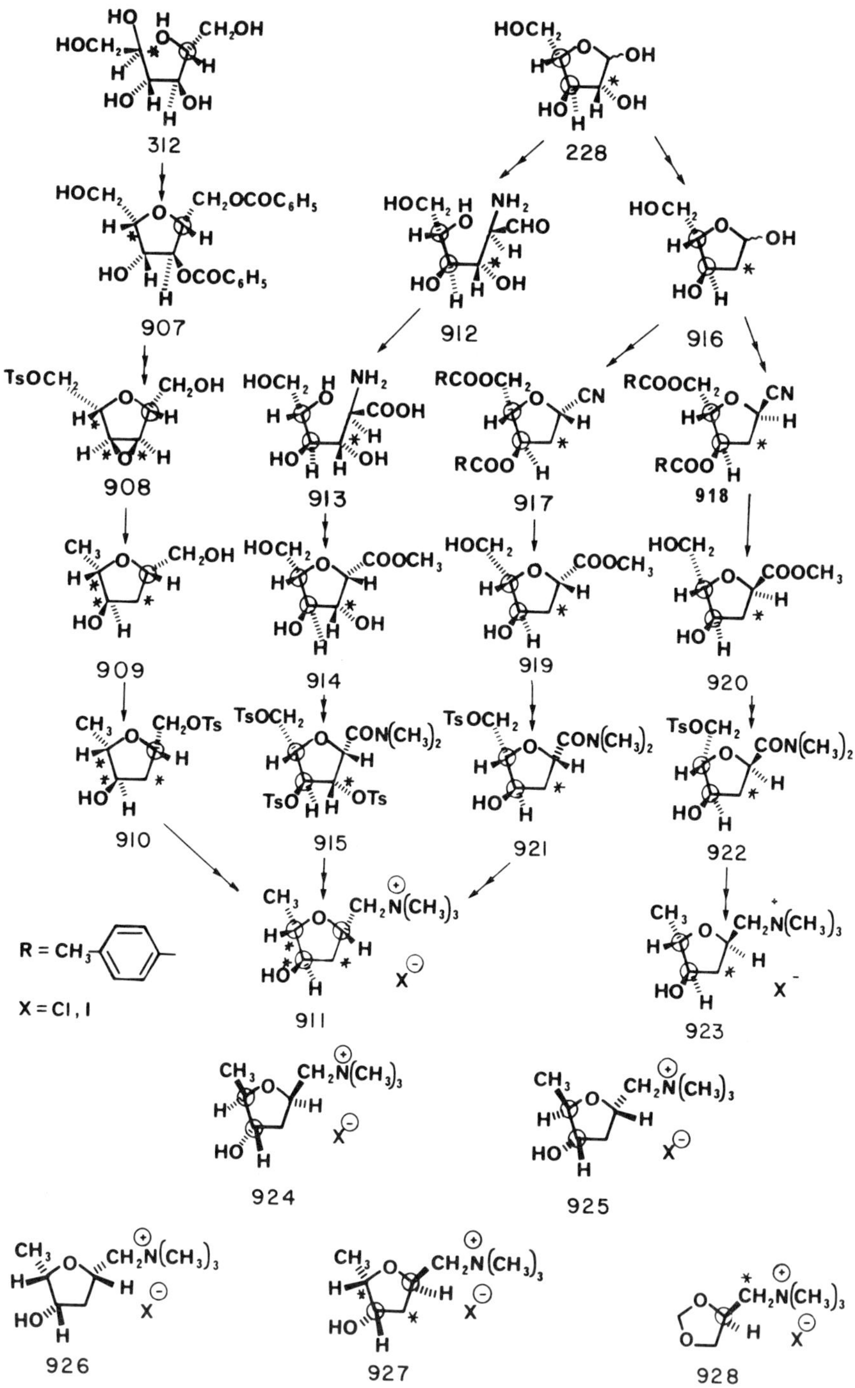

Scheme 56. Muscarines.

begin with a laboriously prepared unnatural sugar. Nonetheless, the Hardegger synthesis (*399*) of **911**, apparently the major natural muscarine, remains a milestone in the development of the use of sugars as chiral starting materials. Thus, the readily available L-arabinose (**228**) was homologated to the unnatural L-glucosamine (**912**). Catalytic air oxidation gave the amino acid **913**, the diazotization of which occurred with internal participation to give after esterification **914**. Ester **914** was converted by treatment with dimethylamine and then tosyl chloride to the tritosylate **915**. Treatment of this compound with lithium aluminum hydride caused a remarkable series of conversions: reductive removal of two of the tosylate groups, cleavage of the third, and reduction of the amide. While the reaction was not completely selective, yielding some C-4 deoxygenated material, L-(+)-muscarine (**911**) could be isolated from the product mixture obtained after methylation.

A second synthesis (*400*) of **911** began with D-mannitol (**312**), which in several steps, including acidic dehydration, was converted to furanose **907**. Tosylation followed by treatment with base gave epoxide **908**. Reduction of this compound with sodium bis(2-methoxyethoxy)aluminum hydride gave a 12 : 1 mixture of **909** and its C-3 regioisomer. After selective tosylation, isomer separation gave **910**, which was readily transformed into **911**.

A third synthesis (*401*) of L-(+)-muscarine (**911**) began, as did Hardegger's, with L-arabinose (**228**), which was transformed in five steps to 2-deoxy-L-ribose (**916**). This material was converted to a mixture of the nitriles **917** and **918**. After separation, nitrile methanolysis gave ester **919** and its mono- and ditolyl esters. Treatment of this mixture with dimethylamine caused the desired ester-to-amide transformation and completed hydrolysis of the tolyl esters. Selective tosylation gave **921**, which after reduction and methylation yielded **911**. Similar reactions performed on **918** gave D-(−)-allomuscarine (**923**). Using 2-deoxy-D-ribose, Huynh-Dinh (*401*) also prepared D-(−)-muscarine (**924**) and L-(+)-allomuscarine (**925**). Compound **924** had been first prepared by Hardegger from D-glucosamine (*402*) or 2-deoxy-D-ribose (*403*) using the methodology outlined above.

The preparation of D-(+)-epiallomuscarine (**927**) in a nonspecific manner from D-glucose has been described (*404*). The muscarine analog **928** was prepared from D-mannitol (*405*).

2. Vicinal Hydroxy Chiral Amines, Hydroxy Chiral Alcohols, and Chiral Hydroxy Amines

The number of compounds fitting into the classifications defined in the title of this section is relatively small. Examples, however, are known

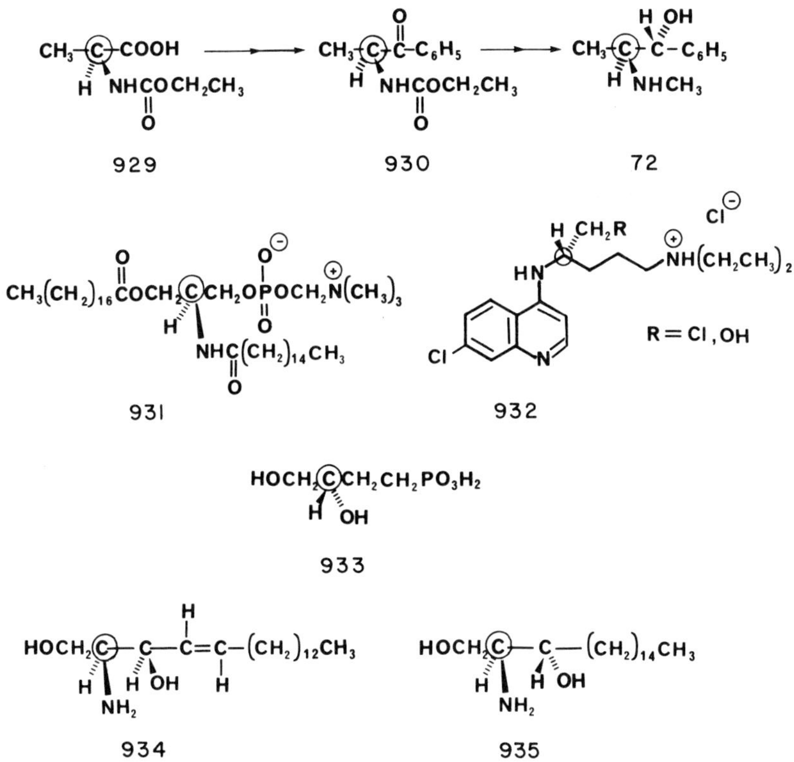

Scheme 57. Amino alcohols and diols.

(see Scheme 57), most being characterized by direct introduction of the preexisting chiral unit. As part of a study of the use of amino acids as chiral educts, Rapoport (*406*) has converted the carbamate **929** from (*S*)-alanine to its acid chloride. Friedel–Crafts reaction of this material or (better) treatment with phenylmagnesium bromide gave ketone **930**. Two-step reduction of this compound gave a 4 : 1 mixture of L-ephedrine (**72**) and the enantiomeric ψ-ephedrine. Other preparations of interest include the phosphatidylcholine analog **931** [from (*S*)-serine] (*407*), antimalarial chloroquine analogs **932** [from (*S*)-glutamic acid] (*408*), isoster **933** of glycerol-3-phosphate [from (*S*)-malic acid] (*409*), and D-*erythro*-sphingosine (**934**) (*410*) and D-*threo*-sphinganine (**935**) (*411*) [both from (*S*)-serine].

3. Chiral Alcohols and Chiral Amines

Picrotoxinin (**946**, Scheme 58) is a chiral alcohol of some interest as an antagonist of γ-aminobutyric acid. This molecule contains eight asymmet-

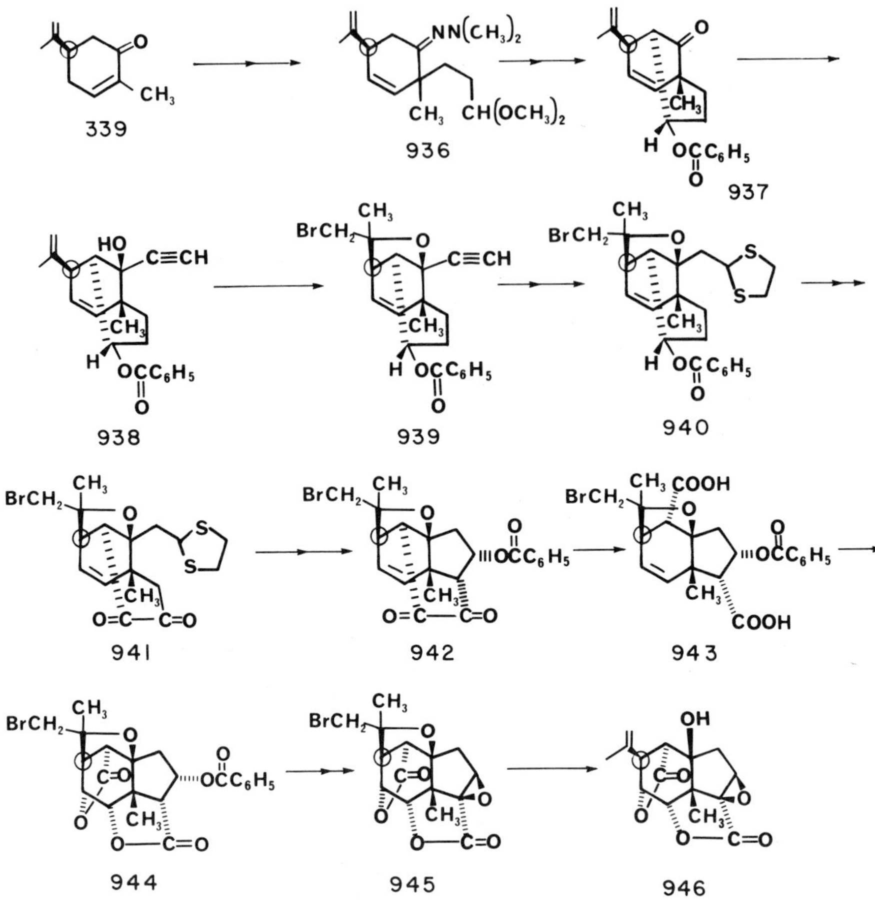

Scheme 58. Picrotoxinin.

ric centers in a very compact array. Its synthesis by Corey (*412*) provides a dramatic illustration of the manner in which stereoselective reactions controlled by one chiral center can lead to a very complex molecule. Interestingly, the second asymmetric center was introduced with only low selectivity. Thus, alkylation of the dimethylhydrazone of (*R*)-(−) carvone (**339**) gave a 6:4 mixture of diastereomers **936**. Hydrazone cleavage, acidic acetal hydrolysis and aldol reaction, benzoylation, and epimer separation gave bicyclic compound **937**. Ethynylation to **938** was followed by treatment with *N*-bromosuccinimide to protect the isopropenyl group and the tertiary alcohol as the bromoether **939**. Construction of the cyclopentane ring in several steps involved hydroboration and protection of the resultant aldehyde (→ **940**), benzoate cleavage and two-step oxidation (→ **941**), and dithiolane cleavage under conditions giving aldolization, followed by benzoylation. Oxidative cleavage of the resultant **942** gave diacid **943**, which was converted to dilactone **944** in essentially quantitative yield by treatment with lead tetraacetate. Elimination of the elements of benzoic acid and epoxidation gave **945**, which was reductively fragmented with zinc to give picrotoxinin.

The cannabinols (Scheme 59) are of major concern as the bioactive constituents of marijuana (for a general review of cannabinoid chemistry, see ref. *17*). The compounds have aroused additional interest since they apparently inhibit the nausea often caused by chemotherapy and may also be effective in inhibiting glaucoma. A facile synthesis of $\Delta^{1,6}$-tetrahydrocannabinol (**949**) and also the $\Delta^{1,2}$ isomer was devised in 1967 by Mechoulam and co-workers (*413*). Condensation of the isomeric verbenol mixture **947** obtained by reduction of (−)-α-pinene with olivetol gave adduct **948**. Rearrangement of this material with boron trifluoride gave **949** with an optical rotation value ~97% of that of the natural mate-

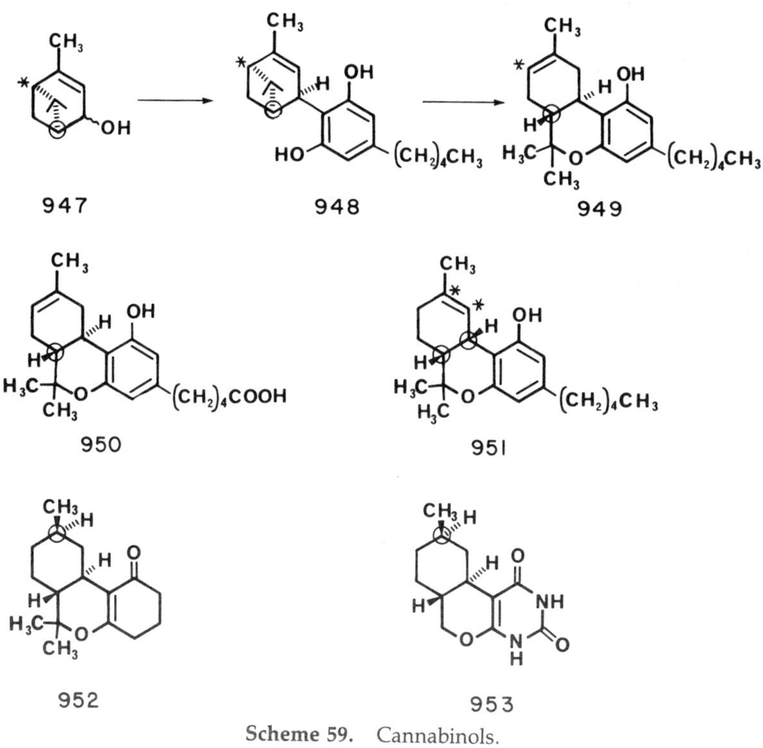

Scheme 59. Cannabinols.

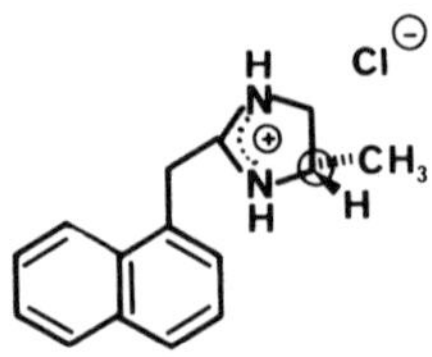

954

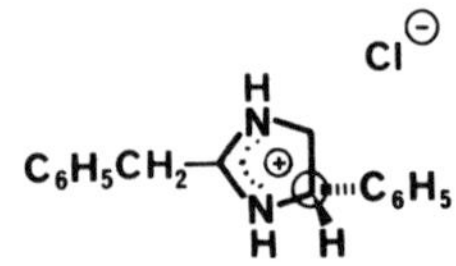

955

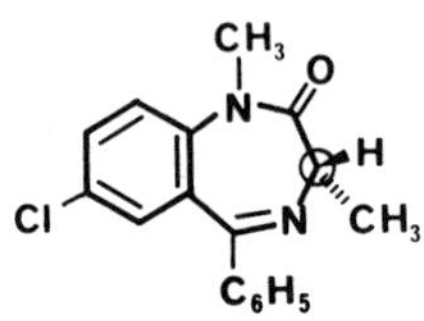

956

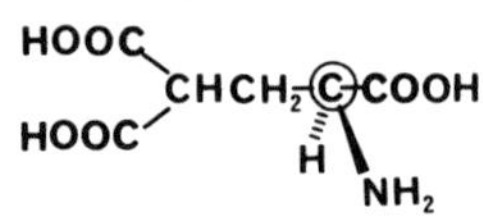

957

Scheme 60. Chiral pharmaceutical agents.

rial. A similar but less specific synthesis of $\Delta^{1,2}$-tetrahydrocannabinol using limonene as chiral source was reported (*414*) during the same year.

Chemistry similar to that described above was used to prepare the carboxylic acid **950** (*415*) from limonene and the cis compound **951** (in low yield) from *trans*-2-carene oxide (*416*). Citronellal was the chiral starting material for the cannabinoid analogs **952** (*417*) and **953** (*418*), which were formed in stereospecific Diels–Alder reactions.

Potential pharmaceuticals having chiral amine functionalities are also known. The syntheses of these compounds were often straightforward. We note the syntheses of the potential adrenergic drugs **954** (*419*) and **955** (*420*) (Scheme 60) from (*S*)-alanine and (*S*)-phenylalanine, respectively, and of the chiral benzodiazepine **956**, again from (*S*)-alanine (*421*). Danishefsky (*422*) employed an elegant synthetic strategy to convert (*S*)-glutamic acid to (*S*)-γ-carboxyglutamic acid (**957**), a constituent of several blood clotting factors.

E. Insect Pheromones and Related Compounds

Insect pheromones are chemical substances that serve as a means of communication between members of a particular species (*2, 16*). These materials can often be isolated in only very small amounts, and their structure determination is accordingly difficult. In addition to the gross structure, a knowledge of geometrical and enantiomeric composition is also required, since it has been established (*2*) that these properties are critical to activity. In fact, some pheromones are *Z*/*E* and/or *R*/*S* mixtures in specific ratios. Total synthesis is often essential in establishing pheromone constitution and configuration. In addition to supplying quantities of natural pheromones for testing purposes, synthesis also allows the preparation of analogs. Finally, if pheromones are to be used extensively for insect pest control, the cost of the active substance will be critical. In this context, the use of commercially available chiral educts for pheromone preparation certainly merits consideration.

The synthesis of pheromones has been reviewed, both in general (*2, 9, 16*) and with emphasis on chiral forms (*10, 15, 22*). I shall take a somewhat different viewpoint, however, to the discussion of pheromone synthesis. Rather than concentrating on the insect species involved, emphasis will be placed on the chiral functionality to be obtained.

Analysis of the structures of chiral insect pheromones indicates that the source of chirality in the majority of examples resides in a secondary alcohol and/or secondary methyl group. The few remaining cases are of secondary alkyl groups larger than methyl (usually isopropenyl) and some examples of tertiary alcohols. The secondary or tertiary alcohol groups are often masked as esters (including lactones), ketals, or epoxides. Nonetheless, the commonality of structural features allows some generalization in synthetic approaches to these substances. The specific starting material and chemical manipulations to be carried out are, of course, determined by the structural features surrounding the chiral center.

1. Pheromones Containing Chiral Secondary Methyl Units

The fact that many pheromones are terpene based, or at least can be conceived as arising by terpene-related biosynthetic pathways, makes terpenes obvious candidates as synthesis starting materials. As noted previously, the readily available citronellols, because of their differentiated bifunctional substitution around the chiral center, have found favor in this regard. One relatively simple pheromone for which (*S*)-citronellol was employed (*423*) is (*R*)-6-methyl-3-octanone (**960**, Scheme 61). Mesylation and reduction of **324** gave the olefin **958**. Oxidative cleavage of the double bond was carried out in three stages (olefin → diol → aldehyde → acid). The resultant acid **959**, as its acid chloride, was reacted with diethylcadmium to give **960**, an ant alarm pheromone. The optical rotation of synthetic **960** was ~80% of that of the natural material; presumably, this represents enantiomeric purity of the starting material. The enantiomer of **960** was also prepared, using (*S*)-2-methyl-1-butanol as the source of chirality.

Closely related in structure to **960** and in biological activity as an ant

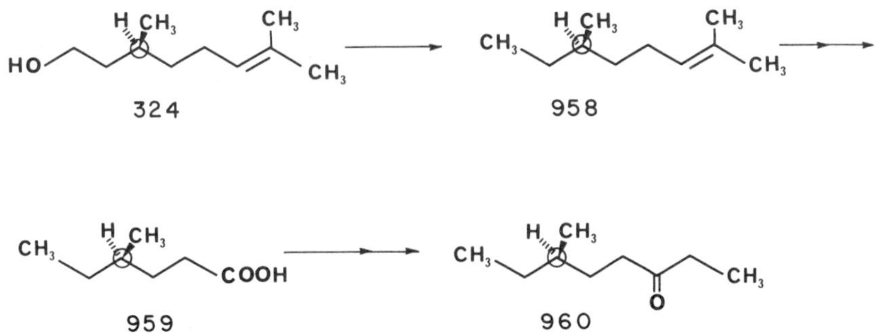

Scheme 61. (*R*)-6-Methyl-3-octanone.

alarm pheromone is manicone (**962**, Scheme 62). Two syntheses (*93, 424*) of the *S* enantiomer of this compound have been reported. Both syntheses proceed via (*S*)-2-methyl-1-butanal (**174**). Mukaiyama (*424*) prepared this material ($[\alpha]_D^{20}$ +31.4°) by oxidation of alcohol **175**. Nakai (*93*), on the other hand, reported that he was unable to do the oxidation without racemization and so prepared **174** ($[\alpha]_D^{25}$ +35.2°) by oxidative decarboxylation of L-isoleucine (**254**) with silver picolinate. The sequences then diverged. Mukaiyama carried out a Lewis acid–catalyzed condensation of **174** with the trimethylsilyl enol ether of 3-pentanone to give the diastereomer mixture **961**. Dehydration with toluenesulfonic acid gave manicone ($[\alpha]_D^{20}$ +45.45°). By condensation of **174** with a different silyl enol ether and correlation after subsequent manipulations with a known compound, Mukaiyama claimed a 97% optical purity for synthetic manicone.

In the Nakai synthesis, aldehyde **174** was treated with the appropriate Grignard reagent to give the allylic alcohol **963**. Esterification with *N,N*-dimethylthiocarbamoyl chloride and thermal sigmatropic rearrangement gave **964**, which was then transformed into manicone of $[\alpha]_D^{20}$ +39.6°. Thus, despite using aldehyde **174** of higher optical purity than that employed by Mukaiyama, Nakai obtained less pure manicone. It is reasonable to assume that partial racemization occurred in the Grignard step (**174** → **963**).

Unfortunately, the optical rotation of naturally occurring manicone has not yet been reported. In addition, neither Mukaiyama nor Nakai used any method other than optical rotation to determine enantiomeric composition. On the basis of the lower rotation of aldehyde **174** used by Mukaiyama and the lower rotating manicone obtained by Nakai, there is a distinct possibility that neither synthesis gave material of high enantiomeric purity. The critical necessity for confirming enantiomeric purities estimated from optical rotations by independent methodology is once again indicated.

Another group of pheromones containing the same chiral terminal sec-

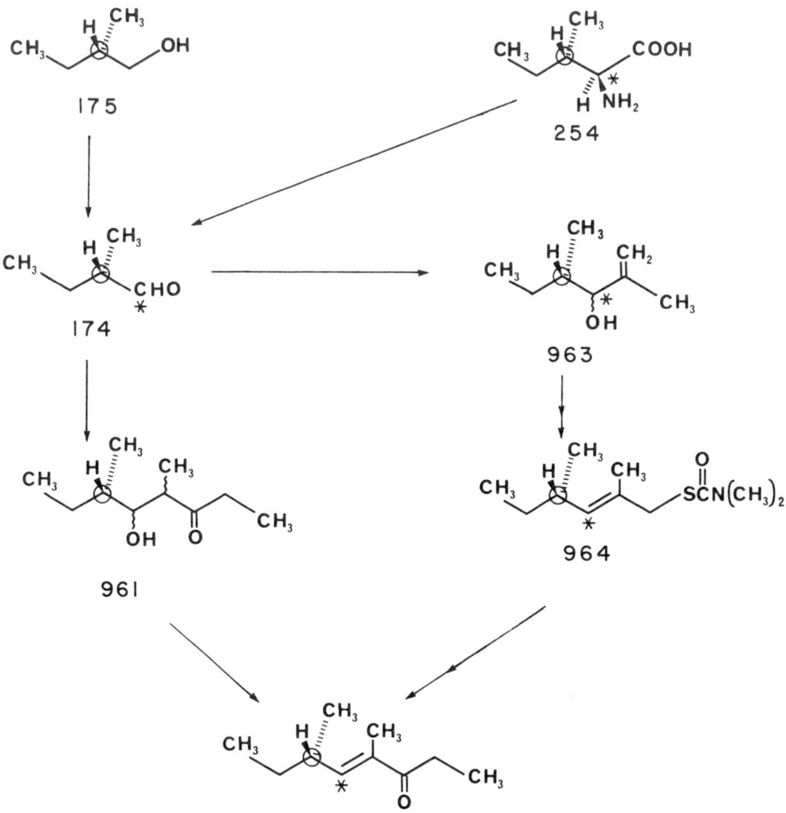

Scheme 62. Manicone.

ondary butyl unit as 6-methyl-3-octanone and manicone is the multicomponent mixture produced by the female dermestid beetle. The gross structure is that of (*R*)-14-methylhexadec-8-ene with a terminal oxidation level of alcohol, aldehyde, or methyl ester. The oxidation state and *Z*/*E* olefin configuration varies from species to species. Early work in this area confirmed the *R* absolute configuration by synthesis from (*S*)-2-methylbutan-1-ol of the unnatural[3] enantiomeric (*S*,*Z*)-14-methylhexadec-8-en-1-ol (*425*, *426*), its *S*,*E* isomer (*426*), the corresponding *S*,*Z* and *S*,*E* aldehydes (*426*), and the (*S*,*Z*)-methyl ester (*425*).

Subsequently, the pheromones themselves were prepared using the citronellols rather than the methylbutanols as sources of chirality (see Scheme 63). The approach taken by Mori (*116*, *429*) produced both *R* and

[3] For some species the native pheromones may have the *S* configuration (see ref. *427*, but see also ref. *428*).

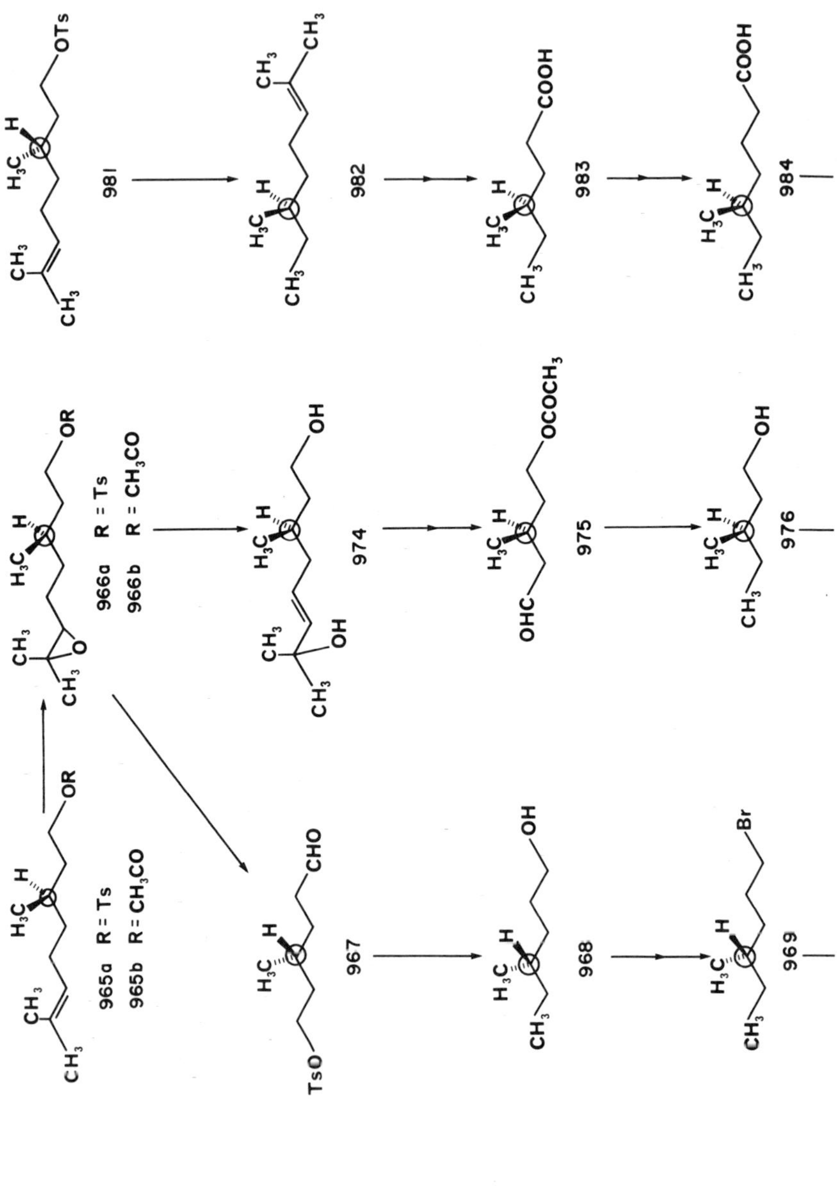
965a R = Ts
965b R = CH3CO
966a R = Ts
966b R = CH3CO
967
968
969
974
975
976
981
982
983
984

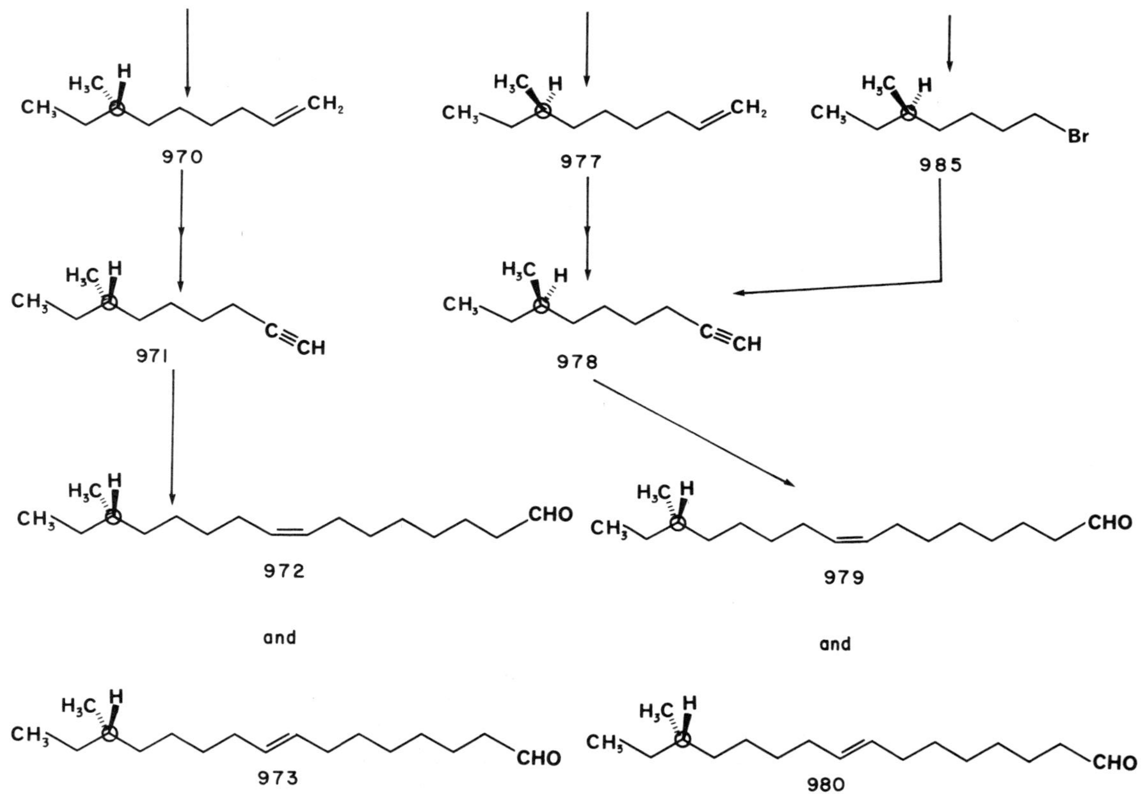

Scheme 63. 14-Methylhexadec-8-enal.

S aldehydes **972**, **973**, **979**, and **980** from (*R*)-(+)-citronellol. Thus, tosylate **965a** and acetate **965b** were epoxidized to **966a** and **966b**, respectively. Oxidative cleavage of **966a** with periodic acid gave aldehyde **967**. Hydride reduction of this material gave alcohol **968**, which was converted via the tosylate to bromide **969**. Coupling of the corresponding Grignard reagent with allyl bromide then gave the *S* olefin **970**.

The enantiomeric *R* olefin **977** was obtained from **966b** by initial treatment with sodium phenylselenide and then hydrogen peroxide. Epoxidation of allylic alcohol **974** and then periodic acid cleavage gave aldehyde **975**. It is worthy of note that compounds **967** and **975** are synthetically equivalent enantiomers whose chirality is dependent on the differing oxidation states of the termini. Wolff–Kishner reduction of **975** gave alcohol **976** which, as its tosylate, was coupled with 3-butenylmagnesium bromide to give *R* olefin **977**.

The olefins **970** and **977** were then converted, via the acetylenes **971** and **978**, to the *S,Z* aldehyde **972** ($[\alpha]_D^{21}$ +6.02°) (*116*), *S,E* aldehyde **973** ($[\alpha]_D^{22}$ +6.18°) (*429*), *R,Z* aldehyde **979** ($[\alpha]_D^{21}$ −5.94°) (*116*), and *R,E* aldehyde **980** ($[\alpha]_D^{21}$ −5.99°) (*429*). Mori estimated these compounds to be of ≥92.5% optical purity based on a value reported (*426*) by Rossi for the *S* alcohol corresponding to **972**. Rossi (*427*) nonetheless was highly critical of the Mori work, claiming on the basis of rotations that many of Mori's compounds were much less optically pure than stated. In the same paper, Rossi converted (*S*)-(−)-citronellol of 84% optical purity to *R,Z* aldehyde **979** ($[\alpha]_D^{25}$ −5.90°) and *R,E* aldehyde **980** ($[\alpha]_D^{25}$ −5.04°), values similar to those obtained by Mori. While Rossi may be correct in criticizing some of the reported rotations, his synthesis (see below) is different enough from Mori's that it is reasonable to assume that both proceeded without racemization. The true enantiomeric purity of the synthetic intermediates and products can only be determined when a method more reliable than optical purity is employed (*430*).

In the Rossi synthesis (*427*), the *S* tosylate **981** was reduced with lithium aluminum hydride to olefin **982**, which was oxidized in three stages to acid **983**. Three-step homologation provided **984**, which was converted in the usual manner to bromide **985**. Reaction of this compound with lithium acetylide gave **978**, which was then taken on to the desired aldehydes. A third synthesis (*431*) of the *R* phermones involved Kolbe coupling of **983**, prepared by Rossi's method, with the appropriate acetylenic acid.

Mori's reply to the Rossi criticism is contained in a report (*428*) of the synthesis of all four diastereomers of 3,11-dimethylnonacosan-2-one and 29-hydroxy-3,11-dimethylnonacosan-2-one pheromones of the German cockroach. In this work, (*R*)-citronellol prepared from (−)-isopulegol and of established enantiomeric purity [92%, by HPLC of the diastereomeric amides formed from the corresponding acid and (*R*)-α-(1-naphthyl)-ethylamine] was used. This same material had been employed in the dermestid beetle phermone work, although as described above, only optical purities were described at that time (*116*).

Mori's approach to the cockroach pheromones (Scheme 64) was once again based on use of the bifunctional nature of citronellol to obtain synthons of either *R* or *S* stereochemistry. (*R*)-Citronellol (**986**) was converted via its tosylate to iodide **987**. Malonic ester two-carbon homologation of **987** gave ester **988**. Reduction and acetylation produced **989**, which was subjected to epoxidation and periodic acid cleavage to give aldehyde **990**. Reaction with a C-15 phosphorane, hydrogenation, and tosylation gave the *R* C-24 segment **991**.

To prepare the *S* enantiomer **995** of **991**, iodide **987** was first treated with hexadecynyllithium to give **992**. Epoxidation and oxidation as for **989**, gave aldehyde **993**. Three-step, one-carbon homologation and acetylene hydrogenation then produced the *S* C-24 unit **995**.

The second chiral center in the pheromone was also prepared from (*R*)-citronellol. Pyrolysis of the derived xanthate gave the diene **996** (*432*). The usual epoxidation and oxidation of the more electron-rich olefin produced aldehyde **997**, which was reduced to alcohol **998**. The *R* bromide **999** was then prepared in the usual manner.

One-carbon homologation of **986** in four steps gave ester **1000**, the epoxidation and oxidation of which led to aldehyde **1001** (*90*). Oxidation to the acid and Hunsdiecker reaction removed one carbon to give bromide **1002**. Olefin formation by oxidation of the derived *O*-nitrophenylselenide produced olefinic acid **1003**. Reduction and bromide formation yielded the *S* compound **1004**. The Grignard reagents from **999** and **1004** were coupled with tosylates **991** and **995**. Hydration and oxidation of the coupling products then gave the desired pheromones. It was shown by rotation and mixed melting point that the natural pheromone is the *S*,*S* diastereomer **1005**. Interestingly, all four diastereomers were equally bioactive. By proper manipulation of intermediates, the second pherome (**1006**), also of *S*,*S* configuration, was obtained.

The intermediates **968** (Scheme 63) and **1002** (Scheme 64) were converted, in a straightforward manner, to the enantiomeric forms of 10-methyldodecyl acetate, a minor component of the pheromone mixture of the smaller tea tortix moth (*433*).

An intermediate from the synthesis of the cockroach pheromones was also employed (*434*) in a preparation of the yellow scale pheromone **1012** (Scheme 65). Modification of the oxidation state of acid **1003** (Scheme 64) gave aldehyde **1008**, reaction of which with the Grignard reagent from bromide **1007** led to the diastereomeric alcohol mixture **1009**. Wittig rear-

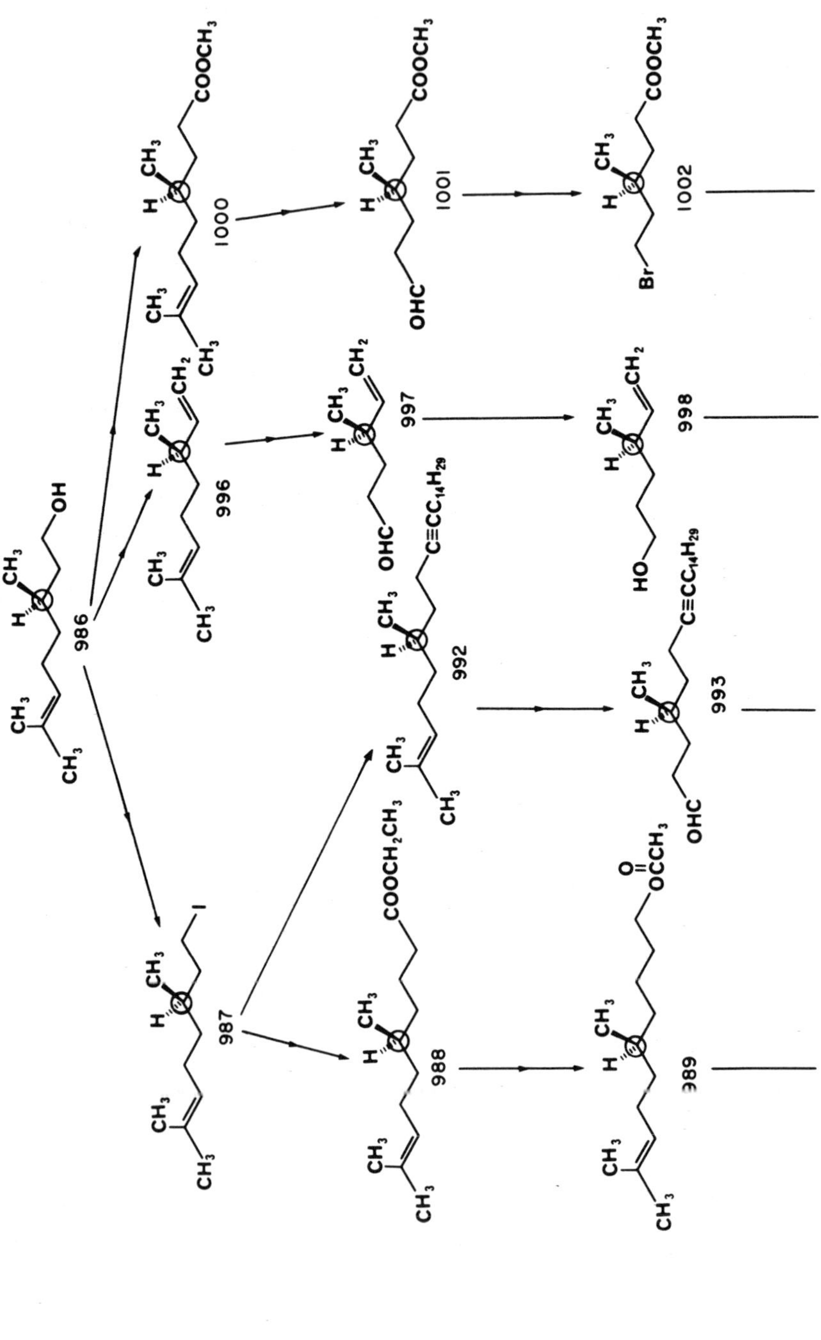
986
987
988
989
992
993
996
997
998
1000
1001
1002

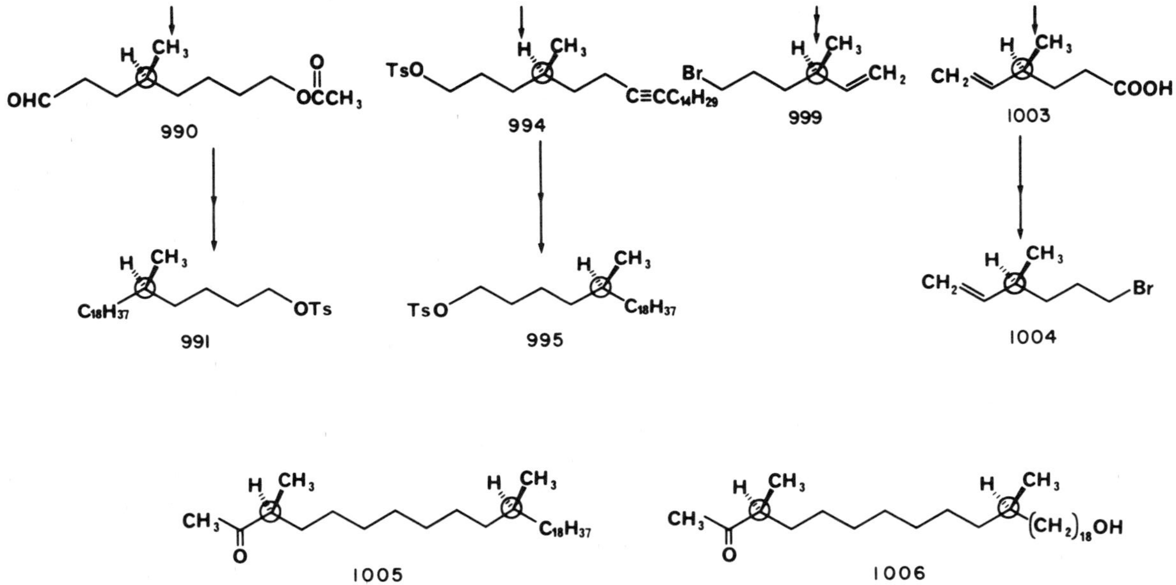

Scheme 64. (3*S*,11*S*)-3,11-Dimethylnonacosan-2-one.

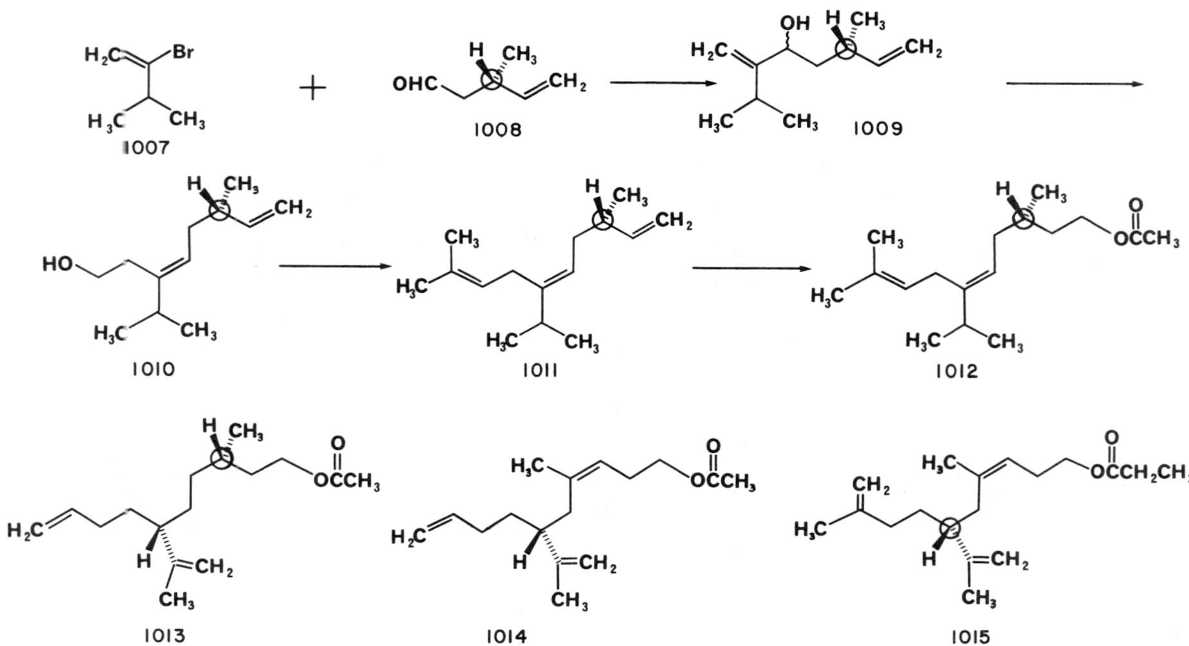

Scheme 65. Scale insect pheromones.

rangement of the derived tri(*n*-butyl)stannyl ether gave the *E* alcohol **1010**. Oxidation and Wittig reaction gave triene **1011**, which was selectively hydroborated, oxidized, and acetylated to **1012**. Unfortunately, this material turned out to be the enantiomer of the natural product. Subsequently, Mori (*435*) also achieved a synthesis of the bioactive *S* enantiomer of **1012**, again from (*R*)-citronellic acid.

The acetate **1013** is one of two pheromones of the California red scale. Two syntheses (*436, 437*) of this material have been described. Both syntheses start from citronellyl derivatives but require diastereomer separations due to nonspecific introduction of the second chiral center. The second California red scale pheromone is **1014**, the preparation of which from (*S*)-carvone has been reported (*438, 439*). Pheromone **1015** of the white peach scale was synthesized (*440*) from (*R*)-limonene.

2. Pheromones Containing Chiral Secondary Methyl and Alcohol Units

Several pheromones contain both chiral secondary methyl units and chiral secondary alcohol units, often in a vicinal relationship. One of the simplest of such compounds is (3*S*,4*S*)-4-methylheptan-3-ol, one of the aggregation pheromones of the smaller European elm bark beetle. The absolute configuration of this compound was established by synthesis (*441*) of the enantiomer and a diastereomer from (*R*)-citronellic acid. A synthesis (Scheme 66) of the pheromone **1023** itself has been described (*442*). The epoxide **1016**, readily derived from D-glucose in several steps, was treated with methylmagnesium chloride. Oxidation of the resultant alcohol gave ketone **1017**. Basic equilibration at C-3 and hydride reduction produced the two chiral centers of the product in the proper orientation. Acetal hydrolysis with acid and reprotection of the aldehyde as a dithiane then gave **1019**. At this point, the excess functionality was removed by lead tetraacetate oxidation. Ethylidenation of the product **1020** gave the olefin mixture **1021**. After protection of the alcohol as its benzyl ether, dithiane hydrolysis produced **1022**. Completion of the synthesis was accomplished by Wittig methylenation and hydrogenation, which saturated both olefinic linkages and cleaved the benzyl ether.

Serricornin, (4*S*,6*S*,7*S*)-4,6-dimethyl-7-hydroxynonan-3-one, is the sex pheromone of the female cigarette beetle. The structure, which can be conceived of as simply a three-carbon acylation product of **1023**, was established (*443, 444*) by synthesis of the (4*S*,6*R*,7*R*)-diastereomer **1024** from D-glucose.

The pine sawfly pheromones **1030** (Scheme 67) also contain the vicinal

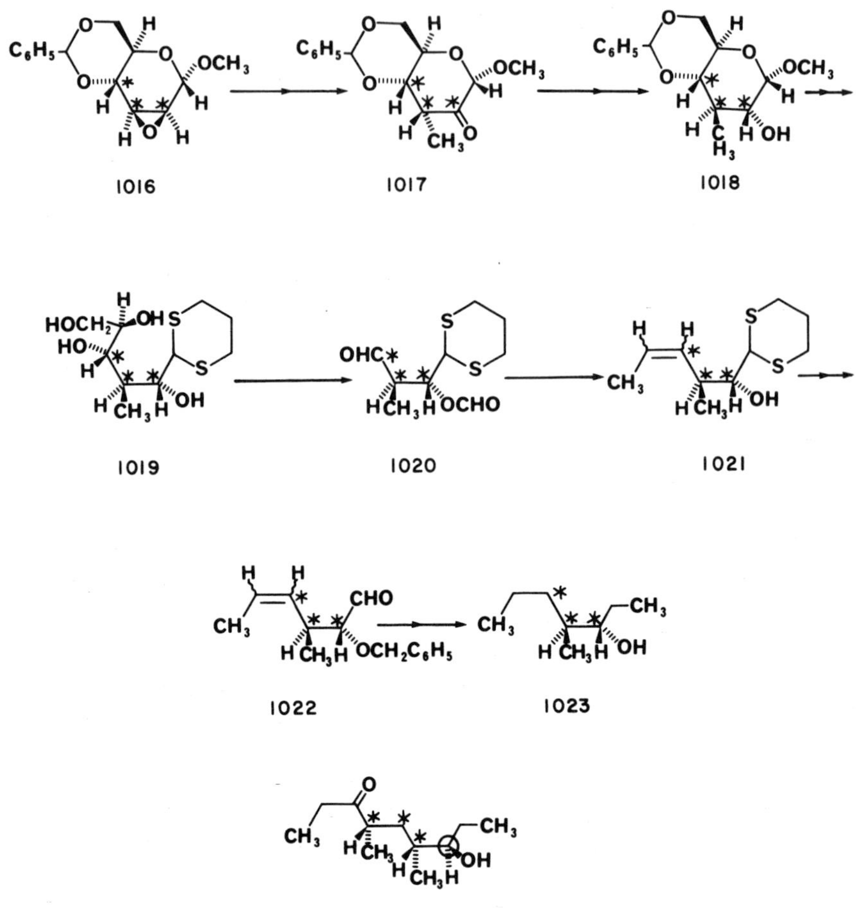

Scheme 66. (3*S*,4*S*)-4-Methylheptan-3-ol.

methyl alcohol functionality. Since the C-2 and C-3 centers were known to be erythro, there were four possible absolute configurations. Mori (*90*) has synthesized all four diastereomers from (*R*)-citronellol and tartaric acid; preparation of the natural 2*S*,3*S*,7*S* diastereomer is shown. The C-7 center was obtained from tosylate **965a** by initial coupling with lithium dihexylcuprate. Ozonolysis of the resulting hydrocarbon **1025** and then reduction gave alcohol **1026**, converted in the usual manner to bromide **1027**.

To prepare the other portion of **1030**, (*R*,*R*)-tartaric acid was converted in several steps via acetonide **1028** to (*S*,*S*)-butanediol (**150**). Reaction of

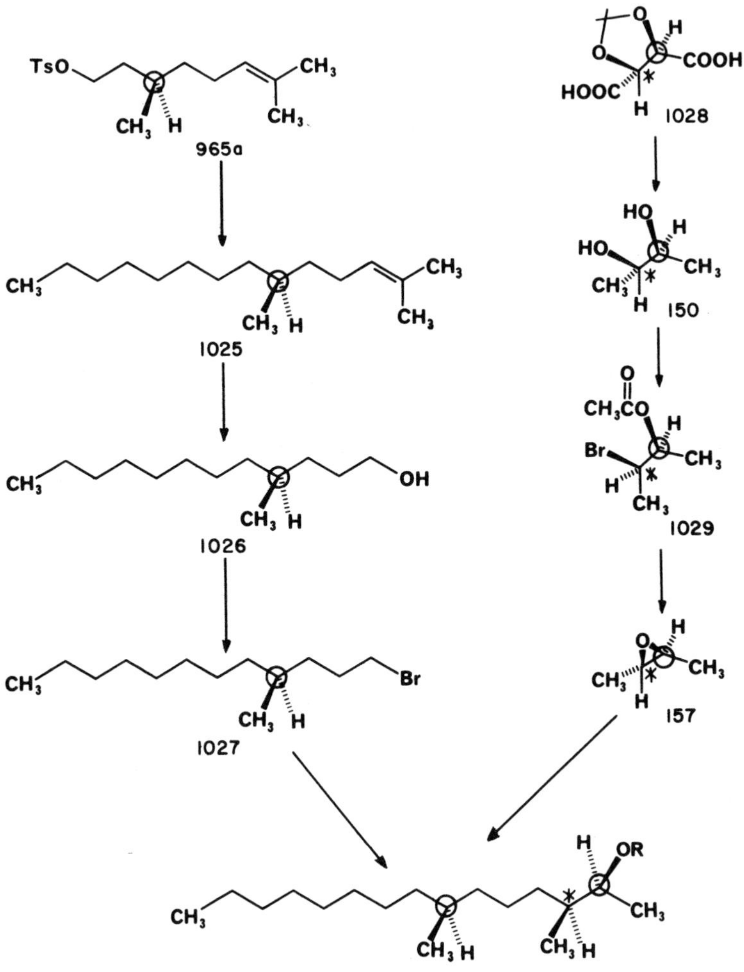

Scheme 67. Pine sawfly pheromones. R = $COCH_3$, $COCH_2CH_3$.

this diol with dry hydrogen bromide (gas) in acetic acid gave **1029**, which on treatment with base gave epoxide **157**. Conversion of bromide **1027** to the lithiocuprate and then reaction with **157** gave, after acylation, the pheromones.

A second synthesis of **1030** has been reported (*445*). This work also employed (*R*,*R*)-tartaric acid in the form of epoxide **157**. The C-7 center was obtained by asymmetric synthesis.

Methodology for conversion of the unnatural L-glucose to (2*S*,5*S*)-5-

hydroxy-2-methylhexanoic acid lactone has been developed in the enantiomeric series (*446*).

3. Pheromones Containing Chiral Secondary Alcohol Units

Several pheromones are relatively simple molecules containing a chiral secondary alcohol. One such example is sulcatol (6-methylhept-5-en-2-ol), the aggregation pheromone of the ambrosia beetle. Syntheses of both *R* and *S* enantiomers of this compound were required, since the natural material is apparently a 35:65 mixture of these materials.

Despite the comparatively uncomplicated structure of sulcatol, three entirely different syntheses of the enantiomers have been reported. In the first synthesis (Scheme 68), done by Mori (*107*), diazotization of (*R*)-glutamic acid gave lactone **1031**, which was then reduced to alcohol **197**. Removal of the excess oxygen atom, via the tosylate and iodide, produced the methyl-substituted lactone **211**. Partial reduction to **1032** with diisobutylaluminum hydride and then Wittig reaction gave (*S*)-sulcatol (**1033**). A parallel series of reactions converted (*S*)-glutamic acid to (*R*)-sulcatol (**1034**).

The second synthesis (*447*) of the sulcatols (Scheme 69) used carbohydrate starting materials and reductive removal of the excess oxygen atoms. To prepare (*S*)-sulcatol, L-fucose was converted in three steps to the protected form **1035**. Acetal cleavage and dimesylation gave **1036**, from which the mesyl groups were reductively eliminated with zinc and sodium iodide. Hydrogenation of the resultant unsaturated sugar then gave **1037**. Acetal hydrolysis and periodiate oxidation gave **1038**, which on Wittig reaction gave (*S*)-sulcatol (**1033**). The route to (*R*)-sulcatol (**1034**) was

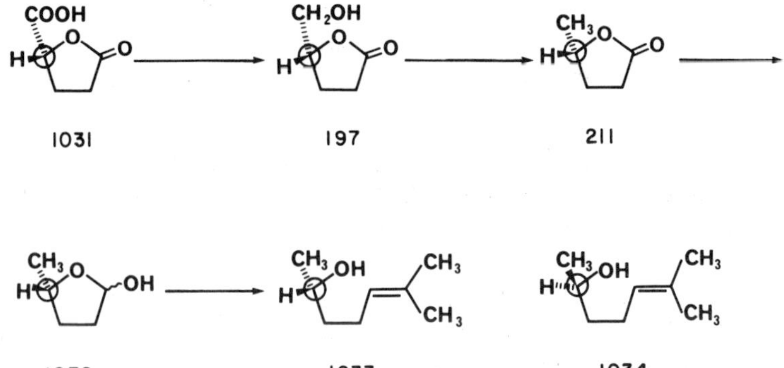

Scheme 68. Sulcatol synthesis I.

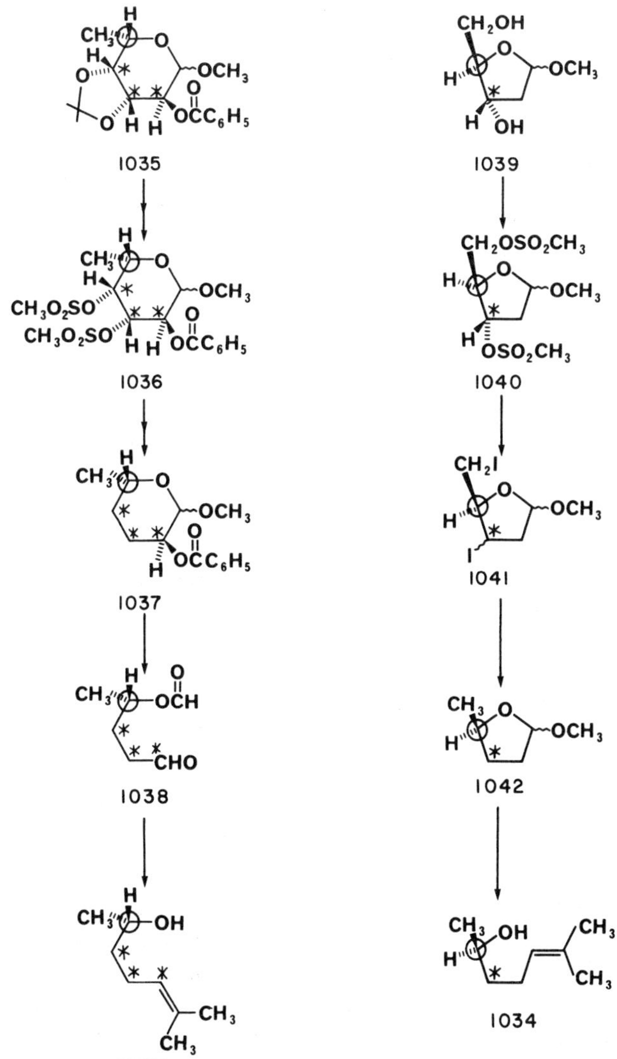

Scheme 69. Sulcatol synthesis II.

somewhat more straightforward. Thus, the protected form **1039** of 2-deoxy-D-ribose was transformed via dimesylate **1040** to the iodide mixture **1041**. Hydrogenolysis over Raney nickel then gave acetal **1042**, which was hydrolyzed and homologated to the desired **1034** (*447*).

Mori (*107*) recognized the possibility of using (*R*)- and (*S*)-2-methyloxirane as sulcatol synthons. However, the unavailability of both enantio-

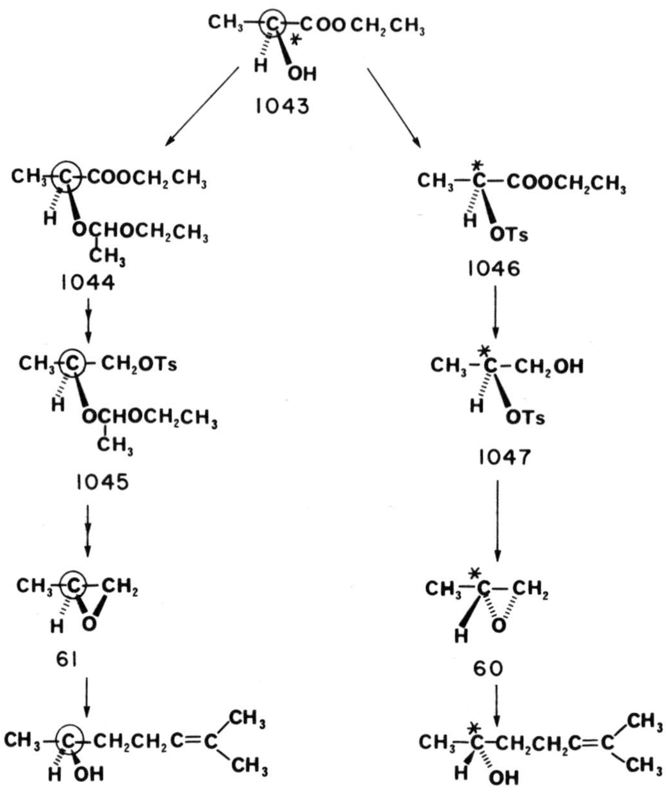

Scheme 70. Sulcatol synthesis III.

mers of known enantiomeric purity had convinced him to develop the glutamic acid route instead. Slessor (*70*) later showed (Scheme 70) that both epoxides, however, could cleanly be obtained from (*S*)-ethyl lactate. The (*S*)-oxirane **61** was prepared, as described by Seebach (*38*), by initial protection of the alcohol with ethyl vinyl ether. Ester reduction and tosylation produced **1045**, which was deprotected with acid and cyclized with base to **61**. Alternatively, Slessor first tosylated **1043** to **1046**, which could be reduced with diborane to **1047**. Treatment of this compound with base gave the (*R*)-oxirane **60**. The sulcatols were then obtained by reaction of **61** and **60** with 3-methyl-2-butenylmagnesium chloride.

The pheromone of the male *Ips* bark beetle consists of three major components: (+)-*cis*-verbenol, (*S*)-(−)-ipsenol (**1052**, Scheme 71), and (*S*)-(+)-ipsdienol (**1057**, Scheme 72). The verbenol was obtained in a straightforward manner from its trans isomer (*448*), that coming from

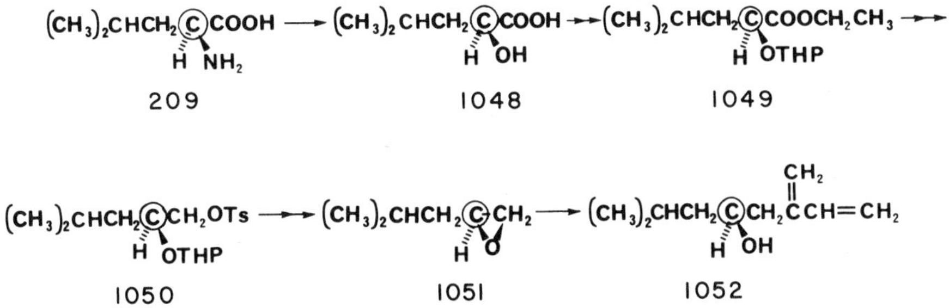

Scheme 71. Ipsenol.

(−)-α-pinene. The preparation of ipsenol was somewhat less obvious (*79*, *449*). Diazotization of (*S*)-leucine (**209**) took place, as usual for amino acids, with retention of configuration. The hydroxy acid **1048** thus obtained was protected as the tetrahydropyranyl ether **1049**, the reduction and tosylation of which gave **1050**. Treatment with acid removed the

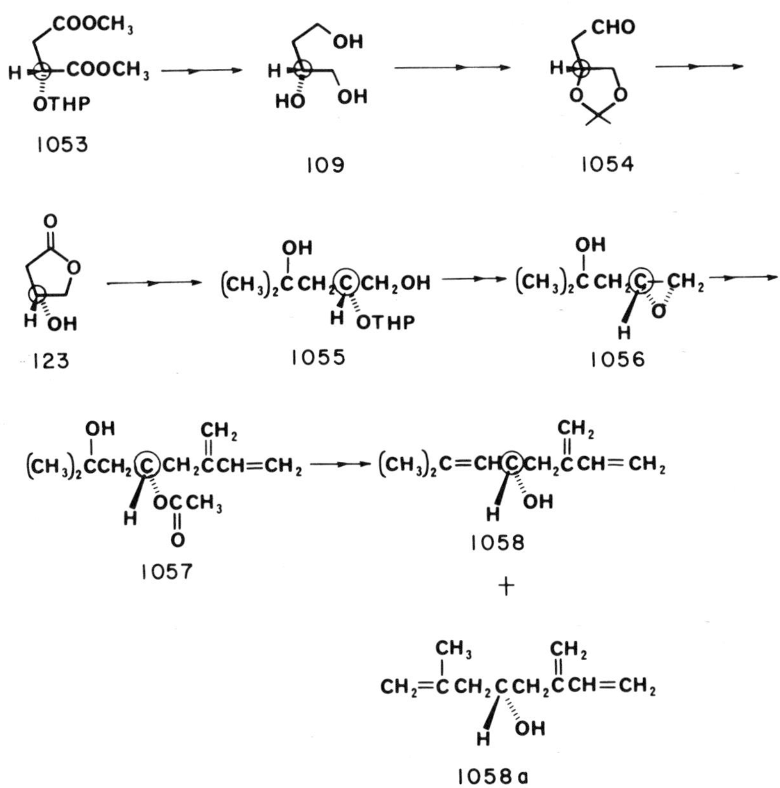

Scheme 72. Ipsdienol.

tetrahydropyranyl group and then basic ring closure gave epoxide **1051**. In the original work (*449*), an elaborate sequence of reactions was used to introduce the diene unit. Later (*79*), it was shown that reaction of **1051** with 2-(1,3-butadienyl)magnesium chloride (*450*) gave ipsenol in one step. Use of (*R*)-leucine gave the inactive *R* enantiomer of **1052**.

The extra degree of unsaturation in ipsdienol (**1058**) made the synthesis of this compound substantially more difficult than that of ipsenol. The first two preparations of this material were mitigated by the use of verbenol of ~80% optical purity (*451*) in one case and by a Wittig reaction on (*R*)-glyceraldehyde acetonide that occurred with substantial racemization (*452*) in the other case. The most satisfactory synthesis of (*S*)-ipsdienol (Scheme 72) is that of Mori (*79*), beginning with (*R*)-malic acid. The protected diester **1053** of this compound was reduced and deprotected to triol **109**. Reprotection as the acetonide and oxidation gave aldehyde **1054**. Further oxidation to the acid level and hydrolysis of the acetonide, which was accompanied by lactonization, produced **123**. Protection of the alcohol and reaction with excess methylmagnesium iodide led to **1055**. The epoxide **1056** was then generated by ether cleavage, selective tosylation, and treatment with base. Reaction, as in the ipsenol synthesis, with 2-(1,3-butadienyl)magnesium chloride (*450*) gave, after acetylation of the product, acetate **1057**. Elimination of the tertiary alcohol with phosphorus oxychloride gave a mixture of olefins. After reductive removal of the acetate groups, separation of ipsdienol (**1058**) and its isomer **1058a** was possible.

4. Pheromones Containing Chiral Secondary Alcohols as Lactones

In some pheromones, the chiral secondary alcohol is present as part of a γ-lactone. Three examples, all of which have been synthesized from glutamic acid, are shown in Scheme 73. The (*R*)-lactone **1031**, obtained (*453*) by diazotization of (*R*)-glutamic acid, could be cleanly reduced with borane methylsulfide to alcohol **197**. Displacement of the derived tosylate with the appropriate lithium dialkylcuprate was also clean, giving (*S*)-γ-caprolactone (**1059**), a dermistid beetle pheromone, and (*Z*)-6-dodecen-4-olide (**1060**), which is not an insect pheromone but rather the "social scent" of the black-tailed deer (*453*). An alternative use of acid **1031** involved Rosenmund reduction of the corresponding acid chloride to give aldehyde **1061**. Wittig reaction then gave the Japanese beetle pheromone **1062** as a 9:1 mixture with its *E* isomer (*454*).

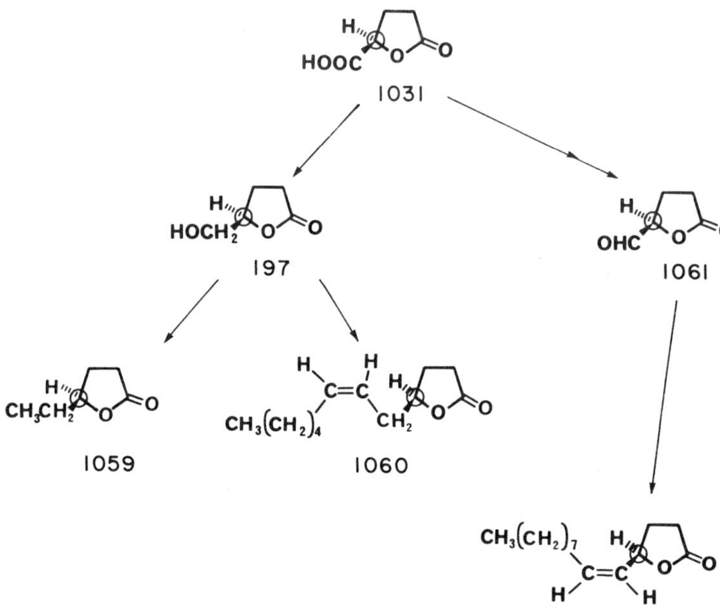

Scheme 73. Lactone pheromones.

The enantiomers of **1059**, **1060**, and **1062** were also prepared using (*S*)-glutamic acid as starting material (*454*).

5. Pheromones Containing Chiral Secondary Alcohols as Ketals

A structurally very interesting group of insect pheromones contains the chiral secondary alcohol as a ketal. There are two types of such compounds: those in which the alcohol is isolated and those in which the alcohol is part of a diol. One example of the former is chalcogran, the aggregation pheromone of a beetle infecting Norway spruce.

Chalcogran has the gross structure 2-ethyl-1,6-dioxaspiro[4.4]nonane. Since the natural product is a mixture at the ketal center, chiral syntheses of this compound have focused on preparation of the equivalent of (7*R*)- or (7*S*)-1,7-dihydroxynonan-4-one, which was then internally dehydrated to chalcogran. Four separate syntheses of the *R* form of chalcogran (**1065** and **1066**) have been reported, each using a different chiral starting material. The first synthesis (Scheme 74), done by Silverstein (*455*), is interest-

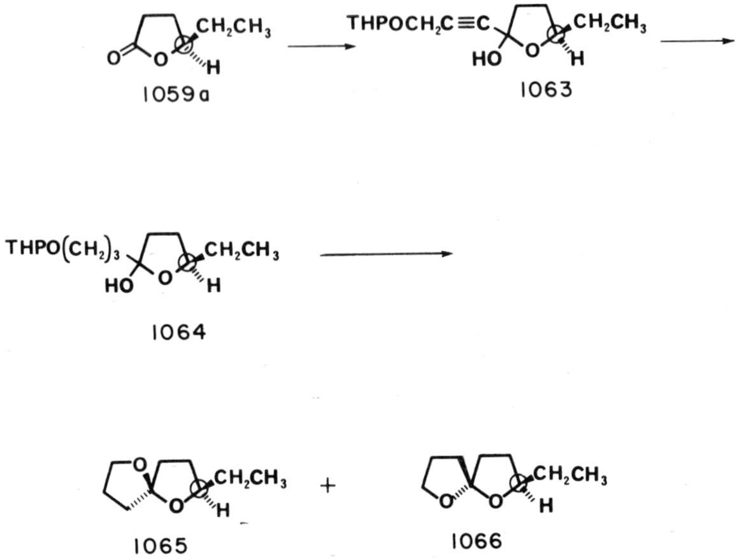

Scheme 74. Chalcogran synthesis I.

ing in that it used one synthetic pheromone to prepare another. (*R*)-γ-Caprolactone (**1059a**) was treated with the lithium derivative of propargyl alcohol tetrahydropyranyl ether to give hemiacetal **1063**. The acetylenic linkage was removed catalytically to give **1064**, which on treatment with acid was converted to chalcogran.

The second approach (*83*) to chalcogran (Scheme 75) began with diazotization of (*R*)-α-aminobutyric acid (**93**). Two-step protection of the product gave ester **1067**, which was reduced and deprotected to give diol **107**. A method somewhat different from usual (and unfortunately, of relatively low yield) was used to convert the diol to *R* epoxide **118**. This method involved treatment of **107** with hydrogen bromide in acetic acid and then base-catalyzed hydrolysis and ring closure of the 92 : 8 mixture of **1068** and **1069**. Terminal alkylation of the dianion of α-acetyl-γ-butyrolactone with epoxide **118** gave, after acidification, the dehydrated lactone **1070** (or its double-bond isomer). Basic hydrolysis and then acidification provided the necessary decarboxylation and ring closure to the usual chalcogran mixture.

The third synthesis (*456, 457*) of chalcogran (Scheme 76) was of an entirely different sort, involving destruction of three of the four chiral centers of D-glucose. For this purpose, the dideoxyacetal **1071** (*458*) was treated with 1,3-propanedithiol and a Lewis acid catalyst to give triol **1072**, which then was protected as acetonide **1073**. Removal of the termi-

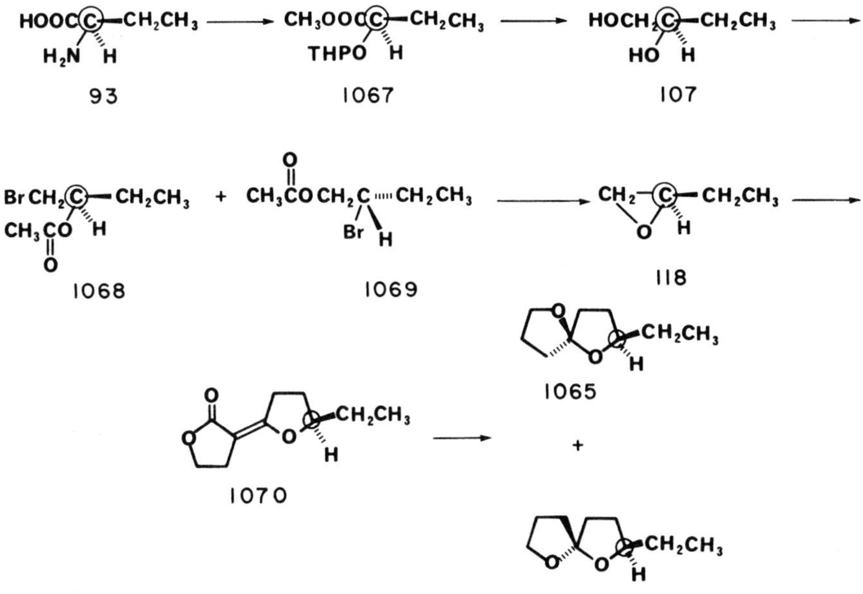

Scheme 75. Chalcogran synthesis II.

nal oxygen by hydride reduction of the tosylate gave **1074**. Dithiane anion formation with butyllithium led to elimination of acetone to give the homoallylic alcohol **1075**, which was reduced with lithium aluminum hydride to **1076**. After protection of the alcohol, alkylation with 3-bromopropanol tetrahydropyranyl ether gave **1077**, a masked form of chalcogran. Removal of the protecting groups then produced the pheromone mixture of **1065** and **1066**.

The fourth synthesis (*75*) of chalcogran (Scheme 77) is similar to the third in its final stages, involving dithiane intermediates. However, the approach to these final intermediates used by Seebach was entirely different. Thus, the O-protected (*S*)-malic acid dimethyl ester **1078** was reduced with hydride to diol **1079** and subsequently converted to dibromide **1080**. Alcohol deprotection and basic epoxide formation gave the key four-carbon unit **104**. Alkylation of the appropriate dithiane with this unit gave **1081**. Epoxide opening with lithium dimethylcuprate provided, in **1082**, the requisite carbon framework. The usual deprotection and acidic cyclization completed the chalcogran synthesis.

Seebach used the same methodology and many of the same intermediates to prepare the wasp pheromones **1083** and **1084** (*75*). A synthesis of the *S* enantiomers of these compounds and of the isomeric **1085** and **1086**,

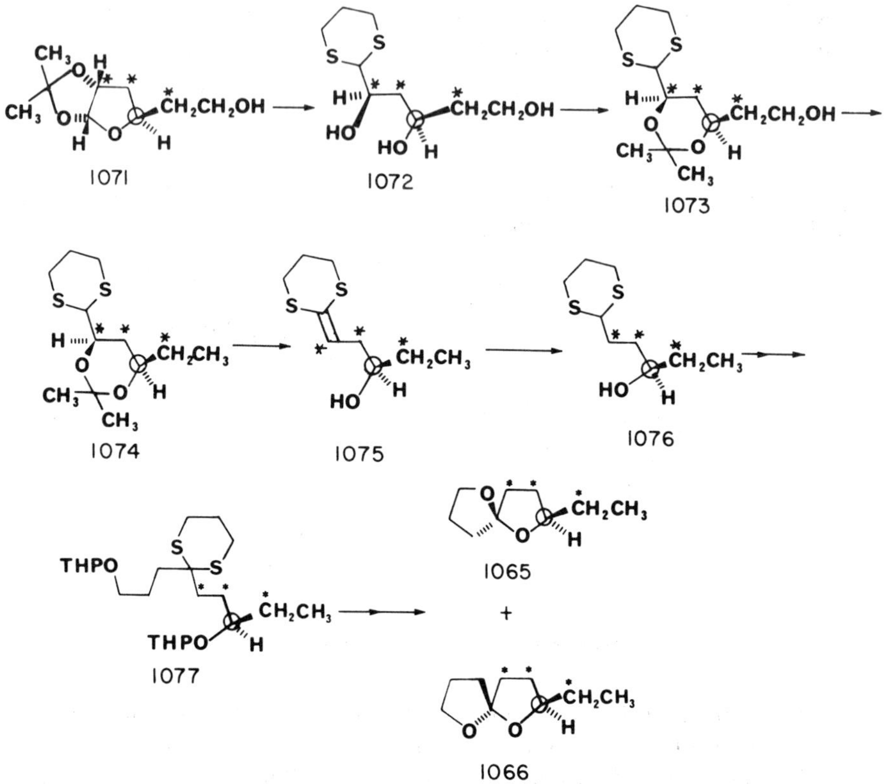

Scheme 76. Chalcogran synthesis III.

also wasp pheromones, was reported (*459*). (*S*)-Lactic acid was the chiral starting material in this work.

The second type of ketal pheromone involves a 1,2-diol unit. One such compound is α-multistriatin (**1094**, Scheme 78), along with (3*S*,4*S*)-4-methylheptan-3-ol (**1023**, Scheme 66) aggregation pheromones of the smaller European elm bark beetle. The first synthesis (*460*) from a chiral precursor involved several reactions of (*R*)-glyceraldehyde acetonide and derivatives that might have occurred with partial racemization. By using chiral NMR shift reagents, Mori could find no evidence for an enantiomer in the product and concluded that his material was enantiomerically pure. However, this NMR analysis was not done on racemic material as a control, and it was later shown (*461*), and acknowledged by Mori (*91*), that this synthesis was not enantiospecific. A second synthesis of α-multistriatin by Kocienski (*432*) also produced material of low rotation. In addition, both the Mori and Kocienski syntheses required separation of

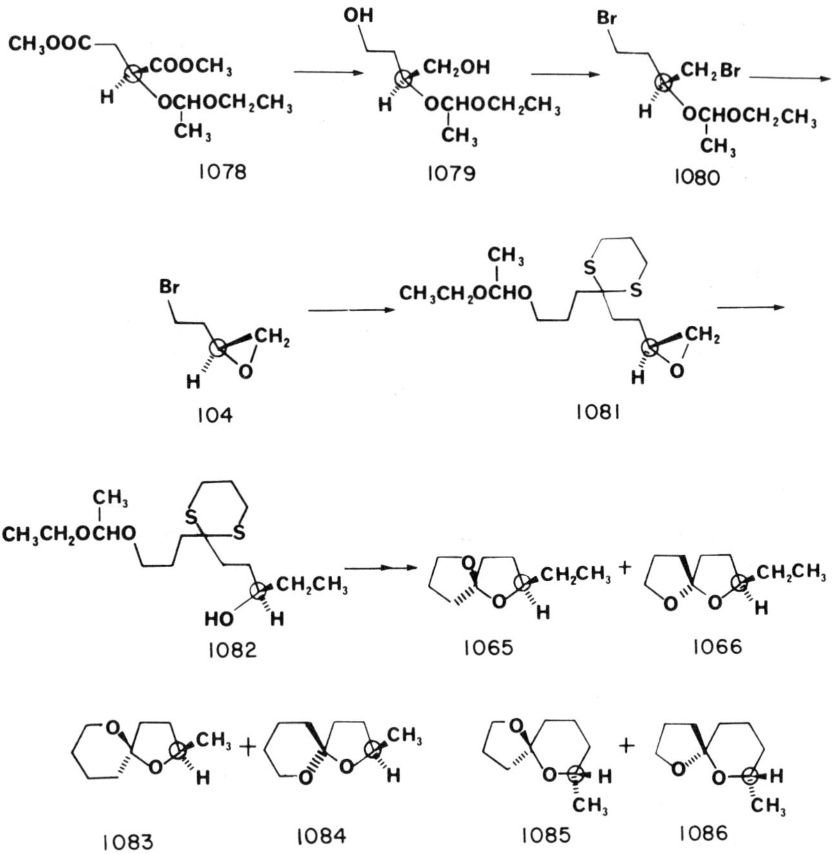

Scheme 77. Chalcogran synthesis IV.

the final product from three diastereomers by preparative gas chromatography.

There are two syntheses of α-multistriatin of preparative value in that they give enantiomerically pure material and, in addition, require no diastereomer separation. Both use D-glucose as starting material, incorporate the same carbons into the final product, and depend on formation of 1,3-diaxial methyl groups on a pyranose ring. Interestingly, the syntheses differ in the order of introduction of these methyl groups.

The first synthesis (Scheme 78), done by Weiler (*462*), involved initial reaction of the readily derived epoxide **523** with lithium dimethylcuprate to give alcohol **1087**. Treatment of the corresponding xanthate with tri-*n*-butylstannane led to deoxygenation, which was followed by selective acid hydrolysis to give **1088**. Introduction of the second methyl group was

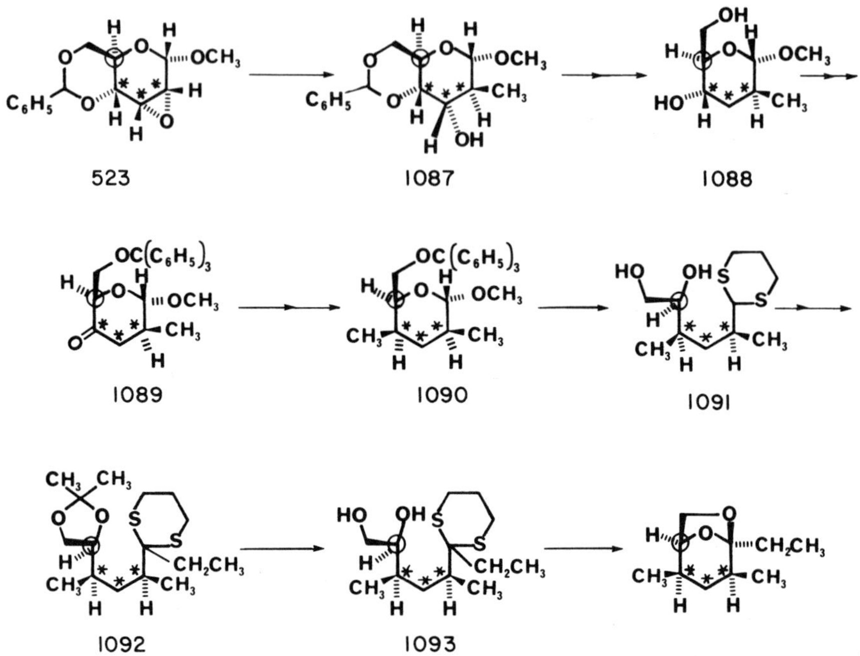

Scheme 78. α-Multistriatin synthesis I.

begun by selective protection of the primary alcohol and oxidation of the secondary alcohol to ketone **1089**. After Wittig methylenation, stereoselective hydrogenation over Wilkinson's catalyst produced **1090**, in which the methyl groups are 1,3-diaxial. At this point, trityl ether cleavage and ring opening was effected with 1,3-propanedithiol and boron trifluoride etherate. After protection of the diol unit of **1091** as an acetonide, the final two carbon atoms were introduced by dithiane anion alkylation, giving **1092**. Rather than simply heat **1092** with strong mineral acid, which would have led to α-multistriatin contaminated with the diastereomeric γ-multistriatin (due to partial epimerization α to the carbonyl before ketalization), Weiler chose a two-step method involving initial deketalization to **1093**. Heating this material in anhydrous acetonitrile with mercuric oxide and mercuric chloride cleanly converted **1093** to the desired α-multistriatin (**1094**), free of diastereomers.

In Fraser-Reid's (*463*) synthesis of α-multistriatin (Scheme 79), the protected form **1095** of D-glucose was hydrogenated to remove the benzylidene group. Tritylation of the primary alcohol and mesylation of the secondary alcohols gave **1096**. Elimination of the mesyloxy groups was

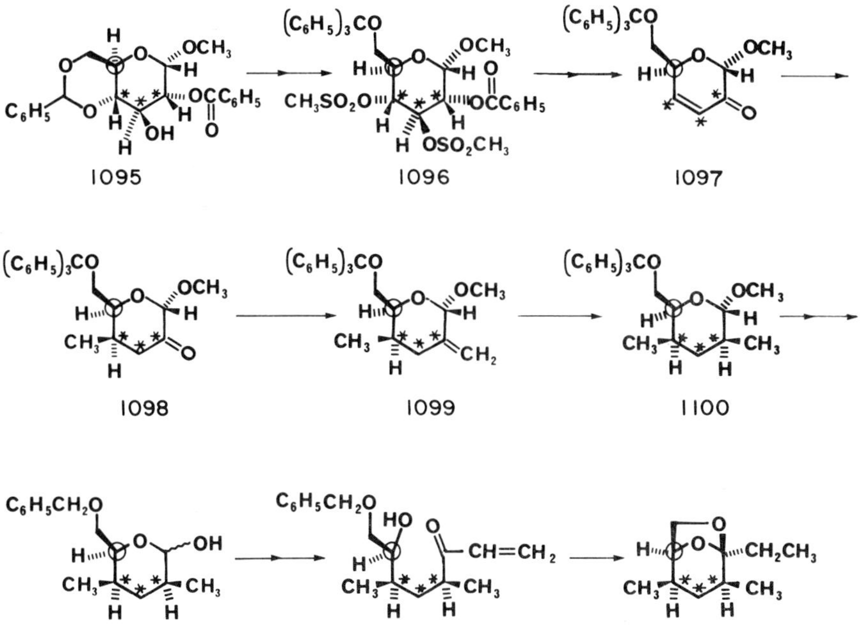

Scheme 79. *α*-Multistriatin synthesis II.

effected with potassium iodide and zinc–copper couple. The product of this reaction was hydrolyzed under basic conditions and oxidized to the enone **1097**. At this time, the first methyl group was introduced with proper (axial) stereochemistry by conjugate addition of lithium dimethylcuprate, giving **1098**. As in the Weiler synthesis, the second methyl group was prepared by hydrogenation of an *exo*-methylenepyranose, in this case **1099**. The hydrogenation was not completely specific, giving under the best conditions a 9 : 1 mixture of **1100** and its C-2 diastereomer. A change of the protecting group at C-6 and acetal hydrolysis to **1101** set the stage for the final conversion to *α*-multistriatin. Treatment of **1101** with vinylmagnesium bromide and manganese dioxide oxidation of the crude product gave the unsaturated ketone **1102**. Catalytic hydrogenation saturated the double bond and cleaved the benzyl ether. Under these conditions, the product spontaneously cyclized to *α*-multistriatin (**1094**).

While *α*-multistriatin is the pheromone of American populations of the smaller European elm bark beetle, (−)-δ-multistriatin (of unknown absolute configuration) apparently is the pheromone of European populations of this insect. Mori (*91*) devised a stereoselective synthesis of both enantiomers of this compound, starting from the appropriate tartaric acids.

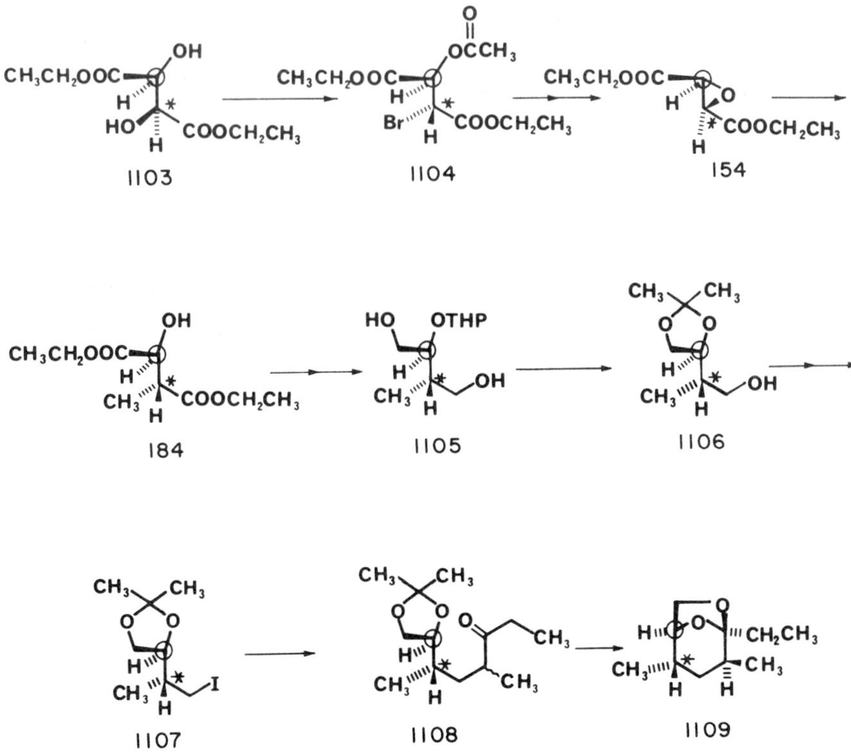

Scheme 80. δ-Multistriatin.

Preparation of the 1*S*,2*S*,4*S*,5*R* enantiomer **1109** is shown in Scheme 80. Treatment of diethyl (*S*,*S*)-tartrate with hydrogen bromide in acetic acid proceeded with inversion of configuration at one center to give **1104**. After acetate hydrolysis, base-catalyzed epoxidation with a second inversion gave the *S*,*S* epoxide **154**. Epoxide opening with lithium dimethylcuprate (third inversion) then gave the alcohol **184**. After the tetrahydropyranyl ether had been formed (for solubility reasons), hydride reduction produced the partially masked triol **1105**, which was easily transformed into acetonide **1106**. Conversion of **1106** via its tosylate to iodide **1107** was followed by alkylation of 3-pentanone with this material. The alkylation product **1108** was a mixture of diastereomers. This proved to be of no consequence in this case, since acidic deketalization and ring closure provided almost exclusively δ-multistriatin (**1109**). This stereochemical outcome is a consequence of equilibration α to the carbonyl and that the other diastereomer (β-multistriatin) is not formed because of an undesirable 1,3-diaxial methyl–methyl interaction.

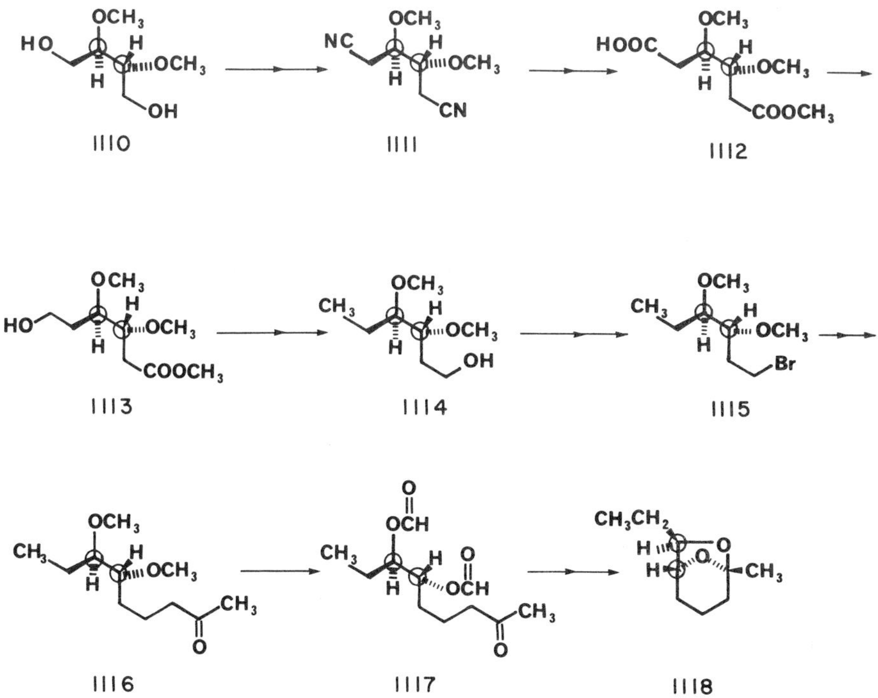

Scheme 81. (1*R*,7*R*)-*exo*-Brevicomin synthesis I.

A second ketal pheromone of the diol type is *exo*-brevicomin, an attractant produced by the western pine beetle. Syntheses of both enantiomers of this compound, from the appropriate tartaric acid, have been reported by Mori (*464*). Scheme 81 delineates the preparation of the biologically active 1*R*,7*R* enantiomer **1118** from (*S*,*S*)-tartaric acid. Thus, hydride reduction of (*S*,*S*)-diethyl tartrate dimethyl ether gave the diol **1110**, which was converted via its ditosylate to the bisnitrile **1111**. Nitrile methanolysis was followed by selective hydrolysis to **1112**, by which process the ends of the molecule were differentiated. The acid function was reduced with diborane to give alcohol **1113**. After this alcohol had been converted to a tosylate, lithium aluminum hydride reduction simultaneously cleaved the tosylate and reduced the ester. The resulting alcohol **1114** was converted in two steps to bromide **1115**. Three-carbon chain elongation by alkylation of ethyl acetoacetate and then saponification and decarboxylation, gave ketone **1116**. At this point, cleavage of the methyl ethers was required. This was a difficult reaction to carry out, but was finally effected in low yield by oxidation to the bisformate **1117** with chromium trioxide. Hydrolysis to the diol and ketalization with acid gave

1118 ($[\alpha]_D^{26}$ +84.1°). Use of (*R*,*R*)-tartaric acid gave the enantiomer ($[\alpha]_D^{24}$ −80.0°).

A second synthesis (*465*) of the unnatural, inactive 1*S*,7*S* material also employed (*R*,*R*)-tartaric acid as starting material. In this work, the overall yield was much higher, but since the product had $[\alpha]_D^{24}$ −67.5°, it was felt (*16, 465*) that inadvertent partial racemization had occurred somewhere in the synthesis. It may be, however, that this is not the case. A third brevicomin synthesis (*466*), in itself of no particular interest since it involved a major diastereomer separation, gave materials of $[\alpha]_D$ +70° and −66°. One must then consider that the rotations reported (*464*) by Mori are not those of pure brevicomins. Since the yield in the **1116**→**1118** set of conversions was only 11%, the possibility that the final products were contaminated by highly rotatory impurities must be acknowledged.

The matter of brevicomin rotations might be considered at an inconclusive rest at this point were it not for a fourth synthesis (*467*) by Fraser-Reid of the 1*R*,7*R* enantiomer of $[\alpha]_D^{23}$ +81.5° (Scheme 82). In this preparation, diacetone glucose (**302**) was benzylated and selectively deprotected to **1119**. After dimesylation (to **1120**), reductive elimination with

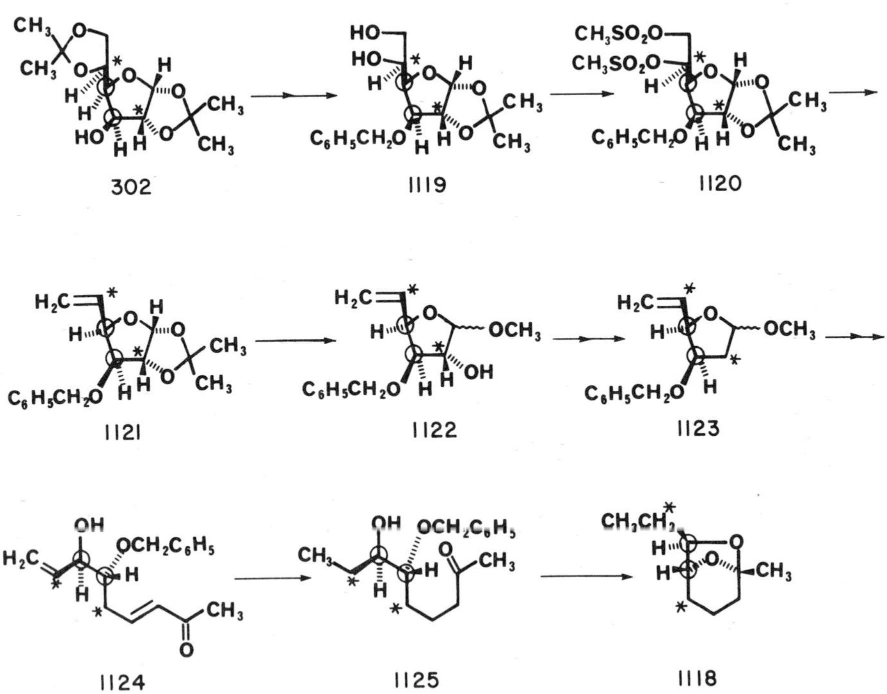

Scheme 82. (1*R*,7*R*)-*exo*-Brevicomin synthesis II.

sodium iodide gave olefin **1121**, thus removing one of the two excess chiral centers of glucose. To remove the second center (at C-2), **1121** was first reprotected as the anomeric mixture **1122**. Reduction of the derived xanthate with tri-*n*-butyltin hydride effected the desired deoxygenation, giving **1123**. At this point, the masked aldehyde was liberated. Reaction with 1-(triphenylphosphoranylidene)-2-propanone provided the unstable enone **1124**. It was found that hydrogenation of this compound in two stages was most efficient. After saturation of the double bonds over Raney nickel, purification of **1125** was carried out. Subsequent hydrogenation over palladium, removing the benzyl group, was accompanied by spontaneous cyclization to *exo*-brevicomin.

With the information at hand, it is not possible to evaluate these four syntheses of *exo*-brevicomin for their ability to provide enantiomerically pure products. Once again, the need for independent methods of determining enantiomeric composition is indicated.

Another ketal pheromone is frontalin, isolated from several species of pine bark beetles. This compound, while structurally similar to those whose synthesis has been described above, differs in that a tertiary rather than a secondary alcohol is part of the diol unit. Two groups have used approaches that are conceptually quite similar to prepare the biologically active 1*S*,5*R* enantiomer **1133** (Scheme 83) of this compound. The methods employed to prepare the tertiary alcohol center were different, however. Ohrui and Emoto (*468*) used the known (*469*) reaction of monomesylate **1126** (readily obtained from D-glucose) with methylmagnesium iodide, which gave alcohol **1127**. This interesting reaction may proceed by internal hydride transfer with concomitant mesylate elimination and ketone formation. The intermediate ketone then reacts, as expected, with Grignard reagent to give the observed product. Protecting-group manipulation converted **1127** to diol **1128**, which was subjected to oxidative cleavage, sodium borohydride reduction, and benzylation to provide **1129**. At this point, hydrolysis, reduction of the liberated aldehyde, and reprotection gave alcohol **1130**. Debenzylation and periodate oxidation gave **1131**, which was immediately condensed with the appropriate Wittig reagent to give ketone **1132**. Removal of the double bond by hydrogenation and treatment of the resulting material with acid converted **1132** to frontalin (**1133**).

The alternate approach to frontalin, as chosen by Fraser-Reid (*470*) involved treatment of the β-epoxide **1016** (*471*) with lithium aluminum hydride and oxidation of the resulting alcohol. Methylenation of the product ketone **1134** then gave **1135**. The correct tertiary alcohol configuration was established by oxymercuration–demercuration. After alcohol protection, ether **1136** could be isolated. Treatment of this compound with acetic

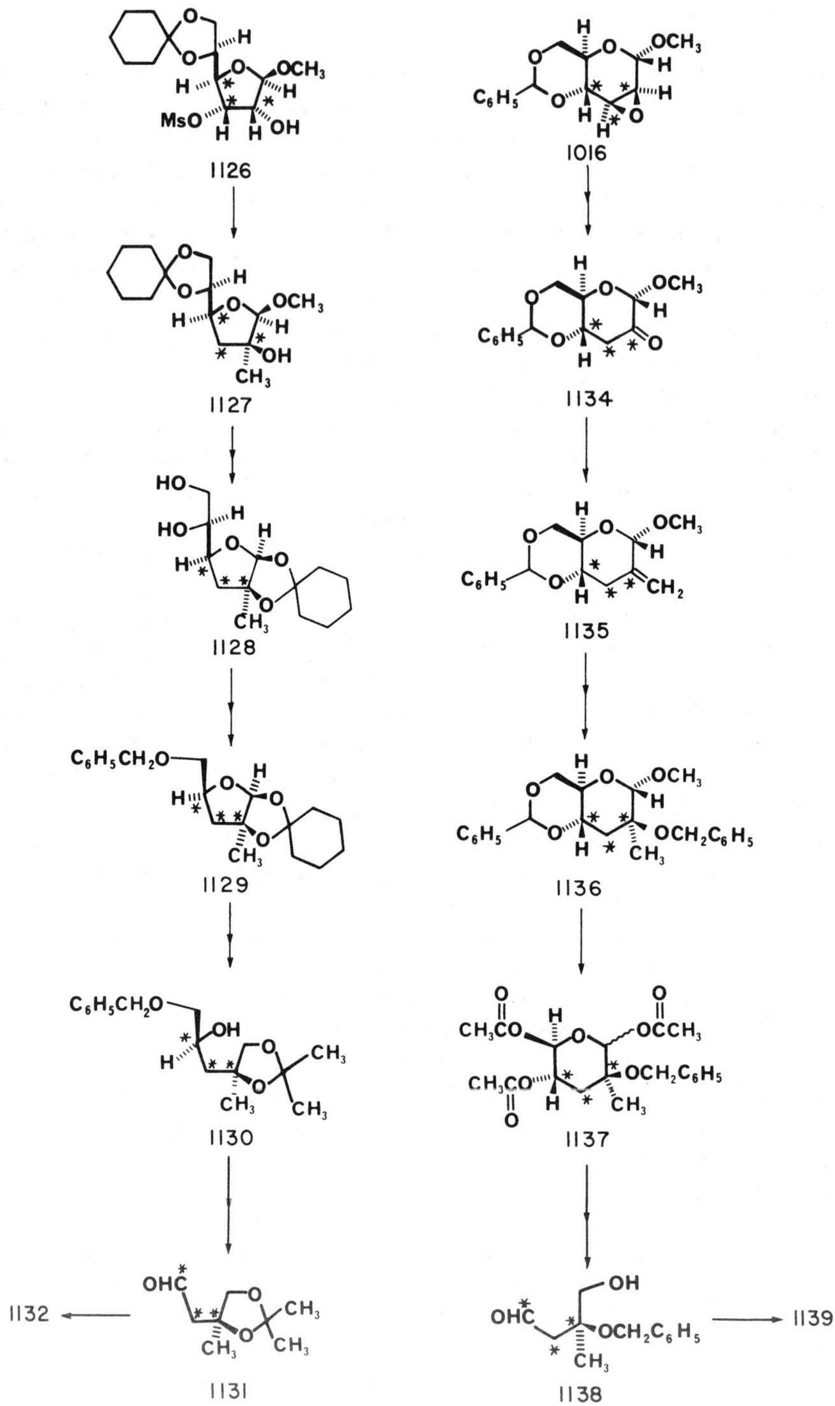

Scheme 83. (1*S*,5*R*)-Frontalin.

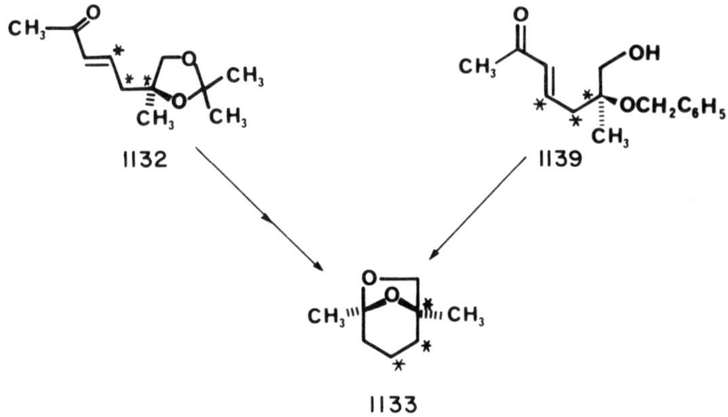

Scheme 83 (*Continued*)

anhydride and boron trifluoride etherate gave the triacetate **1137**, which was not purified. Reaction of this material with sodium borohydride (to cleave the esters and reduce the thus liberated aldehyde) followed by sodium periodate cleavage gave aldehyde **1138**. Chain lengthening, as for **1131**, produced **1139**. At this point, catalytic hydrogenation of this compound led directly to frontalin.

A short synthesis of the unnatural 1*R*,5*S* enantiomer of frontalin from (*R*)-linalool has been described (*472*).

The last ketal pheromone to be discussed, **1148** (Scheme 84), is derived from a 1,3-diol. The methodology employed in preparation of this compound was the same as used (in a slightly different form) to prepare chalcogran (Scheme 76). In the case at hand (*473*), the dichloride **1140**, obtained in good yield from methyl α-D-glucopyranoside by treatment with sulfuryl chloride and then sodium iodide, was reduced with tri-*n*-butyltin hydride to diol **1141**. The usual transacetalization with propane-1,3-dithiol gave **1142**, which was protected in two stages giving **1143**. Treatment of **1143** with butyllithium gave an anion that underwent elimination to **1144**. The olefin thus obtained was reduced by treatment with lithium aluminum hydride. Alkylation of **1145** with propylene oxide and then desulfurization produced the mixture **1146**. That this compound was a mixture was of no consequence, since the next step was oxidation to **1147**. The acidic conditions used to deketalize this compound also caused cyclization to **1148** to occur.

The absolute configuration of the naturally occurring pheromone is not known. As an aid in making this determination, the authors, by selective inversions of intermediates, also prepared the three other enantiomers (*474*).

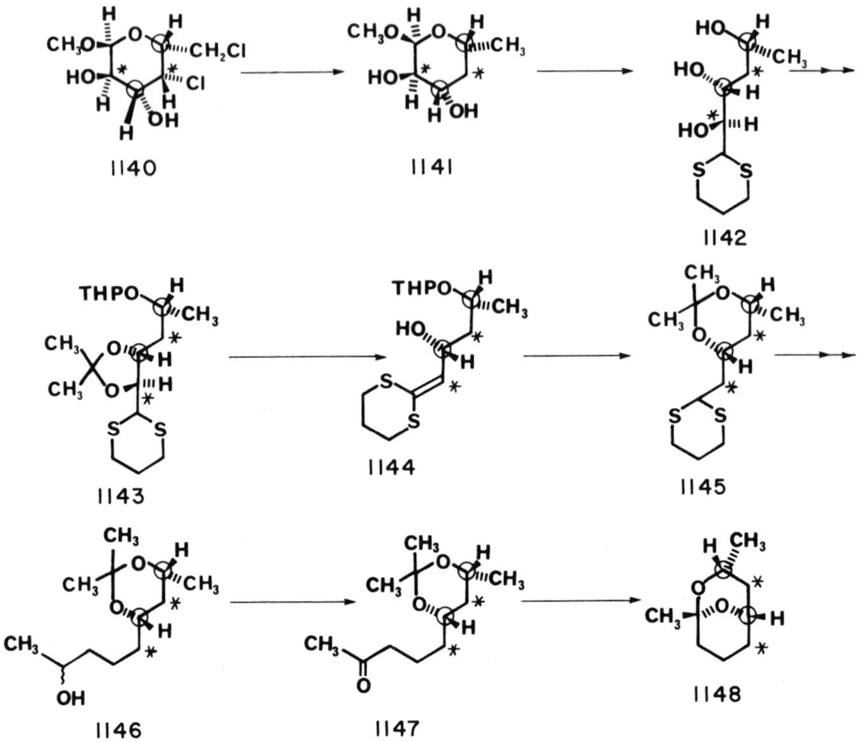

Scheme 84. 1,3-Diol/ketal pheromone.

6. Pheromones Containing Chiral Secondary Alcohols as Epoxides

There is only one pheromone of this type that has been the subject of synthesis from chiral precursors. This compound is dispalure, pheromone of the gypsy moth. Two chiral syntheses of the natural 7*R*,8*S* form **1153** of this compound have been recorded. The first of these (*475*) began with glutamic acid but required a tedious diastereomer separation. A better synthesis was developed by Mori (*476*) (Scheme 85). The alcohol **1113** (Scheme 81), an *exo*-brevicomin intermediate prepared from (*S*,*S*)-tartaric acid, was tosylated and then coupled with di(3-methylbutyl)lithium cuprate to give **1149**. Demethylation with boron trichloride gave directly the lactone **1150**. To introduce the second alkyl chain, **1150** was first converted to the tetrahydropyranyl ether and then partially reduced to lactol **1151**. Reaction of this masked aldehyde with the C-8 Wittig reagent and then saturation of the produced olefin gave **1152**. The epoxide ring of

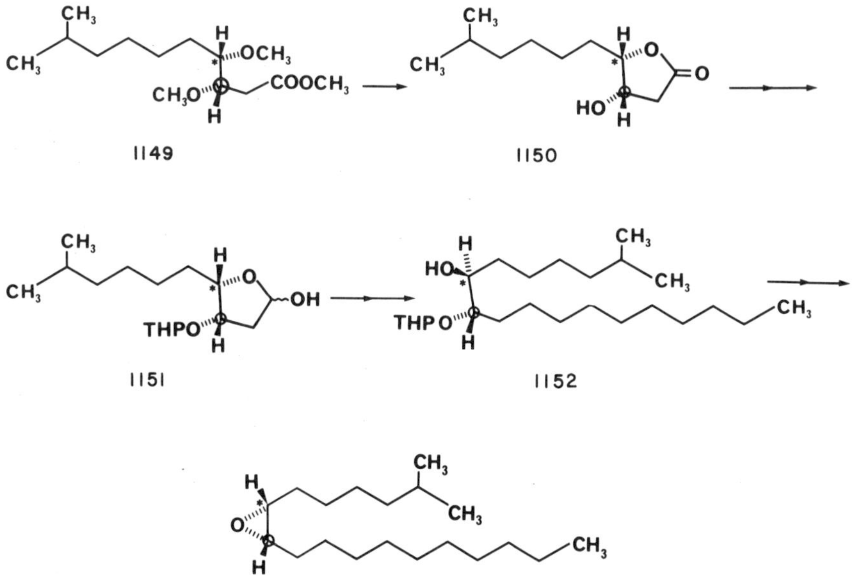

1153

Scheme 85. (7*R*,8*S*)-Dispalure.

1153 was then prepared by the standard sequence of reactions involving tosylation, acidic ether cleavage, and base-catalyzed ring closure.

7. Miscellaneous Insect Pheromones and Related Compounds

In this section are presented syntheses of insect pheromones and related compounds that do not fit into the general structural groups outlined above. Most of these syntheses require little comment. We therefore note the preparation (Scheme 86) of the juvenile hormones juvabione (**1154**) (*477*) and juvabiol (**1155**) (*478*) and the defensive secretions chrysomelidial (**1156**) and plagiolactone (**1157**) (*479*), all from (*R*)-limonene. (*S*)-Carvone was starting material in a preparation of defense secretion **1158** of a west African termite (*480*).

Pinenes were the logical precursors for the cyclobutyl pheromones grandisol **1159** (*481*) of the boll weevil and **1160** (*482*), of the citrus mealybug. (*R*,*R*)-Tartaric acid was converted to a mixture of diasteriomeric stegobinones (**1161**) (*483*), confirming the indicated 2*R*,3*R* configuration of the natural drugstore beetle pheromone. This same starting material also led to anthopleurine (**1162**) (*484*), a sea anenome alarm pheromone.

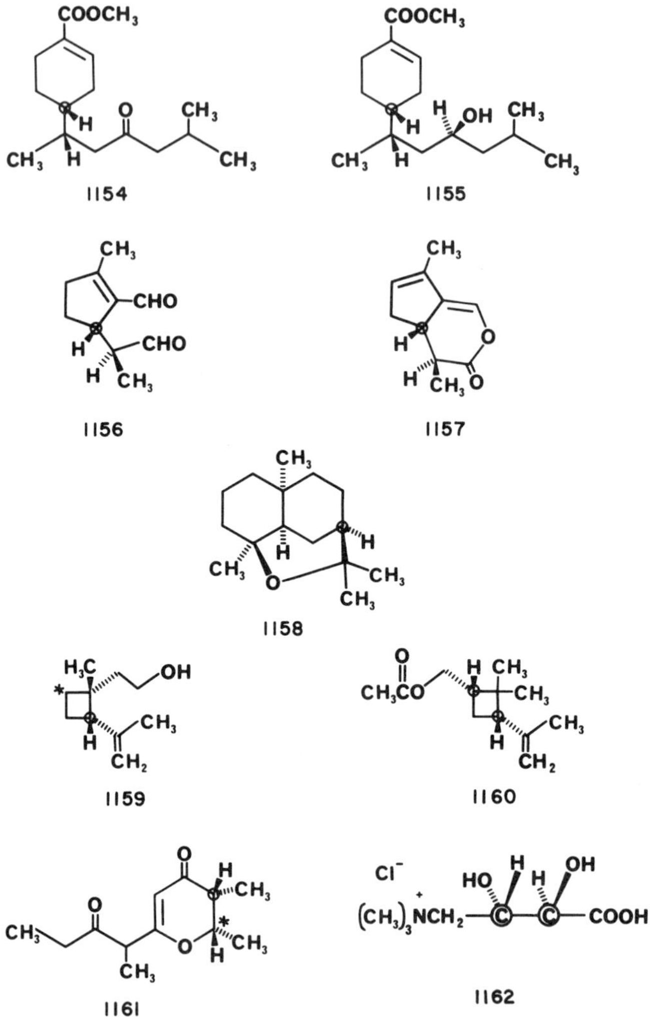

Scheme 86. Miscellaneous pheromones and related compounds.

A structurally unique pheromone (*485*) of a dermestid beetle is the dihydrofuran **1170** (Scheme 87). An unusual synthesis of this material from D-glucose was devised. The readily available ketone **1134** (Scheme 83) was treated with the anion of 2-methyl-1,3-dithiane. Hydrolysis of the product gave the methyl ketone **1163**. Conversion of this functionality to isopropyl was effected by Wittig methylenation and catalytic hydrogenation, which also removed the acetal. The resulting triol **1164** was selectively tosylated and then reduced with lithium aluminum hydride and oxidized to ketone **1165**. Having thus removed one of the excess oxygen

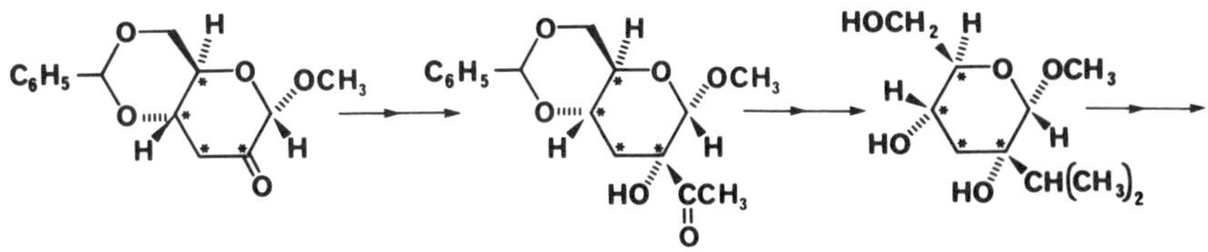

1134 1163 1164

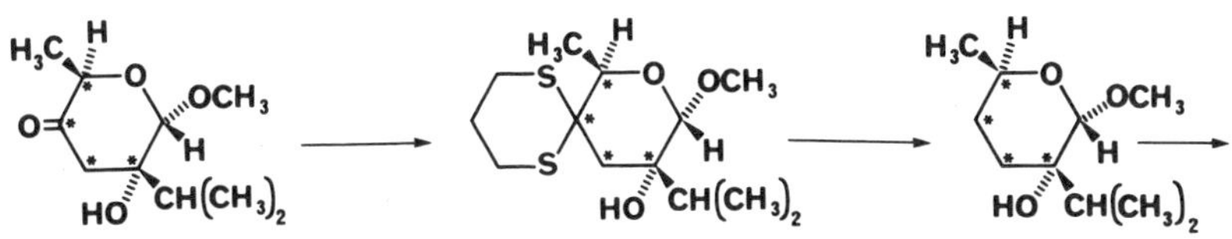

1165 1166 1167

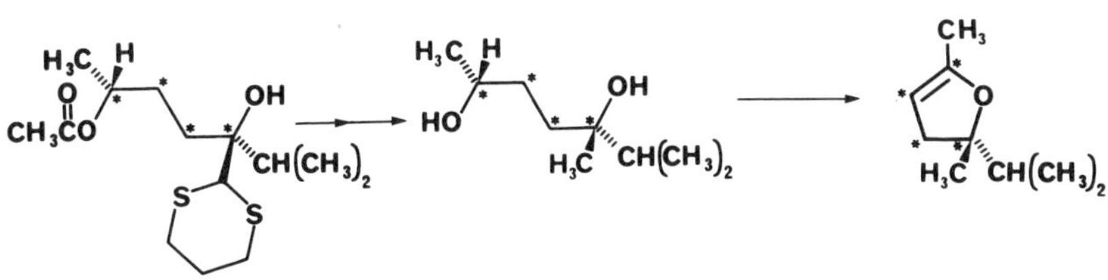

1168 1169 1170

Scheme 87. Dermestid beetle pheromone.

atoms, a second was eliminated by carefully controlled thioketalization (to **1166**) and desulfurization over Raney nickel. A third oxygen was excised by initial thioketalization of **1167** under more forcing conditions. Desulfurization of the resulting **1168** and cleavage of the acetate with methoxide provided diol **1169**. Oxidation of this material with chromium trioxide–pyridine complex in pentane gave a keto alcohol, which under these conditions cyclized with elimination of water, giving the pheromone **1170**.

F. Sugars

As is abundantly evident from the preceding discussion, sugars have found substantial favor as chiral synthetic starting materials. The synthesis of sugars themselves, which usually involves conversion of one sugar into another, is an art which has been practiced almost since the beginnings of organic chemistry. It would be far beyond the scope of this review to attempt to cover the monumental literature in this area.

Although the syntheses of most sugars involve modification of other sugars, there are a few examples in which this is not the case. In these efforts, amino acids have been employed, and the additional asymmetric centers have been created in stereospecific or stereoselective reactions. One good example is Yamada's synthesis (*486, 487*) of D-ribose from (*S*)-glutamic acid (Scheme 88). As in most other syntheses involving glutamic acid, the initial reaction was diazotization to the lactonic acid **1171**. Ester-

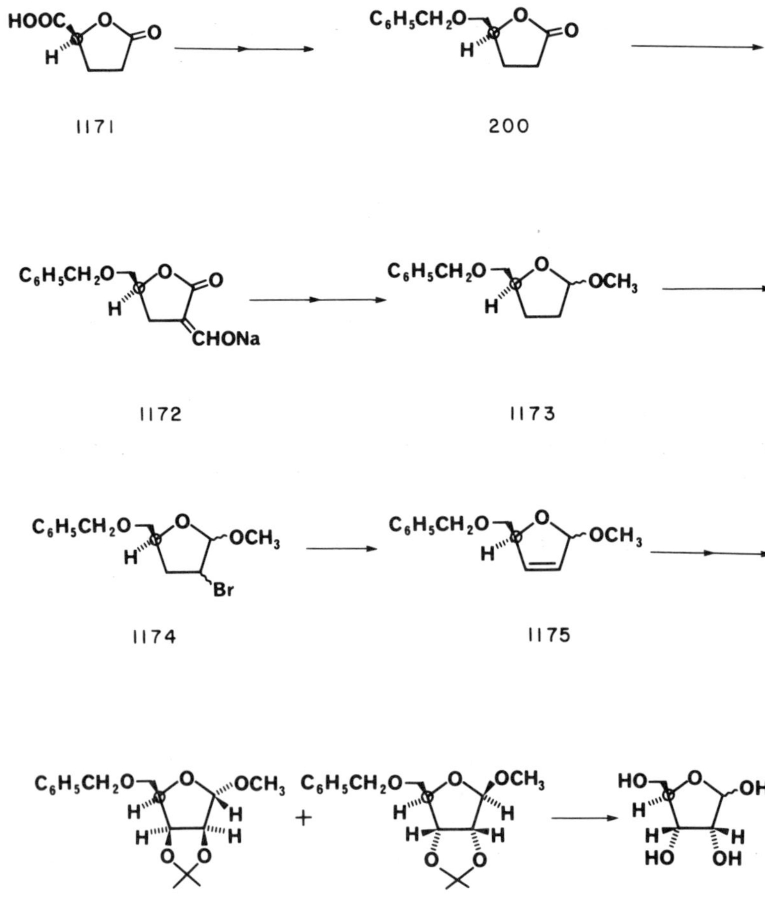

1171 200

1172 1173

1174 1175

1176 1177 235

Scheme 88. D-Ribose.

ification, selective reduction with sodium borohydride, and etherification gave the protected alcohol **200**. Formylation of **200** gave the adduct **1172**, which on treatment with acid underwent ring opening, decarboxylation, and ring reclosure. Protection of the unstable product as its methylfuranoside gave the anomeric mixture **1173**. To introduce the remaining two hydroxyl groups, **1173** was first brominated to **1174** and then dehydrobrominated to **1175**. Osmium tetroxide oxidation of this mixture was stereospecific (trans) with regard to the anomeric center. Treatment of the product diols with acetone gave a 1:2.7 mixture of acetonides of the undesired (**1176**) and desired (**1177**) configurations. Hydrogenolytic deprotection of the latter in the presence of acid led directly to D-ribose (**235**).

Many antibiotics are themselves amino sugars or contain these units as appendages. There has been some effort expended to devise chiral syntheses of these compounds from noncarbohydrate precursors. One such effort (*488*) has led to a synthesis of *N*-benzoyl-L-daunosamine (**1183**, Scheme 89). Diazotization of D-threonine, followed by esterification and ketalization, gave **1178**. Selective reduction at low temperature converted this compound directly to aldehyde **1179**. Reaction with allylmagnesium bromide was selective, giving a 4 : 1 mixture of **1180** and its diastereomer. The mixture was not separable at this stage. Rather, conversion to the tosylate and displacement with azide (with inversion) gave **1181**, also a mixture. Hydride reduction of the azide, deketalization, and benzoylation produced **1182**, which was obtained in diastereomerically pure form by crystallization. Ozonolysis with a reductive workup then completed the

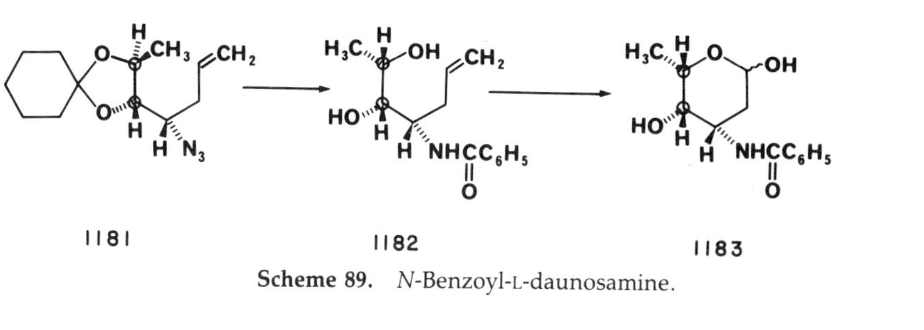

Scheme 89. *N*-Benzoyl-L-daunosamine.

synthesis of **1183**. Using modifications in the methodology, these authors also obtained (*489, 490*) a number of similar amino sugars.

A second synthesis of an amino sugar from amino acid precursors is that of methyl α-L-mycaminoside (**1190**, Scheme 90). This synthesis, done by Yamada (*491*), is a nice illustration of how stereocontrolled elaboration of one chiral center can yield a complex molecule. Thus, diazotization of (*S*)-alanine (**15**) and acetylation of the product gave **1184**. Coupling of the derived acid chloride with the Grignard reagent from propiolaldehyde dimethyl acetal gave the basic sugar C-6 unit **1185**. Semihydrogenation and deacylation produced alcohol **1186**, which on heating with phosphoric acid in carbon tetrachloride was transformed into pyranoside **1187** and its anomer (~2 : 1 ratio). Hydride reduction of isolated **1187** introduced the third chiral center. The produced alcohol **1188** was epoxidized

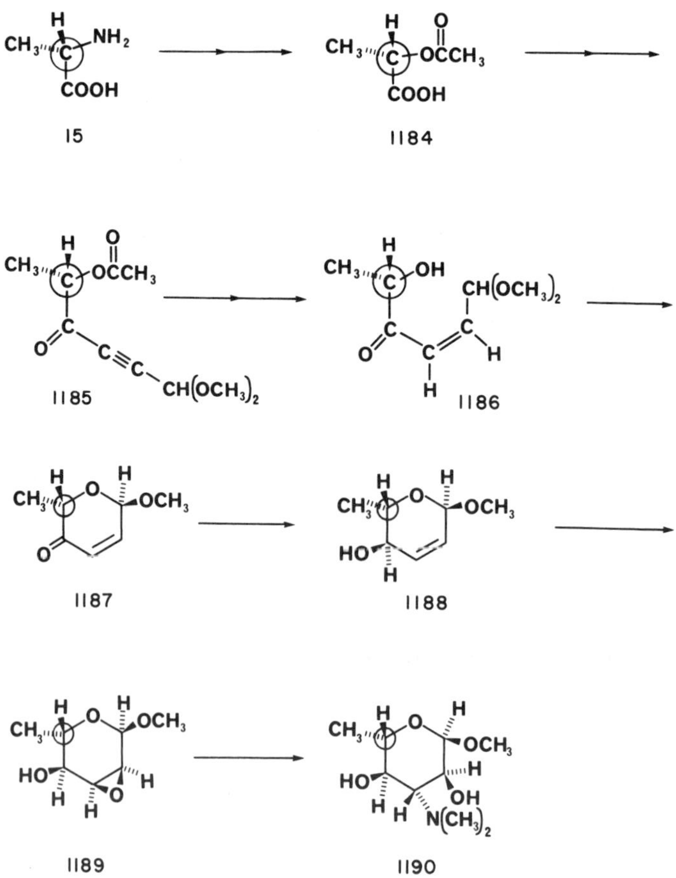

Scheme 90. Methyl α-L-mycaminoside.

to **1189** and then treated with aqueous dimethylamine to give the amino sugar **1190**. The last three reactions all appear to have been stereospecific.

A synthesis of 3-amino-2,3-dideoxytetrose derivatives from (*S*)-aspartic acid has been reported (*492*).

G. Vitamins

Of the 20 or so chemical substances loosely classified (*493*) as vitamins, 8 are chiral molecules: (*a*) vitamin B_2 (riboflavin), (*b*) vitamin B_5 (pantothenic acid), (*c*) vitamin B_{12} (cobalamin), (*d*) vitamin C (ascorbic acid), (*e*) vitamin D_3 (cholecalciferol), (*f*) vitamin E (α-tocopherol), (*g*) vitamin H (biotin), and (*h*) vitamin M or B_c (folic acid). The structures of these materials are shown in Scheme 91.

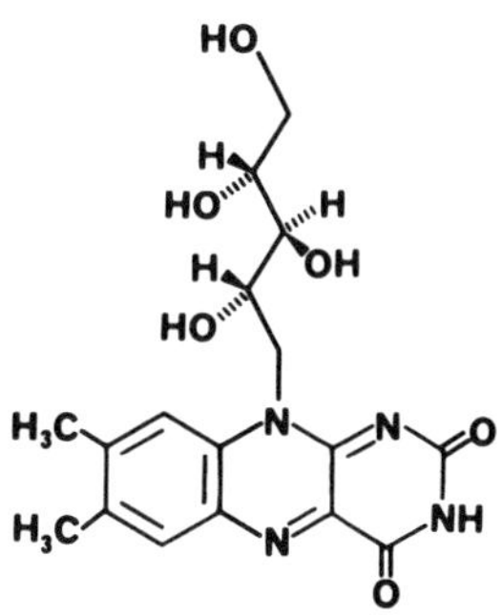

RIBOFLAVIN (1191)

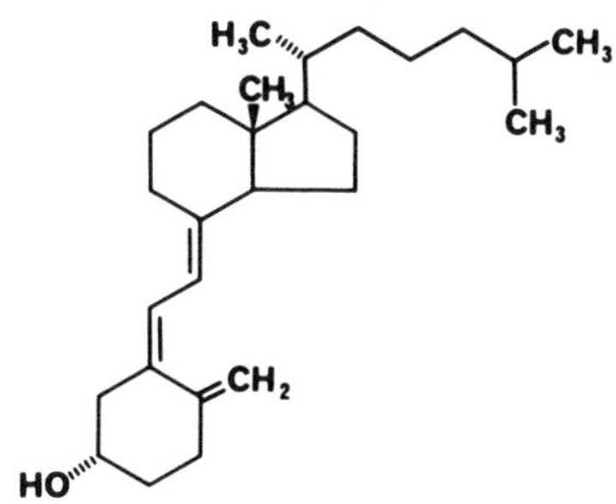

CHOLECALCIFEROL (1192)

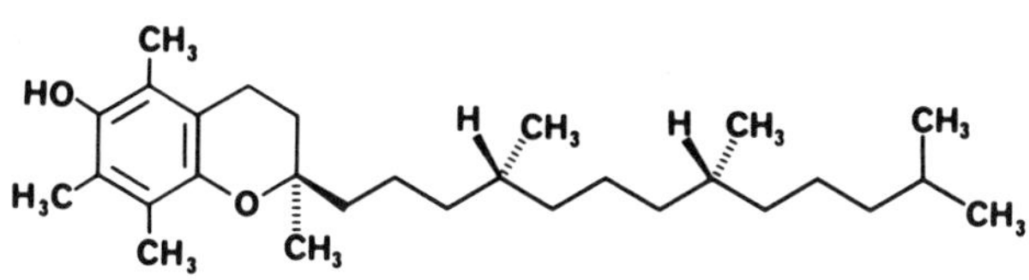

α - TOCOPHEROL (1193)

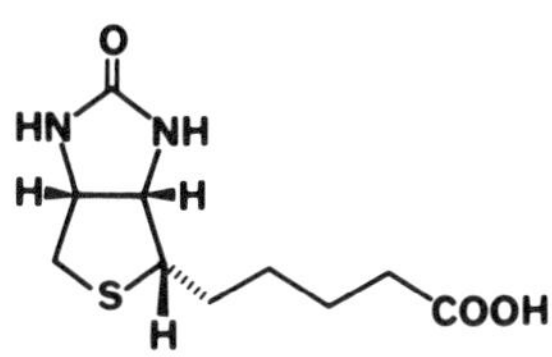

BIOTIN (1194)

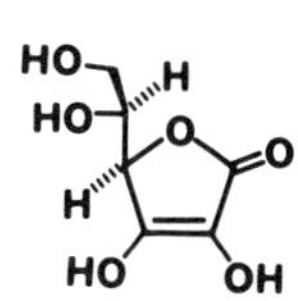

ASCORBIC ACID (1195)

Scheme 91. Chiral vitamins (continued on next page).

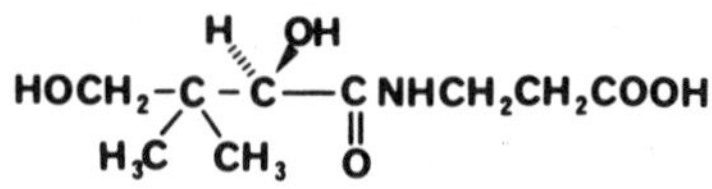

PANTOTHENIC ACID (1196)

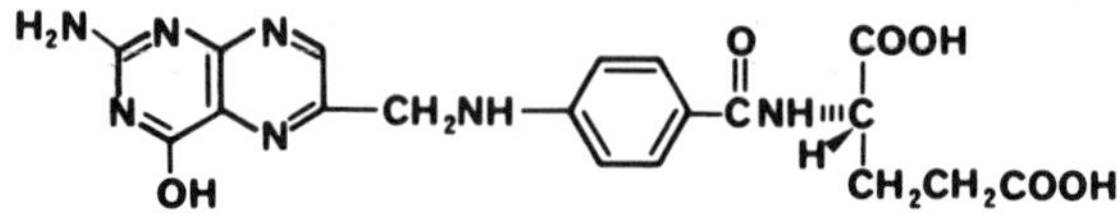

FOLIC ACID (1197)

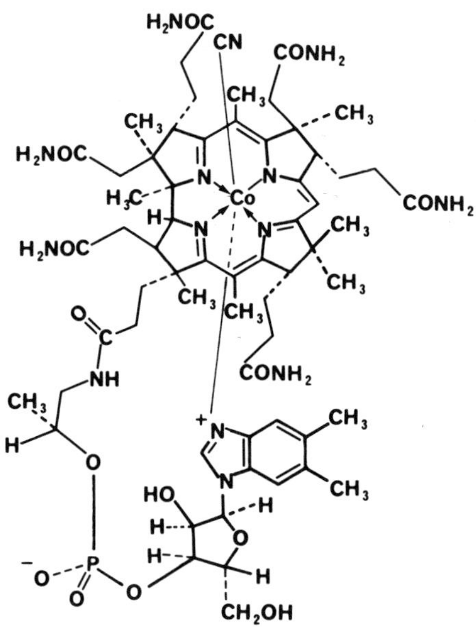

COBALAMIN (1198)

Scheme 91 (*Continued*)

The straightforward inclusion of D-ribose in riboflavin and of (*S*)-glutamic acid in folic acid syntheses as well as the classical preparation of ascorbic acid from D-glucose require no comment in this review. On the other hand, there have been a number of interesting reports in recent years of chiral syntheses of the other five vitamins and their derivatives. These syntheses will be described in the following sections.

1. Pantothenic Acid

Pantothenic acid (**1197**) is generally obtained by reaction of (*R*)-pantolactone with β-alanine. A synthesis of the chiral lactone, which in the past

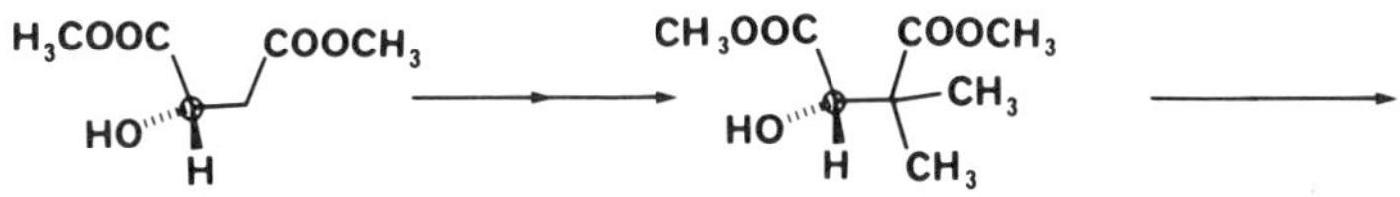

1199 1200

1201 1202

Scheme 92. (*S*)-Pantolactone.

has involved a resolution, thus constitutes a solution to the problem of preparing the vitamin in enantiomeric form. There has been a report (*494*) of the synthesis of the enantiomeric (*S*)-pantolactone (**1202**) from (*S*)-malic acid (*494*) (Scheme 92). This straightforward preparation involved sequential dimethylation of (*S*)-dimethyl malate (**1199**) as its dianion to give **1200**. A highly selective saponification gave the half acid **1201**, which was reduced with lithium triethylborohydride directly to **1202**.

For this synthesis to be of use in preparation of the natural vitamin, (*R*)-malic acid is required. While this compound is available (but expensive), it can also be obtained readily from (*R*,*R*)-tartaric acid (*4*).

2. Cobalamin

If one were asked at this time to pick the single greatest achievement in organic synthesis, it would be very difficult to rationalize a choice other than Eschenmoser and Woodward's synthesis of vitamin B_{12} (**1198**). This work has been amply reviewed (*495, 496*), not least by the chemists themselves (*497–500*), and consequently, the discussion here will be brief.

Due to previous work (*501*), a formal total synthesis of cobalamin could be effected by preparation of cobyric acid (**1220**, Scheme 93). The gross scheme for synthesis of this latter compound involved obtaining synthons corresponding to the four rings in enantiomerically pure form followed by joining of these units. While this project was truly a joint effort, its execution by the two groups varied somewhat in detail. The Woodward variation is described here since it used chiral starting materials for two rings: (−)-camphor for ring D and (+)-camphorquinone for ring C. The correct

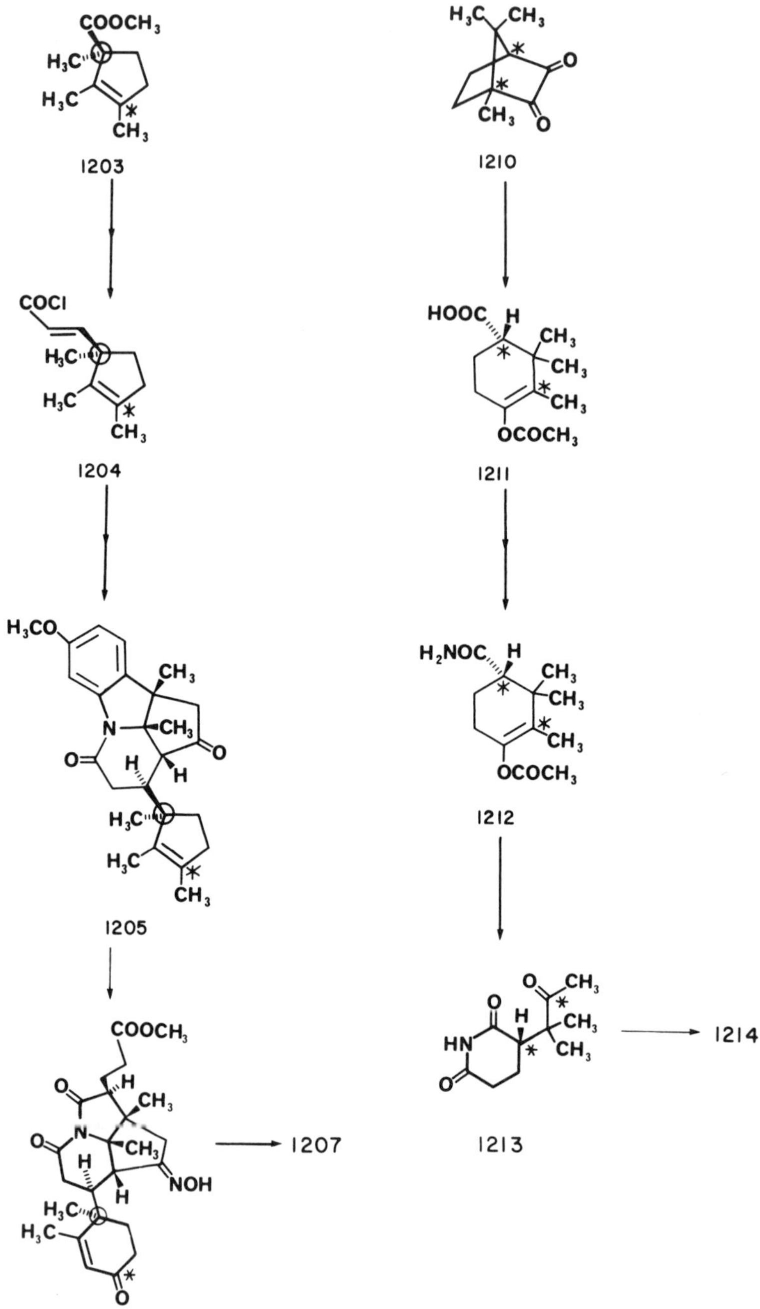

Scheme 93. Cobyric acid.

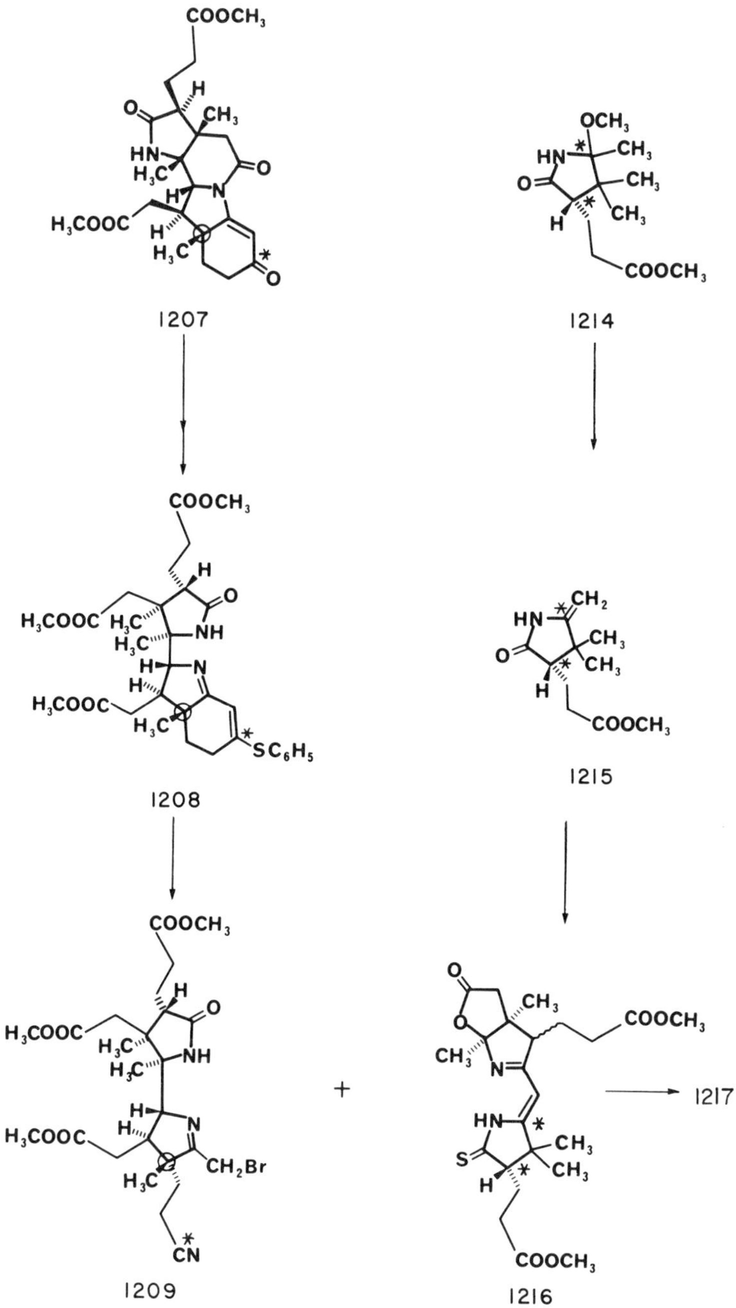

Scheme 93 (*Continued*)

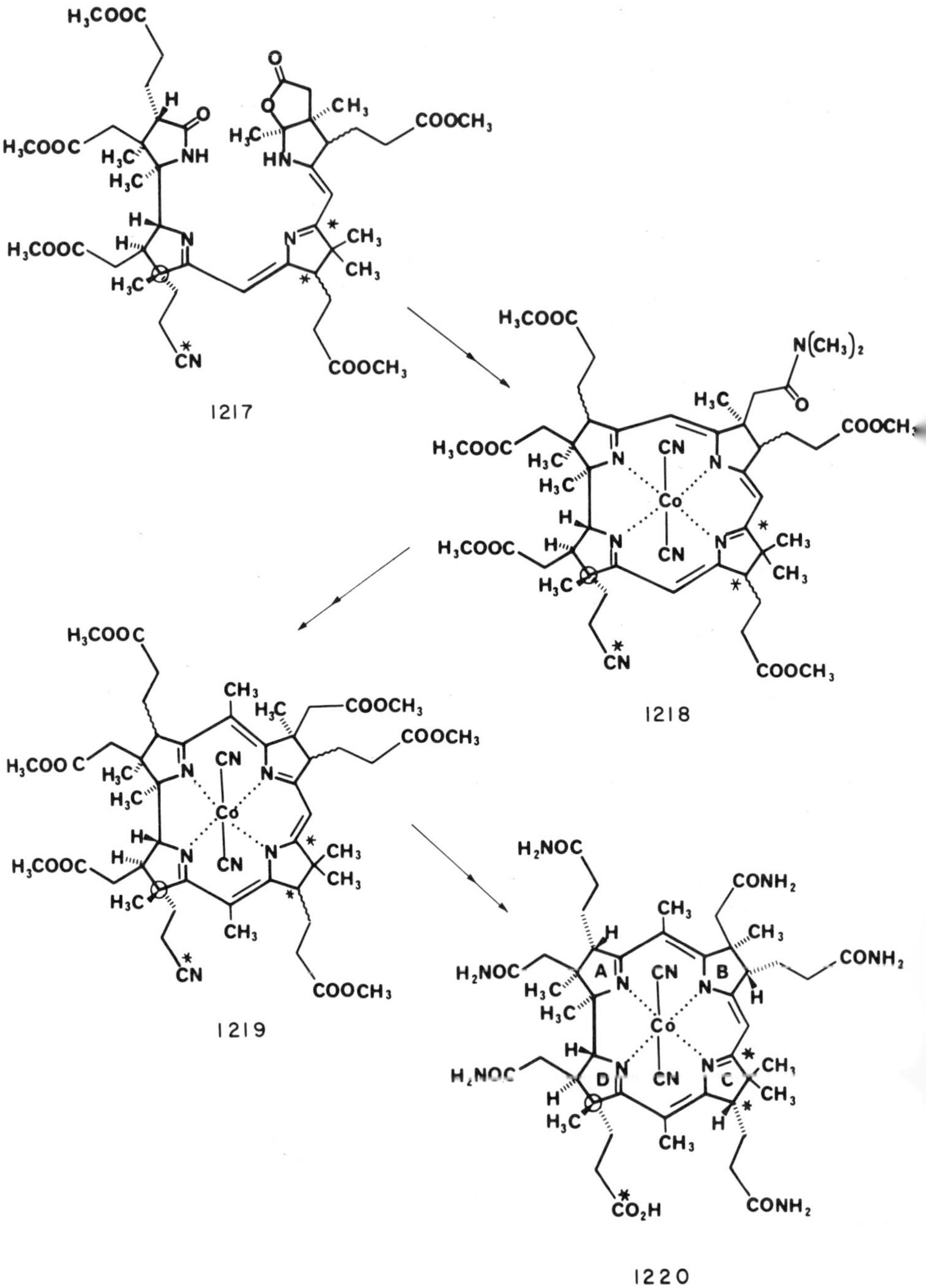

Scheme 93 (*Continued*)

enantiomers of the ring A and B synthons were obtained by resolution and thus are not described here.

The left-half portion (**1209**) of cobyric acid (rings A and D) was obtained by initial conversion, in six steps, of (−)-camphor to methyl laurolenate (**1203**). This compound was homologated to **1204**, which was used to acylate the ring-A precursor. After internal Michael addition, the pentacycle **1205** was obtained. A variety of transformations, including destruction of the aromatic ring and ring expansion of the cyclopentene, produced the enone **1206**. The next set of manipulations involved Beckman-type ring enlargement, opening of the caprolactam ring, oxidative opening of the cyclohexenone ring, and reclosure to the vinylogous amide **1207**. In a relatively straightforward manner, this compound was converted by ring opening to **1208**, which after ozonization and functional group modification gave the desired **1209**.

To prepare the right-half portion (**1216**) of cobyric acid (rings B and C), (+)-camphorquinone (**1210**) was opened with boron trifluoride etherate to the acid **1211**. Conversion to the amide **1212** was followed by ozonization and reductive workup to give the glutarimide **1213**. Treatment with acid gave the rearranged protected aminal **1214**, which was thermally converted to **1215**. Coupling to the ring-B segment (a mixture of diastereomers) gave the requisite **1216**.

The coupling of **1209** and **1216** was effected by a sulfide contraction process. Unfortunately, in this process the ring-C stereochemistry, so carefully nurtured from (+)-camphorquinone and brought to **1216**, was lost, and **1217** was obtained as a mixture of diastereomers at two centers. Introduction of the cobalt atom and ring closure were among the steps involved in converting **1217** to **1218**. In these steps, a third center was partially racemized. After introduction of the last two carbon atoms (**1219**), diastereomer separation and functional group modification was all that was required to complete the cobyric acid preparation.

It is impossible, in the limited space available, to do justice to this magnificent synthesis. Perhaps it is better just to sit back and reflect on the brilliance of the retrosynthetic analysis that led to two camphor derivatives as progenitors of cobyric acid.

3. Biotin

Biotin, as a synthetic target, presents a substantial challenge to the organic chemist. The three adjacent chiral centers on the tetrahydrothiophene ring, in the less stable cis configuration, make careful planning of the synthetic approach mandatory. The obvious relationship of the tetrahydrothiophene ring to the furanose form of sugars makes the latter

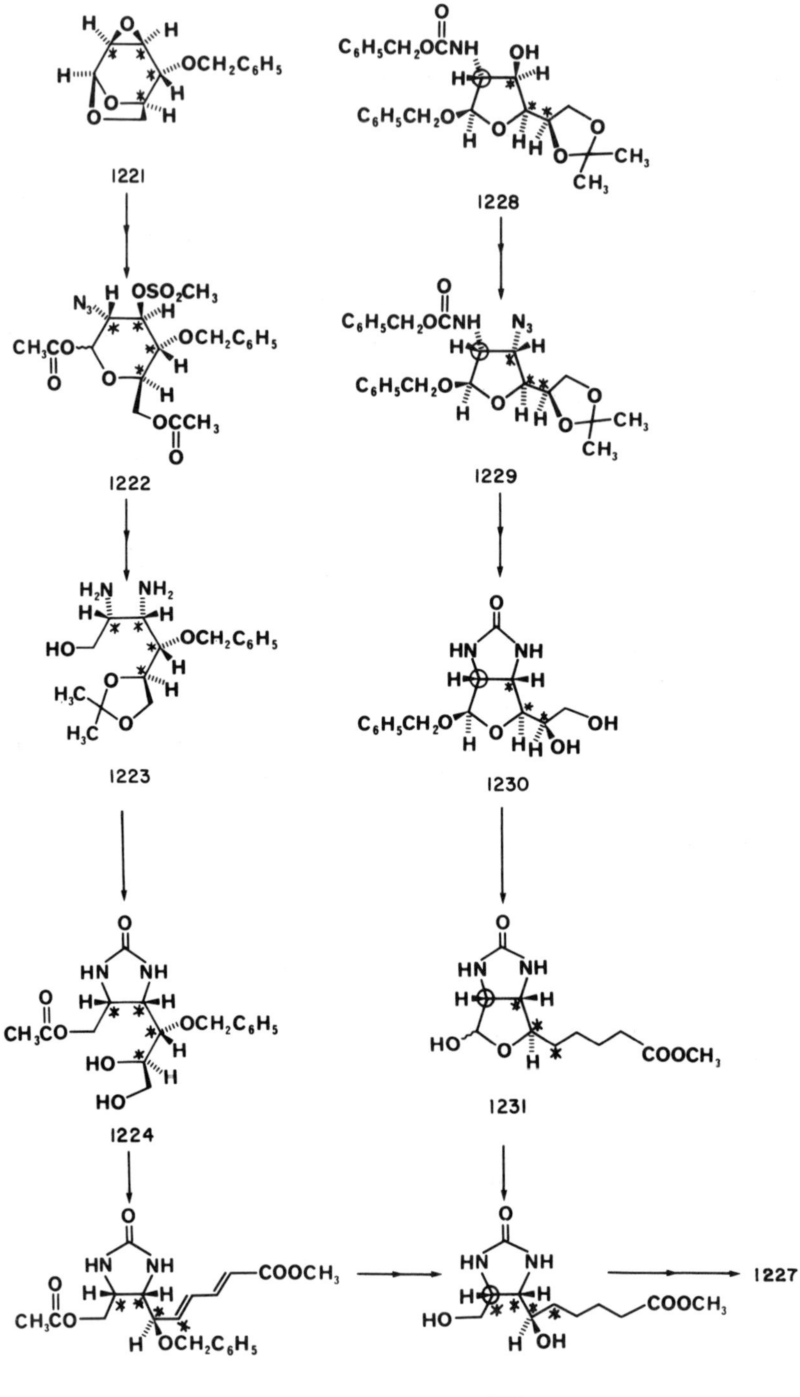

Scheme 94. Biotin syntheses I and II.

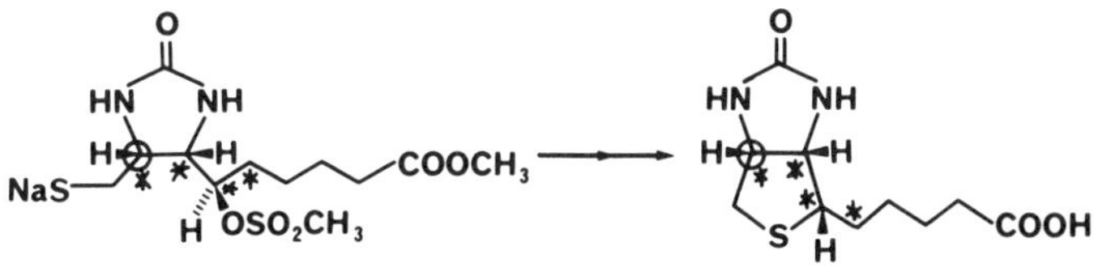

1227 1194

Scheme 94 (*Continued*)

good candidates as starting materials for biotin. Two such syntheses (Scheme 94) proceeded in a "biomimetic" way with formation of the thiophene ring as a final step.

In the first of these syntheses (*502*), anhydroglucose was converted in four straightforward steps to the epoxide **1221**. Ring opening with azide, mesylation, and acetolysis gave **1222**. Solvolytic removal of the acetates, reduction of the thus liberated aldehyde, acetal formation, displacement of the mesylate with azide, and selective hydrogenation of the azido groups gave the diamine **1223**. Reaction with phosgene, selective acetylation and ketal hydrolysis provided **1224**. At this point, the remaining carbon atoms were introduced by periodate cleavage of the diol and Wittig reaction with the aldehyde thus formed. Hydrogenolytic debenzylation and saturation of the olefinic bonds, and then saponification of the acetate, converted **1225** to **1226**. The introduction of the sulfur atom was achieved by bismesylation followed by reaction with a large excess of sodium sulfide. These reactions gave, albeit in only 30% yield (via **1227**), the methyl ester of biotin (**1194**).

A second group (*503*) devised a somewhat shorter synthesis of the diol **1226** and thus of biotin. In this case, instead of using a glucose derivative, which required two inversions at C-2 along with inversions at C-3 and C-4 and destruction of the chirality at C-5, D-glucosamine, with the proper configuration at C-2, was employed. Thus, a several-step protection of D-glucosamine gave the furanose **1228**. The usual tosylation–azide sequence provided **1229**. Selective hydrogenation to reduce only the azide group was effected over Raney nickel. Treatment of the product of this reaction with base caused cyclization and thus avoided the usual use of phosgene for formation of the imidazolidone ring. Hydrolysis of the acetal then gave the diol **1230**. In a sequence similar to that described above, periodate oxidation, Wittig reaction, and hydrogenation (accompanied by debenzylation) gave hemiacetal **1231**. Reduction of this compound with sodium borohydride gave the diol **1226**, which again led to biotin.

A third preparation of biotin (*504*) (chronologically the first) employed many of the same reactions as the other two but used a different synthetic

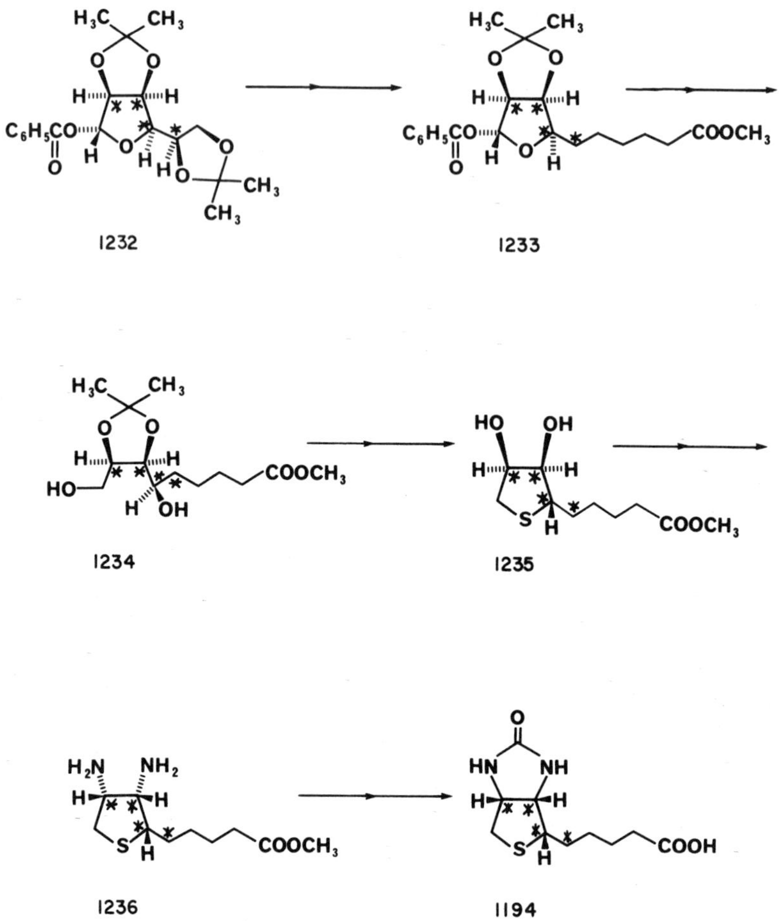

Scheme 95. Biotin synthesis III.

order. In this case (Scheme 95), the tetrahydrothiophene ring was formed prior to introduction of the nitrogen atoms. The benzoylated diketal **1232** of D-mannose was selectively deketalized. The usual sequence (oxidative diol cleavage, Wittig reaction, and hydrogenation) introduced the desired side chain in **1233**. Ester hydrolysis and borohydride reduction gave diol **1234**, which by mesylation, ring formation with sodium sulfide, and ketal hydrolysis was converted to tetrahydrothiophene **1235**. The two nitrogen atoms were introduced by mesylation, displacement with azide, hydrogenation over platinum in acetic anhydride, and basic hydrolysis of the thus formed acetamide groups. The diamine **1236** was converted to biotin by reaction with phosgene and then saponification.

An alternate approach to **1234** from D-arabinose has been described (*505*). While conceptually simpler than the synthesis described in Scheme 95, this work involved a very low yield Wittig reaction and thus is of little value.

The general approaches described above have also been employed in syntheses of *d*-oxybiotin (*506, 507*) and of (+)-desthiobiotin (*508*), both from D-glucose.

In addition to carbohydrate based approaches to biotin, in which inorganic sulfide is incorporated, preparations of this vitamin from the sulfur-containing amino acids cysteine and cystine have been reported (*509, 510*). In the first such sequence (*509*) (Scheme 96), the Hoffmann–La Roche group converted the protected cysteine **1237**, by reduction and oxidation, to aldehyde **1238**. Reaction of this compound with vinylmagnesium chloride gave the alcohol **1239**, which underwent orthoester Claisen rearrangement to **1240**. Bromination of this olefin caused a "dramatic and stereospecific rearrangement [ref. *509*]." The initial bromonium ion apparently suffered internal attack to give the sulfonium cation **1241**. Expulsion of benzaldehyde then gave the observed product (**1242**).

A direct displacement of the bromide in **1242**, or the deacylated amine, by a nitrogen equivalent was projected as the final step in obtaining the biotin stereochemistry. This reaction could not be effected with the desired inversion, however, due to the intermediacy of an aziridinium ion. The solution to this was a double inversion. Thus, heating the amino bromide in acetic acid produced the bromolactam **1243**. This compound underwent displacement with azide to give, along with a distressingly large amount of elimination product, the azido lactam **1244**.

At this point, the stereochemical problems were solved. Catalytic azide reduction and amide opening produced **1245**, which by phosgenation and esterification was converted to **1246**. Using an adaptation of the commercial biotin synthesis, this compound was converted via the thiophanium salt **1247** to the vitamin.

The second amino acid–based synthesis (*510*) of biotin (Scheme 97) used an intramolecular nitrone cycloaddition to provide the correct stereochemistry. Zinc dust reduction of the cystine derivative **1248** in the presence of oxygen occurred with cyclization to yield the olefinic sulfide **1249**. A partial reduction of the ester group and subsequent reaction with *N*-benzylhydroxylamine gave nitrone **1250**. Heating this compound in toluene gave exclusively **1251**, which possessed the requisite stereochemical features. Zinc reduction of the N—O bond and treatment with methyl chloroformate produced urethane **1252**. As a further step in unraveling the cycloaddition product, basic hydrolysis of **1252**, which occurred with recyclization to form the imidazolidinone ring, gave after esterifica-

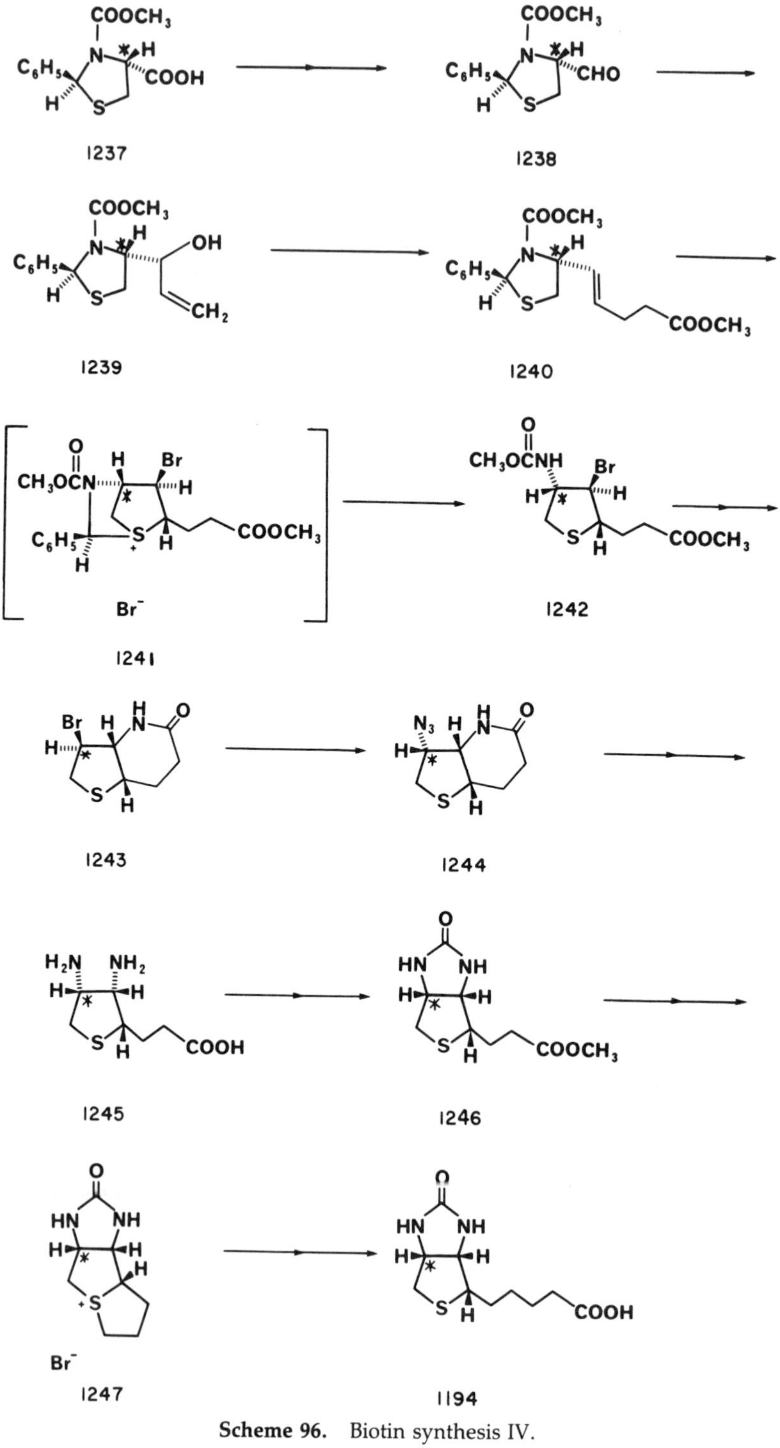

Scheme 96. Biotin synthesis IV.

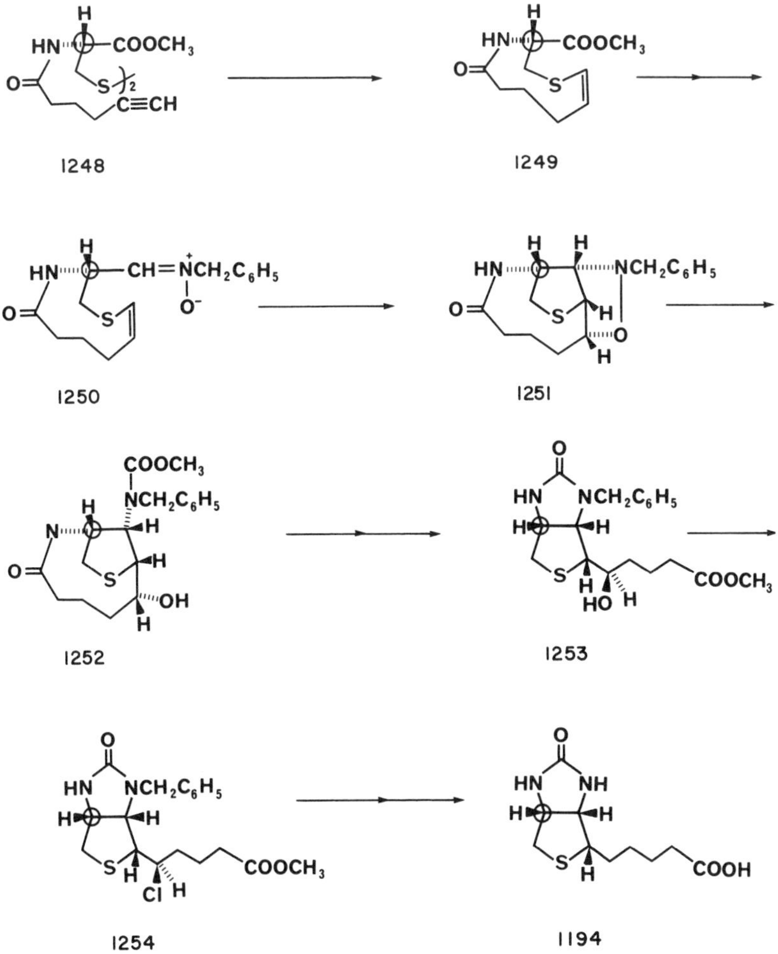

Scheme 97. Biotin synthesis V.

tion the hydroxybiotin **1253**. The extra oxygen was removed by conversion to chloride **1254** and then reduction with sodium borohydride in dimethylformamide. Subsequent acidic debenzylation and ester hydrolysis then gave biotin (**1194**).

4. α-TOCOPHEROL

α-Tocopherol, or vitamin E, is unique among the chiral vitamins in that the material having the "natural" stereochemistry is not alone in possessing bioactivity. With three chiral centers, there are eight diastereomers of α-tocopherol. Although the relative biopotency of (2*R*,4′*R*,8′*R*)-α-tocopherol and the totally synthetic racemic mixture is still in question, it is apparent that the ratio is substantially less than 1.5 : 1, while 8 : 1 would be expected if the other diastereomers were inactive. It is perhaps for this reason that, until recent years, relatively little effort has been put into synthesis of the natural material.

α-Tocopherol possesses two distinct types of chiral units: the tetrasubstituted chroman center and the two trisubstituted chain centers. It should come as no surprise then that syntheses of this molecule to date (*511*) have focused on preparation of chiral chroman one-carbon or two-carbon pieces and coupling of these with the appropriate chiral C-15 or C-14 unit.

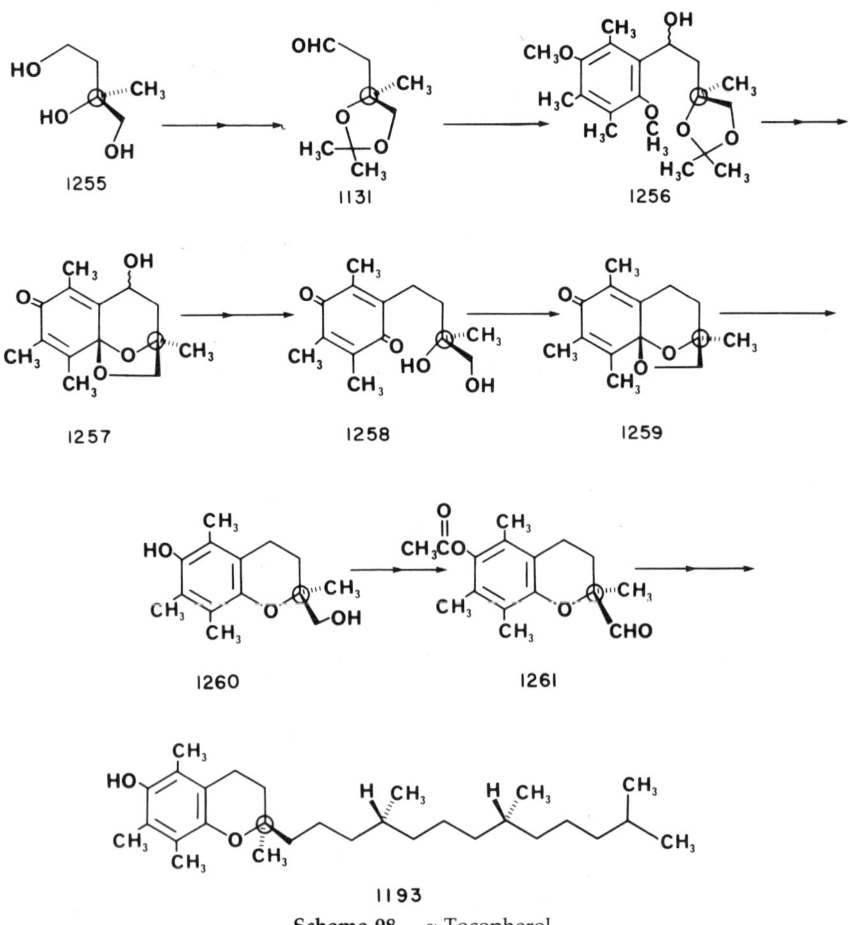

Scheme 98. α-Tocopherol.

One synthesis (*97*) of a chiral chroman (Scheme 98) began with (*S*)-citramalic acid. Reduction to the triol **1255** was followed by acetonide formation and oxidation to aldehyde **1131**. This aldehyde (Scheme 83, see above) has also been obtained from D-glucose. Reaction of the aldehyde with the appropriate aryl Grignard reagent led to the diastereomeric alcohol mixture **1256**. Hydrogenolytic removal of the benzylic alcohol from **1256** could not be effected. Thus, this compound was oxidatively demethylated with ceric ammonium nitrate and then cyclized with acid to ketal **1257**. Hydrogenation of this compound did cause the desired debenzylation and at the same time reduced the quinone. During chromatographic purification of the resulting phenol, oxidation (presumably by air) occurred, giving the known (*555*) **1258**, which was converted to **1259** with acid. Reductive aromatization of **1259** provided the phenolic alcohol **1260**, which was converted in four steps to aldehyde **1261**. Completion of the α-tocopherol (**1193**) synthesis involved Wittig reaction with a C-15 phosphonium salt derived from (*R,R*)-phytol and then hydrogenation to remove the thus generated double bond.

With the exception of the synthesis just described, most preparations of α-tocopherol have relied on a chroman unit whose preparation has included a classical resolution. The chiral secondary methyl groups at C-4′ and C-8′ cannot be so easily obtained by resolution, however, and so alternative approaches have been investigated. One favorite avenue has been via (*R*)-dihydrocitronellic acid, obtained in enantiomerically pure form from (*R*)-pulegone. The syntheses have involved a crossed Kolbe coupling (*512*) and unions with chiral C-5 units derived by microbiological hydroxylation (*34*) and resolution (*513*). In this last work, the C-15 product was further converted to (7*R*,11*R*)-phytol.

An interesting alternative use of (*R*)-pulegone to give the C-15 ester **1268** (Scheme 99) has appeared (*514*). In this work, the pulegone derived ketal **1262** was ozonized to give, after reductive workup, alcohol **1263**. Acidic ketal hydrolysis was accompanied by loss of water. Sodium borohydride reduction of the resultant unsaturated ketone in the presence of Ce^{3+} gave allylic alcohol **1264**. This compound was subjected to a modification of the Cope–Carroll rearrangement, using isovaleryl–Meldrum's acid (**1265**) to form the β-keto ester. This reaction gave ketone **1266**, which was then ozonized. The newly formed aldehyde was oxidized and esterified to **1267**. Removal of the two excess oxygen atoms was effected by desulfurization of the derived bis(thioketal). The overall stereoselectivity of this reaction sequence was not indicated. However, one might worry about the diastereomeric integrity of alcohol **1264** and of possible racemization at C-8′ (α-tocopherol numbering) after establishment of a carbonyl group at C-7′.

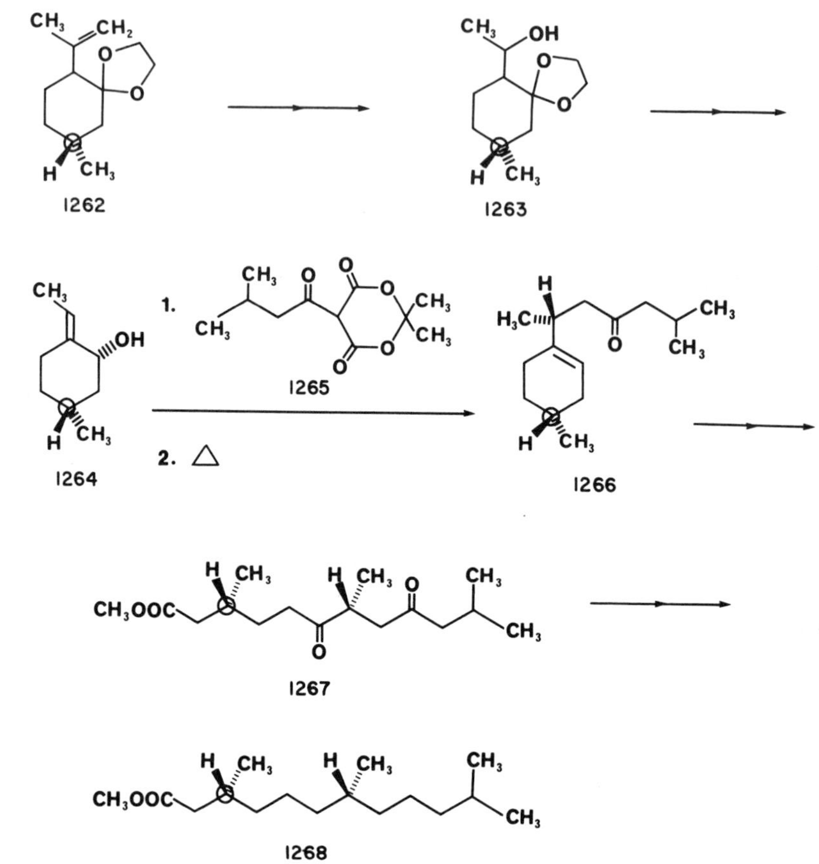

Scheme 99. α-Tocopherol synthon.

Another synthesis of ester **1268** has been described (*515*). This work, designed as a showcase for an elegant palladium-catalyzed allylic transfer of chirality, nonetheless required 19 steps to obtain **1268** from D-glucose.

5. Cholecalciferol

Cholecalciferol, or vitamin D_3 (**1192**), is an item of commerce that is routinely prepared by photolysis and then thermolysis of 7-dehydrocholesterol. Research in this area had apparently come to a stop until reports (*516*) were made, beginning in 1971, of a number of physiologically active metabolites of the compound. These reports regenerated the synthetic interest as a means of obtaining material for biological evaluation and, in some cases, of confirming the absolute stereochemistry of the metabolites. One of the most significant of these metabolites is 1α,25-dihydroxycholecalciferol (**1278**). A total synthesis of this compound has now been reported by a group at Hoffmann–La Roche (*517*) (Scheme 100). The synthesis is marked by the preparation of two chiral pieces, one derived from *d*-carvone and the other by an asymmetric synthesis. Union of these pieces provided the vitamin without resorting to the usual low-yield photolytic step.

Conversion of *d*-carvone to the epoxide **1269** was stereospecific. Horner reaction added the two-carbon (carboethoxy)methylene unit in a geometrically selective (9:1) fashion. The isolated *E* isomer **1270** was opened with sodium acetate in acetic acid to give, after acetylation of the monoacetate, diacetate **1271**. Oxidation of the isopropenyl group in two steps, including a stereospecific (with retention) Baeyer–Villiger rearrangement, gave triacetate **1272**, which was partially hydrolyzed and reprotected to **1273**. Elimination of the elements of water from this molecule was a difficult task that could, nonetheless, be carried out with the unusual reagent, bis[2,2,2-trifluoro-1-phenyl-1-(trifluoromethyl)ethoxy] diphenyl sulfurane. At this point, the olefin geometry needed to be corrected from *E* to *Z*. This was accomplished in a triplet-sensitized photoisomerization. The resultant **1275** was converted, in four straightforward steps, to phosphine oxide **1276**. Reaction (*556*) of the anion of **1276** with ketone **1277**, which had been obtained by elaboration of a product of an asymmetric synthesis, gave, after removal of the silyl protecting groups, the desired vitamin.

The standard cholecalciferol technology, namely, photolysis of the corresponding 7-dehydrocholesterol, was employed to synthesize several minor metabolites of vitamin D_3. In some of these cases, the requisite cholesterol derivative was prepared using a small chiral piece to give the desired side-chain oxidation pattern. Examples include (24*R*),25-dihydroxycholecalciferol (**1279**) from D-glyceric acid (*518*), (25*S*),26-dihydroxycholecalciferol (**1280**) from (*S*)-citramalic acid (*94*) and (23*S*,25*R*)-25-hydroxycholecalciferol 26,23-lactone (**1281**) from (*R*)-citramalic acid (*96*).

H. Miscellaneous Compounds (Including Terpenoids)

There are a number of reported syntheses of chiral molecules from other chiral molecules that fall outside the molecular-target areas described above. A brief overview of some of these syntheses is given in this section.

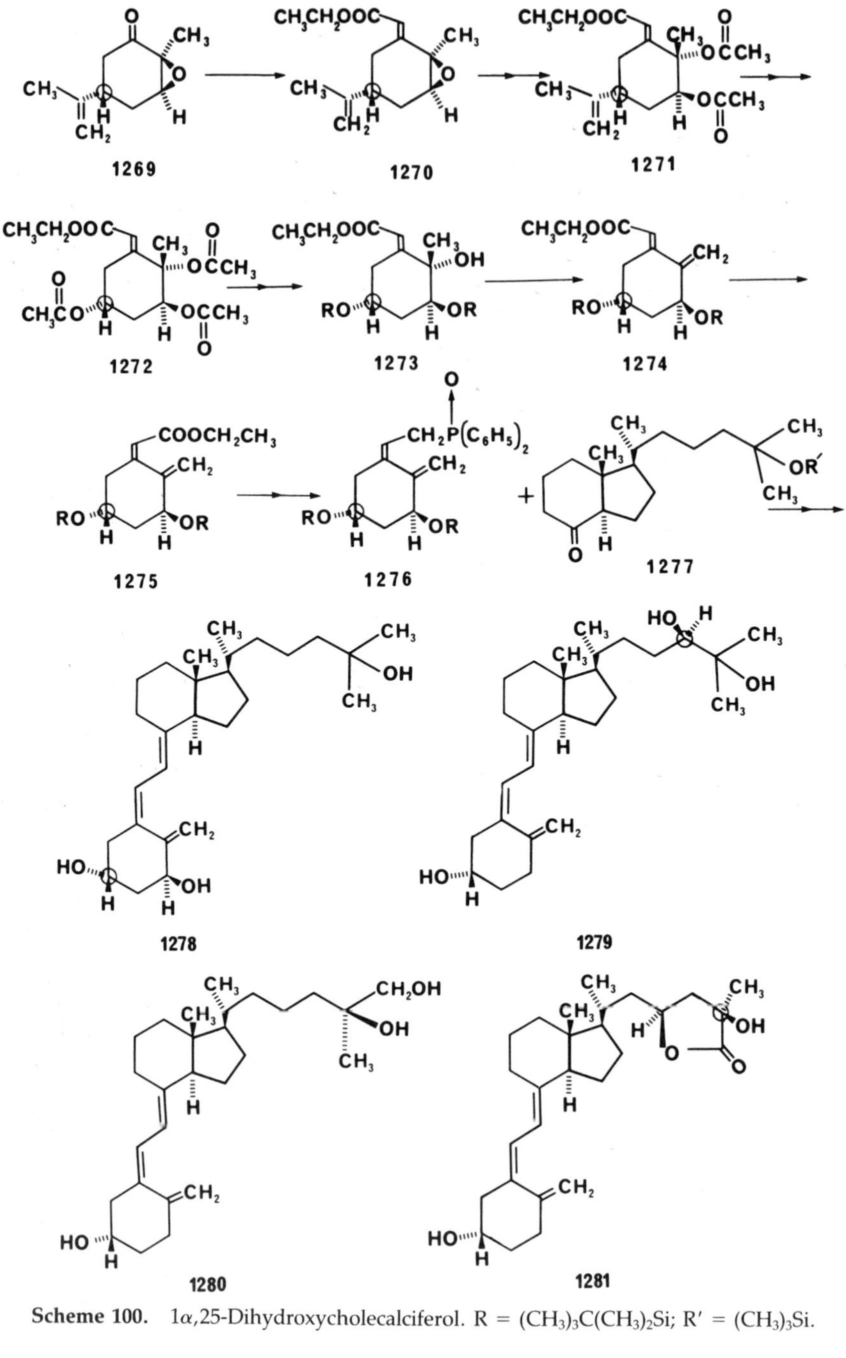

1269, 1270, 1271, 1272, 1273, 1274, 1275, 1276, 1277, 1278, 1279, 1280, 1281

Scheme 100. 1α,25-Dihydroxycholecalciferol. R = $(CH_3)_3C(CH_3)_2Si$; R′ = $(CH_3)_3Si$.

The field of terpenes is one of the first in which organic chemists became synthetically active. Before modern methods of structural elucidation became available, one widely used means to establish the constitution of a terpene was to convert it to another terpene. This process could involve manipulations (degradation and/or construction) in which the original chiral center(s) were alone incorporated into the product or in which these centers were used as a basis for the stereoselective introduction of new centers of asymmetry. Both types of syntheses are quite capably reviewed by ApSimon (*519*). One particularly elegant application of the latter approach is Büchi's (*520*) now classic synthesis of patchouli alcohol, starting from (+)-camphor.

To illustrate the simple degradative approach to a terpenoid, one can consider syntheses of (1*R*,3*R*)-chrysanthemic acid (**1286a**, Scheme 101). This molecule is the acid component of both naturally occurring and most synthetic pyrethrin-type insecticides. There have been at least six reports (*521–526*) of the conversion of Δ^3-carene (**356**) to **1286a**. This multiplicity of effort represents attempts to overcome some difficult and, often, low-yield chemistry. The **356** to **1286a** conversion proceeds in some cases through intermediates **1282** and **1283**. These intermediates have served as starting materials in preparations of the sesquiterpene (−)-taylorione (**1284**) (*527*) and the diterpene (−)-casbene (**1285**) (*528*).

In contrast to the strictly degradative approaches to chrysanthemic acid, one can also note the preparation of this compound from (+)-α-pinene (**364**) (*529*) and from (*S*)-pantolactone (**1202**) (*530*). The preparation of the latter compound from (*S*)-malic acid was described previously (Scheme 92). The epoxide **1287**, derived ultimately from D-glucose, was converted (*531*) to both the *Chrysanthemum* dicarboxylic acid **1286b** and its enantiomer. The methodology employed would clearly be adaptable to the preparation of the parent **1286a**. To complete the chrysanthemic acid picture, one should note the preparation of the analogs **1288** (*532*) from (−)-α-pinene and **1289** (*533*) from thujone.

As in the case of terpenes, the use of one amino acid to prepare another is a well-established methodology. One is particularly impressed by Danishefsky's conversion of (*S*)-glutamic acid to (*S*)-tyrosine (*534*), (*S*)-pretyrosine (arogenate) (*535*), and (*S*)-dopa (*536*). Amino acids have also served as starting materials for preparations of the naturally occurring iron chelators enterobactin (*537*), avenic acid A (*538, 539*), avenic acid B (*540*), and 2′-deoxymugineic acid (*539, 541*), as well as the structurally related nicotianamine (*542*).

As a final example of the inspiration shown by an organic chemist in the synthesis of one chiral molecule from another, one can consider the preparation of the steroid synthon **1295** (Scheme 102) from (−)-camphor (*543*).

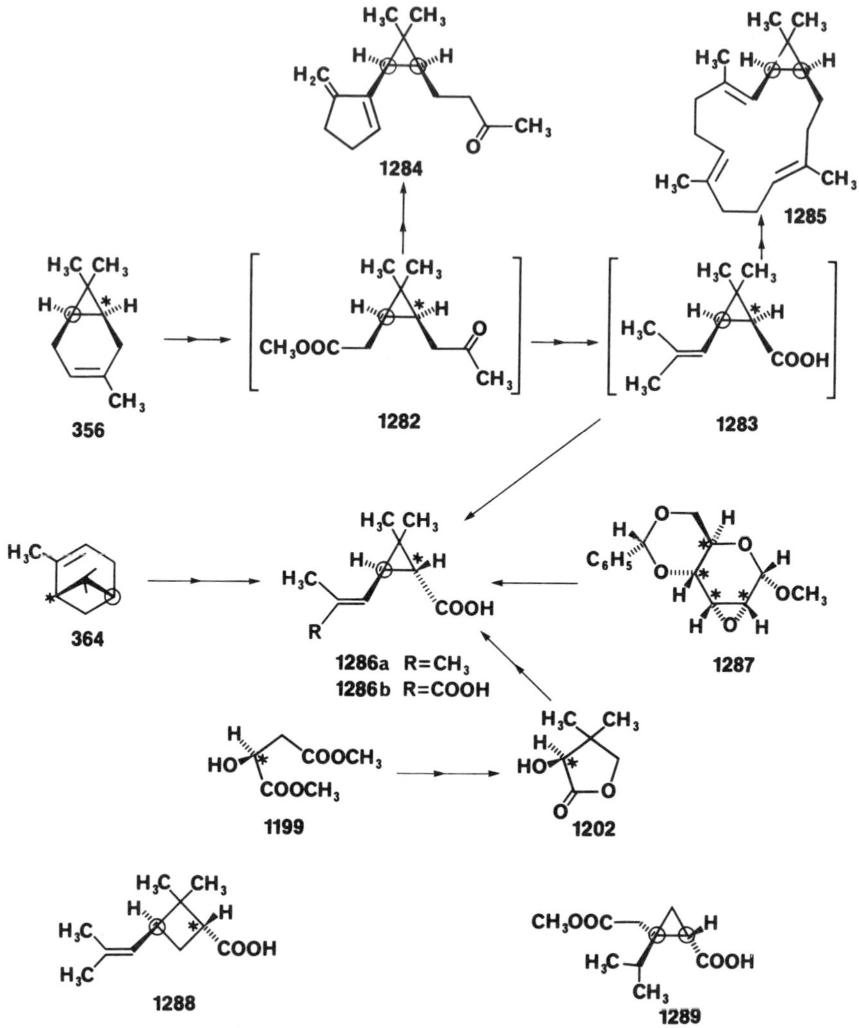

Scheme 101. (1*R*,3*R*)-Chrysanthemic acid.

A three-step preparation gave (−)-π-bromocamphor (**1290**), which by malonate chain lengthening on the corresponding iodide was converted to ketone **1291**. After preparation of the oxime **1292**, Beckmann fragmentation unraveled the bicyclo[2.2.1]heptane system to generate (after completion of a partial double-bond isomerization) nitrile **1293**. Treatment of this compound with base gave the steroid ring C,D compound **1294**. This material was subsequently annulated to the tricyclic compound **1295** in the usual Robinson manner.

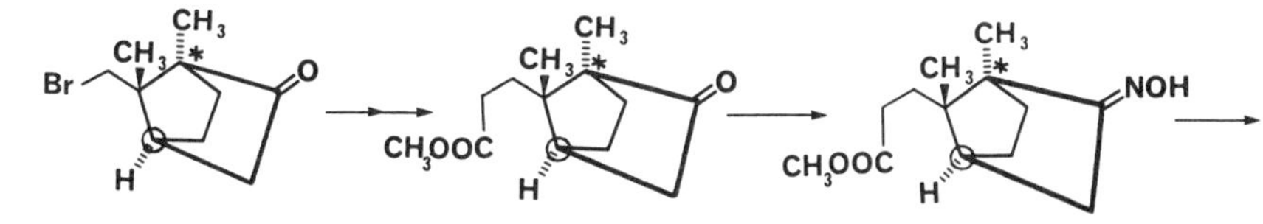

1290 1291 1292

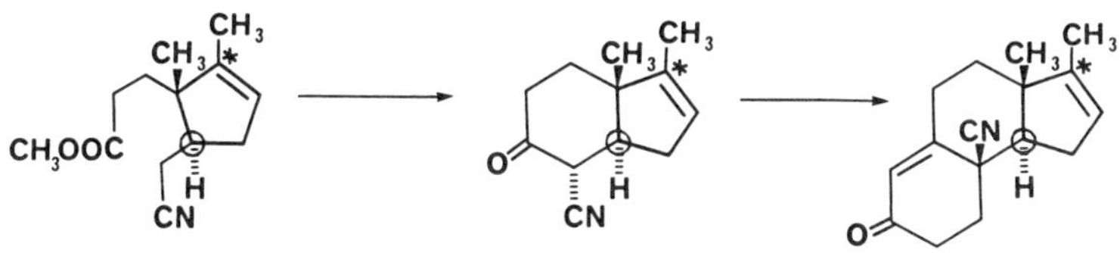

1293 1294 1295

Scheme 102. Steroid synthon.

Addendum

Since the preparation of Section III of this chapter, two major microbially produced compounds have become commercially available. (*S*)-3-Methyl-γ-butyrolactone (**180**) has been used to prepare the isoprenoid chain of (*R*,*R*,*R*)-α-tocopherol (**1193**) (*92*). (*S*)-β-Hydroxyisobutyric acid (**85**) has been employed in a variety of syntheses, including those of (*R*,*R*,*R*)-α-tocopherol (**1193**) (*34, 544*) and all seven possible diastereomers of it (*33*), (*R*)- and (*S*)-muscone (*73*), the polyether ionophores lasalocid A (**775**) (*545*) and monensin (**791**) (*309*), ionophore A-23187 (*546*), two syntheses of the ansa portion of rifamycin S (*547, 548*), the ergot alkaloid paliclavine (*549*), (25*S*),26-dihydroxycholesterol (*550*), a maytansine synthon (*551*), the potent neurotransmitter (−)-domoic acid (*552*), antileukemic (+)-phyllanthocin (*553*), amino acid (*S*)-(2-carboxypropyl)-L-cysteine (*557*), the Prelog–Djerassi lactone (*558*), and segments of the very potent palytoxin (*559, 560*).

References

1. Burger, A. (ed.) (1970). "Medicinal Chemistry," 3rd Ed. Wiley (Interscience), New York.

2. Brand, J. M., Young, J. C., and Silverstein, R. M. (1979). *In* "Progress in the Chemistry of Organic Natural Products" (W. Herz, H. Grisebach, and G. W. Kirby, eds.) Vol. 37, pp. 1–190. Springer-Verlag, New York.

3. Seebach, D., and Kalinowski, H.-O. (1976). *Nachr. Chem. Tech.* **24,** 415–418.
4. Seebach, D., and Hungerbühler, E. (1980). *In* "Modern Synthetic Methods 1980" (R. Scheffold, ed.), pp. 91–171. Salle and Sauerländer, Frankfurt am Main.
5. Jacques, J., Collet, A., and Wilen, S. H. (1981). "Enantiomers, Racemates and Resolutions." Wiley (Interscience), New York.
6. Wilen, S. H. (1972). "Tables of Resolving Agents and Optical Resolutions." University of Notre Dame Press, Notre Dame.
7. Valentine, Jr., D., and Scott, J. W. (1978). *Synthesis* 329–356.
8. Semple, J. E., and Joullié, M. M. (1980). *Heterocycles* **14,** 1825–1870.
9. Henrick, C. A. (1977). *Tetrahedron* **33,** 1845–1889.
10. Rossi, R. (1978). *Synthesis* 413–434.
11. Masamune, S., Bates, G. S., and Corcoran, J. W. (1977). *Angew. Chem. Int. Ed. Engl.* **16,** 585–607.
12. Heathcock, C. H. (1973). *In* "The Total Synthesis of Natural Products" (J. ApSimon, ed.), Vol. 2, pp. 197–558. Wiley (Interscience), New York.
13. ApSimon, J. W., and Hooper, J. W. (1973). *In* "The Total Synthesis of Natural Products" (J. ApSimon, ed.), Vol. 2, pp. 559–640. Wiley (Interscience), New York.
14. Corey, E. J., Niwa, H., Falck, J. R., Mioskowski, C., Arai, Y., and Marfat, A. (1980). *In* "Advances in Prostaglandin and Thromboxane Research" (B. Samuelsson, P. W. Ramwell, and R. Paoletti, ed.), Vol. 6, pp. 19–25. Raven, New York.
15. Mori, K. (1981). *J. Syn. Org. Chem. Jpn.* **39,** 63–72.
16. Mori, K. (1981). *In* "The Total Synthesis of Natural Products" (J. ApSimon, ed.), Vol. 4, pp. 1–183. Wiley (Interscience), New York.
17. Razdan, R. K. (1981). *In* "The Total Synthesis of Natural Products" (J. ApSimon, ed.), Vol. 4, pp. 185–262. Wiley (Interscience), New York.
18. Wierenga, W. (1981). *In* "The Total Synthesis of Natural Products" (J. ApSimon, ed.), Vol. 4, pp. 263–351. Wiley (Interscience), New York.
19. Bindra, J. S. (1981). *In* "The Total Synthesis of Natural Products" (J. ApSimon, ed.), Vol. 4, pp. 353–449. Wiley (Interscience), New York.
20. Szabo, W. A., and Lee, H. T. (1980). *Aldrichimica Acta* **13,** 13–20.
21. (1978). "Carbohydrate Chemistry" (Specialist Periodical Reports), Vol. 11, pp. 228–231. Chemical Society, London. Sequel volumes provide continuing chapters under the heading "The Synthesis of Optically Active Non-carbohydrate Compounds".
22. Smith, L. R., and Williams, H. J. (1979). *J. Chem. Ed.* **56,** 696–698.
23. Fraser-Reid, B. (1975). *Acc. Chem. Res.* **8,** 192–201.
24. Ohrui, H. (1976). *Kagaku To Seibutsu* **14,** 170–172; (1977). *Chem. Abstr.* **86,** 43922.
25. Yamada, S. (1978). *Kagaku Sosetsu* **19,** 163–205; (1980). *Chem. Abstr.* **92,** 164259.
26. Hanessian, S. (1979). *Acc. Chem. Res.* **12,** 159–165.
27. Fraser-Reid, B., and Anderson, R. C. (1980). *In* "Progress in the Chemistry of Organic Natural Products" (W. Herz, H. Grisebach, and G. W. Kirby, eds.), Vol. 39, pp. 1–61. Springer-Verlag, New York.
28. Inch, T. D. (1972). *In* "Advances in Carbohydrate Chemistry and Biochemistry" (R. S. Tipson and D. Horton, eds.), Vol. 27, pp. 191–213. Academic Press, New York.
29. Vasella, A. (1980). *In* "Modern Synthetic Methods 1980" (R. Scheffold, ed.), pp. 173–267. Salle and Sauerländer, Frankfurt am Main.
30. Fraser-Reid, B., Sun, K. M., and Tam, T. F. (1981). *Bull. Soc. Chim. Fr.* **II** 238–246.
31. Drauz, K., Kleeman, A., and Martens, J. (1982). *Angew. Chem. Int. Ed. Engl.* **21,** 584–608.
32. Warren, S. G. (1982). "Organic Synthesis: The Disconnection Approach." Wiley, Sons, New York.

33. Cohen, N., Scott, C. G., Neukom, C., Lopresti, R. J., Weber, G., and Saucy, G. (1981). *Helv. Chim. Acta* **64,** 1158–1173.
34. Cohen, N., Eichel, W. F., Lopresti, R. J., Neukom, C., and Saucy, G. (1976). *J. Org. Chem.* **41,** 3505–3511.
35. Hungerbühler, E., Seebach, D., and Wasmuth, D. (1979). *Angew. Chem. Int. Ed. Engl.* **18,** 958–960.
36. Hungerbühler, E., and Seebach, D. (1981). *Helv. Chim. Acta* **64,** 687–702.
37. Hungerbühler, E., Seebach, D., and Wasmuth, D. (1981). *Helv. Chim. Acta* **64,** 1467–1487.
38. Seuring, B., and Seebach, D. (1977). *Helv. Chim. Acta* **60,** 1175–1181.
39. Seebach, D., and Wasmuth, D. (1980). *Helv. Chim. Acta* **63,** 197–200.
40. Seebach, D., Kalinowski, H.-O., Bastani, B., Crass, G., Daum, H., Dörr, H., Du-Preez, N. P., Ehrig, V., Langer, W., Nüssler, C., Oei, H.-A., and Schmidt, M. (1977). *Helv. Chim. Acta* **60,** 301–325.
41. Brown, H. C., Jadhav, P. K., and Desai, M. C. (1982). *J. Org. Chem.* **47,** 4583–4584.
42. Valentine, Jr., D., Chan, K. K., Scott, C. G., Johnson, K. K., Toth, K., and Saucy, G. (1976). *J. Org. Chem.* **41,** 62–65.
43. Hintzer, K., Koppenhoefer, B., and Schurig, V. (1982). *J. Org. Chem.* **47,** 3850–3854.
44. Wipf, B., Kupfer, E., Bertazzi, R., and Leuenberger, H. G. W. (1983). *Helv. Chim. Acta* **66,** 485–488.
45. Baldwin, J. J., Raab, A. W., Mensler, K., Arison, B. H., and McClure, D. E. (1978). *J. Org. Chem.* **43,** 4876–4878.
46. McClure, D. E., Arison, B. H., and Baldwin, J. J. (1979). *J. Am. Chem. Soc.* **101,** 3666–3668.
47. McClure, D. E., Engelhardt, E. L., Mensler, K., King, S., Saari, W. S., Huff, J. R., and Baldwin, J. J. (1979). *J. Org. Chem.* **44,** 1826–1831.
48. Baldwin, J. J., McClure, D. E., Gross, D. M., and Williams, M. (1982). *J. Med. Chem.* **25,** 931–936.
49. Schurig, V., Koppenhoefer, B., and Buerkle, W. (1980). *J. Org. Chem.* **45,** 538–541.
50. Schurig, V., Koppenhöfer, B., and Bürkle, W. (1978). *Angew. Chem. Int. Ed. Engl.* **17,** 937–939.
51. Koppenhoefer, B., Weber, R., and Schurig, V. (1982). *Synthesis* 316–318.
52. Anisuzzaman, A. K. M., and Owen, L. N. (1967). *J. Chem. Soc.* (*C*) 1021–1026.
53. Takano, S., Goto, E., Hirama, M., and Ogasawara, K. (1981). *Heterocycles* **16,** 381–385.
54. Golding, B. T., and Ioannou, P. V. (1977). *Synthesis* 423–424.
55. Lands, W. E. M., and Zschocke, A. (1965). *J. Lipid Res.* **6,** 324–325.
56. Lok, C. M., Ward, J. P., and van Dorp, D. A. (1976). *Chem. Phys. Lipids* **16,** 115–122.
57. Jung, M. E., and Shaw, T. J. (1980). *J. Am. Chem. Soc.* **102,** 6304–6311.
58. Takano, S., Numata, H., and Ogasawara, K. (1982). *Heterocycles* **19,** 327–328.
59. Baer, E. (1952). *Biochem. Prep.* **2,** 31–38.
60. Fischer, H. O. L., and Baer, E. (1934). *Helv. Chim. Acta* **17,** 622–632.
61. Baker, S. B. (1952). *J. Am. Chem. Soc.* **74,** 827–829.
62. Nelson, W. L., Wennerstrom, J. E., and Sankar, S. R. (1977). *J. Org. Chem.* **42,** 1006–1012.
63. Baker, D. C., and Hawkins, L. D. (1982). *J. Org. Chem.* **47,** 2179–2184.
64. Fischer, H. O. L., and Baer, E. (1936). *Helv. Chim. Acta* **19,** 519–532.
65. Yahiro, N. (1982). *Chem. Lett.* 1479–1480.
66. Minoura, Y., Takebayashi, M., and Price, C. C. (1959). *J. Am. Chem. Soc.* **81,** 4689–4692.

67. Maycock, C. D., and Stoodley, R. J. (1976). *J. Chem. Soc. Chem. Commun.* 234–235.
68. Inch, T. D., and Lewis, G. J. (1971). *Carbohydr. Res.* **16,** 455–458.
69. Hillis, L. R., and Ronald, R. C. (1981). *J. Org. Chem.* **46,** 3348–3349.
70. Johnston, B. D., and Slessor, K. N. (1979). *Can. J. Chem.* **57,** 233–235.
71. Golding, B. T., Hall, D. R., and Sakrikar, S. (1973). *J. Chem. Soc. Perkin I* 1214–1220.
72. Ghirardelli, R. G. (1973). *J. Am. Chem. Soc.* **95,** 4987–4990.
73. Branca, Q., and Fischli, A. (1977). *Helv. Chim. Acta* **60,** 925–944.
74. Barton, D. H. R., Bénéchie, M., Khuong-Huu, F., Potier, P., and Reyna-Pinedo, V. (1982). *Tetrahedron Lett.* **23,** 651–654.
75. Hungerbühler, E., Naef, R., Wasmuth, D., Seebach, D., Loosli, H.-R., and Wehrli, A. (1980). *Helv. Chim. Acta* **63,** 1960–1970.
76. Boger. D. L., and Panek, J. S. (1981). *J. Org. Chem.* **46,** 1208–1210.
77. MacNeil, P. A., Roberts, N. K., and Bosnich, B. (1981). *J. Am. Chem. Soc.* **103,** 2273–2280.
78. Lukeš, R., Jarý, J., and Němec, J. (1962). *Collect. Czech. Chem. Commun.* **27,** 735–741.
79. Mori, K., Takigawa, T., and Matsuo, T. (1979). *Tetrahedron* **35,** 933–940.
80. Meyers, A. I., and Lawson, J. P. (1982). *Tetrahedron Lett.* **23,** 4883–4886.
81. Corey, E. J., Shirahama, H., Yamamoto, H., Terashima, S., Venkateswarlu, A., and Schaaf, T. K. (1971). *J. Am. Chem. Soc.* **93,** 1490–1491.
82. Corey, E. J., Niwa, H., and Knolle, J. (1978). *J. Am. Chem. Soc.* **100,** 1942–1943.
83. Mori, K., Sasaki, M., Tamada, S., Suguro, T., and Masuda, S. (1979). *Tetrahedron* **35,** 1601–1605.
84. Schmidt, U., Talbiersky, J., Bartkowiak, F., and Wild, J. (1980). *Angew. Chem. Int. Ed. Engl.* **19,** 198–199.
85. Shieh, H.-M., and Prestwich, G. D. (1981). *J. Org. Chem.* **46,** 4319–4321.
86. Collum, D. B., McDonald, III, J. H., and Still, W. C. (1980). *J. Am. Chem. Soc.* **102,** 2118–2120.
87. Corey, E. J., Marfat, A., and Hoover, D. J. (1981). *Tetrahedron Lett.* **22,** 1587–1590.
88. Fujita, K., Nakai, H., Kobayashi, S., Inoue, K., Nojima, S., and Ohno, M. (1982). *Tetrahedron Lett.* **23,** 3507–3510.
89. Nicolaou, K. C., Papahatjis, D. P., Claremon, D. A., and Dolle, III, R. E. (1981). *J. Am. Chem. Soc.* **103,** 6967–6969.
90. Mori, K., and Tamada, S. (1979). *Tetrahedron* **35,** 1279–1284.
91. Mori, K., and Iwasawa, H. (1980). *Tetrahedron* **36,** 87–90.
92. Schmid, M., and Barner, R. (1979). *Helv. Chim. Acta* **62,** 464–473.
93. Mimura, T., Kimura, Y., and Nakai, T. (1979). *Chem. Lett.* 1361–1364.
94. Barner, R., Hübscher, J., Daly, J. J., and Schönholzer, P. (1981). *Helv. Chim. Acta* **64,** 915–938.
95. Yamada, S., Nakayama, K., and Takayama, H. (1981). *Tetrahedron Lett.* **22,** 2591–2594.
96. Yamada, S., Nakayama, K., and Takayama, H. (1981). *Chem. Pharm. Bull.* **29,** 2393–2396.
97. Barner, R., and Schmid, M. (1979). *Helv. Chim. Acta* **62,** 2384–2399.
98. Ogura, K., Yamashita, M., and Tsuchihashi, G. (1976). *Tetrahedron Lett.* 759–762.
99. Eguchi, C., and Kakuta, A. (1974). *Bull. Chem. Soc. Jpn.* **47,** 1704–1708.
100. Taniguchi, M., Koga, K., and Yamada, S. (1974). *Tetrahedron* **30,** 3547–3552.
101. Takano, S., Goto, E., Hirama, M., and Ogasawara, K. (1981). *Heterocycles* **16,** 951–954.

102. Takano, S., Yonaga, M., and Ogasawara, K. (1981). *Synthesis* 265–266.
103. Camps, P., Font, J., and Ponsati, O. (1981). *Tetrahedron Lett.* **22,** 1471–1472.
104. Tomioka, K., Ishiguro, T., and Koga, K. (1979). *J. Chem. Soc. Chem. Commun.* 652–653.
105. Tomioka, K., Ishiguro, T., and Koga, K. (1980). *Tetrahedron Lett.* **21,** 2973–2976.
106. Deanne, C. C., and Inch, T. D. (1969). *J. Chem. Soc. Chem. Commun.* 813.
107. Mori, K. (1975). *Tetrahedron* **31,** 3011–3012.
108. Vigneron, J. P., Méric, R., Larchevêque, M., Debal, A., Kunesch, G., Zagatti, P., and Gallois, M. (1982). *Tetrahedron Lett.* **23,** 5051–5054.
109. Kitahara, T., Mori, K., and Matsui, M. (1979). *Tetrahedron Lett.* 3021–3024.
110. Červinka, O., and Hub, L. (1968). *Collect. Czech. Chem. Commun.* **33,** 2927–2932.
111. Shibuya, H., Yamamoto, Y., Miura, I., and Kitagawa, I. (1982). *Heterocycles* **17,** 215–218.
112. Ireland, R. E., McGarvey, G. J., Anderson, R. C., Badoud, R., Fitzsimmons, B., and Thaisrivongs, S. (1980). *J. Am. Chem. Soc.* **102,** 6178–6180.
113. Takano, S., Masuda, K., and Ogasawara, K. (1981). *Heterocycles* **16,** 1509–1513.
114. Kisfaludy, L., Korenczki, F., and Kathó, A. (1982). *Synthesis* 163.
115. Schreiber, S. L. (1980). *J. Am. Chem. Soc.* **102,** 6163–6165.
116. Mori, K., Suguro, T., and Uchida, M. (1978). *Tetrahedron* **34,** 3119–3123.
117. Yamada, S., and Akimoto, H. (1969). *Tetrahedron Lett.* 3105–3108.
118. Yamada, S., Tomioka, K., and Koga, K. (1976). *Tetrahedron Lett.* 61–64.
119. Akimoto, H., Okamura, K., Yui, M., Shioiri, T., Kuramoto, M., Kikugawa, Y., and Yamada, S. (1974). *Chem. Pharm. Bull.* **22,** 2614–2623.
120. Ungemach, F., DiPierro, M., Weber, R., and Cook, J. M. (1981). *J. Org. Chem.* **46,** 164–168.
121. Brossi, A., Focella, A., and Teitel, S. (1973). *J. Med. Chem.* **16,** 418–420.
122. Soerens, D., Sandrin, J., Ungemach, F., Mokry, P., Wu, G. S., Yamanaka, E., Hutchins, L., DiPierro, M., and Cook, J. M. (1979). *J. Org. Chem.* **44,** 535–545.
123. van Tamelen, E. E., and Dorschel, C. (1976). *Bioorg. Chem.* **5,** 203–228.
124. Yamada, S., Konda, M., and Shioiri, T. (1972). *Tetrahedron Lett.* 2215–2218.
125. Konda, M., Shioiri, T., and Yamada, S. (1975). *Chem. Pharm. Bull.* **23,** 1025–1031.
126. Konda, M., Oh-ishi, T., and Yamada, S. (1977). *Chem. Pharm. Bull.* **25,** 69–74.
127. Konda, M., Shioiri, T., and Yamada, S. (1975). *Chem. Pharm. Bull.* **23,** 1063–1076.
128. Barton, D. H. R., Bhakuni, D. S., James, R., and Kirby, G. W. (1967). *J. Chem. Soc. C* 128–132.
129. Takano, S., Chiba, K., Yonaga, M., and Ogasawara, K. (1980). *J. Chem. Soc. Chem. Commun.* 616–617.
130. Takano, S., Yonaga, M., and Ogasawara, K. (1981). *J. Chem. Soc. Chem. Commun.* 1153–1155.
131. Takano, S., Yonaga, M., Chiba, K., and Ogasawara, K. (1980). *Tetrahedron Lett.* **21,** 3697–3700.
132. Neuss, N., Gorman, M., Hargrove, W., Cone, N. J., Biemann, K., Büchi, G., and Manning, R. E. (1964). *J. Am. Chem. Soc.* **86,** 1440–1442.
133. Takano, S., Kasahara, C., and Ogasawara, K. (1981). *J. Chem. Soc. Chem. Commun.* 637–638.
134. Takano, S., Tamura, N., and Ogasawara, K. (1981). *J. Chem. Soc. Chem. Commun.* 1155–1156.
135. Yamada, S., Tomioka, K., and Koga, K. (1976). *Tetrahedron Lett.* 57–60.
136. Tomioka, K., Koga, K., and Yamada, S. (1977). *Chem. Pharm. Bull.* **25,** 2681–2688.

137. Shimizu, K., Tomioka, K., Yamada, S., and Koga, K. (1977). *Heterocycles* **25,** 277–282.
138. Shimizu, K., Tomioka, K., Yamada, S., and Koga, K. (1978). *Chem. Pharm. Bull.* **26,** 3765–3771.
139. Kametani, T., Takagi, N., Toyota, M., Honda, T., and Fukumoto, K. (1981). *Heterocycles* **16,** 591–594.
140. Kutney, J. P., Eigendorf, G. K., Matsue, H., Murai, A., Tanaka, K., Sung, W. L., Wada, K., and Worth, B. R. (1978). *J. Am. Chem. Soc.* **100,** 938–943.
141. Dörnyei, G., and Sźantay, C. (1976). *Acta Chim. Acad. Sci. Hung.* **89,** 161–172.
142. Robins, D. J., and Sakdarat, S. (1981). *J. Chem. Soc. Perkin* **I** 909–913.
143. Overman, L. E., and Bell, K. L. (1981). *J. Am. Chem. Soc.* **103,** 1851–1853.
144. Babidge, P. J., Massy-Westropp, R. A., Pyne, S. G., Shiengthong, D., Ungphakorn, A., and Veerachat, G. (1980). *Aust. J. Chem.* **33,** 1841–1845.
145. Russel, J. H., and Hunziker, H. (1969). *Tetrahedron Lett.* 4035–4036.
146. Faber, L., and Wiegrebe, W. (1973). *Helv. Chim. Acta* **56,** 2882–2884.
147. Fujimoto, R., and Kishi, Y. (1981). *Tetrahedron Lett.* **22,** 4197–4198.
148. Ritchie, R., and Saxton, J. E. (1975). *J. Chem. Soc. Chem. Commun.* 611–612.
149. Kametani, T., Kanaya, N., and Ihara, M. (1981). *J. Chem. Soc. Perkin* **I** 959–963.
150. Ritchie, R., and Saxton, J. E. (1981). *Tetrahedron* **37,** 4295–4303.
151. Kametani, T., Kanaya, N., and Ihara, M. (1980). *J. Am. Chem. Soc.* **102,** 3974–3975.
152. Takamatsu, N., Inoue, S., and Kishi, Y. (1971). *Tetrahedron Lett.* 4661–4664.
153. Takamatsu, N., Inoue, S., and Kishi, Y. (1971). *Tetrahedron Lett.* 4665–4668.
154. Takano, S., Imamura, Y., and Ogasawara, K. (1981). *Tetrahedron Lett.* **22,** 4479–4482.
155. Aketa, K., Terashima, S., and Yamada, S. (1976). *Chem. Pharm. Bull.* **24,** 621–631.
156. Saitoh, Y., Moriyama, Y., Hirota, H., Takahashi, T., and Khuong-huu, Q. (1981). *Bull. Chem. Soc. Jpn.* **54,** 488–492.
157. Macmillan J. G., Springer, J. P., Clardy, J., Cole, R. J., and Kirksey, J. W. (1976). *J. Am. Chem. Soc.* **98,** 246–247.
158. Habermehl, G., Andres, H., Miyahara, K., Witkop, B., and Daly, J. W. (1976). *Liebigs Ann. Chem.* 1577–1583.
159. Oppolzer, W., and Flaskamp, E. (1977). *Helv. Chim. Acta* **60,** 204–207.
160. Oppolzer, W., and Petrzilka, M. (1978). *Helv. Chim. Acta* **61,** 2755–2762.
161. Allinger, N. L., and Riew, C. K. (1975). *J. Org. Chem.* **40,** 1316–1321.
162. Rebek, Jr., J., and Shue, Y. K. (1980). *J. Am. Chem. Soc.* **102,** 5426–5427.
163. Samuelsson, B. (1982). Angew. Chem. Int. Ed. Engl. **21,** 902–910.
164. Roberts. S. M., and Scheinmann, F. (eds.) (1982). "New Synthetic Routes to Prostaglandins and Thromboxanes," Academic Press, New York.
165. Borgeat, P., and Sirois, P. (1981). *J. Med. Chem.* **24,** 121–126.
166. Kuehl, Jr., F. A., and Egan, R. W. (1980). *Science* **210,** 978–984.
167. Corey, E. J., Schaaf, T. K., Huber, W., Koelliker, U., and Weinshenker, N. M. (1970). *J. Am. Chem. Soc.* **92,** 397–398.
168. Johnson, F., Paul, K. G., Favara, D., Ciabatti, R., and Guzzi, U. (1982). *J. Am. Chem. Soc.* **104,** 2190–2198.
169. Corey, E. J., Albonico, S. M., Koelliker, U., Schaaf, T. K., and Varma, R. K. (1971). *J. Am. Chem. Soc.* **93,** 1491–1493.
170. Corey, E. J., Nicolaou, K. C., and Beames, D. J. (1974). *Tetrahedron Lett.* 2439–2440.
171. Ferrier, R. J., and Prasit, P. (1981). *J. Chem. Soc. Chem. Commun.* 983–985.
172. Stork, G., and Raucher, S. (1976). *J. Am. Chem. Soc.* **98,** 1583–1584.

173. Stork, G., Takahashi, T., Kawamoto, I., and Suzuki, T. (1978). *J. Am. Chem. Soc.* **100,** 8272–8273.
174. Stork, G., and Takahashi, T. (1977). *J. Am. Chem. Soc.* **99,** 1275–1276.
175. Naruto, M., Ohno, K., and Naruse, N. (1978). *Chem. Lett.* 1419–1422.
176. Naruto, M., Ohno, K., Naruse, N., and Takeuchi, H. (1979). *Tetrahedron Lett.* 251–254.
177. Naruto, M., Ohno, K., Naruse, N., and Takeuchi, H. (1978). *Chem. Lett.* 1423–1424.
178. Berkowitz, W. F., Sasson, I., Sampathkumar, P. S., Hrabie, J., Choudhry, S., and Pierce, D. (1979). *Tetrahedron Lett.* 1641–1644.
179. Berkowitz, W. F., Choudhry, S. C., and Hrabie, J. A. (1982). *J. Org. Chem.* **47,** 824–829.
180. Weinges, K., von der Eltz, H., and Tran-Viet, D. (1980). *Angew. Chem. Int. Ed. Engl.* **19,** 628.
181. Mitscher, L. A., Clark, III, G. W., Bokelman, G., Showalter, H. D. H., Shirahata, K., and Hudson, P. B. (1977). *Heterocycles* **7,** 779–786.
182. Clark, III, G. W., Hammesfahr, P. D., Hudson, P. B., Mitscher, L. A., Shirahata, K., Sulko, J., and Veysoglu, T. (1979). *Heterocycles* **13,** 163–167.
183. Mitscher, L. A., Clark, III, G. W., and Hudson, P. B. (1978). *Tetrahedron Lett.* 2553–2556.
184. Stork, G., and Isobe, M. (1975). *J. Am. Chem. Soc.* **97,** 6260–6261.
185. Elliott, J. D., Hetmanski, M., Stoodley, R. J. (1981). *J. Chem. Soc. Perkin* **I** 1782–1789.
186. Horton, D., and Machinami, T. (1981). *J. Chem. Soc. Chem. Commun.* 88–90.
187. Ohrui, H., and Kuzuhara, H. (1980). *Agric. Biol. Chem.* **44,** 907–912.
188. Fringuelli, F., Patumi, M., and Taticchi, A. (1978). *Gazz. Chim. Ital.* **108,** 487–490.
189. Barry, J., and Kagan, H. B. (1981). *Synthesis* 453–455.
190. Lourens, G. J., and Koekemoer, J. M. (1975). *Tetrahedron Lett.* 3719–3722.
191. Arndt, R. R., Koekemoer, J. M., Lourens, G. J., and Venter, E. M. (1981). *S. Afr. J. Chem.* **34,** 121–127.
192. Hanessian, S., Dextraze, P., Fougerousse, A., and Guindon, Y. (1974). *Tetrahedron Lett.* 3983–3986.
193. Lourens, G. J., and Koekemoer, J. M. (1975). *Tetrahedron Lett.* 3715–3718.
194. Boessenkool, I. K., and Lourens, G. J. (1980). *S. Afr. J. Chem.* **33,** 112–115.
195. Barrière, J.-C., Chiaroni, A., Cléophax, J., Géro, S. D., Riche, C., and Vuilhorgne, M. (1981). *Helv. Chim. Acta* **64,** 1140–1144.
196. Castellanos, L., Gateau-Olesker, A., Panne-Jacolot, F., Cleophax, J., and Gero, S. D. (1981). *Tetrahedron* **37,** 1691–1696.
197. Saijo, S., Wada, M., Himizu, J., and Ishida, A. (1980). *Chem. Pharm. Bull.* **28,** 1449–1458.
198. Fujimoto, Y., Yadav, J. S., and Sih, C. J. (1980). *Tetrahedron Lett.* **21,** 2481–2482.
199. Nelson, N. A., and Jackson, R. W. (1976). *Tetrahedron Lett.* 3275–3278.
200. Corey, E. J., Shibasaki, M., and Knolle, J. (1977). *Tetrahedron Lett.* 1625–1626.
201. Hernandez, O. (1978). *Tetrahedron Lett.* 219–222.
202. Ohrui, H., and Emoto, S. (1977). *Agric. Biol. Chem.* **41,** 1773–1778.
203. Hanessian, S., and Lavallee, P. (1977). *Can. J. Chem.* **55,** 562–565.
204. Hanessian, S., and Lavallee, P. (1981). *Can. J. Chem.* **59,** 870–877.
205. Kelly, A. G., and Roberts, J. S. (1980). *J. Chem. Soc. Chem. Commun.* 228–229.
206. Schaaf, T. K., Bussolotti, D. L., and Parry, M. J. (1981). *J. Am. Chem. Soc.* **103,** 6502–6505.

207. Nicolaou, K. C., Magolda, R. L., and Claremon, D. A. (1980). *Adv. Prostaglandin Thromboxane Res.* **6,** 481–483.

208. Ansell, M. F., Caton, M. P. L., Palfreyman, M. N., and Stuttle, K. A. J. (1979). *Tetrahedron Lett.* 4497–4498.

209. Fried, J., Barton, J., Kittisopikul, S., Needleman, P., and Wyche, A. (1980). *Adv. Prostaglandin Thromboxane Res.* **6,** 427–436.

210. Just, G., Luthe, C., and Oh, H. (1980). *Tetrahedron Lett.* **21,** 1001–1004.

211. Just, G., and Crosilla, D. (1980). *Can. J. Chem.* **58,** 2349–2357.

212. Just, G., and Oh, H. (1981). *Can. J. Chem.* **59,** 2729–2736.

213. Just, G., and Luthe, C. (1980). *Can. J. Chem.* **58,** 1799–1805.

214. Just, G., and Luthe, C. (1980). *Can. J. Chem.* **58,** 2286–2294.

215. Corey, E. J., Clark, D. A., Goto, G., Marfat, A., and Mioskowski, C. (1980). *J. Am. Chem. Soc.* **102,** 1436–1439.

216. a. Cohen, N., Banner, B. L., and Lopresti, R. J. (1980). *Tetrahedron Lett.* **21,** 4163–4166; b. Cohen, N., Banner, B. L., Lopresti, R. J., Wong, F., Rosenberger, M., Liu, Y.-Y., Thom, E., and Liebman, A. A. (1983). *J. Am. Chem. Soc.* **105,** 3661–3672.

217. Rokach, J., Young, R. N., Kakushima, M., Lau, C.-K., Seguin, R., Frenette, R., and Guindon, Y. (1981). *Tetrahedron Lett.* **22,** 979–982.

218. Rokach, J., Zamboni, R., Lau, C.-K., and Guindon, Y. (1981). *Tetrahedron Lett.* **22,** 2759–2762.

219. Rokach, J., Lau, C.-K., Zamboni, R., and Guindon, Y. (1981). *Tetrahedron Lett.* **22,** 2763–2766.

220. Corey, E. J., Clark, D. A., Marfat, A., and Goto, G. (1980). *Tetrahedron Lett.* **21,** 3143–3146.

221. Marriott, D. P., and Bantick, J. R. (1981). *Tetrahedron Lett.* **22,** 3657–3658.

222. Corey, E. J., and Goto, G. (1980). *Tetrahedron Lett.* **21,** 3463–3466.

223. Corey, E. J., Marfat, A., Goto, G., and Brion, F. (1980). *J. Am. Chem. Soc.* **102,** 7984–7985.

224. Corey, E. J., Marfat, A., Munroe, J., Kim, K. S., Hopkins, P. B., and Brion, F. (1981). *Tetrahedron Lett.* **22,** 1077–1080.

225. Corey, E. J., Hopkins, P. B., Munroe, J. E., Marfat, A., and Hashimoto, S. (1980). *J. Am. Chem. Soc.* **102,** 7986–7987.

226. Corey, E. J., Marfat, A., and Laguzza, B. C. (1981). *Tetrahedron Lett.* **22,** 3339–3342.

227. Corey, E. J., and Kang, J. (1981). *J. Am. Chem. Soc.* **103,** 4618–4619.

228. Sheehan, J. C., and Henery-Logan, K. R. (1959). *J. Am. Chem. Soc.* **81,** 3089–3094.

229. Woodward, R. B., Heusler, K., Gosteli, J., Naegeli, P., Oppolzer, W., Ramage, R., Ranganathan, S., and Vorbrüggen, H. (1966). *J. Am. Chem. Soc.* **88,** 852–853.

230. Woodward, R. B. (1966). *Science* **153,** 487–493.

231. Salzmann, T. N., Ratcliffe, R. W., Christensen, B. G., and Bouffard, F. A. (1980). *J. Am. Chem. Soc.* **102,** 6161–6163. See also Reider, P. J., and Grabowski, E. J. J. (1982). *Tetrahedron Lett.* **23,** 2293–2296.

232. Ikota, N., Shibata, H., and Koga, K. (1980). *Heterocycles* **14,** 1077–1080.

233. Salzmann, T. N., Ratcliffe, R. W., and Christensen, B. G. (1980). *Tetrahedron Lett.* **21,** 1193–1196.

234. Karady, S., Amato, J. S., Reamer, R. A., and Weinstock, L. M. (1981). *J. Am. Chem. Soc.* **103,** 6765–6767.

235. Shiozaki, M., Ishida, N., Hiraoka, T., and Yanagisawa, H. (1981). *Tetrahedron Lett.* **22,** 5205–5208.

236. Baldwin, J. E., Christie, M. A., Haber, S. B., and Kruse, L. I. (1976). *J. Am. Chem. Soc.* **98,** 3045–3047.

237. Baldwin, J. E., Au, A., Christie, M., Haber, S. B., and Hesson, D. (1975). *J. Am. Chem. Soc.* **97,** 5957–5958.

238. Baldwin, J. E., and Christie, M. A. (1978). *J. Am. Chem. Soc.* **100,** 4597–4599.

239. Nakatsuka, S., Tanino, H., and Kishi, Y. (1975). *J. Am. Chem. Soc.* **97,** 5008–5010.

240. Nakatsuka, S., Tanino, H., and Kishi, Y. (1975). *J. Am. Chem. Soc.* **97,** 5010–5012.

241. Koppel, G. A., McShane, L., Jose, F., and Cooper, R. D. G. (1978). *J. Am. Chem. Soc.* **100,** 3933–3935.

242. Cooper, R. D. G., Jose, F., McShane, L., and Koppel, G. A. (1978). *Tetrahedron Lett.* 2243–2246.

243. Mattingly, P. G., Kerwin, Jr., J. F., and Miller, M. J. (1979). *J. Am. Chem. Soc.* **101,** 3983–3985.

244. Mattingly, P. G., and Miller, M. J. (1980). *J. Org. Chem.* **45,** 410–415.

245. Miller, M. J., Mattingly, P. G., Morrison, M. A., Kerwin, Jr., J. F. (1980). *J. Am. Chem. Soc.* **102,** 7026–7032.

246. Bose, A. K., Sahu, D. P., and Manhas, M. S. (1981). *J. Org. Chem.* **46,** 1229–1230.

247. Mattingly, P. G., and Miller, M. J. (1981). *J. Org. Chem.* **46,** 1557–1564.

248. Townsend, C. A., and Nguyen, L. T. (1981). *J. Am. Chem. Soc.* **103,** 4582–4583.

249. Townsend, C. A., and Nguyen, L. T. (1982). *Tetrahedron Lett.* **23,** 4859–4862.

250. Kamiya, T., Oku, T., Nakaguchi, O., Takeno, H., and Hashimoto, M. (1978). *Tetrahedron Lett.* 5119–5122.

251. Kamiya, T., Hashimoto, M., Nakaguchi, O., and Oku, T. (1979). *Tetrahedron* **35,** 323–328.

252. Floyd, D. M., Fritz, A. W. and Cimarusti, C. M. (1982). *J. Org. Chem.* **47,** 176–178.

253. Tenneson, S. M., and Belleau, B. (1980). *Can. J. Chem.* **58,** 1605–1607.

254. Seebach, D., Seuring, B., Kalinowski, H.-O., Lubosch, W., and Renger, B. (1977). *Angew. Chem. Int. Ed. Engl.* **16,** 264–265.

255. Seuring, B., and Seebach, D. (1978). *Liebigs Ann. Chem.* 2044–2073.

256. Mali, R. S., Pohmakotr, M., Weidmann, B., and Seebach, D. (1981). *Liebigs Ann. Chem.* 2272–2284.

257. Seidel, W., and Seebach, D. (1982). *Tetrahedron Lett.* **23,** 159–162.

258. Utimoto, K., Uchida, K., Yamaya, M., and Nozaki, H. (1977). *Tetrahedron Lett.* 3641–3642.

259. Schmidt, U., Gombos, J., Haslinger, E., and Zak, H. (1976). *Chem. Ber.* **109,** 2628–2644.

260. Gombos, J., Haslinger, E., and Schmidt, U. (1976). *Chem. Ber.* **109,** 2645–2647.

261. Sun, K. M., and Fraser-Reid, B. (1980). *Can. J. Chem.* **58,** 2732–2735.

262. Ireland, R. E., and Vevert, J.-P. (1981). *Can. J. Chem.* **59,** 572–583.

263. Hanessian, S., and Frenette, R. (1979). *Tetrahedron Lett.* 3391–3394.

264. Stork, G., Nakahara, Y., Nakahara, Y., and Greenlee, W. J. (1978). *J. Am. Chem. Soc.* **100,** 7775–7777.

265. Tatsuta, K., Nakagawa, A., Maniwa, S., and Kinoshita, M. (1980). *Tetrahedron Lett.* **21,** 1479–1482.

266. Tatsuta, K., Amemiya, Y., Mamiwa, S., and Kinoshita, M. (1980). *Tetrahedron Lett.* **21,** 2837–2840.

267. Tatsuta, K., Yamauchi, T., and Kinoshita, M. (1978). *Bull. Chem. Soc. Jpn.* **51,** 3035–3038.

268. Tatsuta, K., Tanaka, A., Fujimoto, K., and Kinoshita, M. (1977). *J. Am. Chem. Soc.* **99,** 5826–5827.

269. Freiberg, L. A., Egan, R. S., and Washburn, W. H. (1974). *J. Org. Chem.* **39,** 2474–2475.

270. Nicolaou, K. C., Pavia, M. R., and Seitz, S. P. (1979). *Tetrahedron Lett.* 2327–2330.

271. Nicolaou, K. C., Pavia, M. R., and Seitz, S. P. (1981). *J. Am. Chem. Soc.* **103,** 1224–1226.

272. Nicolaou, K. C., Seitz, S. P., and Pavia, M. R. (1981). *J. Am. Chem. Soc.* **103,** 1222–1224.

273. Ziegler, F. E., Gilligan, P. J., and Chakraborty, U. R. (1979). *Tetrahedron Lett.* 3371–3374.

274. Ziegler, F. E., and Gilligan, P. J. (1981). *J. Org. Chem.* **46,** 3874–3880.

275. Tatsuta, K., Amemiya, Y., Kanemura, Y., and Kinoshita, M. (1981). *Tetrahedron Lett.* **22,** 3997–4000.

276. Nakata, M., Ikeyama, Y., Takao, H., and Kinoshita, M. (1980). *Bull. Chem. Soc. Jpn.* **53,** 3252–3258.

277. Rosenthal, A., and Sprinzl, M. (1969). *Can. J. Chem.* **47,** 4477–4481.

278. Tatsuta, K., Amemiya, Y., Kanemura, Y., Takahashi, H., and Kinoshita, M. (1982). *Tetrahedron Lett.* **23,** 3375–3378.

279. Hanessian, S., and Rancourt, G. (1977). *Pure Appl. Chem.* **49,** 1201–1214.

280. Hanessian, S., and Rancourt, G. (1977). *Can. J. Chem.* **55,** 1111–1113.

281. Hanessian, S., Rancourt, G., and Guindon, Y. (1978). *Can. J. Chem.* **56,** 1843–1846.

282. Redlich, H., and Neumann, H.-J., (1981). *Chem. Ber.* **114,** 2029–2050.

283. Masamune, S., Kim, C. U., Wilson, K. E., Spessard, G. O., Georghiou, P. E., and Bates, G. S. (1975). *J. Am. Chem. Soc.* **97,** 3512–3513.

284. Masamune, S., Yamamoto, H., Kamata, S., and Fukuzawa, A. (1975). *J. Am. Chem. Soc.* **97,** 3513–3515.

285. Jarosz, S., and Fraser-Reid, B. (1981). *Tetrahedron Lett.* **22,** 2533–2534.

286. Ireland, R. E., and Daub, J. P. (1981). *J. Org. Chem.* **46,** 479–485.

287. Isobe, M., Ichikawa, Y., and Goto, T. (1981). *Tetrahedron Lett.* **22,** 4287–4290.

288. Costa, S. S., Lagrange, A., Olesker, A., Lukacs, G., and Thang, T. T. (1980). *J. Chem. Soc. Chem. Commun.* 721–723.

289. Aburaki, S., and Kinoshita, M. (1979). *Bull. Chem. Soc. Jpn.* **52,** 198–203.

290. Aburaki, S., Konishi, N., and Kinoshita, M. (1975). *Bull. Chem. Soc. Jpn.* **48,** 1254–1259.

291. Kinoshita, M., Aburaki, S., Wada, M., and Umezawa, S. (1973). *Bull. Chem. Soc. Jpn.* **46,** 1279–1287.

292. Corey, E. J., Weigel, L. O., Chamberlin, A. R., and Lipshutz, B. (1980). *J. Am. Chem. Soc.* **102,** 1439–1441.

293. Corey, E. J., Weigel, L. O., Chamberlin, A. R., Cho, H., and Hua, D.H. (1980). *J. Am. Chem. Soc.* **102,** 6613–6615.

294. Isobe, M., Ichikawa, Y., Kitamura, M., and Goto, T. (1981). *Chem. Letters* 457–460.

295. Barton, D. H. R., Géro, S. D., and Maycock, C. D. (1980). *J. Chem. Soc. Chem. Commun.* 1089–1091.

296. Ho, P.-T. (1980). *Can. J. Chem.* **58,** 858–860.

297. Pan, B.-C., Zhang, H., Pan, C.-Y., Shu, Y., Wang, Z.-F., and Gao, Y.-S. (1980). *Hua Hseuh Hsueh Pao* **38,** 502–506; (1981). *Chem. Abstr.* **94,** 174940.

298. Bonjouklian, R., and Ganem, B. (1979). *Carbohydr. Res.* **76,** 245–251.

299. Meyers, A. I., and Amos, R. A. (1980). *J. Am. Chem. Soc.* **102,** 870–872.

300. Kinoshita, M., and Mariyama, S. (1975). *Bull. Chem. Soc. Jpn.* **48,** 2081–2083.

301. Kinoshita, M., Hamazaki, H., and Awamura, M. (1978). *Bull. Chem. Soc. Jpn.* **51,** 3595–3598.

302. Ireland, R. E., Thaisrivongs, S., and Wilcox, C. S. (1980). *J. Am. Chem. Soc.* **102,** 1155–1157.

303. Ireland, R. E., Thaisrivongs, S., Vanier, N., and Wilcox, C. S. (1980). *J. Org. Chem.* **45,** 48–61.
304. Ireland, R. E., Wilcox, C. S., and Thaisrivongs, S. (1978). *J. Org. Chem.* **43,** 786–787.
305. Nakata, T., Schmid, G., Vranesic, B., Okigawa, M., Smith-Palmer, T., and Kishi, Y. (1978). *J. Am. Chem. Soc.* **100,** 2933–2935.
306. Ireland, R. E., and Wilcox, C. S. (1980). *J. Org. Chem.* **45,** 197–202.
307. Ireland, R. E., Wuts, P. G. M., and Ernst, B. (1981). *J. Am. Chem. Soc.* **103,** 3205–3207.
308. Collum, D. B., McDonald, III, J. H., and Still, W. C. (1980). *J. Am. Chem. Soc.* **102,** 2117–2118.
309. Collum, D. B., McDonald, III, J. H., and Still, W. C. (1980). *J. Am. Chem. Soc.* **102,** 2120–2121.
310. Fukuyama, T., Akasaka, K., Karanewsky, D. S., Wang, C.-L. J., Schmid, G., and Kishi, Y. (1979). *J. Am. Chem. Soc.* **101,** 262–263.
311. Nicolaou, K. C., Claremon, D. A., Papahatjis, D. P., and Magolda, R. L. (1981). *J. Am. Chem. Soc.* **103,** 6969–6971.
312. Hanessian, S., Tyler, P. C., Demailly, G., and Chapleur, Y. (1981). *J. Am. Chem. Soc.* **103,** 6243–6246.
313. Nakahara, Y., Beppu, K., and Ogawa, T. (1981). *Tetrahedron Lett.* **22,** 3197–3200.
314. Hanessian, S., Tyler, P. C., and Chapleur, Y. (1981). *Tetrahedron Lett.* **22,** 4583–4586.
315. Tomioka, K., Mizuguchi, H., and Koga, K. (1978). *Tetrahedron Lett.* 4687–4690.
316. Tomioka, K., and Koga, K. (1979). *Tetrahedron Lett.* 3315–3318.
317. Tomioka, K., and Koga, K. (1979). *Heterocycles* **12,** 1523–1528.
318. Robin, J. P., Gringore, O., and Brown, E. (1980). *Tetrahedron Lett.* **21,** 2709–2712.
319. Anderson, R. C., and Fraser-Reid, B. (1975). *J. Am. Chem. Soc.* **97,** 3870–3871.
320. Ohrui, H., and Emoto, S. (1975). *Tetrahedron Lett.* 3657–3660.
321. Rosenthal, A., and Nguyen, L. (1969). *J. Org. Chem.* **34,** 1029–1034.
322. Anderson, R. C., and Fraser-Reid, B. (1977). *Tetrahedron Lett.* 2865–2868.
323. Anderson, R. C., and Fraser-Reid, B. (1978). *Tetrahedron Lett.* 3233–3236.
324. Ohrui, H., Sueda, N., and Kuzuhara, H. (1981). *Nippon Kagaku Kaishi* **5,** 769–775; (1981). *Chem. Abstr.* **95,** 169612.
325. Tam, T. F., and Fraser-Reid, B. (1980). *J. Chem. Soc. Chem. Commun.* 556–558.
326. Watanabe, M., and Yoshikoshi, A. (1980). *Chem. Lett.* 1315–1318.
327. Fujimoto, Y., Miura, H., Shimizu, T., and Tatsuno, T. (1980). *Tetrahedron Lett.* **21,** 3409–3412.
328. Tanaka, A., and Yamashita, K. (1981). *Chem. Lett.* 319–322.
329. Bloomer, J. L., and Kappler, F. E. (1976). *J. Chem. Soc. Perkin* **I** 1485–1491.
330. Hecht, S. M., Rupprecht, K. M., and Jacobs, P. M. (1979). *J. Am. Chem. Soc.* **101,** 3982–3983.
331. Arai, H., Hagmann, W. K., Suguna, H., and Hecht, S. M. (1980). *J. Am. Chem. Soc.* **102,** 6631–6633.
332. Ohgi, T., and Hecht, S. M. (1981). *J. Org. Chem.* **46,** 1232–1234.
333. Umezawa, Y., Morishima, H., Saito, S., Takita, T., Umezawa, H., Kobayashi, S., Otsuka, M., Narita, M., and Ohno, M. (1980). *J. Am. Chem. Soc.* **102,** 6630–6631.
334. Yoshioka, T., Hara, T., Takita, T., and Umezawa, H. (1974). *J. Antibiotics* **27,** 356–357.
335. Narita, M., Otsuka, M., Kobayashi, S., Ohno, M., Umezawa, Y., Morishma, H., Saito, S., Takita, T., and Umezawa, H. (1982). *Tetrahedron Lett.* **23,** 525–528.

336. Takita, T., Umezawa, Y., Saito, S., Morishima, H., Umezawa, H., Muraoka, Y., Suzuki, M., Otsuka, M., Kobayashi, S., and Ohno, M. (1981). *Tetrahedron Lett.* **22,** 671–674.
337. Saito, S., Umezawa, Y., Morishima, H., Takita, T., Umezawa, H., Nirita, M., Otsuka, M., Kobayashi, S., and Ohno, M. (1982). *Tetrahedron Lett.* **23,** 529–532.
338. Takita, T., Umezawa, Y., Saito, S., Morishima, H., Naganawa, H., Umezawa, H., Tsuchiya, T., Miyake, T., Kageyama, S., Umezawa, S., Muraoka, Y., Suzuki, M., Otsuka, M., Narita, M., Kobayashi, S., and Ohno, M. (1982). *Tetrahedron Lett.* **23,** 521–524.
339. Aoyagi, Y., Katano, K., Suguna, H., Primeau, J., Chang, L.-H., and Hecht, S. M. (1982). *J. Am. Chem. Soc.* **104,** 5537–5538.
340. Tsuchiya, T., Miyake, T., Kageyama, S., Umezawa, S., Umezawa, H., and Takita, T. (1981). *Tetrahedron Lett.* **22,** 1413–1416.
341. Pozsgay, V., Ohgi, T., and Hecht, S. M. (1981). *J. Org. Chem.* **46,** 3761–3763.
342. Shibahara, S., Kondo, S., Maeda, K., Umezawa, H., and Ohno, M. (1972). *J. Am. Chem. Soc.* **94,** 4353–4354.
343. Streicher, W., and Reinshagen, H. (1980). *Carbohydr. Res.* **83,** 383–388.
344. Ogawa, T., Katano, K., and Matsui, M. (1980). *Tetrahedron* **36,** 2727–2733.
345. Ohrui, H., Kuzuhara, H., and Emoto, S. (1971). *Tetrahedron Lett.* 4267–4270.
346. Kuzuhara, H., and Emoto, S. (1973). *Tetrahedron Lett.* 5051–5054.
347. Kuzuhara, H., Ohrui, H., and Emoto, S. (1973). *Tetrahedron Lett.* 5055–5058.
348. Kuzuhara, H., Kimura, M., and Emoto, S. (1975). *Carbohydr. Res.* **45,** 245–255.
349. Hanessian, S., and Roy, R. (1979). *J. Am. Chem. Soc.* **101,** 5839–5841.
350. Hanessian, S., and Roy, R. (1979). *Jpn. J. Antibiot.* **32** (Suppl.), S73–S90; (1980). *Chem. Abstr.* **93,** 132390.
351. Wong, C. M., Buccini, J., Chang, I., Raa, J. T., and Schwenk, R. (1969). *Can. J. Chem.* **47,** 2421–2424.
352. Wong, C. M., Buccini, J., and Raa, J. T. (1968). *Can. J. Chem.* **46,** 3091–3094.
353. Felner, I., and Schenker, K. (1970). *Helv. Chim. Acta* **53,** 754–763.
354. Verheyden, J. P. H., Richardson, A. C., Bhatt, R. S., Grant, B. D., Fitch, W. L., and Moffatt, J. G. (1978). *Pure Appl. Chem.* **50,** 1363–1383.
355. Kakinuma, K., Ōtake, N., and Yonehara, H. (1980). *Tetrahedron Lett.* **21,** 167–168.
356. Leimgruber, W., Batcho, A. D., and Czajkowski, R. C. (1968). *J. Am. Chem. Soc.* **90,** 5641–5643.
357. Rebek, Jr., J., and Gehret, J.-C. E. (1977). *Tetrahedron Lett.* 3027–3028.
358. Rebek, Jr., J., and Shaber, S. (1981). *Heterocycles* **15,** 161–163.
359. Semple, J. E., Wang, P. C., Lysenko, Z., and Joullié, M. M. (1980). *J. Am. Chem. Soc.* **102,** 7505–7510.
360. Just, G., and Payette, D. R. (1980). *Tetrahedron Lett.* **21,** 3219–3222.
361. Silverman, R. B., and Holladay, M. W. (1981). *J. Am. Chem. Soc.* **103,** 7357–7358.
362. Kusumoto, S., Tsuji, S., and Shiba, T. (1974). *Bull. Chem. Soc. Jpn.* **47,** 2690–2695.
363. Goto, T., and Ohgi, T. (1974). *Tetrahedron Lett.* 1413–1416.
364. Kusumoto, S., Tsuji, S., Shima, K., and Shiba, T. (1976). *Bull. Chem. Soc. Jpn.* **49,** 3611–3614.
365. Yamada, S., Koga, K., and Matsuo, H. (1963). *Chem. Pharm. Bull.* **11,** 1140–1145.
366. Ottenheijm, H. C. J., Liskamp, R. M. J., van Nispen, S. P. J. M., Boots, H. A., and Tijhuis, M. W. (1981). *J. Org. Chem.* **46,** 3273–3283.
367. Bastian, G., Bessodes, M., Panzica, R. P., Abushanab, E., Chen, S.-F., Stoeckler, J. D., and Parks, Jr., R. E. (1981). *J. Med. Chem.* **24,** 1383–1385.
368. Thiem, J., and Wessel, H.-P. (1980). *Tetrahedron Lett.* **21,** 3571–3574.

369. Dodd, J. H., and Weinreb, S. M. (1979). *Tetrahedron Lett.* 3593–3596.
370. Thiem, J., and Wessel, H.-P. (1981). *Liebigs Ann. Chem.* 2216–2227.
371. Mincher, D. J., and Shaw, G. (1981). *J. Chem. Soc. Chem. Commun.* 508–510.
372. Fraser-Reid, B., and Walker, D. L. (1980). *Can. J. Chem.* **58,** 2694–2702.
373. van Tamelen, E. E., Demers, J. P., Taylor, E. G., and Koller, K. (1980). *J. Am. Chem. Soc.* **102,** 5424–5425.
374. Laguzza, B. C., and Ganem, B. (1981). *Tetrahedron Lett.* **22,** 1483–1486.
375. Iwakawa, M., Pinto, B. M., and Szarek, W. A. (1978). *Can. J. Chem.* **56,** 326–335.
376. Feit, P. W. (1964). *J. Med. Chem.* **7,** 14–17.
377. Greene, A. E. (1980). *J. Am. Chem. Soc.* **102,** 5337–5343.
378. Secrist, III, J. A., and Barnes, K. D. (1980). *J. Org. Chem.* **45,** 4526–4528.
379. Ohzeki, M., Mizoguchi, T., Koga, K., and Yamada, S. (1977). *Chem. Pharm Bull.* **25,** 2676–2680.
380. Just, G., and Potvin, P. (1980). *Can. J. Chem.* **58,** 2173–2177.
381. Barnes, N. J., Davidson, A. H., Hughes, L. R., Procter, G., and Rajcoomar, V. (1981). *Tetrahedron Lett.* **22,** 1751–1754.
382. Tomioka, K., and Koga, K. (1980). *Tetrahedron Lett.* **21,** 2321–2324.
383. Tulshian, D. B., and Fraser-Reid, B. (1981). *J. Am. Chem. Soc.* **103,** 474–475.
384. Tam, T. F., and Fraser-Reid, B. (1980). *J. Org. Chem.* **45,** 1344–1346.
385. Pyne, S. G., Hensel, M. J., and Fuchs, P. L. (1982). *J. Am. Chem. Soc.* **104,** 5719–5728.
386. Nakata, M., Takao, H., Ikeyama, Y., Sakai, T., Tatsuta, K., and Kinoshita, M. (1981). *Bull. Chem. Soc. Jpn.* **54,** 1749–1756.
387. Ohrui, H., and Emoto, S. (1978). *Tetrahedron Lett.* 2095–2098.
388. Pougny, J.-R., and Sinaÿ, P. (1978). *Tetrahedron Lett.* 3301–3304.
389. Sueda, N., Ohrui, H., and Kuzuhara, H. (1979). *Tetrahedron Lett.* 2039–2042.
390. Pietraszkiewicz, M., and Sinaÿ, P. (1979). *Tetrahedron Lett.* 4741–4744.
391. Cutting, W. (1967). "Handbook of Pharmacology," 3rd ed. Appleton-Century-Crofts, New York.
392. Danilewicz, J. C., and Kemp, J. E. G. (1973). *J. Med. Chem.* **16,** 168–169.
393. Caroon, J. M., Clark, R. D., Kluge, A. F., Nelson, J. T., Strosberg, A. M., Unger, S. H., Michel, A. D., and Whiting, R. L. (1981). *J. Med. Chem.* **24,** 1320–1328.
394. Weinstock, L. M., Mulvey, D. M., and Tull, R. (1976). *J. Org. Chem.* **41,** 3121–3124.
395. Nelson, W. L., Wennerstrom, J. E., Dyer, D. C., and Engel, M. (1977). *J. Med. Chem.* **20,** 880–885.
396. Nelson, W. L., and Wennerstrom, J. E. (1976). *J. Chem. Soc. Chem. Commun.* 921–922.
397. Delalande, S. A. (1981). Fr. Patent No. 2459800.
398. Tsuda, Y., Yoshimoto, K., and Nishikawa, T. (1981). *Chem. Pharm. Bull.* **29,** 3593–3600.
399. Hardegger, E., and Lohse, F. (1957). *Helv. Chim. Acta* **40,** 2383–2389.
400. Mubarak, A. M., and Brown, D. M. (1982). *J. Chem. Soc. Perkin* **I** 809–813.
401. Pochet, S., Huynh-Dinh, T. (1982). *J. Org. Chem.* **47,** 193–198.
402. Cox, H. C., Hardegger, E., Kögl, F., Liechti, P., Lohse, F., and Salemink, C. A. (1958). *Helv. Chim. Acta* **41,** 229–234.
403. Hardegger, E., Furter, H., and Kiss, J. (1958). *Helv. Chim. Acta* **41,** 2401–2410.
404. Wang, P.-C., and Joullié, M. M. (1980). *J. Org. Chem.* **45,** 5359–5363.
405. Triggle, D. J., and Belleau, B. (1962). *Can. J. Chem.* **40,** 1201–1215.
406. Buckley, III, T. F., and Rapoport, H. (1981). *J. Am. Chem. Soc.* **103,** 6157–6163.
407. Chandrakumar, N. S., and Hajdu, J. (1981). *Tetrahedron Lett.* **22,** 2949–2952.

408. Witiak, D. T., Grattan, D. A., Heaslip, R. J., and Rahwan, R. G. (1981). *J. Med. Chem.* **24,** 712–717.
409. Tang, K.-C., Tropp, B. E., and Engel, R. (1978). *Tetrahedron* **34,** 2873–2878.
410. Newman, H. (1973). *J. Am. Chem. Soc.* **95,** 4098–4099.
411. Newman, H. (1974). *J. Org. Chem.* **39,** 100–103.
412. Corey, E. J., and Pearce, H. L. (1979). *J. Am. Chem. Soc.* **101,** 5841–5843.
413. Mechoulam, R., Braun, P., and Gaoni, Y. (1967). *J. Am. Chem. Soc.* **89,** 4552–4554.
414. Petrzilka, T., Haefliger, W., Sikemeier, C., Ohloff, G., and Eschenmoser, A. (1967). *Helv. Chim. Acta* **50,** 719–723.
415. Franke, I., and Binder, M. (1980). *Helv. Chim. Acta* **63,** 2508–2514.
416. Uliss, D. B., Razdan, R. K., Dalzell, H. C., and Handrick, G. R. (1977). *Tetrahedron 33,* 2055–2059.
417. Tietze, L.-F., von Kiedrowski, G., Harms, K., Clegg, W., and Sheldrick, G. (1980). *Angew. Chem. Int. Ed. Eng.* **19,** 134–135.
418. Tietze, L.-F., and von Kiedrowski, G. (1981). *Tetrahedron Lett.* **22,** 219–222.
419. Miller, D. D., Hsu, F.-L., Ruffolo, Jr., R. R., and Patil, P. N. (1976). *J. Med. Chem.* **19,** 1382–1384.
420. Hsu, F.-L., Hamada, A., Booher, M. E., Fuder, H., Patil, P. N., and Miller, D. D. (1980). *J. Med. Chem.* **23,** 1232–1235.
421. Šunjić, V., Kajfež, F., Štromar, I., Blažević, N., and Kolbah, D. (1973). *J. Heterocycl. Chem.* **10,** 591–599.
422. Danishefsky, S., Berman, E., Clizbe, L. A., and Hirama, M. (1979). *J. Am. Chem. Soc.* **101,** 4385–4386.
423. Rossi, R., and Salvadori, P. A. (1979). *Synthesis* 209–210.
424. Banno, K., and Mukaiyama, T. (1976). *Chem. Lett.* 279–282.
425. Mori, K. (1974). *Tetrahedron* **30,** 3817–3820.
426. Rossi, R., and Carpita, A. (1977). *Tetrahedron* **33,** 2447–2450.
427. Rossi, R., Salvadori, P. A., Carpita, A., and Niccoli, A. (1979). *Tetrahedron* **35,** 2039–2042.
428. Mori, K., Masuda, S., and Suguro, T. (1981). *Tetrahedron* **37,** 1329–1340.
429. Suguro, T., and Mori, K. (1979). *Agric. Biol. Chem.* **43,** 409–410.
430. Mori, K., Kuwahara, S., Levinson, H. Z., and Levinson, A. R. (1982). *Tetrahedron* **38,** 2291–2297.
431. Jensen, U., and Schäfer, H.-J. (1981). *Chem. Ber.* **114,** 292–297.
432. Cernigliaro, G. J., and Kocienski, P. J. (1977). *J. Org. Chem.* **42,** 3622–3624.
433. Suguro, T., and Mori, K. (1979). *Agric. Biol. Chem.* **43,** 869–870.
434. Masuda, S., Kuwahara, S., Suguro, T., and Mori, K. (1981). *Agric. Biol. Chem.* **45,** 2515–2520.
435. Mori, K., and Kuwahara, S. (1982). *Tetrahedron* **38,** 521–525.
436. Snider, B. B., and Rodini, D. (1978). *Tetrahedron Lett.* 1399–1400.
437. Anderson, R. J., Adams, K. G., Chinn, H. R., and Henrick, C. A. (1980). *J. Org. Chem.* **45,** 2229–2236.
438. Roelofs, W. L., Gieselmann, M. J., Cardé, A. M., Tashiro, H., Moreno, D. S., Henrick, C. A., and Anderson, R. J. (1977). *Nature* **267,** 698–699.
439. Roelofs, W., Gieselmann, M., Cardé, A., Tashiro, H., Moreno, D. S., Henrick, C. A., and Anderson, R. J. (1978). *J. Chem. Ecol.* **4,** 211–224.
440. Heath, R. R., Doolittle, R. E., Sonnet, P. E., and Tumlinson, J. H. (1980). *J. Org. Chem.* **45,** 2910–2912.
441. Mori, K. (1977). *Tetrahedron* **33,** 289–294.
442. Pougny, J.-R., and Sinaÿ, P. (1982). *J. Chem. Res. (M)* 0186–0196.

443. Mori, K., Nomi, H., Chuman, T., Kohno, M., Kato, K., and Noguchi, M. (1981). *Tetrahedron Lett.* **22,** 1127–1130.
444. Mori, M., Chuman, T., Kohno, M., Kato, K., Noguchi, M., Nomi, H., and Mori, K. (1982). *Tetrahedron Lett.* **23,** 667–670.
445. Byström, S., Högberg, H.-E., and Norin, T. (1981). *Tetrahedron* **37,** 2249–2254.
446. Hanessian, S., Demailly, G., Chapleur, Y., and Leger, S. (1981). *J. Chem. Soc. Chem. Commun.* 1125–1126.
447. Schuler, H. R., and Slessor, K. N. (1977). *Can. J. Chem.* **55,** 3280–3287.
448. Mori, K., Mizumachi, N., and Matsui, M. (1976). *Agric. Biol. Chem.* **40,** 1611–1615.
449. Mori, K. (1976). *Tetrahedron* **32,** 1101–1106.
450. Kondo, K., Dobashi, S., and Matsumoto, M. (1976). *Chem. Lett.* 1077–1080.
451. Ohloff, G., and Giersch, W. (1977). *Helv. Chim. Acta* **60,** 1496–1500.
452. Mori, K. (1976). *Tetrahedron Lett.* 1609–1612.
453. Ravid, U., Silverstein, R. M., and Smith, L. R. (1978). *Tetrahedron* **34,** 1449–1452.
454. Doolittle, R. E., Tumlinson, J. H., Proveaux, A. T., and Heath, R. R. (1980). *J. Chem. Ecol.* **6,** 473–485.
455. Smith, L. R., Williams, H. J., and Silverstein, R. M. (1978). *Tetrahedron Lett.* 3231–3232.
456. Redlich, H., and Francke, W. (1980). *Angew. Chem. Int. Ed. Engl.* **19,** 630–631.
457. Redlich, H. (1982). *Liebigs Ann. Chem.* 708–716.
458. Barton, D. H. R., and McCombie, S. W. (1975). *J. Chem. Soc. Perkin* **I** 1574–1585.
459. Hintzer, K., Weber, R., and Schurig, V. (1981). *Tetrahedron Lett.* **22,** 55–58.
460. Mori, K. (1976). *Tetrahedron* **32,** 1979–1981.
461. Sum, P.-E., and Weiler, L. (1978). *Can. J. Chem.* **56,** 2700–2702.
462. Sum, P.-E., and Weiler, L. (1982). *Can. J. Chem.* **60,** 327–334.
463. Plaumann, D. E., Fitzsimmons, B. J., Ritchie, B. M., and Fraser-Reid, B. (1982). *J. Org. Chem.* **47,** 941–946.
464. Mori, K. (1974). *Tetrahedron* **30,** 4223–4227.
465. Meyer, H. H. (1977). *Liebigs Ann. Chem.* 732–736.
466. Bernardi, R., Fuganti, C., and Grasselli, P. (1981). *Tetrahedron Lett.* **22,** 4021–4024.
467. Sherk, A. E., and Fraser-Reid, B. (1982). *J. Org. Chem.* **47,** 932–935.
468. Ohrui, H., and Emoto, S. (1976). *Agric. Biol. Chem.* **40,** 2267–2270.
469. Kawana, M., and Emoto, S. (1975). *Tetrahedron Lett.* 3395–3398.
470. Jarosz, S., Hicks, D. R., and Fraser-Reid, B. (1982). *J. Org. Chem.* **47,** 935–940.
471. Hicks, D. R., and Fraser-Reid, B. (1974). *Synthesis* 203.
472. Magnus, P., and Roy, G. (1978). *J. Chem. Soc. Chem. Commun.* 297–298.
473. Redlich, H., Schneider, B., and Francke, W. (1980). *Tetrahedron Lett.* **21,** 3009–3012.
474. Redlich, H., Schneider, B., and Francke, W. (1980). *Tetrahedron Lett.* **21,** 3013–3016.
475. Iwaki, S., Marumo, S., Saito, T., Yamada, M., and Katagiri, K. (1974). *J. Am. Chem. Soc.* **96,** 7842–7844.
476. Mori, K., Takigawa, T., and Matsui, M. (1979). *Tetrahedron* **35,** 833–837.
477. Pawson, B. A., Cheung, H.-C., Gurbaxani, S., and Saucy, G. (1970). *J. Am. Chem. Soc.* **92,** 336–343.
478. Williams, D. R., and Phillips, J. G. (1981). *J. Org. Chem.* **46,** 5452–5454.
479. Meinwald, J., and Jones, T. H. (1978). *J. Am. Chem. Soc.* **100,** 1883–1886.
480. Baker, R., Evans, D. A., and McDowell, P. G. (1977). *J. Chem. Soc. Chem. Commun.* 111.
481. Hobbs, P. D., and Magnus, P. D. (1976). *J. Am. Chem. Soc.* **98,** 4594–4600.

482. Bierl-Leonhardt, B. A., Moreno, D. S., Schwarz, M., Fagerlund, J., and Plimmer, J. R. (1981). *Tetrahedron Lett.* **22,** 389–392.
483. Mori, K., Ebata, T., and Sakakibara, M. (1981). *Tetrahedron* **37,** 709–713.
484. Musich, J. A., and Rapoport, H. (1978). *J. Am. Chem. Soc.* **100,** 4865–4872.
485. Redlich, H., and Xiang-jun, J. (1982). *Liebigs Ann. Chem.* 717–722.
486. Taniguchi, M., Koga, K., and Yamada, S. (1974). *Tetrahedron* **30,** 3547–3552.
487. Taniguchi, M., Koga, K., and Yamada, S. (1974). *Chem. Pharm. Bull.* **22,** 2318–2323.
488. Fuganti, C., Grasselli, P., and Pedrocchi-Fantoni, G. (1981). *Tetrahedron Lett.* **22,** 4017–4020.
489. Fronza, G., Fuganti, C., Grasselli, P., and Marinoni, G. (1979). *Tetrahedron Lett.* 3883–3386.
490. Fronza, G., Fuganti, C., and Grasselli, P. (1980). *Tetrahedron Lett.* **21,** 2999–3000.
491. Koga, K., Yamada, S., Yoh, M., and Mizoguchi, T. (1974). *Carbohydr. Res.* **36,** C9–C11.
492. David, S., and Veyrières, A. (1970). *Carbohydr. Res.* **13,** 203–209.
493. Marks, J. [n.d.] "A Guide to the Vitamins: Their Role in Health and Disease." Medical and Technical Publishing, Lancaster, England.
494. Wasmuth, D., Arigoni, D., and Seebach, D. (1982). *Helv. Chim. Acta* **65,** 344–352.
495. Bindra, J. S., and Bindra, R. (1975). "Creativity in Organic Synthesis," Vol. 1, pp. 297–314. Academic Press, New York.
496. Stevens, R. V. (1982). *In* "B_{12}" (D. Dolphin, ed.), Vol. 1, pp. 169–200. Wiley (Interscience), New York.
497. Woodward, R. B. (1968). *Pure Appl. Chem.* **17,** 519–547.
498. Woodward, R. B. (1971). *Pure Appl. Chem.* **25,** 283–304.
499. Woodward, R. B. (1973). *Pure Appl. Chem.* **33,** 145–177.
500. Eschenmoser, A., and Wintner, C. E. (1977). *Science* **196,** 1410–1420.
501. Friedrich, W., Gross, G., Bernhauer, K., and Zeller, P. (1960). *Helv. Chim. Acta* **43,** 704–712.
502. Ogawa, T., Kawano, T., and Matsui, M. (1977). *Carbohydr. Res.* **57,** C31–C35.
503. Ohrui, H., Sueda, N., and Emoto, S. (1978). *Agric. Biol. Chem.* **42,** 865–868.
504. Ohrui, H., and Emoto, S. (1975). *Tetrahedron Lett.* 2765–2766.
505. Vogel, F. G. M., Paust, J., and Nürrenbach, A. (1980). *Liebigs Ann. Chem.* 1972–1977.
506. Ohrui, H., Kuzuhara, H., and Emoto, S. (1970). *Agric. Biol. Chem.* **34,** 375–380.
507. Ohrui, H., Kuzuhara, H., and Emoto, S. (1971). *Agric. Biol. Chem.* **35,** 752–755.
508. Kuzuhara, H., Ohrui, H., and Emoto, S. (1971). *Agric. Biol. Chem.* **35,** 8–17.
509. Confalone, P. N., Pizzolato, G., Baggiolini, E. G., Lollar, D., and Uskoković, M. R. (1975). *J. Am. Chem. Soc.* **97,** 5936–5938.
510. Baggiolini, E. G., Lee, H. L., Pizzolato, G., and Uskoković, M. R. (1982). *J. Am. Chem. Soc.* **104,** 6460–6462.
511. a. Kasparek, S. (1980). *In* "Vitamin E: A Comprehensive Treatise" (L. Machlin, ed.), pp. 7–65. Dekker, New York. b. Saucy, G., and Cohen, N. (1981). *In* "Studies in Organic Chemistry" (Z. Yoshida, ed.), Vol. 6, pp. 1–65. Elsevier, Amsterdam.
512. Takahashi, J., Mori, K., and Matsui, M. (1979). *Agric. Biol. Chem.* **43,** 1605–1606.
513. Fujisawa, T., Sato, T., Kawara, T., and Ohashi, K. (1981). *Tetrahedron Lett.* **22,** 4823–4826.
514. Koreeda, M., and Brown, L. (1982). Abstracts, 183rd. American Chemical Society Meeting, Las Vegas, Nevada, O41.
515. Trost, B. M., and Klun, T. P. (1981). *J. Am. Chem. Soc.* **103,** 1864–1865.

516. DeLuca, H. F., Paaren, H. G., and Schnoes, H. U. (1979). *Top. Curr. Chem.* **83,** 1–65.
517. Baggiolini, E. G., Iacobelli, J. A., Hennessey, B. M., and Uskoković, M. R. (1982). *J. Am. Chem. Soc.* **104,** 2945–2948.
518. Takayama, H., Ohmori, M., and Yamada, S. (1980). *Tetrahedron Lett.* **21,** 5027–5028.
519. ApSimon J. (ed) (1973/1983). "The Total Synthesis of Natural Products", Vol. 2 and Vol. 5. Wiley (Interscience), New York.
520. Büchi, G., MacLeod, Jr., W. D., and Padilla, O. J. (1964). *J. Am. Chem. Soc.* **86,** 4438–4444. See also Näf, F., Decorzant, R., Giersch, W., and Ohloff, G. (1981). *Helv. Chim. Acta* **64,** 1387–1397.
521. Matsui, M., Yoshioka, H., Sakamoto, H., Yamada, Y., and Kitahara, T. (1967). *Agric. Biol. Chem.* **31,** 33–39.
522. Cocker, W., St. J. Lauder, H., and Shannon, P. V. R. (1972). *J. Chem. Soc. Chem. Commun.* 684–685.
523. Gopichand, Y., Khanra, A. S., Mitra, R. B., and Chakravarti, K. K. (1975). *Ind. J. Chem.* **13,** 433–436.
524. Khanra, A. S., and Mitra, R. B. (1976). *Ind. J. Chem.* **14B,** 716–718.
525. Ho, T.-L., and Din, Z. U. (1980). *Synth. Comm.* **10,** 921–927.
526. Mane, M., Gore, K. G., and Kulkarni, G. H. (1980). *Ind. J. Chem.* **198,** 605–607.
527. Nakayama, M., Ohira, S., Shinke, S., Matsushita, Y., Matsuo, A., and Hayashi, S. (1979). *Chem. Lett.* 1245–1246.
528. Crombie, L, Kneen, G., Pattenden, G., and Whybrow, D. (1980). *J. Chem. Soc. Perkin* **I** 1711–1717.
529. Mitra, R. B., and Khanra, A. S. (1977). *Synth. Comm.* **7,** 245–250.
530. Matsuo, T., Mori, K., and Matsui, M. (1976). *Tetrahedron Lett.* 1979–1982.
531. Fitzsimmons, B. J., and Fraser-Reid, B. (1979). *J. Am. Chem. Soc.* **101,** 6122–6124.
532. Scharf, H.-D., Kalkoff, H., and Janus, J. (1979). *Tetrahedron* **35,** 2513–2521.
533. Kutney, J. P., Choudhury, M. K., Decesare, J. M., Jacobs, H., Singh, A. K., and Worth, B. R. (1981). *Can. J. Chem.* **59,** 3162–3167.
534. Danishefsky, S., Morris, J., and Clizbe, L. A. (1981). *Heterocycles* **15,** 1205–1207.
535. Danishefsky, S., Morris, J., and Clizbe, L. A. (1981). *J. Am. Chem. Soc.* **103,** 1602–1604.
536. Danishefsky, S., and Craig, T. A. (1981). *Tetrahedron* **37,** 4081–4086.
537. Rastetter, W. H., Erickson, T. J., and Venuti, M. C. (1981). *J. Org. Chem.* **46,** 3579–3590.
538. Ohfune, Y., and Nomoto, K. (1981). *Chem. Lett.* 827–828.
539. Fushiya, S., Sato, Y., Nakatsuyama, S., Kanuma, N., and Nozoe, S. (1981). *Chem. Lett.* 909–912.
540. Fushiya, S., Sato, Y., Nozoe, S. (1980). *Chem. Lett.* 1215–1218.
541. Ohfune, Y., Tomita, M., and Nomoto, K. (1981). *J. Am. Chem. Soc.* **103,** 2409–2410.
542. Fushiya, S., Nakatsuyama, S., Sato, Y., and Nozoe, S. (1981). *Heterocycles* **15,** 819–822.
543. Stevens, R. V., and Gaeta, F. C. A. (1977). *J. Am. Chem. Soc.* **99,** 6105–6106.
544. Cohen, N., Eichel, W. F., Lopresti, R. J., Neukom, C., and Saucy, G. (1976). *J. Org. Chem.* **41,** 3512–3515.
545. Nakata, T., and Kishi, Y. (1978). *Tetrahedron Lett.* 2745–2748.
546. Evans, D. A., Sacks, C. E., Kleschick, W. A., and Taber, T. R. (1979). *J. Am. Chem. Soc.* **101,** 6789–6791.
547. Nagaoka, H., and Kishi, Y. (1981). *Tetrahedron* **37,** 3873–3888.

548. Masamune, S., Imperiali, B., and Garvey, D. S. (1982). *J. Am. Chem. Soc.* **104,** 5528–5531.
549. Kozikowski, A. P., and Chen, Y.-Y. (1981). *J. Org. Chem.* **46,** 5248–5250.
550. Byon, C.-Y., Gut, M., and Toome, V. (1981). *J. Org. Chem.* **46,** 3901–3903.
551. Meyers, A. I., and Hudspeth, J. P. (1981). *Tetrahedron Lett.* **22,** 3925–3928.
552. Ohfune, Y., and Tomita, M. (1982). *J. Am. Chem. Soc.* **104,** 3511–3513.
553. McGuirk, P. R., and Collum, D. B. (1982). *J. Am. Chem. Soc.* **104,** 4496–4497.
554. Takano, S., Seya, K., Goto, E., Hirama, M., and Ogasawara, K. (1983). *Synthesis* 116–117.
555. Cohen, N., Lopresti, R. J., and Saucy, G. (1979). *J. Am. Chem. Soc.* **101,** 6710–6716.
556. Lythgoe, B. (1980). *Chem. Soc. Rev.* **9,** 449–475.
557. Parry, R. J., and Naidu, M. V. (1983). *Tetrahedron Lett.* **24,** 1133–1134.
558. Schlessinger, R. H., and Poss, M. A. (1982). *J. Am. Chem. Soc.* **104,** 357–358.
559. Klein, L. L., McWhorter, Jr., W. W., Ko, S. S., Pfaff, K.-P., Kishi, Y., Uemura, D., and Hirata, Y. (1982). *J. Am. Chem. Soc.* **104,** 7362–7364.
560. Ko, S. S., Finan, J. M., Yonaga, M., Kishi, Y., Uemura, D., and Hirata, Y. (1982). *J. Am. Chem. Soc.* **104,** 7364–7367.

2

Optical Activation and Utilization of Compounds Containing Chiral Sulfur Centers

Michael R. Barbachyn
Carl R. Johnson
Department of Chemistry
Wayne State University
Detroit, Michigan

I. Introduction

The preparation and chemistry of chiral organosulfur reagents have received much attention since the early 1960s. Interest in this area was initially concerned with the mechanism of nucleophilic substitution at tri- and tetracoordinate sulfur and, more recently, with the exceptional versatility these compounds exhibit in asymmetric reactions. The present chapter focuses on those compounds having demonstrated or potential use as

ISBN 0-12-507704-1

reagents in asymmetric synthesis. This review will first highlight methods for the optical activation of chiral organosulfur reagents—especially sulfoxides, sulfonium salts, and sulfoximines and their derivatives—with emphasis placed on those procedures furnishing compounds of high optical purity. Uses of these optically active organosulfur compounds as reagents for asymmetric synthesis will then be discussed. A review that covers a wider range of chiral organosulfur compounds has been published (*1*). Several other general reviews pertinent to this area are available (*2a–g*).

II. The Optical Activation of Chiral Organosulfur Compounds

A. Sulfoxides

A number of techniques have been developed for the preparation of optically active sulfoxides. Approaches that have been used include (*1, 2*) (*a*) classical resolution of racemic sulfoxides containing a resolving handle (e.g., hydroxyl, amino, and acidic groups), (*b*) formation and separation of optically active diastereomeric transition-metal complexes, (*c*) direct resolution of racemic sulfoxides by chromatography over optically activated stationary phases, (*d*) optical enrichment by incorporation into inclusion compounds with a chiral host molecule, (*e*) asymmetric oxidation of sulfides, (*f*) kinetic resolution by partial oxidation or reduction with chiral reagents, and (*g*) stereospecific nucleophilic substitution on a variety of optically active sulfinyl compounds.

Harrison, Kenyon, and Phillips (*3*) first employed the classical resolution method in 1926 for the optical activation of 4-aminophenyl 4-methylphenyl sulfoxide (**1**) and 3-carboxyphenyl methyl sulfoxide (**2**) via the

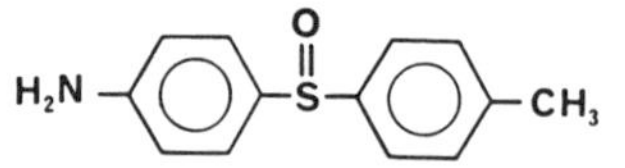

1

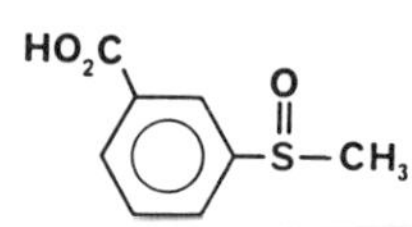

2

separation of diastereomeric salts of *d*-camphorsulfonic acid and brucine, respectively. Subsequently, a number of examples (*4a–d*) of this resolution scheme have appeared, including one (*5*) where a phosphonic acid group was used in the resolution of the sulfoxide **3**.

The optical resolution of 1,3-dithiane 1-oxide by (*a*) addition of the lithio derivative to (+)-camphor, (*b*) separation of the diastereomeric

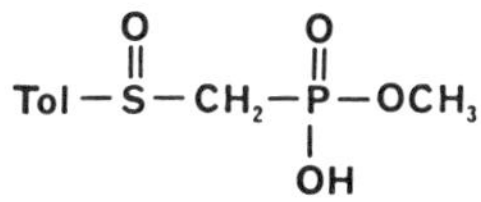

3

Tol = p-CH_3—C_6H_4—

adducts **4** and **5**, and (*c*) base-catalyzed retroaddition of each has been accomplished. (*6*)

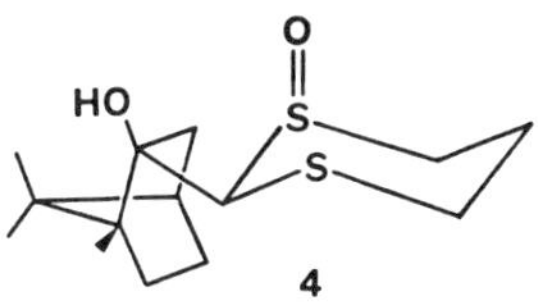

4

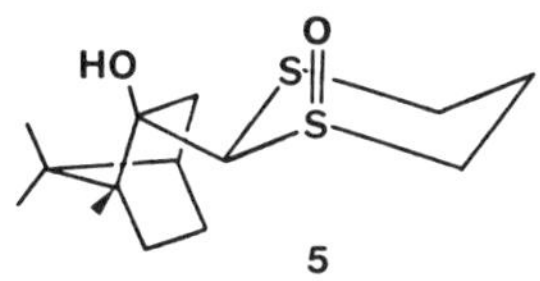

5

Optically active transition-metal complexes have become increasingly important in the resolution of chiral sulfoxides. Backer and Keuning resolved 2,5-dithiaspiro-[3.3]heptane 2,5-dioxide with a chiral cobalt complex in 1934 (*7*). An elegant resolution of ethyl *p*-tolyl sulfoxide was accomplished by the fractional recrystallization of diastereomeric complexes derived from (+)- and (−)-*trans*-dichloroethylene(α-methylbenzylamine)platinum(II) and subsequent release of the optically pure sulfoxide with aqueous sodium cyanide (*8*). In a more recent development, alkylaryl sulfoxides were resolved by high-performance liquid chromatography (HPLC) of diastereomeric platinum complexes derived from *trans*-chloro(*N*,*N*-dimethyl-D-phenylglycine)(ethylene)platinum(II) (*9*). Interestingly, the faster eluting component was always the (*R*)-sulfoxide complex.

The development of chiral stationary phases for the direct chromatographic separation of enantiomeric sulfoxides has received a great deal of attention. Montanari and co-workers utilized an activated α-lactose stationary phase to partially resolve racemic unsaturated vinyl disulfoxides in 1959 (*10*). The bonding of a (*R*)-2,2,2-trifluoro-1-(9-anthryl)ethanol moiety (**6**) to silica gel has provided a stationary phase useful in separation of sulfoxide enantiomers using HPLC (*11a,b*).

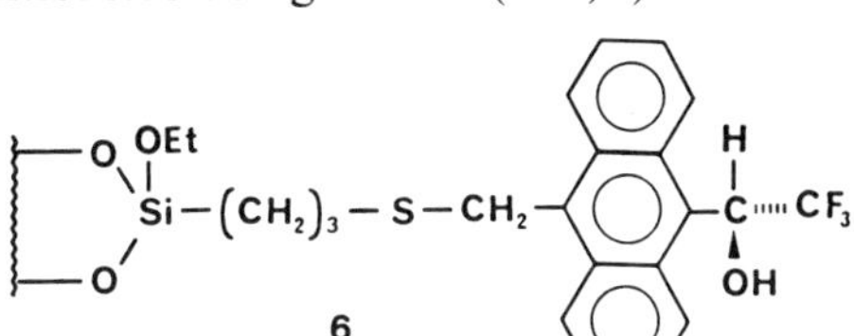

6

Another nonclassical approach to the optical resolution of chiral sulfoxides involves the formation of inclusion complexes from racemic sulfoxides (guests) and β-cyclodextrin (host) (*12*). The precipitated inclusion compounds can be decomposed to release the optically active sulfoxides, albeit of low optical purity [up to 14.5% enantiomeric excess (ee) after one inclusion process].

The asymmetric oxidation of sulfides using chiral peracids was first described in 1960 (*13a–c*). In these oxidations the degree of asymmetric induction is generally very low (0–8% ee), and the absolute configuration of the sulfoxide product is largely unpredictable (*14a,b*). Davis reported the use of optically active 2-sulfonyloxaziridines (e.g., **7**) for the asym-

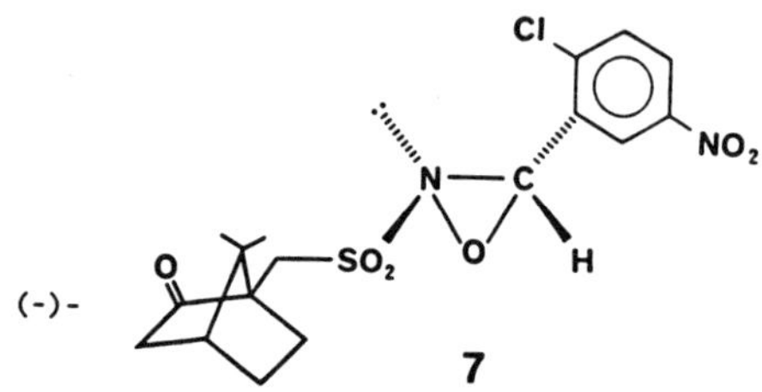

7

metric oxidation of sulfides (*15*). The observed enantioselectivity (up to 46% ee) is the highest reported for a chiral oxidizing agent, and additionally, the absolute configuration of the optically active sulfoxides can be reliably predicted using a chiral recognition model.

An alternative optical activation scheme employing a chiral catalyst for the asymmetric oxidation of sulfides to chiral sulfoxides has had some success. While earlier results (*16a,b*) were disappointing (<10% ee), a more recently developed protein-mediated two-phase sodium metaperiodate oxidation has afforded optical purities of up to 81% on a small scale (*17a–c*). An electrochemical asymmetric oxidation of cyclohexyl phenyl sulfide on poly(L-valine)-coated platinum electrodes has been reported (*18*). The product (−)-cyclohexyl phenyl sulfoxide was obtained in 54% optical yield.

The oxidation of unsymmetric sulfides by microbes is an interesting method for the preparation of optically active sulfoxides (Table 1). Henbest studied the oxidation of alkyl *p*-tolyl sulfides with *Aspergillus niger* and observed optical purities of 32% (methyl *p*-tolyl sulfoxide) to 99% (*tert*-butyl *p*-tolyl sulfoxide) (*19*). The variation in asymmetric induction indicates the high sensitivity toward substrate structure. *Mortierella isabellina* NRRL 1757 and *Helminthosporium* sp. NRRL 4671 were used to stereospecifically oxidize ethyl and methyl *p*-tolyl sulfide to the corresponding optically pure sulfoxides in good chemical yield (*20*).

Optically active sulfoxides have also been prepared from resolved heterosulfonium salts obtained from the corresponding sulfides. Johnson first employed this route in 1972 (*21*). Benzyl *p*-tolyl sulfide was treated

TABLE 1
Microbial Oxidation of Sulfides to Sulfoxides

R	R′	Microbe[a]	Chemical yield (%)	Optical purity(%)	Rotation sign	Reference
Tol	*t*Bu	A	25	99	+	*19*
Tol	*i*Pr	A	12	70	+	*19*
Tol	*n*Bu	A	14	32	+	*19*
Tol	Me	A	48	35	+	*19*
$PhCH_2$	*t*Bu	A	65	91	−	*19*
$PhCH_2$	Ph	A	14	26	−	*19*
Tol	Me	B	60	100	+	*20*
Tol	Me	C	50	100	−	*20*

[a] A, *Aspergillus niger;* B, *Mortierella isabellina* NRRL 1757; C, *Helminthosporium* sp. NRRL 4671.

sequentially with 1-chlorobenzotriazole, *l*-menthol, and silver tetrafluoroborate to afford the optically active (+)-menthoxysulfonium salt **8**, which after recrystallization and basic hydrolysis (*22*) furnished (−)-benzyl *p*-tolyl sulfoxide (87% optically pure). A number of related reports have since appeared (*23a,b,c*) including one in which the resolved aminosulfonium salt **9** was converted via oxidation and deimination to optically pure (−)-methyl *p*-tolyl sulfoxide (*24*). Similarly, Oae reported the stereoselective deimination of (+)-*S*-methyl-*S*-phenylsulfoximine to (−)-methyl phenyl sulfoxide utilizing *p*-toluenesulfonyl nitrite (*25*) in an extension of Cram's work (*26a–c*).

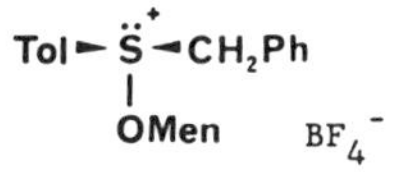

8

Men = *l*-menthyl

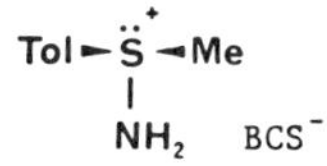

9

BCS = bromocamphorsulfonate

Racemic sulfoxides have been kinetically resolved by partial oxidation (*27a,b*) or reduction (*28a–c*) with a deficiency of chiral reagent. The degree of asymmetric induction is generally low (<5%), although a few examples have attained higher optical purities. One of these, a Pummerer reaction of racemic sulfoxides with optically active α-phenylbutyric acid chloride and base, afforded unreduced sulfoxides in good optical purity (up to 70%) (*29*). The lithio derivative of racemic methyl *p*-tolyl sulfoxide undergoes an asymmetric addition to a deficiency of various (−)-menthyl carboxylates (*30*). The resultant optically active β-ketosulfoxides were obtained in up to 71.5% optical yield. Optically activated *p*-tolyl vinyl sulfoxide has been obtained in the asymmetric elimination of *β*-halogenoethyl *p*-tolyl sulfoxides with chiral amine bases (up to 25% ee) (*31*).

The method of choice for the preparation of optically active sulfoxides (e.g., **11**) remains the Andersen synthesis (*32a–d*), in which an organometallic reagent is reacted with an optically active sulfinate ester (e.g., **10**) [Eq. (1)]. The Andersen synthesis, announced in 1962, was pivotal in stimulating both synthetic and mechanistic studies with chiral nonracemic organosulfur compounds. The optically pure sulfinate **10** had been prepared first by Phillips in 1925 by recrystallization of the mixture of diastereomers obtained by reaction of *p*-toluenesulfinyl chloride with *l*-menthol (*33a,b*). Herbrandson and Dickerson found that hydrogen chloride equilibrated the diasteromeric mixture and were able to obtain a yield of more than 90% of the less soluble diastereomer **10** (*33a,b*).

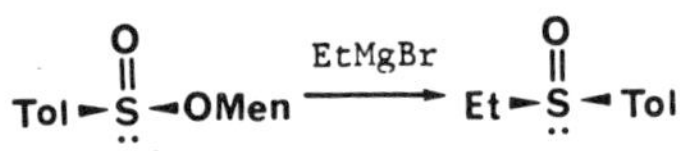

10 **11** (1)

As the example indicates, the Grignard reaction proceeds stereospecifically with inversion of configuration at sulfur (*32a–d, 34*). It was subsequently observed that lithium dialkylcuprates afforded higher yields and cleaner products than the ethereal Grignard reagents (*35a*) that were originally used. Mikolajczyk then noted a further improvement in the reaction by using Grignard reagents in benzene (*35b*). Additional nucleophiles that have been reported include the magnesium enolate of *tert*-butyl acetate (*36*), alkyl- and arylthiomethyllithium reagents (*37*), and methyllithium (*38a,b*) (Table 2). While resolved menthyl sulfinates (*39a,b*) are the most common substrates for the Andersen synthesis, a number of other optically active sulfinate esters have been utilized. These include diastereomeric cholesteryl methanesulfinates (*40a,b*) and sulfinates derived from carbohydrates (*41*).

An extension of the Andersen synthesis has appeared in which an optically active arenesulfinamide is substituted for the usual sulfinate [Eq. (2)] (*42a,b*). Optically activated aryl methyl sulfoxides were obtained upon treatment with methyllithium.

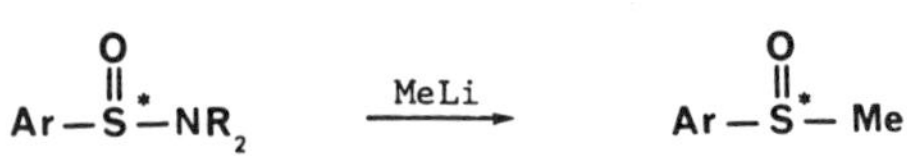

(2)

Wudl and Lee employed the resolved oxathiazolidine **12** derived from *l*-ephedrine to prepare optically active sulfoxides (*43*). Sequential treatment with a Grignard reagent (selective S—O bond cleavage) and then an alkyllithium reagent (S—N bond cleavage) affords the chiral sulfoxide

TABLE 2
Andersen Synthesis of Optically Active Sulfoxides from (−)-(*S*)-Menthyl *p*-toluenesulfinate

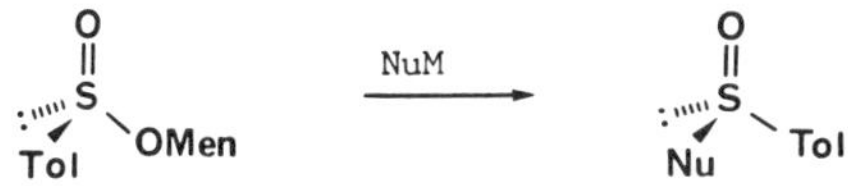

NuM/solvent	Chemical yield (%)	Optical purity (%)	Rotation sign	Reference
Me_2CuLi/Et_2O	55	85.1	+	*35a*
$MeMgBr/Et_2O$	61	85.0	+	*35b*
MeMgBr/PhH	82	89.5	+	*35b*
$EtMgBr/Et_2O$	62	91.5	+	*32*
EtMgBr/PhH	92	97.5	+	*35b*
*i*PrMgBr/Et_2O	40	98.2	+	*35b*
*i*PrMgBr/PhH	79	97.7	+	*35b*
Ph_2CuLi/Et_2O	59	99.1	+	*35a*
PhMgBr/PhH	88	91.0	+	*35b*
$(MeO)_2P(O)CH_2Li/THF$	70	96.6	+	*5*
tBuO_2CCH_2MgBr/Et_2O	60	100.0	+	*36*
$PhSCH_2Li$/THF, DABCO	70	100.0	+	*37*

[Eq. (3)]. Since both nucleophilic groups add with inversion of configuration, the configuration of the sulfoxide is dependent on the order of their addition.

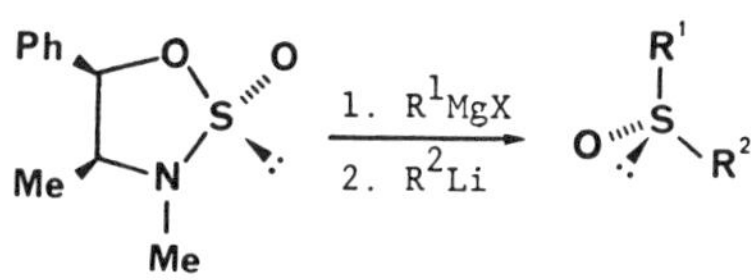

12

(3)

The Andersen synthesis is an excellent method for the preparation of optically active alkyl aryl sulfoxides but is less satisfactory for the synthesis of resolved unsymmetric dialkyl sulfoxides. Johnson developed an exchange reaction in which optically pure alkyl aryl sulfoxides are reacted with appropriate alkyllithium reagents to afford, after quenching, optically active dialkyl sulfoxides with inversion of configuration at sulfur [Eq. (4)] (*44*).

(4)

Optically active bis(arylsulfinyl)ethanes have been prepared via the copper(II)-catalyzed oxidative dimerization of the lithio derivatives of optically pure aryl methyl sulfoxides [Eq. (5)] (*45*).

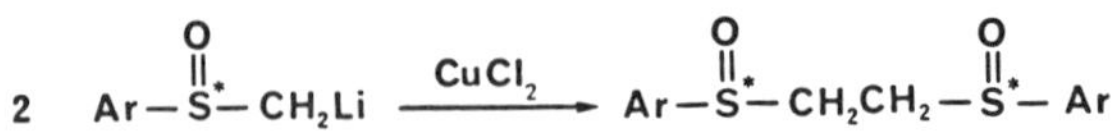

(5)

B. Sulfonium Salts

The first optically active organosulfur compounds to be prepared were the sulfonium salts **13** (*46*) and **14** (*47*), which were resolved via fractional

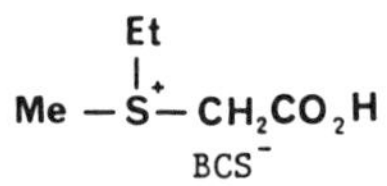

13

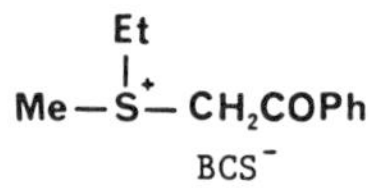

14

crystallization of their diastereomeric bromocamphorsulfonate salts. A number of additional examples have appeared (*2a–h, 48, 49a–e*), and a convenient procedure for this classical resolution method has been described (*50*). The resolved sulfonium salts can racemize at sulfur if the counterion is a good nucleophile (e.g., iodide) in which case reversible S_N2 attack of the nucleophile at an α-carbon is observed (*51*). Other modes of racemization that have been implicated include pyramidal inversion (*49a,c*) and reversible dissociation via an S_N1 mechanism into sulfide and a carbonium ion (*49b*). Increased aromatic substitution enhances the ease of configurational change at sulfur, and optically active triarylsulfonium salts are unknown (*52*).

During the course of their studies on the inversion of sulfur configuration, Johnson and McCants found that resolved sulfoxides underwent O-alkylation with triethyloxonium tetrafluoroborate to afford optically pure ethoxysulfonium salts. Hydrolysis of these salts with aqueous base leads to the starting sulfoxide with an inverted configuration [Eq. (6)] (*22*). The treatment of optically pure alkoxyalkyl-*p*-tolylsulfonium salts with alkyl Grignard or alkylcadmium reagents provides optically active dialkyl-*p*-tolylsulfonium salts with inversion of configuration at sulfur (Table 3) (*53a,b*). Some racemization is observed under the reaction conditions, and the synthesis is limited to dialkylarylsulfonium salts if optically active material is desired.

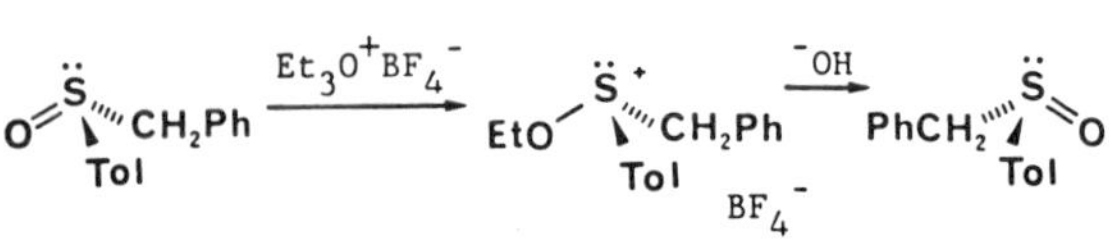

(6)

TABLE 3
Dialkyl-*p*-tolylsulfonium Salts from (*R*)-Ethoxyalkyl-*p*-tolylsulfonium Salts

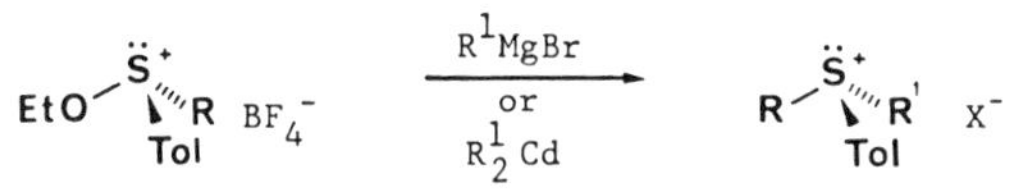

R	R′	X^-	Organometallic	Chemical yield (%)	Rotation sign
Et	Me	Ph_4B^-	MeMgBr	9	+
Et	Me	Br^-	Me_2Cd	72	+
Me	Et	Br^-	Et_2Cd	25	–
Me	Et	$TNBS^{-}$ [a]	Et_2Cd	57	–
Me	*n*Bu	$TNBS^-$	*n*Bu_2Cd	10	–
Me	*n*Bu	Ph_4B^-	*n*Bu_2Cd	10	–
*n*Bu	Me	$TNBS^-$	Me_2Cd	75	+

[a] $TNBS^-$, 2,4,6-Trinitrobenzenesulfonate.

In an extension of Johnson's work, appropriate chiral sulfinamides have been reacted with trifluoromethanesulfonate to afford the chiral alkoxyaminosulfonium salts **15** (*54*) and **16** (isolated as the tetraphenylborate) (*55*).

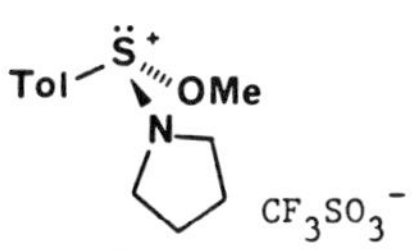

15

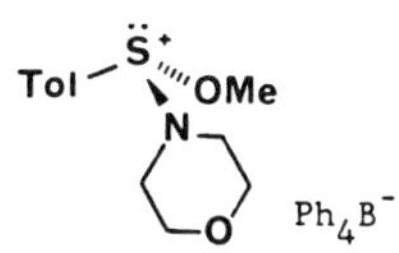

16

Optically active aminosulfonium salts have been prepared by three methods. In the first route, the *N*-tosylsulfilimine **17** is N-alkylated (*56a,b*) with trimethyloxonium tetrafluoroborate to furnish the optically active aminosulfonium salt **18** [Eq. (7)] (*57a,b*). Optically pure sulfoxides

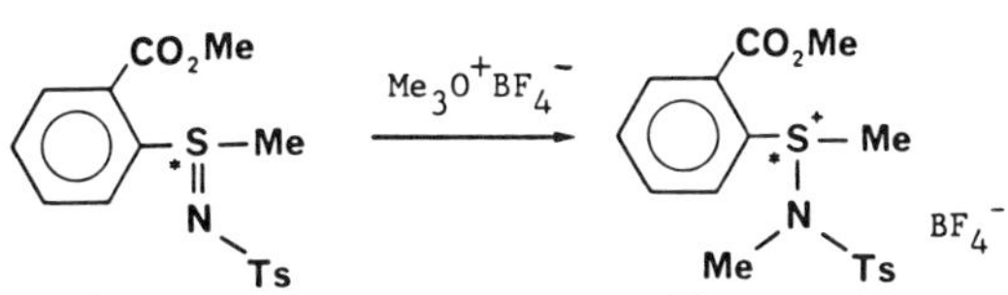

17 **18** (7)

have been treated with *N,N*-diethylaminosulfinyl tetrafluoroborate to afford optically active aminosulfonium salts [Eq. (8)] (*58*). The reaction proceeds largely with retention of configuration at sulfur (up to 83.5% stereospecific) when dialkyl and alkyl aryl sulfoxides are used. Diaryl

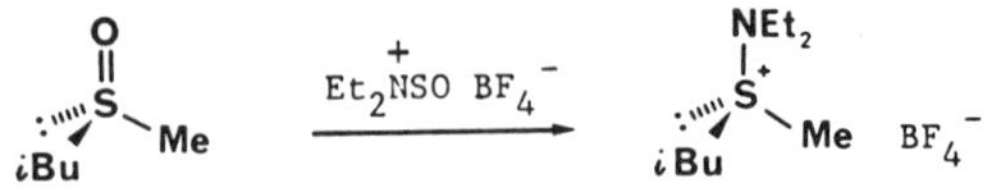

(8)

sulfoxides afford only racemic aminosulfonium salts. A third approach is based on the reaction of methyl *p*-tolyl sulfide with the chiral amination reagent **19** (*24*). Fractional recrystallization of the diastereomeric aminosulfonium salts affords optically pure **9**.

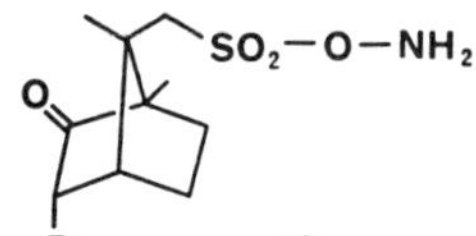

19

The preparation of optically active menthoxysulfonium salts (*21*) **8** (mentioned previously) has been extended to chiral aminomenthoxysulfonium salts such as **20** [Eq. (9)] (*55*). Subsequent nucleophilic substitution reactions with appropriate amines or alcohols have provided optically active diaminosulfonium and alkoxyaminosulfonium salts, respectively.

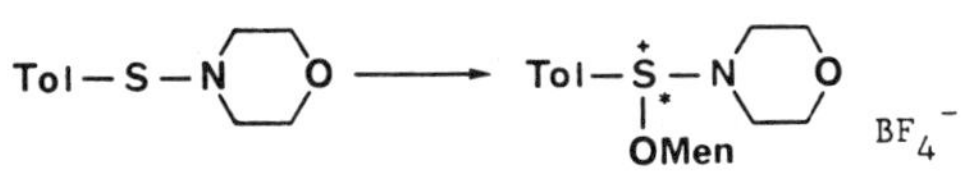

20

(9)

C. Oxosulfonium Salts

The first optically active tetracoordinate oxosulfonium salt, (*R*)-(*N,N*-dimethylamino)methyl-*p*-tolyloxosulfonium tetrafluoroborate (**21**), was prepared by Johnson and Schroeck in 1969 via the exhaustive methylation of (*R*)-*S*-methyl-*S*-*p*-tolylsulfoximine with trimethyloxonium tetrafluoroborate [Eq. (10)] (*59a,b*). Other optically active examples with varied substitution at sulfur and nitrogen were noted (*59b*).

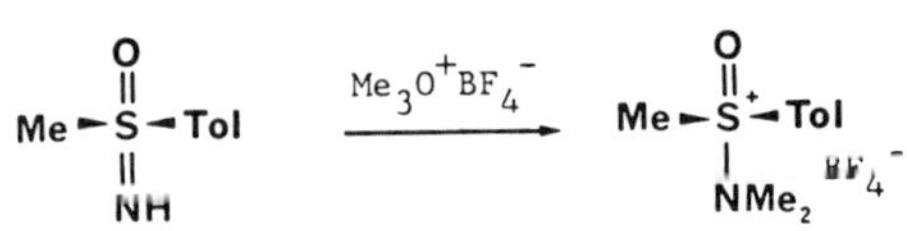

21

(10)

Kobayashi described three successful routes to obtain optically active ethylmethylphenyloxosulfonium perchlorate (**22**). In the first of these (*60*), ethylmethylphenyloxosulfonium mercuritriiodide was treated with (+)-10-camphorsulfonic acid to afford diastereomeric salts. After separation by fractional recrystallization and ion exchange, the enantiomeric oxosulfonium salts **22** were obtained. The S-alkylation of optically active

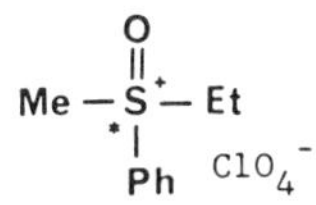

22

ethyl phenyl sulfoxide with methyl iodide–mercuric iodide followed by ion exchange afforded optically active **22** (*61*). Finally, the stereospecific oxidation of optically active ethylmethylphenylsulfonium perchlorate with perbenzoic acid also afforded the desired optical isomers (*62*).

D. Sulfoximines

Sulfoximines can be considered as aza analogs of sulfones. The chemistry of sulfones and sulfoximines have much in common, but the chemistry of sulfoximines is enriched by the presence of the nitrogen group, which renders appropriate molecules chiral and which provides a basic site in the molecule and an additional site for structural variation. Approaches that have been used for the preparation of optically activated sulfoximines include (*a*) direct classical resolution of racemic sulfoximines via their basic nitrogen site or other resolution handle, (*b*) trapping of "nitrenes" with optically active sulfoxides, (*c*) stereospecific oxidation of optically active sulfilimines, and (*d*) nucleophilic substitution of optically active sulfonimidoyl fluorides and sulfonimidoates with organometallic reagents. A number of reviews on the preparation and uses of sulfoximines have appeared (*1, 2a–h, 63a,b,c, 64*).

Examples of racemic sulfoximines resolved by the fractional recrystallization of diastereomeric salts include *S*-(3-carboxyphenyl)-*S*-methyl-*N*-(phenylsulfonyl)sulfoximine (**23**) (*65*), using α-methylbenzylamine, and *S*-methyl-*S*-phenylsulfoximine (**24**) (*59b, 66*) and *S*-mesityl-*S*-methylsulfoximine (**25**) (*59b*), using (+)-10-camphorsulfonic acid. Barash de-

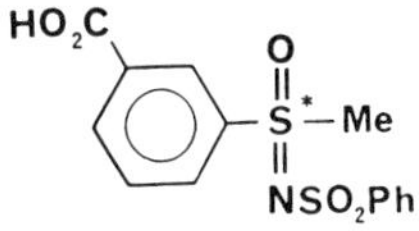

23

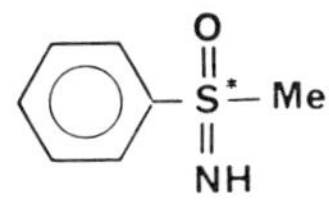

24

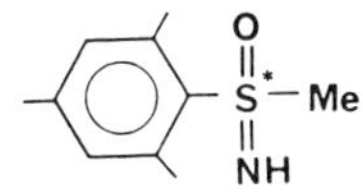

25

scribed the synthesis and separation of the diastereomeric sulfoximines **26** (*67*).

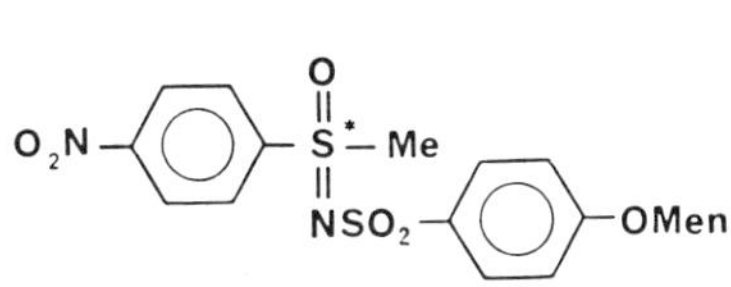

26

In a reaction displaying a high degree of stereospecificity, optically active sulfoxides react with arylsulfonyl azides in the presence of copper to afford optically active *N*-arylsulfonylsulfoximines in high yield and with retention of configuration at sulfur [Eq. (11)] (*26a,b, 59a,b, 68a–c*). The copper-promoted reaction of resolved sulfoxides with Chloramine-T proceeds in an analogous manner to furnish optically active *N*-tosylsulfoximines (*69*).

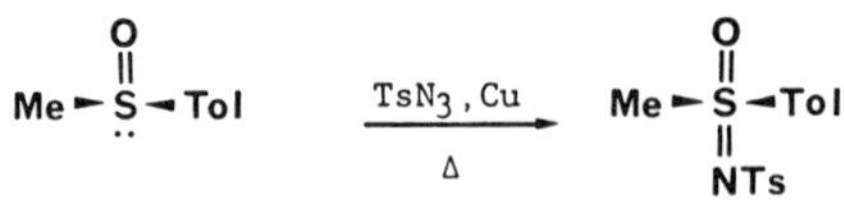

(11)

The *N*-arylsulfonyl group is readily removed with strong acid (*70*), provided that the S-substituents of the sulfoximine are not good carbocation precursors (e.g., benzyl, *tert*-butyl, etc.), in which case S—C bond heterolysis occurs. A variety of N-substituted optically active sulfoximines have been prepared from the NH compounds. N-Chlorination occurs with sodium hypochlorite (*26a,b*), monomethylation can be achieved under Eschweiler–Clarke conditions (*26c, 71a,b*), and dimethylation occurs with trimethyloxonium tetrafluoroborate (*59a,b*) (see above).

In related chemistry, optically active *N*-phthalimidosulfoximines have been prepared in a stereospecific manner with retention of configuration by the reaction of optically active sulfoxides with *N*-aminophthalimide and lead tetraacetate [Eq. (12)] (*72a,b*).

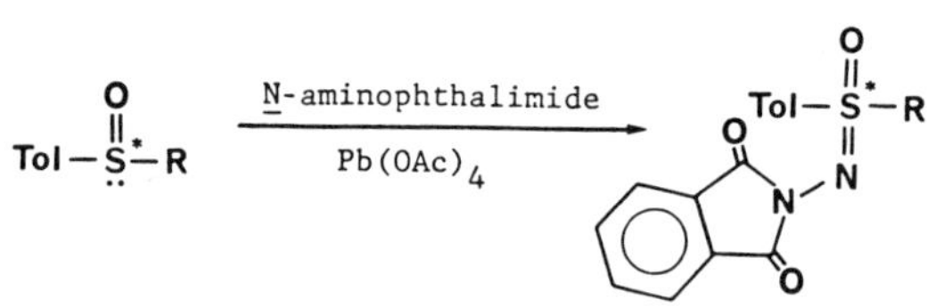

(12)

A simple one-step method for the production of optically active sulfoximines from optically active sulfoxides has been described by Johnson (*73*). Imination with *O*-mesitylsulfonylhydroxylamine (MSH) occurs with retention of configuration to afford the corresponding "free" sulfoximines in good yield and high enantiomeric purity (Table 4). This procedure is superior to the hydrazoic acid method (*74*), in which partial racemization of the starting sulfoxide is observed (*59b*). It also has advantages over the hydrolysis of optically active *N*-tosylsulfoximines with strong acid, since this reaction often fails or results in extensive decomposition.

Potassium permanganate (*68b, 70, 75*) and *m*-chloroperoxybenzoate anion (*26a, 76*) cleanly effect oxidation of optically active *N*-tosylsulfilimines to optically active sulfoximines with retention of configuration at

TABLE 4
Reactions of Sulfoxides with *O*-Mesitylsulfonylhydroxylamine[a]

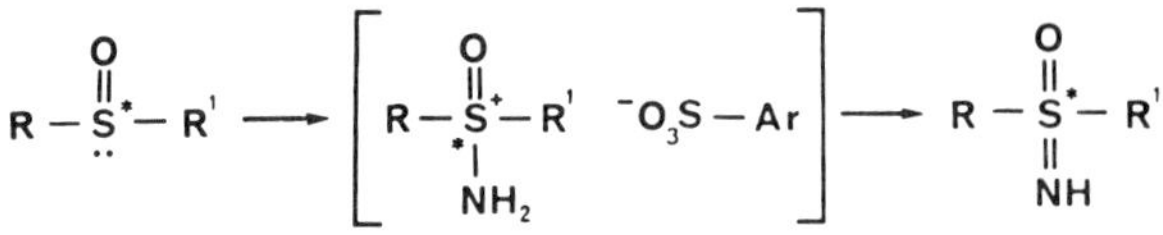

		Sulfoxide		Sulfoxime		
R	R′	Absolute config.	Optical purity (%)	Absolute config.	Optical purity (%)	Chemical yield (%)
Me	Tol	*R*	99.0	*R*	98.5	80
Et	Tol	*R*	100.0	*R*	99.0	70
*i*Pr	Tol	*R*	100.0	*R*	99.0	79
*n*Bu	Tol	*R*	100.0	*R*	99.0	77
$PhCH_2$	Tol	*R*	93.0	*R*	92.0	60
Ph	Tol	*R*	99.5	*R*	99.0	19
Ph	Me	*S*	94.0	*S*	93.5	70
Me	*n*Bu	*R*	92.0	*R*	91.5	78

[a] MSH. From (*73*).

sulfur [Eq. (13)]. Other oxidants that have been utilized for the oxidation of sulfilimines to sulfoximines include sodium metaperiodate (*77*), ruthenium tetroxide (*78*), alkaline hydrogen peroxide (*79*), and sodium hypochlorite under phase-transfer conditions (*80*).

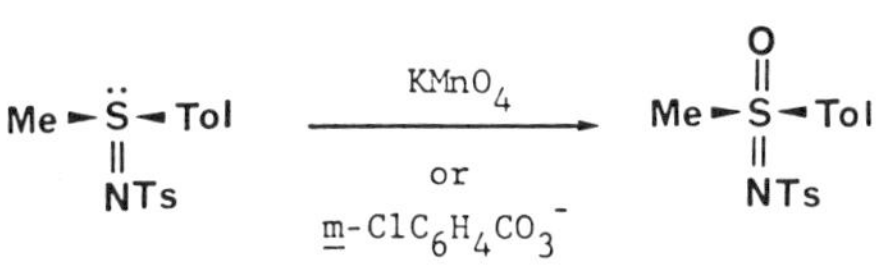

(13)

In related chemistry, Oae described the alkaline hydrolysis of *N*-chloro-*S*-(2-methoxyphenyl)-*S*-phenylsulfilimine (**27**) to the sulfoximine **28** with nearly complete retention of configuration at sulfur (*81*).

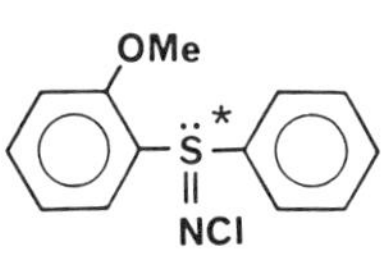

27

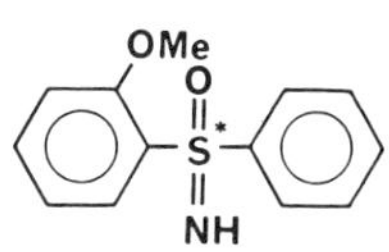

28

The requisite optically active sulfilimines for the above oxidations are readily available from the corresponding optically pure sulfoxides via a variety of iminating reagents (*26a, 82a–g*). While the majority of these iminations proceed with retention of configuration at sulfur, Cram and Day found that *N*-sulfinyl-*p*-toluenesulfonamide reacts with (*R*)-methyl-*p*-tolyl sulfoxide in the presence of pyridine with inversion (98% stereospecific) [Eq. (14)] (*82a*). Alternatively, optically active sulfilimines have

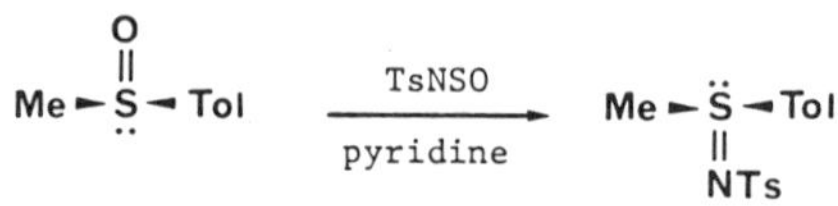

(14)

been prepared by the reaction of a Grignard reagent with diastereomeric menthyl *p*-toluenesulfinimidoates (*83a,b*), by the treatment of menthoxysulfonium salts with Chloramine-T (*23a, 84*), and by classical resolution of racemic sulfilimines via separation of diastereomeric salts (*85a,b*).

In their studies on nucleophilic substitution at sulfur in sulfonimidoyl compounds, Johnson and co-workers developed a method for the stereospecific synthesis of optically active sulfonimidoyl chlorides (*71b, 86a,b, 87*). The oxidation of (*S*)-*N*-methylbenzenesulfinamide (**29**) with chlorine, *N*-chlorobenzotriazole, or *tert*-butyl hypochlorite (*88*) proceeds with retention of configuration at sulfur to afford (*R*)-*N*-methylbenzenesulfonimidoyl chloride (**30**) [Eq. (15)] (*71b, 87*). Substitution of the chlo-

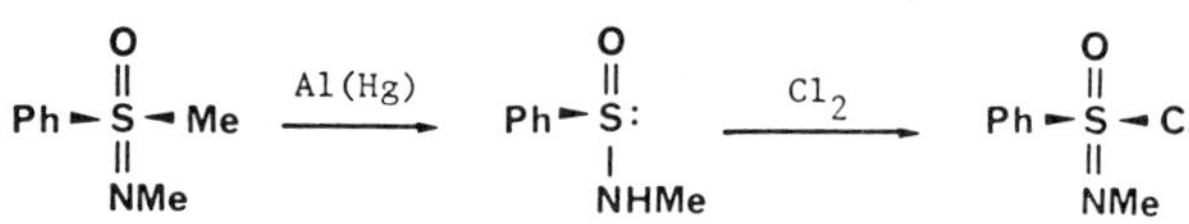

(15)

29 30

ride by a variety of nucleophiles is successful except for reaction with organometallic reagents, which generally results in reduction to sulfinamides rather than substitution to yield sulfoximines (*86a,b*). This problem was circumvented by prior conversion of the sulfonimidoyl chlorides to sulfonimidoates (*71b, 87*). In this example sulfonimidoyl chloride **30** was treated with excess sodium phenoxide to give (*S*)-phenyl-*N*-methylbenzenesulfonimidoate (**31**) with inversion of configuration. This ester was reacted with an excess of methyllithium to furnish a good yield of (*S*)-*N*-*S*-dimethyl-*S*-phenylsulfoximine (**32**) with inversion of configuration at sulfur [Eq. (16)].

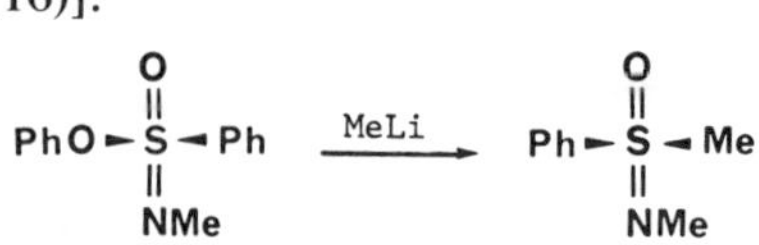

(16)

31 32

Alternatively, the treatment of sulfonimidoyl chlorides with fluoride ion (sodium fluoride in acetonitrile, tetrabutylammonium fluoride in acetonitrile, or potassium fluoride in acetonitrile containing 18-crown-6) results in good to excellent yields of the corresponding sulfonimidoyl fluorides (*89*). These compounds react with primary alkyllithium reagents to afford the corresponding sulfoximines in high yield. Especially interesting were the chromatographically separable diastereomeric sulfoximines **33** derived from (−)-1-phenylethylamine.

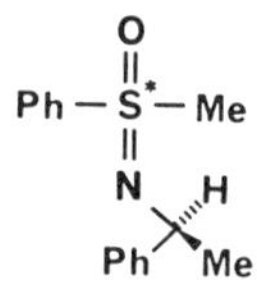

33

III. Applications of Optically Active Organosulfur Reagents in Asymmetric Synthesis

Optically active tri- and tetracoordinate sulfur moieties are readily available with varied oxidation states and diverse substitution patterns. This flexibility imparts extensive synthetic possibilities to these systems. In this section, examples illustrating the effectiveness of optically activated sulfur functional groups in asymmetric synthesis are described. Several excellent reviews on this subject, especially with regard to chiral sulfoxides, are available (*1, 2a–b, 90–92*).

A. Sulfoxides

The first example of an optically active sulfoxide providing asymmetric induction was reported by Goldberg and Sahli in 1965 (*93a,b*). Thermolysis of the sulfoxide (*S*)-**34** afforded (*S*)-4-methylcyclohexene (**35**) in up to 70% optical purity [Eq. (17)].

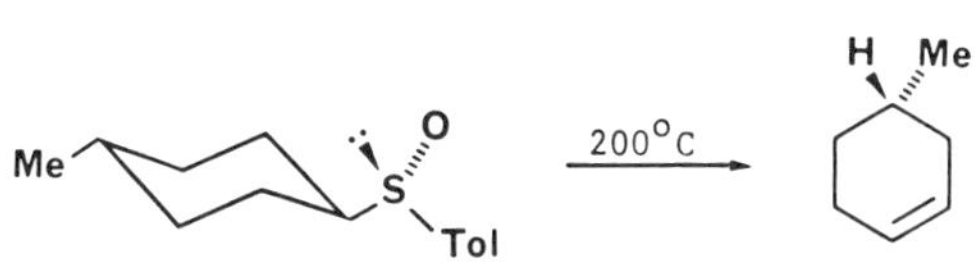

(17)

34 **35**

The Pummerer reaction of optically active sulfoxides generally proceeds with modest asymmetric induction (<30% ee) (*94a–d*). However, an example in which dicyclohexylcarbodiimide (DCC) was used as an acetic acid scavenger provided an increase in the enantioselectivity (up to 70% ee) [Eq. (18)] (*95*).

$$\text{Tol}-\overset{\text{O}}{\overset{\|}{\text{S}}}{}^{*}-\text{CH}_2\text{CO}_2\text{Et} \xrightarrow[\text{DCC}]{\text{Ac}_2\text{O}} \text{Tol}-\text{S}-\underset{\text{OAc}}{\underset{|}{\overset{*}{\text{C}}\text{H}}}-\text{CO}_2\text{Et} \qquad (18)$$

Asymmetric induction has also been observed in the reversible [2,3]sigmatropic rearrangement of optically active allyl aryl sulfoxides to sulfenate esters. If the reaction is performed in the presence of a suitable thiophile (e.g., trimethyl phosphite), then the corresponding optically active allylic alcohols are obtained. In an elegant study by Hoffman and coworkers (*96a–c*), it was revealed that optically active acyclic 3-(*E*)-substituted allylic sulfoxides rearranged with low enantiomeric selectivity but that the corresponding 3-(*Z*)-substituted substrates (e.g., **36**) afforded allylic alcohols (e.g., **37**) in higher optical yields (up to 42%) [Eq. (19)].

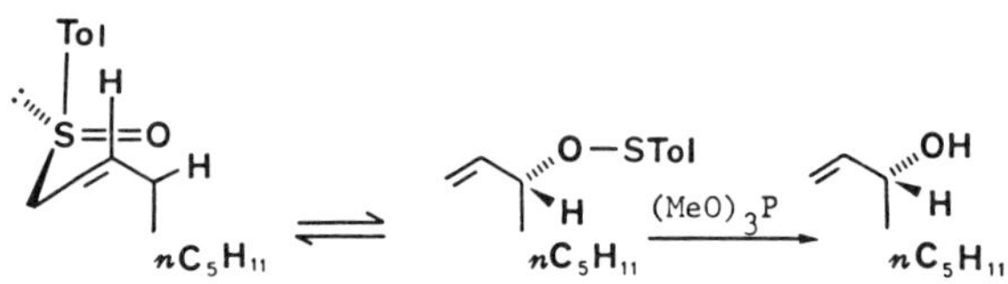

36 37 (19)

Additionally, an enantiomeric excess of greater than 90% was obtained in the [2,3]sigmatropic rearrangement of allylic sulfoxide **38** to the spirocyclic alcohol **39** [Eq. (20)] (*96a–c*).

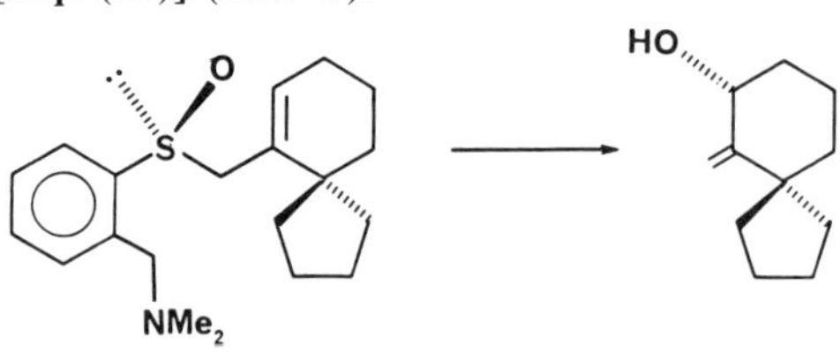

38 39 (20)

High asymmetric induction has been observed in the 1,3-dipolar cycloaddition of (*R*)-*p*-tolyl vinyl sulfoxide with acyclic nitrones (*97*). When *C*,*N*-diphenylnitrone **40** was utilized, the 3-phenyl-4-(*p*-tolylsulfinyl)isoxazolidine **41** was obtained in greater than 90% ee [Eq. (21)].

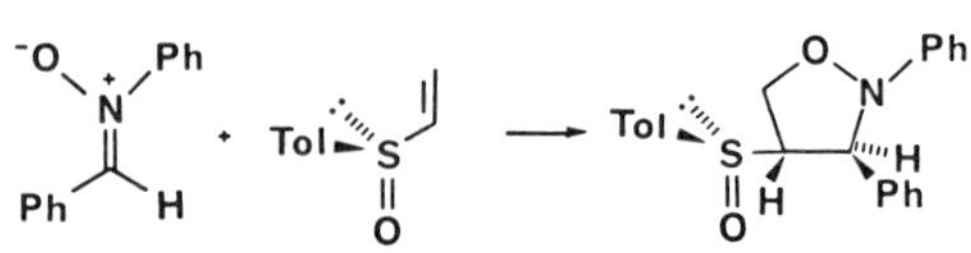

40 41 (21)

Optically pure vinyl sulfoxides have also been used as receptors for conjugate nucleophilic additions. Significant asymmetric induction (60–74% ee) has been observed on the addition of piperidine (*98*) and diethyl sodiomalonate (*99*) to (*R*)-*cis*-propenyl *p*-tolyl sulfoxide and (*R*)-*trans*-β-styryl *p*-tolyl sulfoxide, respectively. The conjugate addition of organometallic reagents to enantiomerically pure α-carbonyl α,β-ethylenic sulfoxides has been reported by Posner and co-workers (*91, 100a–c*). For example, the conjugate addition of 6-methoxy-2-naphthylmagnesium bromide to the optically pure (*S*)-cyclopentenone sulfoxide **42** followed by methylation of the intermediate enolate ion furnishes the sulfoxide **43** with nearly complete asymmetric induction (>98% ee) [Eq. (22)]. Presumably,

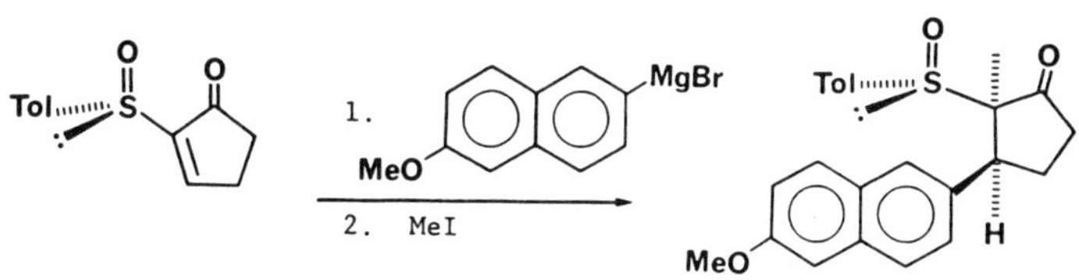

42 **43**

(22)

the remarkable enantioselectivity obtained reflects metal chelation of the carbonyl and sulfoxide, which provides an asymmetric environment for the nucleophile by locking the chiral sulfoxide into one conformation.

The electrophilic addition of bromine to (*R*)-*p*-tolyl vinyl sulfoxide has been examined, and a low level of asymmetric induction has been observed (32 % ee) (*98*). Much higher induction was found with the addition of hypobromous acid to (*R*)-*trans*-2-phenylethenyl *p*-tolyl sulfoxide. The bromohydrins **44** and **45** were formed in a ratio of 9 : 1 [Eq. (23)] (*101*).

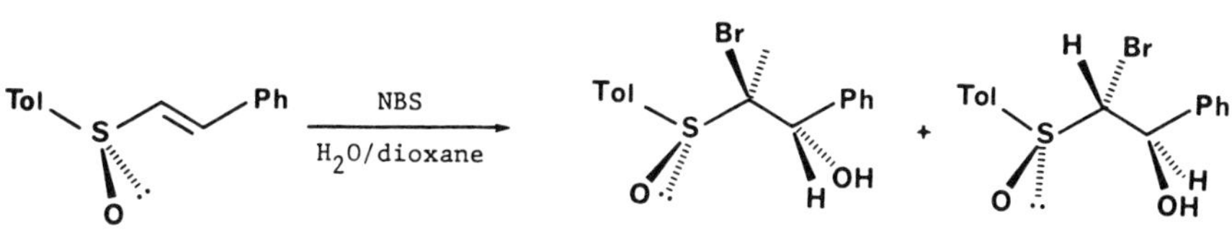

44 **45**

(23)

Optically active alkyl aryl and dialkyl sulfoxides react with electrophilic halogenating reagents to furnish optically active α-halosulfoxides (*102*). Generally, only one of the two possible diastereomeric products is formed, and the reaction proceeds with either retention or inversion of configuration at sulfur, depending on the substrate structure and reaction conditions. The α-bromination of (*S*)-benzyl methyl sulfoxide affords two regioisomeric products in the ratio 1.5 : 1 (*103*). The major product, benzyl α-bromomethyl sulfoxide (**46**), was obtained with retention of configuration at sulfur, and the minor product, α-bromobenzyl methyl sulfoxide (**47**), with inversion [Eq. (24)]. The sulfoxide **47** was reduced with zinc to (*R*)-benzyl methyl sulfoxide (48% optically pure).

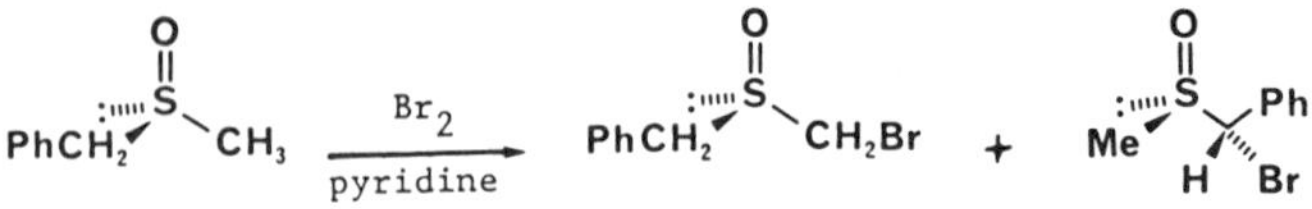

46 **47**

(24)

The ability of sulfoxides to stabilize negative charge on an adjacent carbon atom has prompted the development of synthetic procedures based on optically active α-sulfinylcarbanions (*92*). The first additions of optically active α-sulfinylcarbanions to carbonyl compounds were independently reported by Johnson (*104*) and Durst (*105*) in 1971. The diastereomeric β-hydroxysulfoxides obtained have been separated and subsequently converted to optically active epoxides (>90% optical purity).

Detailed studies on the addition of α-sulfinylcarbanions derived from optically active alkyl aryl sulfoxides to aldehydes and ketones indicate a low level of asymmetric induction (diastereomeric ratios from 1 : 1 to 2.3 : 1) (*106a–d*). However, when (*R*)-methyl *p*-tolyl sulfoxide was added as its lithio derivative to *N*-benzylideneaniline, the β-aminosulfoxide **48** was obtained as a single diastereomer [Eq. (25)] (*106b*).

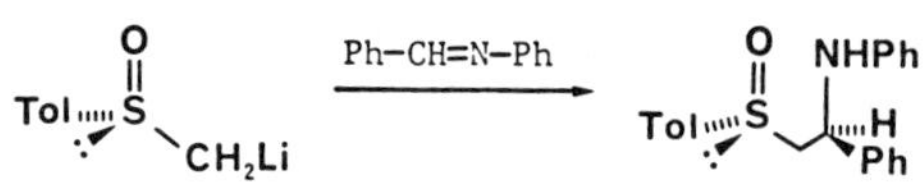

48

(25)

In related chemistry, an optically active acetate enolate equivalent has been described by Solladié (*107, 108a–c*). The aldol condensation of (*R*)-α-sulfinylacetate enolate **49** (M = MgX) with various aldehydes and ke-

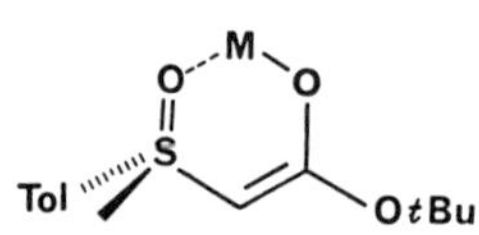

49

tones was found to occur with high stereoselectivity. The degree of asymmetric induction directly correlates with substituent size differences on the carbonyl substrate (*107*).

The successful application of this asymmetric aldol condensation methodology to the synthesis of naturally occurring substances (17–86% ee) has been reported (*20, 108c, 109*).

Optically active sulfoxide-stabilized carbanions have also been reacted with α,β-unsaturated carbonyl compounds. For example, the α-lithio-sulfoxide **50** has been described as a novel chiral acyl anion equivalent (*110a,b*). The conjugate addition of **50** to cyclopentenone **51** occurs with a high degree of asymmetric induction to afford, after deoxygenation of the sulfoxide, the prostaglandin precursor **52** (92% ee) [Eq. (26)]. Similarly, the chiral sodiosulfoxide **49** (M = Na) has also been utilized in conjugate

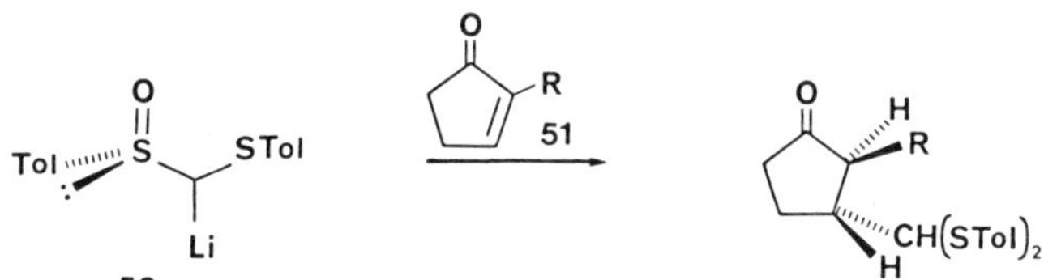

50 52

R = $(CH_2)_6CO_2Me$ (26)

additions, with prochiral α,β-unsaturated esters as the substrates (12–24% ee) (*111*).

Asymmetric induction has been observed in the reduction of optically pure (*R*)-β-ketosulfoxide **53** to the diastereomeric alcohols (*R*,*R*)-**54** and (*S*,*R*)-**55** (*112*). When tetra-*n*-butylammonium borohydride was utilized, diastereomer **54** was the major product (ratio 85 : 15). In contrast, diisobutylaluminum hydride afforded more of diastereomer **55** (ratio 78 : 22) [Eq. (27)]. Subsequent desulfurization with Raney nickel provided enantiomeric methyl carbinols.

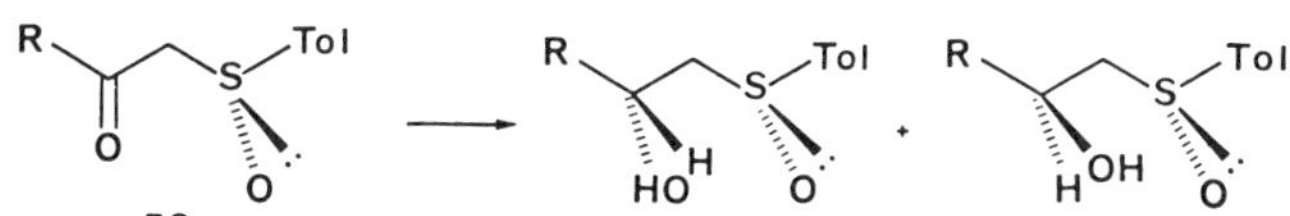

53 54 55

R = Et, Ph, *n*-C_9H_{19} (27)

B. Sulfonium Salts

The most useful applications of sulfonium salts to asymmetric synthesis involve the formation and reaction of optically active sulfonium ylides. Unfortunately, sulfonium ylides are more likely to racemize than their sulfonium salt precursors. Despite this limitation, several examples of optical induction via optically active sulfonium ylides have been noted.

The [2,3]sigmatropic rearrangement of (*S*)-1-adamantylallylethylsulfonium tetrafluoroborate (**56**) under ylide-generating conditions provided the optically active rearranged product (*R*)-1-adamantyl 2-pent-4-enyl sulfide (**57**) with a minimum of 94% asymmetric induction [Eq. (28)] (*49e*). The efficient transfer of chirality from sulfur to carbon reflects the concerted nature of the reaction.

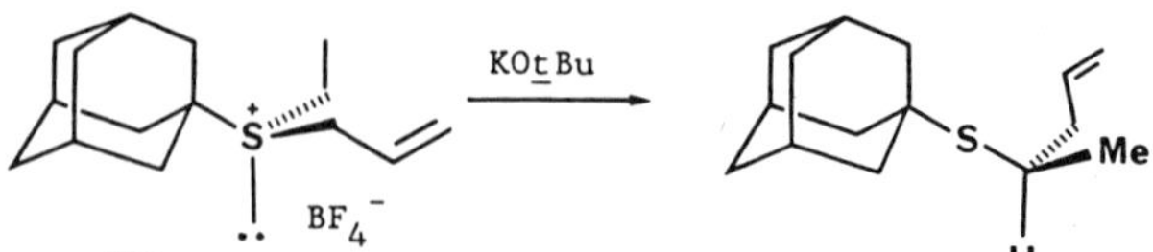

56 57 (28)

Modest asymmetric induction has been observed in the deprotonation of the achiral sulfonium salt **58** with optically pure bases (*113*). The resulting optically active sulfonium ylide undergoes a concerted [2,3]sigmatropic rearrangement to furnish the optically active sulfide **59** (5–12% ee) [Eq. (29)].

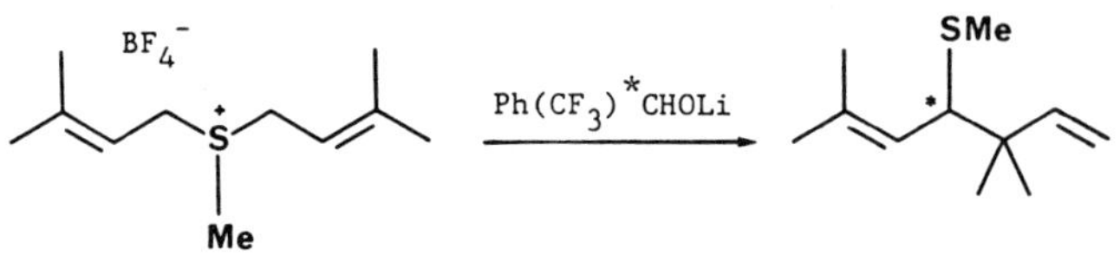

58 59 (29)

C. Oxosulfonium Salts

Ylides derived from oxosulfonium salts are considerably more stable than their sulfonium counterparts. In fact, (dialkylamino)aryloxosulfonium alkylides have long-term stability; in a sealed tube the methylide **60** remains unchanged after 2 months at room temperature (*64*).

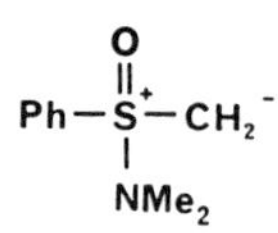

60

Johnson and co-workers reported the use of optically active (dialkylamino)aryloxosulfonium alkylides as asymmetric nucleophilic alkylidene-transfer reagents (*59a,b, 64, 114a,b*). These ylides react with aldehydes and ketones to give epoxides, and with electrophilic alkenes to give cyclopropanes; optical purities of 7–43% have been observed. Additionally, ylides with great structural variation are available from the requisite optically active oxosulfonium salts (**61**). Representative examples of asymmetric induction by these ylides are shown in Table 5.

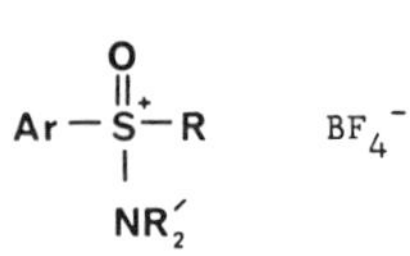

61

61	Ar	R	R′	Absolute config.
a	Tol	Me	Me	*R*
b	Tol	Et	Me	*R*
c	Ph	Me	Me	*S*
d	Ph	Me	Et	*S*
e	2-Mesityl	Me	Me	*R*
f	Ph	Cyclopropyl	Me	*S*

TABLE 5
Reactions of Optically Active Oxosulfonium Ylides with Electrophilic Substrates

Substrate	Oxosulfonium salt	Product	Absolute config.	Optical purity (%)
Benzaldehyde	**61a**	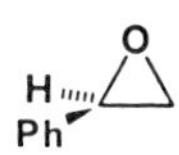	*R*	20.0
trans-Methyl-cinnamate	**61a**		1*S*,2*S*	30.4
Methyl acrylate	**61b**		1*S*,2S	43.2
trans-Methyl-crotonate	**61c**	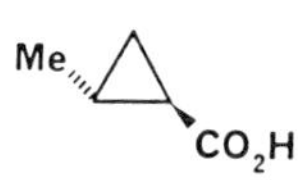	1*R*,2*R*	11.9
trans-Benzal-acetophenone	**61e**	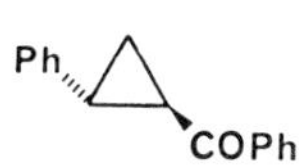	1*R*,2*R*	36.5
trans-Benzal-acetophenone	**61a**		1*S*,2*S*	35.3
trans-Benzal-acetophenone	**61f**	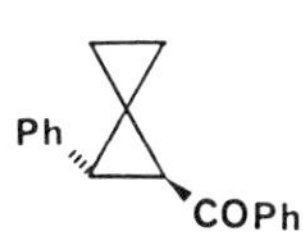	—	—

In related chemistry, the preparation and reactions of (*S*)-(dimethylamino)phenyl(*trans*-2-phenylethenyl)oxosulfonium tetrafluoroborate (**62**) have been described (*115a,b*). When this salt is treated with appropriate dibasic nucleophiles, an asymmetric ethylenation of the nucleophile is observed. Methyl cyanoacetate affords, in 82% yield, (1*S*,2*R*)-methyl 1-cyano-2-phenylcyclopropane carboxylate (**63**) (25.5% optically pure) and methyl nitroacetate gives, in 95% yield, 3-(methoxycarbonyl)-4-phenylisoxazoline 2-oxide (**64**) (33% ee).

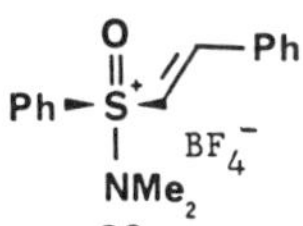

62

63

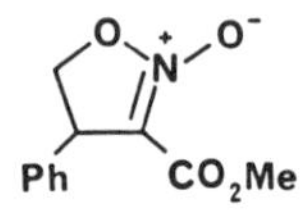

64

D. Sulfoximines

The development of sulfoximines as reagents for asymmetric synthesis has been a central research theme in the authors' laboratory. In 1968, Johnson and Schroeck (*59a,b*) reported the asymmetric synthesis of optically active epoxides and cyclopropanes via chiral oxosulfonium ylides derived from salts of optically pure sulfoximines (see above). Subsequently, the anions of optically pure *N*-(*p*-tolylsulfonyl)sulfoximines have been described as alternative nucleophilic alkylidene-transfer reagents useful for the asymmetric synthesis of epoxides, aziridines, and cyclopropanes (*68c, 116a,b*). For example, the lithio derivative of (*R*)-*S*-methyl-*S*-phenyl-*N*-(*p*-tolylsulfonyl)sulfoximine (**65**) was reacted with *trans*-benzalacetophenone to give a 75% yield of (1*R*,2*R*)-*trans*-1-benzoyl-2-phenylcyclopropane (**66**) (49% optically pure) [Eq. (30)].

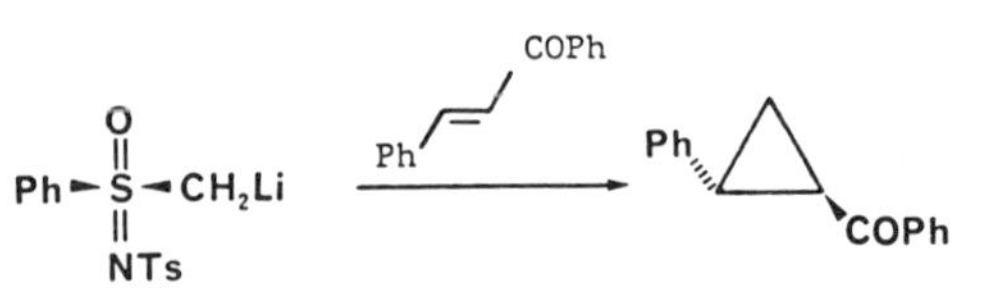

65 66 (30)

In an extension of this methodology, Welch and co-workers obtained 2,2-disubstituted oxetanes from ketones on treatment with 3 equiv of *S*-methyl-*S*-(sodiomethyl)-*N*-(*p*-tolylsulfonyl)sulfoximine (**67**) (*117*). Initial methylene transfer produces an intermediate epoxide that is ring-expanded to the oxetane. Application of this reaction to 4-*tert*-butylcyclohexanone affords a single oxetane (**68**) in 69% yield [Eq. (31)]. While only

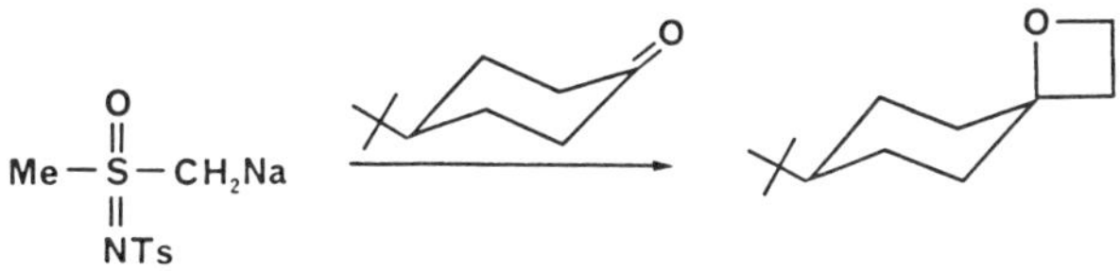

67 68 (31)

achiral alkylidene-transfer reagents were utilized, in principle this reaction is amenable to the asymmetric synthesis of oxetanes.

Borane complexes of optically active β-hydroxysulfoximines have been developed as highly reactive asymmetric reducing agents of prochiral ketones (*118*). The necessary β-hydroxysulfoximines have been prepared by the addition of (+)-(*S*)-*N*,*S*-dimethyl-*S*-phenylsulfoximine (**69**) as the lithium derivative to prochiral or chiral ketones, followed by chromatographic separation of the diastereomeric adducts, for example, **70a** (faster eluting) and **70b** (slower eluting). Similarly, the β-hydroxysulfoximine **71** was prepared as the major adduct from **69** and *l*-menthone. A toluene

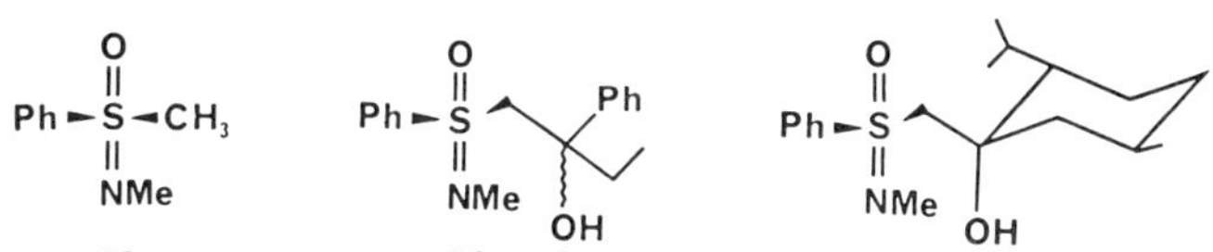

69 70a, 70b 71

solution of an adduct has been treated sequentially with diborane gas and a prochiral ketone to furnish, after workup, the optically active alcohols (Table 6).

Another related sulfoximine-mediated synthesis of optically active alcohols is based on the aluminum amalgam or Raney nickel reduction of optically active β-hydroxysulfoximines (*119a,b*). This method works well for β-hydroxysulfoximines derived from prochiral ketones because the diastereomers are generally separable by medium-pressure liquid chromatography on silica gel. Subsequent desulfurization of the separated adducts affords enantiomeric tertiary alcohols of high optical purity (~100%) [Eq. (32)]. Aldehyde adducts are directly reduced because of the

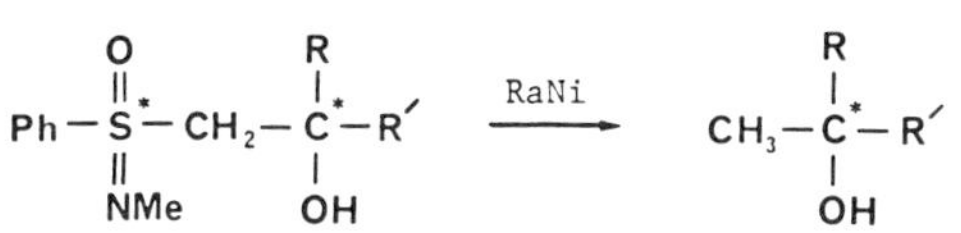

(32)

poor chromatographic response of the diastereomers. The optical purities of the secondary alcohols obtained (25–46%) reflect the degree of asymmetric induction in adduct formation.

In related chemistry, (*S*)-β-ketosulfoximines have been prepared by condensation of the lithio derivative of (+)-(*S*)-**69** with esters (*120*) or

TABLE 6
Asymmetric Reduction of Ketones with
β-Hydroxysulfoximine–Borane Complexes[a]

$$\text{Ligand*—BH}_3 \xrightarrow[\text{toluene, }-78°\text{C}]{\text{RCOR}^1} \xrightarrow{\text{H}_3\text{O}^+} \text{R*CH(OH)R}^1$$

R	R′	Ligand*	Chemical yield (%)	Optical purity (%)	Absolute config.
Ph	Me	**70a**	80	60	*S*
Ph	Me	**70b**	92	57	*R*
Ph	Me	**71**	66	82	*S*
Ph	Et	**70a**	66	33	*S*
Ph	Et	**70b**	74	74	*R*
Ph	Et	**71**	48	29	*S*
Ph	*n*Pr	**70b**	81	70	*R*
Ph	*i*Pr	**70b**	53	8	*R*
Me	*t*Bu	**70b**	60	22	*R*

[a] From (*118*).

nitriles (*121*). Cinquini and co-workers examined the reduction of these substances with sodium borohydride and found asymmetric-induction levels of 0 to 50% (*120*). Johnson and Stark observed higher levels of asymmetric induction when gaseous diborane was utilized in toluene at −78°C (*121*). The resulting diastereomeric β-hydroxysulfoximines are obtained in ratios of 3 : 1 to 9 : 1. Raney nickel desulfurization then affords optically active secondary alcohols with optical purities in the range 18–69% [Eq. (33)].

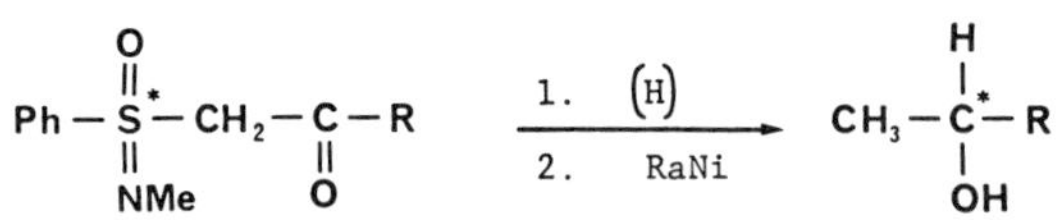

(33)

R = Ph, *t*Bu, *i*Bu, $(CH_2)_5CH_3$

The preparation of (+)-(*S*,*S*)- and (−)-(*S*,*R*)-*N*-methyl-*S*-phenyl-*S*-(*p*-tolylsulfinylmethyl)sulfoximine (**72**) has been described (*122*). It was found that alkylation of these compounds under phase transfer conditions proceeds in a stereoselective manner. Control experiments indicate that the sulfinyl group is largely controlling the stereochemical outcome of the reaction. Furthermore, (+)-(*S*,*S*)-**72** is alkylated stereospecifically in all cases, suggesting that in this diastereomer the two sulfur moieties work together to provide double asymmetric induction (Table 7).

Another synthetically useful transformation of the versatile β-hydroxysulfoximine moieties is their reductive elimination to yield alkenes on treatment with aluminum amalgam in a mixture of tetrahydrofuran, water,

TABLE 7
Alkylation of **73** with Alkyl Halides under Phase-Transfer Conditions

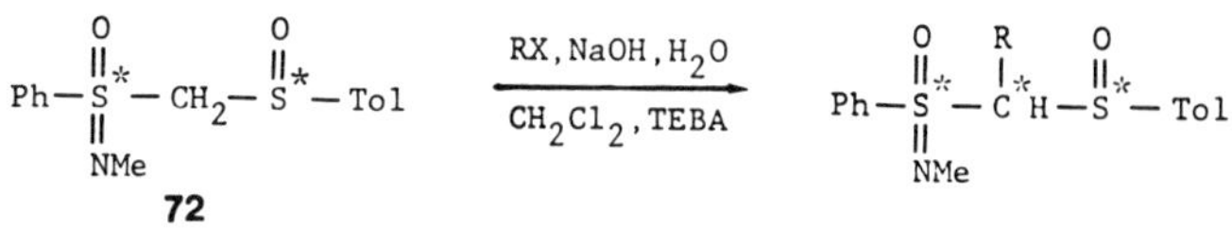

TEBA = benzyltriethylammonium chloride

Absolute config. of **72**	Alkyl halide	Chemical yield (%)	Diastereomeric ratio
(*S,S*)	$H_2C{=}CCH_2Br$	93	100 : 0
(*S,R*)	$H_2C{=}CCH_2Br$	87	83 : 17
(*S,S*)	$PhCH_2Br$	77	100 : 0
(*S,R*)	$PhCH_2Br$	79	80 : 20
(*S,S*)	$HC{\equiv}CCH_2Br$	91	100 : 0
(*S,R*)	$HC{\equiv}CCH_2Br$	73	100 : 0
(*S,S*)	EtBr	80	100 : 0
(*S,R*)	EtBr	82	80 : 20

and acetic acid (*123a,b,c*). The use of optically active β-hydroxysulfoximines allows for ketone methylenation with concomitant optical resolution. A successful application of this procedure has been described (*124*). Treatment of racemic ketone **73** (derived in four steps from 3-methyl-2-cyclohexene-1-one) with (+)-(*S*)-(*N*-methylphenylsulfonimidoyl)methyllithium affords two diastereomeric adducts that can be separated by flash chromatography. Reductive elimination of each affords optically pure β-panasinsene (**74**) of natural and unnatural absolute configuration [Eq. (34)].

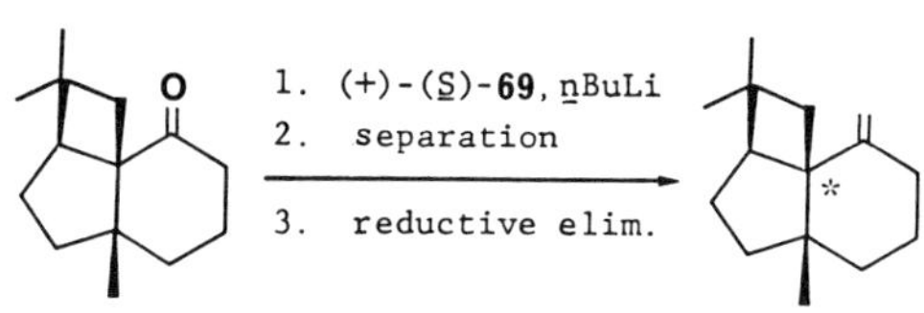

73 **74** (34)

A new ketone resolution technique based on the reversible addition of optically pure *N,S*-dimethyl-*S*-phenylsulfoximine to selected chiral ketones has been reported (*125a,b*). The diastereomeric ketone adducts are separated by medium-pressure or flash liquid chromatography over silica gel and then thermolyzed (80–120°C) to generate the optically active ketone and the recoverable optically pure sulfoximine [Eq. (35) and Table 8]. Best results are obtained with substituted cycloalkanones and acyclic ketones bearing chelating substituents near the carbonyl, since these com-

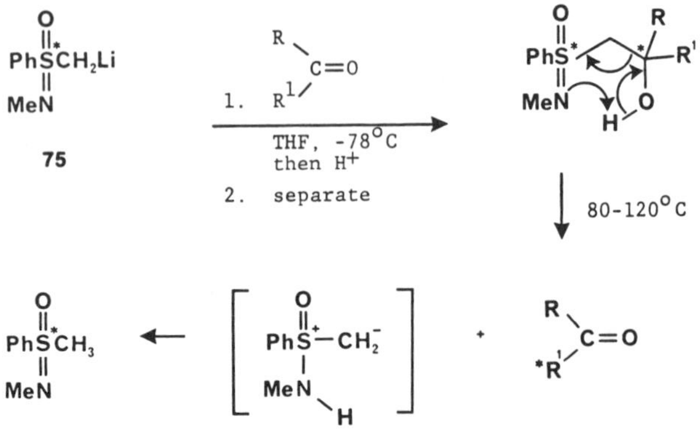

(35)

TABLE 8
Results of the Sulfoximine-Mediated Ketone Resolution[a]

Ketone	Yield of diastereomer (%)	Yield of thermolysis (%)	$[\alpha]_D$ (deg.): temp (°C), conc., solvent
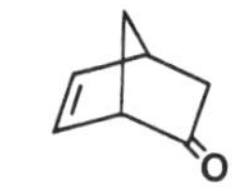	**I** 44	—	—
	II 35	50	−1135.7: 28, 0.7, $CHCl_3$
	**I** 25	80	−64.3: 25, 0.64, $CHCl_3$
	II 23	98	+59.6: 25, 1.3, $CHCl_3$
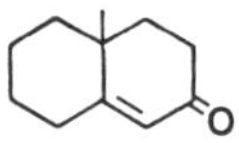	**I** 42	98	−207.0: 24, 1, EtOH
	II 21	84	+208.3: 24, 1, EtOH
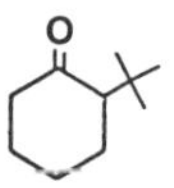	**I** 56	93	+36.0: 24.5, 1, MeOH
	II 12	98	−35.3: 24.5, 1, MeOH
	III 20	73	−35.4: 24.5, 1, MeOH
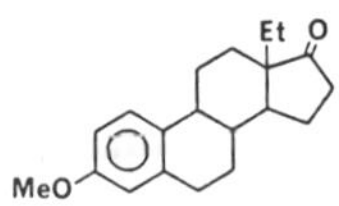	**I** 52	86	+106.4: 25, 1, MeOH/$CHCl_3$
	II 40	61	−104.4: 25, 1, MeOH/$CHCl_3$
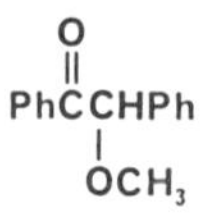	**I** 38	96	+52.8: 19, 0.6, PhH
	II 8		
	III 31	93	−51.5: 20, 0.6, PhH

[a] From (*125a, b*).

pounds provide better diasterofacc selectivity in the sulfoximine addition. The success of the resolution is also dependent on the separability of the adducts. Generally, β-hydroxysulfoximines are highly responsive to separation by chromatography.

Optically pure (*N*-methylphenylsulfonimidoyl)methyllithium (**75**) adds to the carbonyl of prochiral enones to afford two diastereomeric allylic or homoallylic alcohols that are generally separable by chromatography. A cyclopropyl ketone optical-activation method based on Simmons–Smith cyclopropanation of these resolved β-hydroxysulfoximines has been reported [Eq. (36) and Table 9] (*126*). The hydroxyl group was found to

TABLE 9
Optically Active Cyclopropyl Ketones Produced According to Eq. (36)

Cyclopropyl ketone	$[\alpha]_D^{25}$ (deg.): conc. of $CHCl_3$
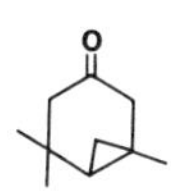	+150.1: 1.04 −147.6: 1.00
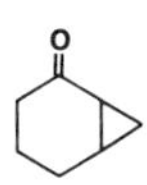	+15.3: 2.04 −15.5: 1.28
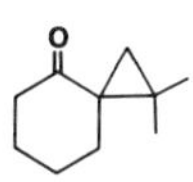	+47.3: 0.51 −50.8: 1.31
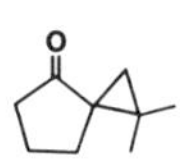	+162.3: 0.63 −171.9: 1.11
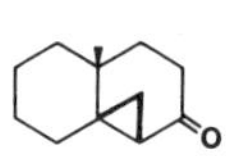	+97.7: 1.76 −95.3: 1.47
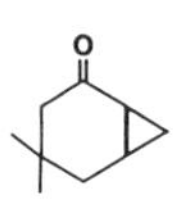	+215.3: 1.08 −214.5: 1.02
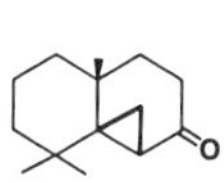	+70.5: 0.78 −72.2: 1.06

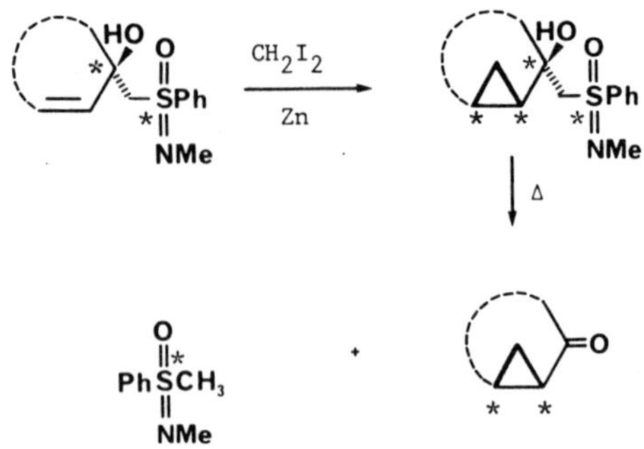

(36)

direct the stereochemical outcome of the methylenation. An application of this methodology to the total synthesis of (+)- and (−)-thujopsene (**76**) has been described.

TABLE 10
Optically Active Dihydroxy Ketones Produced According to Eq. (37)[a]

Dihydroxy ketone	$[\alpha]_D^{25}$ (deg.): conc., solvent
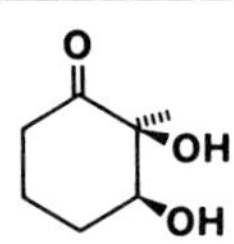	+0.7: 1.01, $CHCl_3$ −0.8: 1.06, $CHCl_3$
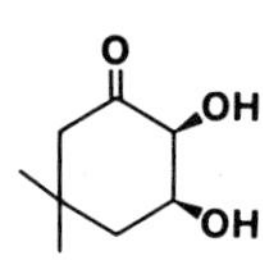	+21.6: 0.36, $CHCl_3$ −21.2: 1.08, $CHCl_3$
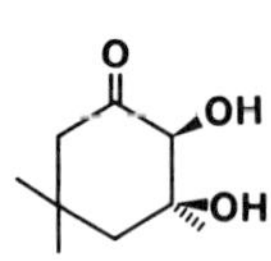	+28.6: 1.05, $CHCl_3$ −29.3: 1.09, $CHCl_3$
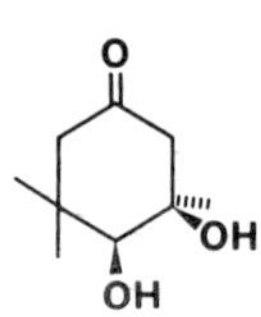	+24.0: 0.92, $CHCl_3$ −24.2: 0.39, $CHCl_3$

[a] From (*127*).

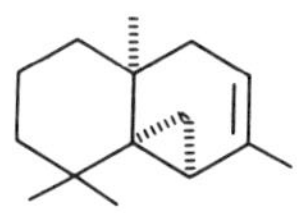

76

In closely related chemistry, the methylimino group of resolved "allylic" β-hydroxysulfoximines has been found to facilitate and direct osmium tetroxide cis hydroxylations of the carbon–carbon double bond with high diastereoface selectivity (*127*). The reaction is successful under both stoichiometric and catalytic conditions, and subsequent thermolysis (see above) affords high yields of optically active dihydroxy ketones [Eq. (37) and Table 10]. An application of this procedure to the synthesis of

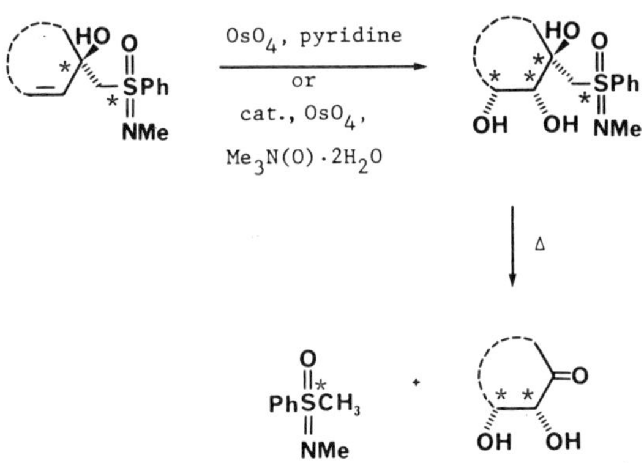

(37)

optically pure (+)- and (−)-pentenomycin precursors **77** (*128*) has been completed.

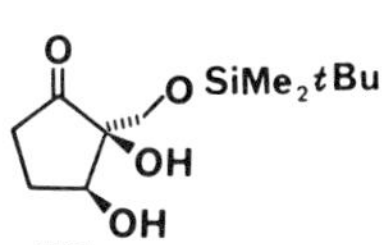

77

Acknowledgment

The work described in this chapter from the authors' laboratory has been supported by grants from the National Science Foundation.

References

1. Mikołajczyk, M., and Drabowicz, J. (1982). "Topics in Stereochemistry" (N. L. Allinger, E. L. Eliel, and S. H. Wilen, eds.), Vol. 13, pp. 333–468. Wiley (Interscience), New York.

2. a. Nudelman, A. (1971). *Int. J. Sulfur Chem B* **6,** 1; b. *idem* (1972). *ibid.* **7,** 241; c. *idem* (1976). *Phosphorus Sulfur* **2,** 51; d. *idem* (1980). *ibid.* **9,** 1; e. Shriner, R. L., Adams, R., and Marvel, C. S. (1943). "Organic Chemistry" (H. Gilman, ed.), Vol. 1, pp. 214–488. Wiley, New York; f. Laur, P. H. (1972). "Sulfur in Organic and Inorganic Chemistry" (A. Senning, ed.), Vol. 3, pp. 91–274. Dekker, New York; g. Wilen, S. H. (1972). "Tables of Resolving Agents and Optical Resolutions" (E. L. Eliel, ed.). University of Notre Dame Press; h. Ingold, C. K. (1971). *Int. J. Sulfur Chem. B* **6,** 65.

3. Harrison, P. W. B., Kenyon, J., and Phillips, H. (1926). *J. Chem. Soc.* 2079.

4. a. Piechulek, W., and Suszko, J. (1933). *Rocz. Chem.* **13,** 520; b. Janczewski, M., and Suszko, J. (1952). *ibid.* **26,** 394; c. Janczewski, M., and Mariarczyk, H. (1977). *ibid.* **51,** 891; d. Huszthy, P., Kapovits, I., Kucsman, A., and Radic, L. (1978). *Tetrahedron Lett.* 1853.

5. Mikolajczyk, M., Midura, W., Grzejszczak, S., Zatorski, A. and Chefczynska, A. (1978). *J. Org. Chem.* **43,** 473.

6. Bryan, R. F., Carey, F. A., Dailey, Jr., O. D., Maher, R. J., and Miller, R. W. (1978). *J. Org. Chem.* **43,** 90.

7. Backer, H. J., and Keuning, K. J. (1934). *Rec. Trav. Chim.* **53,** 798.

8. Cope, A. C., and Caress, E. A. (1966). *J. Am. Chem. Soc.* **88,** 1711.

9. Goldman, M., Kustanovich, Z., Weinstein, S., Tishbee, A., and Gil-Av, E. (1982). *J. Am. Chem. Soc.* **104,** 1093.

10. Farina, G., Montanari, F., and Negrini, A. (1959). *Gazz. Chim. Ital.* **89,** 1548.

11. a. Pirkle, W. H., and House, D. W. (1979). *J. Org. Chem.* **44,** 1957; b. Pirkle, W. H., House, D. W., and Finn, J. M. (1980). *J. Chromatogr.* **192,** 143.

12. Mikołajczyk, M., and Drabowicz, J. (1978). *J. Am. Chem. Soc.* **100,** 2510.

13. a. Mayr, A., Montanari, F., and Tramontini, M. (1960). *Gazz. Chim. Ital.* **90,** 739; b. Balenovic, K., Bregant, N., and Francetic, D. (1960). *Tetrahedron Lett.* 20; c. Maccioni, A., Montanari, F., Secci, M., and Tramontini, M. (1961). *Tetrahedron Lett.* 607.

14. a. Mislow, K., Green, M. M., Laur, P., Melillo, J. T., and Ternay, Jr., A. L. (1965). *J. Am. Chem. Soc.* **87,** 1958; b. Mislow, K., Green, M. M., and Raban, M., *ibid.* **87,** 2761.

15. Davis, F. A., Jenkins, Jr., R. H., Awad, S. B., Stinger, O. D., Watson, W. H., and Galloy, J. (1983). *J. Am. Chem. Soc.* **104,** 5412.

16. a. Higuchi, T., Pitman, I. H., and Gensch, K. H. (1966). *J. Am. Chem. Soc.* **88,** 5676; b. DiFuria, F., Modena, G., and Curci, R. (1976). *Tetrahedron Lett.* 4637.

17. a. Sugimoto, T., Kokubo, T., Miyazaki, J., Tanimoto, S., and Okano, M. (1979). *J. Chem. Soc. Chem. Commun.* 402; b. Sugimoto, T., Kokubo, T., Miyazaki, J., Tanimoto, S., and Okano, M. (1981). *Bioorg. Chem.* **10,** 311; c. Ogura, K., Fujita, M., and Iida, H. (1980). *Tetrahedron Lett.* **21,** 2233.

18. Komori, T., and Nonaka, T. (1983). *J. Am. Chem. Soc.* **105,** 5690.

19. Auret, B. J., Boyd, D. R., Henbest, H. B., and Ross, S. J. (1968). *J. Chem. Soc. C* 2371.

20. a. Abushanab, E., Reed, D., Suzuki, F., and Sih, C. J. (1978). *Tetrahedron Lett.* 3415; b. May, S. W., and Phillips, R. S. (1980). *J. Am. Chem. Soc.* **102,** 5981.

21. Johnson, C. R., Bacon, C. C., and Kingsbury, W. D. (1972). *Tetrahedron Lett.* 501.
22. Johnson, C. R., and McCants, Jr., D. (1965). *J. Am. Chem. Soc.* **87,** 5404.
23. a. Moriyama, M., Oae, S., Numata, T., and Furukawa, N. (1976). *Chem. Ind. (London)* 163; b. Kinoshita, M., Sato, Y., and Kunieda, N. (1974). *Chem. Lett.* 377; c. Sato, Y., Kunieda, N., and Kinoshita, M. (1976). *Chem. Lett.* 563.
24. Kjaer, A., and Malver, O. (1982). *Tetrahedron Lett.* **23,** 2687.
25. Oae, S., Iida, K., and Takata, T. (1981). *Tetrahedron Lett.* **22,** 573.
26. a. Rayner, D. R., von Schriltz, D. M., Day, J., and Cram, D. J. (1968). *J. Am. Chem. Soc.* **90,** 2721; b. Cram, D. J., Day, J., Rayner, D. R., von Schriltz, D. M., Duchamp, D. J., Garwood, D. C. (1970). *ibid.* **92,** 7369; c. Williams, T. R., Booms, R. E., and Cram, D. J. (1971). *ibid.* **93,** 7338.
27. a. Kobayashi, M., and Yobe, A. (1967). *Bull. Chem. Soc. Jpn.* **40,** 224; b. Folli, U., Iarossi, D., and Montanari, F. (1968). *J. Chem. Soc. C* 1372.
28. a. Balenovic, K., and Bregant, N. (1964). *Chem. Ind. (London)* 1577; b. Mikołajczyk, M., and Para, M. (1969). *J. Chem. Soc. Chem. Commun.* 1192; c. Mikołajczyk, M., and Leitloff, M. (1975). *Ups. Khim.* **44,** 1419; d. *idem* (1975). *Russ. Chem. Rev. Engl. Transl.* **44,** 670.
29. Juge, S., and Kagan, H. B. (1975). *Tetrahedron Lett.* 2733.
30. Kuneida, N., Motoki, H., and Kinoshita, M. (1978). *Chem. Lett.* 713.
31. Marchese, G., Naso, F., and Ronzini, L. (1974). *J. Chem. Soc. Chem. Commun.* 830.
32. a. Andersen, K. K. (1962). *Tetrahedron Lett.* 93; b. Andersen, K. K. (1964). *J. Org. Chem.* **29,** 1953; c. Andersen, K. K., Gaffield, W., Papanikolaou, N. E., Foley, J. W., and Perkins, R. I. (1964). *J. Am. Chem. Soc.* **86,** 5637; d. Andersen, K. K. (1971). *Int. J. Sulfur Chem.* **6,** 69.
33. a. Phillips, H. (1925). *J. Chem. Soc.* **127,** 2552; b. Herbrandson, H. F., and Dickerson, R. T. (1959). *J. Am. Chem. Soc.* **81,** 4102.
34. Axelrod, M., Bickart, P., Jacobus, J., Green, M. M., and Mislow, K. (1968). *J. Am. Chem. Soc.* **90,** 4835.
35. a. Harpp, D. N., Vines, S. M., Montillier, J. P., and Chan, T. H. (1976). *J. Org. Chem.* **41,** 3987; b. Drabowicz, J., Bujnicki, B., and Mikołajczyk, M. (1982). *J. Org. Chem.* **47,** 3325.
36. Mioskowski, C., and Solladié, G. (1975). *Tetrahedron Lett.* 3341.
37. Colombo, L., Gennari, C., and Narisano, E. (1978). *Tetrahedron Lett.* 3861.
38. a. Jacobus, J., and Mislow, K. (1967). *J. Am. Chem. Soc.* **89,** 5228; b. Colonna, S., Giovini, R., and Montanari, F. (1968). *Chem. Commun.* 865.
39. For improved preparations see: a. Estep, R. E., and Tavares, D. J. (1973). *Int. J. Sulfur Chem.* **8,** 279; b. Harpp, D. N., Friedlander, B. T., Larsen, C., Steliou, K., and Stockton, A. (1978). *J. Org. Chem.* **43,** 3481.
40. a. O'Brien, J. B. (1968). Ph.D. Dissertation, University of New Hampshire, Durham, N. H.; b. Drabowicz, J. (1974). Ph.D. Dissertation, Centre of Molecular and Macromolecular Studies, Polish Academy of Sciences, Department of Organic Sulfur Compounds, Lodz, Poland.
41. Ridley, D. D., and Smal, M. A. (1981). *J. Chem. Soc. Chem. Commun.* 505.
42. a. Colonna, S., Giovini, R., and Montanari, F. (1968). *Chem. Commun.* 865; b. Jacobus, J., and Mislow, K. (1968). *Chem. Commun.* 253.
43. Wudl, F., and Lee, T. B. K. (1973). *J. Am. Chem. Soc.* **95,** 6349.
44. Lockard, J. P., Schroeck, C. W., and Johnson, C. R. (1973). *Synthesis* 485.
45. Maryanoff, C. A., Maryanoff, B. E., Tang, R., and Mislow, K. (1973). *J. Am. Chem. Soc.* **95,** 5839.
46. Pope, W. J., and Peachey, S. J. (1900). *J. Chem. Soc.* **77,** 1072.

47. Smiles, S. (1900). *J. Chem. Soc.* **77,** 1174.
48. For a review see: Andersen, K. K. (1981). "The Chemistry of the Sulphonium Group" (C. J. M. Stirling, ed.), Part 1, pp. 229–266. Wiley (Interscience), New York.
49. For example see: a. Darwish, D., and Tourigny, G. (1966). *J. Am. Chem. Soc.* **88,** 4303; b. Darwish, D., Hui, S. H., and Tomilson, R. (1968). *J. Am. Chem. Soc.* **90,** 5631; c. Scartazzini, R., and Mislow, K. (1967). *Tetrahedron Lett.* 2719; d. Garbesi, A., Corsi, N., and Fava, A. (1970). *Helv. Chim. Acta* **53,** 1499; e. Trost, B. M., and Hammen, R. F. (1973). *J. Am. Chem. Soc.* **95,** 962.
50. Campbell, S. J., and Darwish, D. (1974). *Can. J. Chem.* **52,** 2953.
51. For example see: Cook, M. J., Dorn, H., and Katritzky, A. R. (1968). *J. Chem. Soc. B* 1468.
52. For example see: Darwish, D., and Scott, C. E. (1973). *Can. J. Chem.* **51,** 3647.
53. a. Andersen, K. K., Caret, R. L., and Ladd, D. L. (1976). *J. Org. Chem.* **41,** 3096; b. Andersen, K. K., Caret, R. L., and Karup-Nielsen, I. (1974). *J. Am. Chem. Soc.* **96,** 8026.
54. Mikołajczyk, M., Bujnicki, B., and Drabowicz, J. (1977). *Bull. Acad. Pol. Sci. Ser. Chem.* **25,** 267.
55. Okuma, K., Minato, H., and Kobayashi, M. (1980). *Bull. Chem. Soc. Jpn.* **53,** 435.
56. a. Johnson, C. R., Rigau, J. J., Haake, M., McCants, Jr., D., Keiser, J. E., and Geertsema, A. (1968). *Tetrahedron Lett.* 3719; b. Johnson, C. R., and Rigau, J. J. (1968). *J. Org. Chem.* **33,** 4340.
57. a. Bohman, O., and Allenmark, S. (1973). *Chem. Scripta* **4,** 202; b. Darwish, D., and Datta, S. K. (1974). *Tetrahedron* **30,** 1155.
58. Drabowicz, J., Bujnicki, B., and Mikołajczyk, M. (1981). *J. Org. Chem.* **46,** 2788.
59. a. Johnson, C. R., and Schroeck, C. W. (1968). *J. Am. Chem. Soc.* **90,** 6852; b. (1973). *ibid.* **95,** 7418.
60. Kobayashi, M., Kamiyama, K., Minato, H., Oishi, Y., Takada, Y., and Hattori, Y. (1971). *J. Chem. Soc. D* 1577.
61. Kamiyama, K., Minato, H., and Kobayashi, M. (1973). *Bull. Chem. Soc. Jpn.* **46,** 3895.
62. Kobayashi, M. (1980). Work presented at the Tenth International Symposium on Organic Sulfur Chemistry, Riga, U.S.S.R.
63. a. Kennewell, P. D., and Taylor, J. B. (1975). *Chem. Soc. Rev.* **4,** 189; b. Kennewell, P. D. (1980). *ibid.* **9,** 477; c. Johnson, C. R. (1979). "Comprehensive Organic Chemistry" (N. D. Jones, ed.), Vol. 3, Chapter 11, pp. 223–232. Pergamon, New York.
64. Johnson, C. R. (1973). *Acc. Chem. Res.* **6,** 341.
65. Kresze, G., and Wustrow, B. (1962). *Chem. Ber.* **95,** 2652.
66. a. Fusco, R., and Tericoni, F. (1965). *Chem. Ind. (Milan)* **47,** 61; b. *idem* (1965). *Chem. Abstr.* **62,** 10357h.
67. Barash, M. (1960). *Nature* **187,** 591.
68. a. Kwart, H., and Khan, A. A. (1967). *J. Am. Chem. Soc.* **89,** 1950; b. Sabol, M. A., Davenport, R. W., and Andersen, K. K. (1968). *Tetrahedron Lett.* 2159; c. Johnson, C. R., Kirchhoff, R. A., Reischer, R. J., and Katekar, G. F. (1973). *J. Am. Chem. Soc.* **95,** 4287.
69. Moriyama, M., Numata, T., and Oae, S. (1974). *Chem. Prep. Proc. Int.* **6,** 207.
70. Bentley, H. R., and Whitehead, J. K. (1950). *J. Chem. Soc.* 2081.
71. a. Schroeck, C. W. (1971). Ph.D. Dissertation, Wayne State University, Detroit, Michigan; b. Jonsson, E. U., and Johnson, C. R. (1971). *J. Am. Chem. Soc.* **93,** 5308.
72. a. Colonna, S., and Stirling, C. J. M. (1971). *J. Chem. Soc. Chem. Commun.* 1591; b. Colonna, S., and Stirling, C. J. M. (1974). *J. Chem. Soc. Perkin Trans.* **I,** 2120.

73. Johnson, C. R., Kirchhoff, R. A., and Corkins, H. G. (1974). *J. Org. Chem.* **39,** 2458.
74. Whitehead, J. R., and Bentley, H. R. (1952). *J. Chem. Soc.* 1572; b. Satzinger, G., and Stoss, P. (1970). *Arzneim.-Forsch.* **20,** 1214.
75. Gilchrist, T. L., and Moody, C. J. (1977). *Chem. Rev.* **77,** 409.
76. Huang, S.-L., and Swern, D. (1979). *J. Org. Chem.* **44,** 2510.
77. Stoss, P., and Satzinger, G. (1974). *Tetrahedron Lett.* 1973.
78. Veale, H. S., Levin, J., and Swern, D. (1978). *Tetrahedron Lett.* 503.
79. Johnson, C. R., and Kirchhoff, R. A. (1979). *J. Org. Chem.* **44,** 2280.
80. Furukawa, N., Akutagawa, K., Yoshimura, T., and Oae, S. (1981). *Tetrahedron Lett.* **22,** 3989.
81. Akasaka, T., Yoshimura, T., Furukawa, N., and Oae, S. (1978). *Chem. Lett.* 417.
82. a. Day, J., and Cram, D. J. (1965). *J. Am. Chem. Soc.* **87,** 4398; b. Garwood, D. C., and Cram, D. J. (1970). *ibid.* **92,** 4575; c. Yamagishi, F. G., Rayer, D. R., Zwicker, E. T., and Cram, D. J. (1973). *ibid.* **95,** 1916; d. Christensen, B. W., and Kjaer, A. (1969). *J. Chem. Soc. Chem. Commun. 934;* e. Christensen, B. W. (1971). *ibid.* 597; f. Maricich, T. J., and Hoffman, V. L. (1972). *Tetrahedron Lett.* 5309; g. Maricich, T. J., and Hoffman, V. L. (1974). *J. Am. Chem. Soc.* **96,** 7770.
83. a. Nudelman, A. (1969). *Diss. Abstr.* **29,** 1049B; b. *idem* (1970). *Chem. Abstr.* **72,** 131905h.
84. Moriyama, M., Yoshimura, T., Furukawa, N., Numata, T., and Oae, S. (1976). *Tetrahedron* **32,** 3003.
85. a. Clarke, S. G., Kenyon, J., and Phillips, H. (1927). *J. Chem. Soc.* 188; b. Holloway, J., Kenyon, J., and Phillips, H. (1928). *ibid.* 3000. For a more recent example see: Bohman, O., and Allenmark, S. (1973). *Chem. Scripta* **4,** 202.
86. a. Jonsson, E. U., Bacon, C. C., and Johnson, C. R. (1971). *J. Am. Chem. Soc.* **93,** 5306; b. Johnson, C. R., Jonsson, E. U., and Bacon, C. C. (1979). *J. Org. Chem.* **44,** 2055.
87. Johnson, C. R., Jonsson, E. U., and Wambsgans, A. (1979). *J. Org. Chem.* **44,** 2061.
88. Johnson, C. R., and Wambsgans, A. (1979). *J. Org. Chem.* **44,** 2278.
89. Johnson, C. R., Bis, K. G., Cantillo, J. H., Meanwell, N. A., Reinhard, M. F. D., Zeller, J. R., and Vonk, G. P. (1983). *J. Org. Chem.* **48,** 1.
90. Colonna, S., Annunziata, R., and Cinquini, M. (1981). *Phosphorus Sulfur* **10,** 197.
91. Posner, G. H., Mallamo, J. P., Miura, K., and Hulce, M. (1981). *Pure Appl. Chem.* **53,** 2307.
92. Solladié, G. (1981). *Synthesis* 185.
93. a. Goldberg, S. I., and Sahli, M. S. (1965). *Tetrahedron Lett.* 4441; b. Goldberg, S.I., and Sahli, M. S. (1967). *J. Org. Chem.* **32,** 2059.
94. For example see: a. Stridsberg, B., and Allenmark, S. (1976). *Acta Chem. Scand.* **30,** 219; b. Numata, T., and Oae, S. (1977). *Tetrahedron Lett.* 1337; c. Numata, T., Itoh, O., and Oae, S. (1977). *Chem. Lett.* 909; d. Mikołajczyk, M., Zatorski, A., Grzejszczak, S., Costisella, B., and Midura, W. (1978). *J. Org. Chem.* **43,** 2518.
95. Numata, T., Itoh, O., and Oae, S. (1979). *Tetrahedron Lett.* 1869.
96. a. Goldmann, S., Hoffmann, R. W., Maak, N., and Geueke, K. J. (1980). *Chem. Ber.* **113,** 831; b. Hoffmann, R. W., Goldmann, S., Gerlach, R., and Maak, N. (1980). *ibid.* **113,** 845; c. Hoffmann, R. W., Gerlach, R., and Goldmann, S. (1980). *ibid.* **113,** 856.
97. Koizumi, T., Hirai, H., and Yoshii, E. (1982). *J. Org. Chem.* **47,** 4004.
98. Abbott, D. J., Colonna, S., Stirling, C. J. M. (1976). *J. Chem. Soc. Perkin Trans.* **I,** 492.
99. Tsuchihashi, G., Mitamura, S., and Ogura, K. (1976). *Tetrahedron Lett.* 855.

100. a. Posner, G. H. (1983). "Asymmetric Synthesis" (J. D. Morrison, ed.), Vol. 2, Ch. 8. Academic Press, New York; b. Posner, G. H., Mallamo, J. P., and Miura, K. (1981). *J. Am. Chem. Soc.* **103,** 2886; c. Posner, G. H., Hulce, M., Mallamo, J. P., Drexler, S. A., and Clardy, J. (1981). *J. Org. Chem.* **46,** 5244.

101. Tsuchihashi, G., Mitamura, S., and Ogura, K. (1974). *Tetrahedron Lett.* 455.

102. Calzavara, P., Cinquini, M., Colonna, S., Fornasier, R., and Montanari, F. (1973). *J. Am. Chem. Soc.* **95,** 7431.

103. Cinquini, M., Colonna, S., and Montanari, F. (1974). *J. Chem. Soc. Perkin Trans.* **I,** 1719.

104. Johnson, C. R., and Schroeck, C. W. (1971). *J. Am. Chem. Soc.* **93,** 5303.

105. Durst, T., Viau, R., Van Den Elzen, R., and Nguyen, C. H. (1971). *J. Chem. Soc. Chem. Commun.* 1334.

106. a. Tsuchihashi, G., Iriuchijima, S., and Ishibashi, M. (1972). *Tetrahedron Lett.* 4605; b. Kingsbury, C. A. (1972). *J. Org. Chem.* **37,** 102; c. Kunieda, N., Kinoshita, M., and Nokami, J. (1977). *Chem. Lett.* 289; d. Tsuchihashi, G., Iriuchijima, S., and Maniwa, K. (1973). *Tetrahedron Lett.* 3389.

107. For a review on the addition of various chiral nucleophiles to aldehydes and ketones see Solladié, G. (1983). "Asymmetric Synthesis" (J. D. Morrison, ed.), Vol. 2, Ch. 6. Academic Press, New York.

108. a. Mioskowski, C., and Solladié, G. (1977). *J. Chem. Soc. Chem. Commun.* 162; b. Mioskowski, C., and Solladié, G. (1980). *Tetrahedron* **36,** 227; c. Solladié, G., and Matloubi-Moghadam, F. (1982). *J. Org. Chem.* **47,** 91.

109. Corey, E. J., Weigel, L. O., Chamberlin, A. R., Cho, H., and Hua, D. H. (1980). *J. Am. Chem. Soc.* **102,** 6613.

110. a. Colombo, Z., Gennari, C., Resnati, G., and Scolastico, C. (1981). *Synthesis* 74; b. Colombo, Z., Gennari, C., Resnati, G., and Scolastico, C. (1981). *J. Chem. Soc. Perkin Trans.* **I,** 1284.

111. Matloubi, F., and Solladie, G. (1979). *Tetrahedron Lett.* 2141.

112. Solladié, G., Greck, C., and Demailly, G. (1982). *Tetrahedron Lett.* **23,** 5047.

113. Trost, B. M., and Biddlecom, W. G. (1973). *J. Org. Chem.* **38,** 3438.

114. a. Johnson, C. R., and Janiga, E. R. (1973). *J. Am. Chem. Soc.* **95,** 7692; b. Johnson, C. R., Schroeck, C. W., and Shanklin, J. R. (1973). *ibid.* **95,** 7424.

115. a. Johnson, C. R., and Lockard, J. P. *Tetrahedron Lett.* 1971, 4589; b. Johnson, C. R., Lockard, J. P., and Kennedy, E. R. (1980). *J. Org. Chem.* **45,** 264.

116. a. Johnson, C. R., and Katekar, G. F. (1970). *J. Am. Chem. Soc.* **92,** 5753; b. Johnson, C. R., Katekar, G. F., Huxol, R. F., and Janiga, E. R. (1971). *ibid.* **93,** 3771.

117. Welch, S. C., Rao, A. S. C. P., Lyon, J. T., and Assercq, J. (1983). *J. Am. Chem. Soc.* **105,** 252.

118. Johnson, C. R., and Stark, Jr., C. J. (1979). *Tetrahedron Lett.* 4713.

119. a. Schroeck, C. W., and Johnson, C. R. (1971). *J. Am. Chem. Soc.* **93,** 5305; b. Johnson, C. R., and Stark, Jr., C. J. (1982). *J. Org. Chem.* **47,** 1193.

120. Annunziata, R., Cinquini, M., and Cozzi, F. (1981). *J. Chem. Soc. Perkin Trans.* **I,** 1109.

121. Johnson, C. R., and Stark, Jr., C. J. (1982). *J. Org. Chem.* **47,** 1196.

122. Annunziata, R., and Cinquini, M. (1982). *Synthesis* 767.

123. a. Johnson, C. R., and Kirchhoff, R. A. (1979). *J. Am. Chem. Soc.* **101,** 3602; b. Johnson, C. R., Shanklin, J. R., and Kirchhoff, R. A. (1973). *ibid.* **95,** 6462; C. Morton, Jr., D. R., and Brokaw, F. C. (1979). *J. Org. Chem.* **44,** 2880.

124. Johnson, C. R., and Meanwell, N. A. (1981). *J. Am. Chem. Soc.* **103,** 7667.

125. a. Johnson, C. R., and Zeller, J. R. (1982). *J. Am. Chem. Soc.* **104,** 4021; b. Johnson, C. R., and Zeller, J. R. (1984). *Tetrahedron,* **40,** 1225.

126. Johnson, C. R., and Barbachyn, M. R. (1982). *J. Am. Chem. Soc.* **104,** 4290.

127. Johnson, C. R., and Barbachyn, M. R. (1984). *J. Am. Chem. Soc.* **106,** 2459.

128. Smith, III, A. B., Branca, S. J., Pilla, N. N., and Guaciaro, M. A. (1982). *J. Org. Chem.* **47,** 1855.

3

Preparation of the Enantiomers of Compounds Containing Chiral Phosphorus Centers

Donald Valentine, Jr.
Santa Clara, California

I. Introduction

Appropriately substituted three-, four-, and five-coordinate organophosphorus compounds contain chiral phosphorus atoms. Many such compounds can be prepared in optically active form. Enantiomers of P-chiral

ISBN 0-12-507704-1

compounds are of interest for three main reasons. First, there is intrinsic interest in preparation of P-chiral systems and in the study of the many stereoselective transformations of these compounds. Second, many P-chiral organophosphorus compounds exhibit interesting biological activities that may be specific to one enantiomer. Third, P-chiral compounds have achieved considerable notoriety and are of practical value as ligands in catalysts for asymmetric organic syntheses.

This chapter is intended to serve as a practical guide to the preparation of new P-chiral compounds by either resolution or asymmetric synthesis. Considerable progress has been made in the preparation and use of P-chiral compounds. Use of newer methods is stressed here; however, an attempt is also made to provide an overview. Section I of this chapter comprises a brief survey of P-chiral chemistry, including a summary of synthetic methods, of readily available P-chiral systems whose absolute configuration is known, and of factors influencing the stereochemical results obtained in reactions of P-chiral compounds. Section II is concerned with preparation of P-chiral phosphines and related substances. Section III is concerned with the preparation of various P-chiral thio acids and related substances.

TABLE 1
Some Readily Available P-Chiral Compounds

Structure	Configuration	Method	References
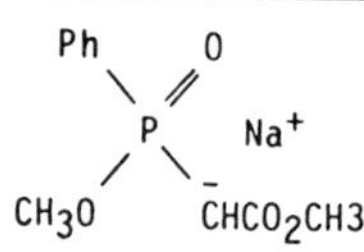	(*S*)-(−)	Resolution	*50, 145*
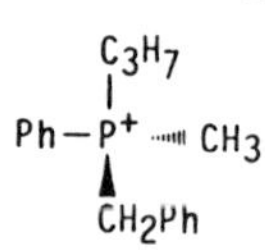	(*S*)-(+)	Resolution[a]	*82, 87, 170*
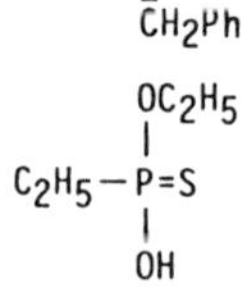	—	Resolution[b]	*1*
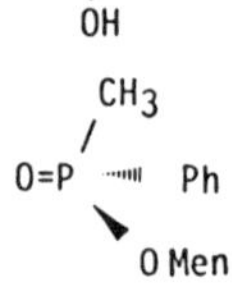	(*S*)-(−)	Separation of diastereomers	*109*

[a] Many similar systems have been resolved (*81*).

[b] Many similar systems have been resolved (*156*). Others are available by asymmetric syntheses described in Section III.

Many transformations of P-chiral compounds are known to be stereospecific or highly stereoselective under at least some conditions. It is often possible to convert one of the readily available P-chiral compounds of known absolute configuration, by a series of steps known to be stereospecific, to desired target molecules. Such syntheses offer synthetic flexibility and convenience, since both product enantiomers can often be obtained from a single precursor. They offer the further advantage that the absolute configuration of the products is correlated with that of the precursor. Use of Table 1 (which shows available P-chiral substances of known absolute configuration) and Schemes 1 and 2 (which show typical stereochemical results for many interconversions of P-chiral compounds) will often be helpful in suggesting routes to particular P-chiral targets.

Many of the reactions in Schemes 1 and 2 involve nucleophilic displacements at phosphorus in tetracoordinate, pentavalent phosphorus compounds. Study of nucleophilic displacements at phosphorus has revealed a rich variety of stereochemical possibilities. In some systems, nucleophilic displacements are stereospecific or highly stereoselective, proceeding with either predominant inversion or predominant retention. In other systems, only marginal stereoselectivity is observed, and the stereo-

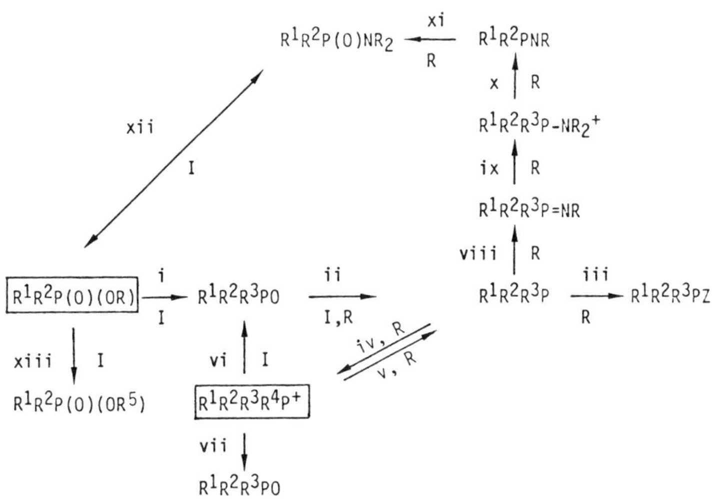

Scheme 1. Some stereoselective reactions of P-chiral organophosphorus compounds. Boxed compounds are primary sources of P-chirality. I, Inversion; R, retention; *Q*, variable or racemization. These results are for acyclic systems only (chiral P is not part of ring system). Conditions: *i*, RMgX. *ii*, Si_2Cl_6(I) or $PhSiH_3$(R). *iii*, Z = O, S, Se. For Z = O, use H_2O_2; for Z = S, S_8; for Z = Se, Se. *iv*, RX. *v*, Cathodic reduction. *vi*, NaOH, R^4 = benzyl gives clean inversion. Variable results in other systems. *vii*, (1) base; (2) $R_2C{=}O$. Wittig reactions occurs with retention at phosphorus. *viii*, RN_3. *ix*, RX. *x*, Cathodic reduction. *xi*, Hydrogen peroxide. *xii*, NaOR; $LiNR_2$. *xiii*, (1) Methyl triflate; (2) $NaOR^5$.

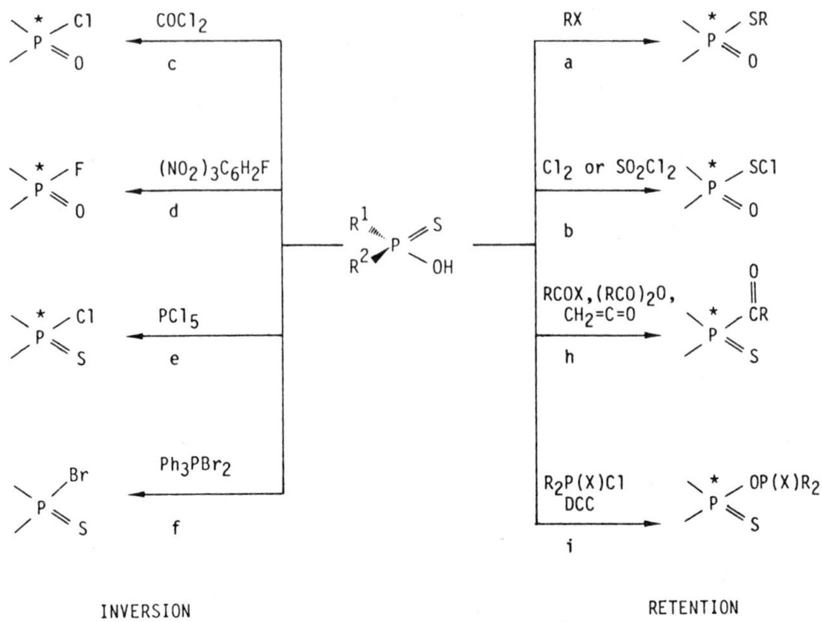

Scheme 2. Stereoselective transformations of phosphorus thioacids. Results may differ in cyclic versus acyclic systems (*62*). *a*, (*3, 11, 55, 143*). *b*, (*111, 142, 143*). *c*, (*3, 54, 138*). *d*, (*17, 156b*). *e*, (*51, 52, 139, 153*). *f*, (*140, 141*). *g*, (*56, 144–148*). *h*, (*144–146, 154*).

chemical results depend on reaction conditions, steric and electronic properties of the nucleophile, the leaving group, and other phosphorus ligands. It has proved a considerable challenge to develop a consistent explanation for the diverse stereochemical behavior of P-chiral compounds.

The stereochemistry of organophosphorus compounds has been extensively investigated; full discussion is beyond the scope of this chapter but many good sources are available (*59, 61, 62, 80, 135, 156, 197*). The following paragraphs are intended only to provide an orientation for synthetic chemists. References are made where appropriate to discussion of specific points of interest.

Reactions resulting in inversion of configuration can occur by an S_N2P mechanism or through trigonalbipyramidal (TBP) intermediates arising from attack of the nucleophile opposite the leaving group, such that both the nucleophile and the leaving group occupy apical positions. The leaving group must depart by Berry pseudorotation (*13*) before ligand reorganization or turnstile mechanisms (*186*) can operate. Scheme 3a illustrates the inversion process. The TBP concept (*53, 186, 192*) also explains retention of configuration, as shown in Scheme 3b. The leaving group attains

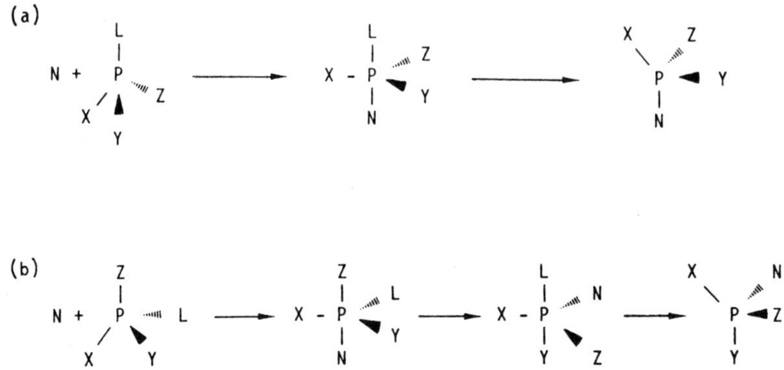

Scheme 3. (a) Nucleophilic attack at phosphorus with inversion of configuration. (b) Nucleophilic attack at phosphorus with retention of configuration. Adapted from (*62*).

an apical position not opposite to the nucleophile, which occupies a basal position following ligand reorganization of the initially formed intermediate by pseudorotation or turnstile mechanism. Further assumptions are necessary to explain why some reactions occur with inversion and others with retention. One idea is that the nucleophile enters opposite the most "apicophilic" group (an apicophilic group is one that favors an apical position). If the apicophilic group is also a good leaving group, inversion occurs. Otherwise, ligand reorganization occurs, resulting in retention of configuration (*16, 19, 35, 167*). The problem is to identify those features that characterize good leaving groups and groups of high apicophilicity. Analysis of cyclic systems is particularly difficult because bond strengths may differ from bond strengths in acyclic analogs, and the presence of the ring may influence the geometry of intermediates (*38, 53, 62, 197*). Detailed general discussions of the TBP concept (*43, 124*) and of the stereochemical features of cyclic systems (*62*) are available.

II. Chiral Tertiary Phosphines and Related Substances

A principal current application of P-chiral phosphines is their use as chiral ligands in catalysts for asymmetric synthesis. High enantiomeric excess (ee) is required of these chiral ligands. In addition, P-chiral phosphines compete with other phosphines in which an enantiomerically pure organic moiety is ligated to achiral phosphorus. It is necessary to consider both whether the P-chiral phosphine offers advantages as a ligand and whether a feasible method exists to prepare required amounts of the phosphine in high enantiomeric excess. Many syntheses of P-chiral tertiary phosphines have been described. None of these methods is general. Unfortunately, features that make particular phosphines attractive as ligands (e.g., steric bulk and other functionality) may make application of common methods difficult. Careful consideration of several methods is advised before embarking on particular syntheses.

A. *Epimerization of Chiral Phosphorus Centers*

Syntheses of P-chiral tertiary phosphines must take into account the moderate epimeric stability of these substances. Tertiary phosphines have, in the absence of special effects (*8–10, 41, 119, 197*), barriers to pyramidal inversion of approximately 30–35 kcal/mol and may be obtained as enantiomers that interconvert slowly at or near ambient temperature (*7, 8, 80, 81, 118, 125, 129, 157*). Above approximately 80–100°C, however, reaction (1) is typically fairly rapid.

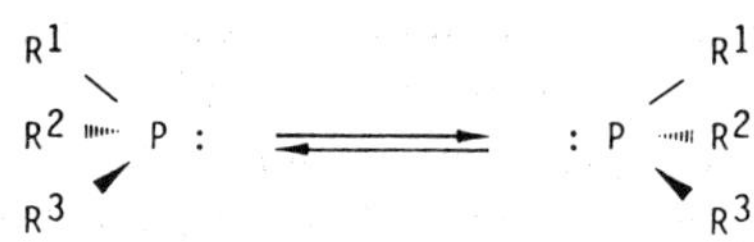

(1)

Racemization of P-chiral phosphinous esters R_2POR occurs with measurable rates even at room temperature. Racemization of these substances is acid catalyzed and may involve exchange of OR groups rather than pyramidal inversion (*79*). Tertiary phosphine oxides, $O{=}PR^1R^2R^3$, usually have high epimeric stability but are racemized by reagents such as $LiAlH_4$ (*76*). Quaternary phosphonium salts are also relatively stable toward epimerization at phosphorus. In addition, phosphinic acid esters and amides, $R^1R^2P(O)(OR)$ and $R^1R^2P(O)(NR_2)$, are relatively stable toward epimerization, although processes in which the symmetric anion $R^1R^2PO_2^-$ is formed lead to racemization.

B. *Preparations of Chiral Tertiary Phosphines*

This section discusses the modern synthetic methods to obtain P-chiral phosphines, encompassing most aspects of the preparations of phosphine oxides, phosphonium salts, and phosphinates. The most important preparative methods to obtain P-chiral tertiary phosphines are (*a*) the phosphinate ester method (*197*) (Section II,B,1); (*b*) the phosphonium salt method (*80*) (Section II,B,2); (*c*) resolutions of enantiomers using dias-

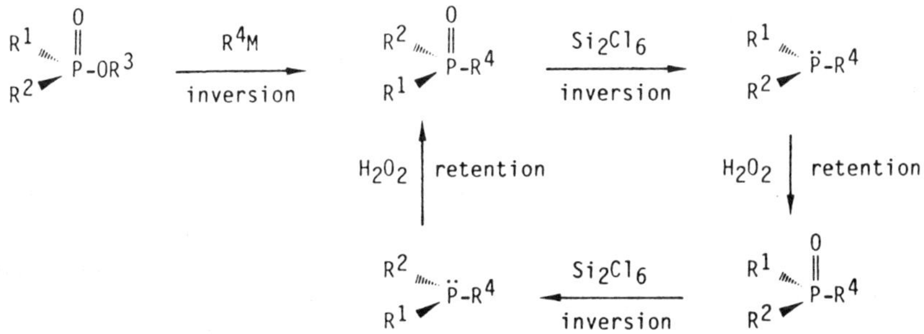

Scheme 4. P-Chiral tertiary phosphines from P-chiral tertiary phosphinate esters (*197*).

teromeric metal complexes (*184*) (Section II,B,3); and (*d*) emerging methods (Section II,B,4).

1. Phosphinate Ester Method

Synthesis of P-chiral tertiary phosphine oxides using P-chiral phosphinate esters as shown in Scheme 4 was developed by Mislow *et al.* (*197*). The method is conceptually similar to the synthesis of S-chiral sulfoxides from S-chiral sulfinate esters (*6*). The phosphinate ester route to P-chiral phosphines is both flexible and, in some cases, adaptable for practical use. Successful use of the method requires that a P-chiral phosphinate ester of high enantiomeric purity be obtainable; that it be possible to convert the enantiomeric phosphinate stereospecifically to a P-chiral phosphine oxide; and that it be possible to reduce the P-chiral phosphine oxide stereospecifically to the desired P-chiral phosphine enantiomer. Problems may arise in each stage of the process.

Phosphinate esters containing chiral phosphorus can be obtained by unexceptional methods, of which reactions (2) and (3) provide one exam-

$$\mathrm{hPCl_2} \xrightarrow{\mathrm{CH_3OH,\ py}} \mathrm{PhP(OCH_3)_2} \xrightarrow[\text{Arbusov}]{\mathrm{CH_3I},\ \text{heat}} \mathrm{(CH_3)PhPO(OCH_3)} \xrightarrow{\mathrm{PCl_5}} \mathrm{Ph{-}P(Cl)(CH_3){=}O} \qquad (2)$$

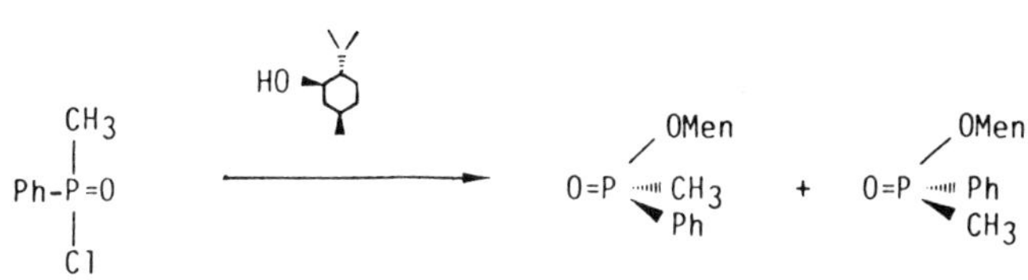

(3)

1 (Sp)-**2** (Rp)-**3**

ple (*197*). Reactions similar to reaction (2) employing menthol (*108, 109, 196*) or cholesterol (*163*) have been used to generate mixtures of diastereomeric phosphinates. It is also possible to convert P-chiral thiophosphinates to P-chiral phosphinates, as outlined in Section III,A. In addition, P-chiral phosphinates can also be obtained from carbohydrate precursors as indicated in Scheme 5 (*27, 59*). This reaction is discussed in more detail in Section III,C. Synthesis of phosphinates from phosphinyl chlorides often produces equal or nearly equal mixtures of diastereomers from which neither diastereomer can readily be obtained in pure form. In the carbohydrate-based route outlined in Scheme 5, the diastereomers are geometric isomers, which can be easier to separate to obtain pure epi-

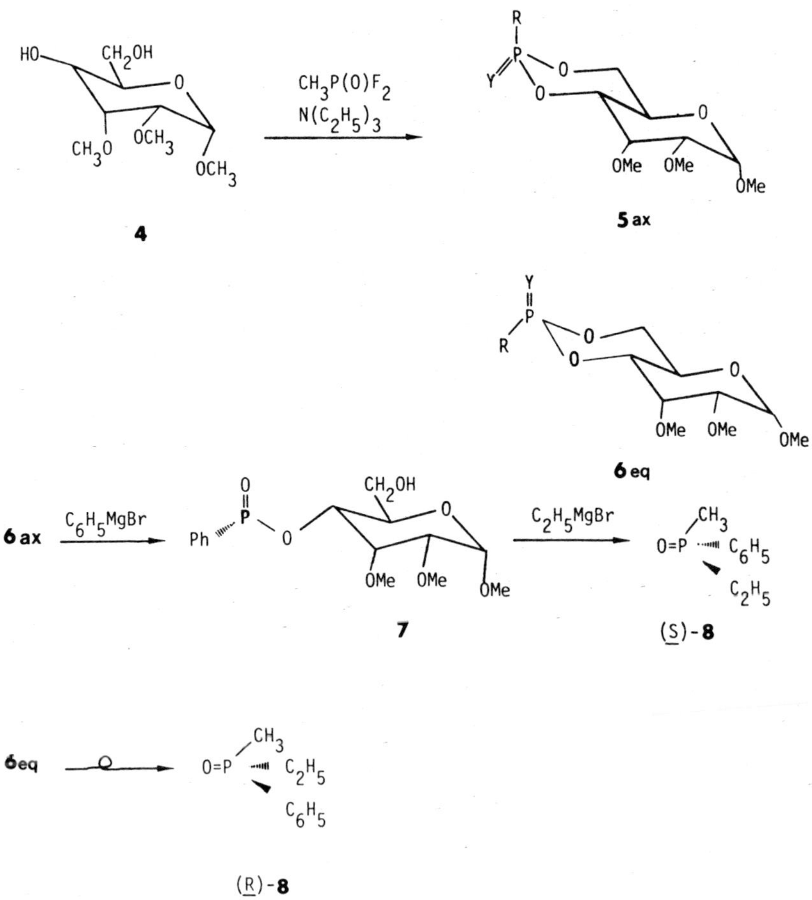

Scheme 5. Use of carbohydrates to prepare P-chiral phosphinates of high epimeric purity.

mers. Finally, the single precursor **4** can provide both tertiary phosphine oxide enantiomers if the order of addition of the Grignard reagents is reversed. The known absolute configuration of the carbohydrate then establishes the absolute configuration of the products (*59*).

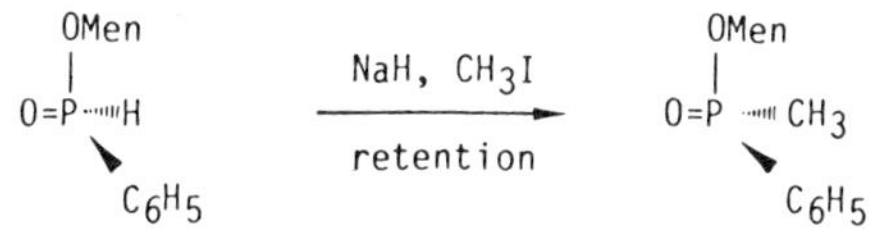

9

(4)

Another possibility is stereospecific alkylation of menthylphenylphosphinate (**9**) (*47, 48*). Many chiral phosphinates are potentially available from **9** via reaction (4). It is also possible to exchange alkoxy groups in phosphinates and phosphinothioates (*36*):

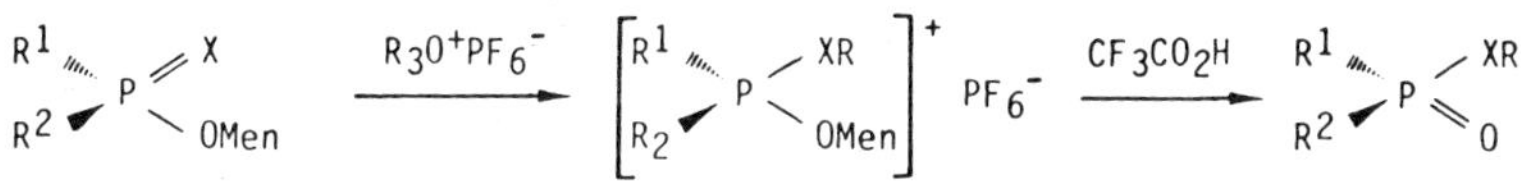

10 **11** **12**

(5)

where X = S or O, R = CH_3 or C_2H_5, the preferred counterion is PF_6^-, and the configuration of **12** is inverted compared to **10** (*30–36, 77*).

Grignard reagents react stereospecifically with phosphinates to give phosphine oxides with inversion of phosphorus configuration (*108, 109*). The reaction may give racemic products or fail completely under the drastic reaction conditions required when there are bulky groups on either phosphorus or the Grignard reagent. Unfortunately, the systems having bulky substituents may be the most interesting for applications. Organolithium reagents are more reactive but epimerization at phosphorus may occur (*94, 121, 181, 197*).

Stereospecific reduction of P-chiral tertiary phosphine oxides to P-chiral tertiary phosphines, with *inversion* of phosphorus configuration, is easily accomplished on a small scale using reaction (6) (*162*).

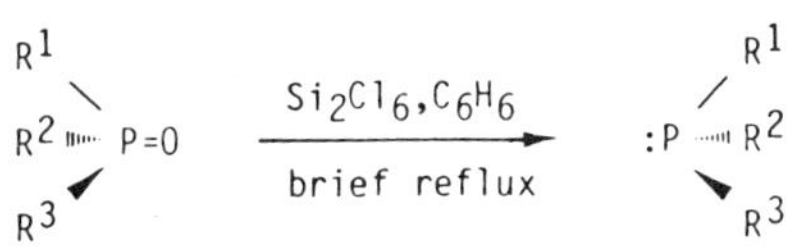

(6)

The Si_2Cl_6 should be freshly distilled. This distillation should be carried out on small scale with adequate safety protection, as explosions have been reported. The shortest possible reaction time should be used in carrying out reaction (6) because Si_2Cl_6 will racemize P-chiral tertiary phosphines. On a larger scale, the cost of Si_2Cl_6 and possible difficulties in its handling make this reagent unattractive. Reagent systems of the composition $HSiCl_3$—NR_3, where R = C_2H_5 or n-C_4H_9, are a less expensive and more easily handled alternative (*83, 162*) that also converts P-chiral tertiary phosphines oxides to the corresponding phosphines with inversion of phosphorus configuration, probably because Si_2Cl_6 is generated *in situ* (*162, 197*). Investigation of different amines, which give different characteristic reflux temperatures of the reaction mixtures, is recommended in order to determine the reaction temperature that gives the most stereoselective reduction.

Stereospecific reduction of $O{=}PR^1R^2R^3$ to $:PR^1R^2R^3$ with *retention* of phosphorus configuration is possible but is usually less attractive than reduction with inversion. Trichlorosilane has been used, but some phosphine oxides have been found to give racemic phosphine products. The more expensive phenylsilane $PhSiH_3$ is a better reagent for reduction of phosphine oxides with retention at phosphorus (*133*) but also does not always give stereospecific reduction (*188*). The lack of reliable reagents for the reduction of phosphine oxides with retention of phosphorus configuration creates the need to have a phosphinate ester having the same phosphorus configuration as that desired in the phosphine product. If this is not possible, it may be feasible, as indicated in Scheme 1, to reduce $O{=}PR^1R^2R^3$ with inversion, reoxidize $:PR^1R^2R^3$ with H_2O_2, with complete retention (*81*), and again reduce $O{=}PR^1R^2R^3$, now of inverted configuration, with inversion. Another option is to convert phosphine oxide to phosphine sulfide (retention) and reduce the phosphine sulfide with Si_2Cl_6 (retention) (*196*).

The phosphinate route has been used to prepare the commercially valuable ligands cyclohexylanisylmethylphosphine (CAMP) (**16**) and 1,2-bis-(*o*-anisylphenylphosphino)ethane (DIPAMP) (**17**) (*102–104, 191*). The synthesis outlined in Scheme 6 could present difficulties because **2** and **3** are not easy to separate. Furthermore, the more readily obtained phosphinate **2** has the same configuration as *R*-(−)-**16**, which cannot be obtained in optically active form by reduction of the corresponding phosphine oxide **14** with $HSiCl_3$. The second problem (but not the first) is solvable in principle by use of unnatural menthol. In the author's opinion, however, the commercial preparation of **16** and **17** is more likely achieved by the modified synthesis shown in Scheme 7, which has been described by Knowles (*104*).

Reaction of **18**, which is racemic but presumably readily epimerized in the presence of base, with menthol has been reported to give **19** and **20** in a 4 : 1 ratio. The diastereomeric excess of **19** is readily crystallized from the reaction mixture. The ester **19** has the correct phosphorus configuration for conversion by successive reductions to the desired (*R,R*)-(−)-**17**, which is the ligand used to obtain *S* amino acids in rhodium phosphine complex–catalyzed asymmetric hydrogenations (*187*). It is now apparent

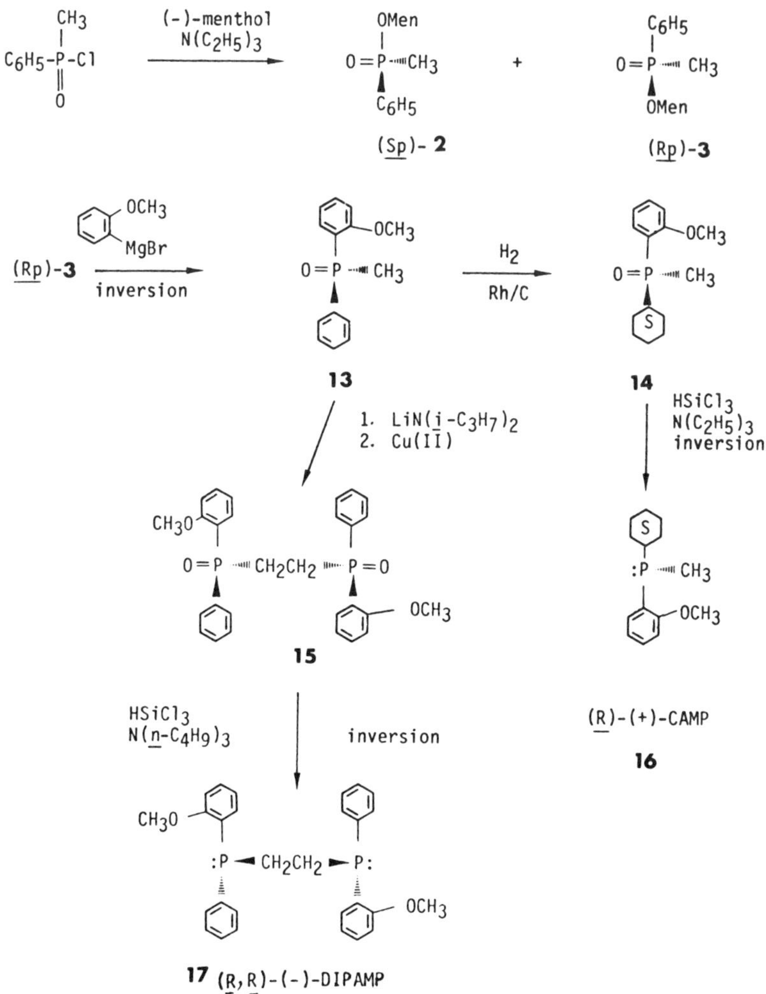

Scheme 6. Syntheses of (*R*)-(+)-CAMP and (*R,R*)-(−)-DIPAMP from menthylmethylphosphinates (*102–104, 191*).

that very low catalyst loadings (a few parts per million) can be used with rhodium DIPAMP (and certain other rhodium chiral phosphine) complex asymmetric hydrogenation catalysts. Consequently, any synthesis that can be used to make multikilo amounts of the correct phosphinate diastereomer will suffice.

A key feature in both Schemes 6 and 7 is dimerization of **13** to give **15** (*191*). The dimerization procedure (*97, 98*) was first used to make chiral bisphosphine systems by Mislow's group (*134*). Reduction of **15** to **17** was

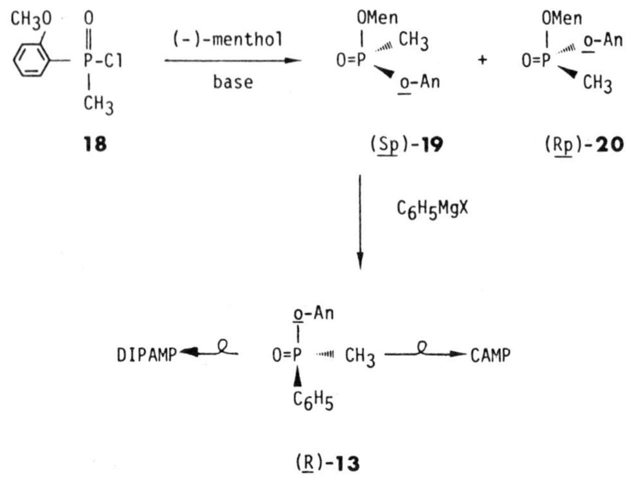

Scheme 7. Alternative synthesis of DIPAMP using *o*-anisylmethylphosphinates (*104*).

achieved using $HSiC_3N(n\text{-}C_4H_9)_3$ (*191*). A small amount of meso isomer is formed, which can be crystallized away from the desired (*R*,*R*)-(−)-**17** diastereomer.

2. Phosphonium Salt Method

Separations of diastereomeric quaternary phosphonium dibenzoyltartrates (*82, 113*) were combined by Horner (*80*) with stereospecific conversions of P-chiral phosphonium cations into phosphine oxides (alkali hydroxide induced cleavage, which produces inversion in favorable cases) or phosphines (cathodic reduction, which occurs with retention). The resulting "phosphonium salt method" (Scheme 8) was the first practical way to prepare P-chiral tertiary phosphines. This method has gradually lost importance as a result of development of more general and convenient alternatives.

An improved method to obtain chiral benzyl quaternary phosphonium salts has been reported (*200*). Treatment of $PhMeRPCH_2Ph^+X^-$(X = Br, Cl; R = *n*-propyl, isopropyl, l-naphthyl) with NaNH gave PhMeRPCHPh, which with tartaric acid R′COOH gave $PhMeRPCH_2Ph^+OOCR^-$, whose diastereomers were separated by fractional crystallization and treated with methyl bromide to give (*R*)- and (*S*)-$PhMeRPCH_2Ph^+Br^-$.

Quaternary phosphonium cations are reduced in good yields at the mercury cathode to give tertiary phosphines according to reaction (7)

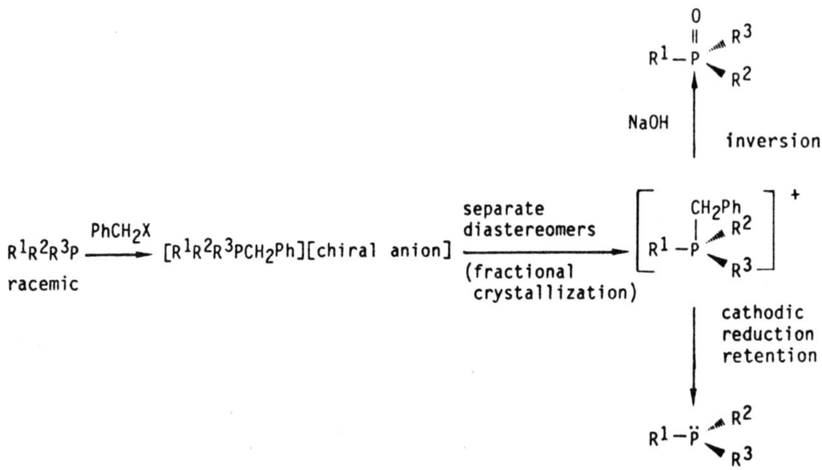

Scheme 8. Chiral tertiary phosphines from chiral quaternary phosphonium salts (*80*).

(*87*). Presumably, one-electron reduction of R_4P^+ is followed by expulsion of R·, leading to formation of R_3P and RH, as in reaction (8) (*80*).

$$R_4P^+ + 2\,e + 2\,H^+ \longrightarrow R_3P + RH \tag{7}$$

$$R_4P^+ + e^- \longrightarrow R_4P\cdot \longrightarrow R_3P\colon + R\cdot \longrightarrow RH \tag{8}$$

Horner (*71, 80*) established the following order of increasing tendency for groups to be lost:

$$(C_6H_5)_3C,\ 4\text{-}CH_3OC_6H_4 < CH_3 < 4\text{-}CH_3C_6H_4$$
$$< 4\text{-}i\text{-}C_3H_7C_6H_4 < C_6H_5 < C_2H_5 < 4\text{-}CF_3C_6H_4$$
$$< 4\text{-}C_6H_5C_6H_4 < n\text{-}C_4H_9 < i\text{-}C_3H_7 < 2\text{-naphthyl}$$
$$< t\text{-}C_4H_9 < 2\text{-}HOCH_2CH_2 < H_2C{=}CH{-}CH_2,\ C_6H_5CH_2$$

Synthetically useful preparations are usually limited to benzylphosphonium salts. Predominant loss of benzyl occurs on either cathodic cleavage or alkali hydroxide–induced cleavage to a phosphine oxide, to give a single product in good yield. Configuration is preserved in cathodic cleavage (*80*) [reaction (9)].

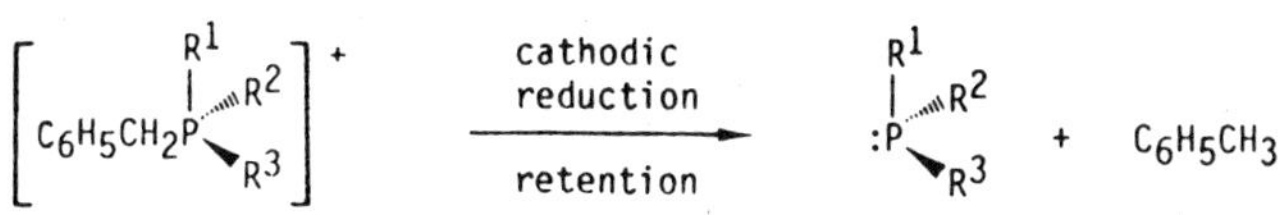

(9)

Alkali hydroxide–induced cleavage of benzyl (*126*) and allyl (*92b*) phosphonium cations occurs with loss of benzyl or allyl to give, in favorable cases, phosphine oxides with 100% inversion of phosphorus configuration

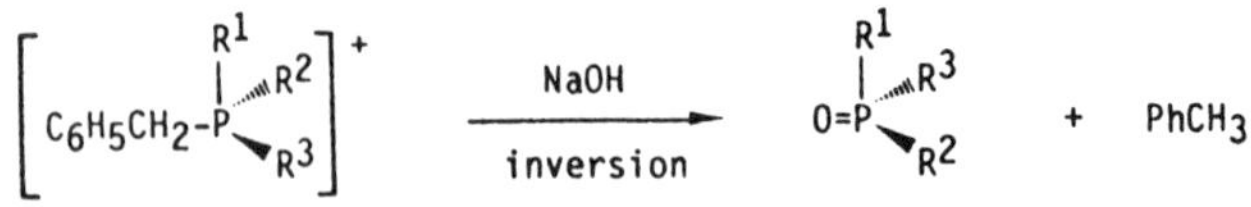

(10)

[reaction (10)]. This reaction represents the ideal case. Bulky substituents such as *tert*-butyl influence the stereochemistry of cleavage of benzyl groups, and the resulting phosphine oxide may be formed with retention and/or partial racemization, depending on both the phosphorus ligands and the reaction conditions (*128a–f*). Certain aryl groups may compete with benzyl as the leaving group (*128d*). Many aryl groups are lost with retention of phosphorus configuration (*127*). Cyanolysis occurs with 100% retention of phosphorus configuration and yields the corresponding phos-

$R^1R^2R^3R^4P^+$

Enantiomers are usually obtained by separations of dibenzoyltartrate salts

cathodic cleavage →	$R^1R^2R^3P$:	Predominant retention when R^4 = benzyl or allyl; when neither group is present, mixtures of products are obtained
cyanolysis →	$R^1R^2R^3P$:	Predominant retention when R^4 = allyl; limited experience is available
$LiAlH_4$ →	$R^1R^2R^3P$:	Mainly retention, with accompanying racemization; limited experience; phosphine oxides are racemized by reagent
NaOH, H_2O →	$R^1R^2R^3P{=}O$	Complete or nearly complete inversion of configuration when R^4 = benzyl or allyl; if bulky groups are present (e.g. *tert*-butyl), either retention or inversion with reduced stereoselectivity may be observed; other R^4 groups such as aryl leave with retention and partial racemization
NaOR, ROH →	$R^1R^2R^3P{=}O$	Same as NaOH and H_2O except that lower stereoselectivities are observed

Scheme 9. Cleavage of phosphonium cations to give phosphines and phosphine oxides (*80, 125*).

phine (*92a*). The reaction of phosphonium cations with $LiAlH_4$ also gives phosphines with retention of configuration, but partial racemization occurs.

The many possible stereochemical results on alkali hydroxide and other reagent-induced phosphonium cation cleavages are summarized in Scheme 9 and Table 2. The stereochemistry of phosphonium-cation cleavages has been reviewed (*125*). The diverse stereochemistry observed in different phosphonium-cation cleavages provides considerable fodder for the mechanistic chemist but tends to reduce the value of the phosphonium salt method for the synthetic chemist.

TABLE 2
Some Representative Results of Alkali Hydroxide–Induced Cleavage of Phosphonium Cations to Phosphine Oxides[a]

Phosphonium salt $R^1R^2R^3R^4P^+X^-$					Leaving group	Reaction medium[b]	Stereochemical result[c]
R^1	R^2	R^3	R^4	X^-			
CH_3	Ph	CH_2Ph	*p*-Anisyl	Br^-	R^3	1	87% inversion
						2	93% inversion
CH_3	Ph	CH_2Ph	$C_{10}H_7$	Br^-	R^3	1	70% inversion
						2	64% inversion
CH_3	*n*-C_3H_7	CH_2Ph	Cyclohexyl	Br^-	R^3	1	86% retention
						2	61% inversion
CH_3	*n*-C_3H_7	C_2H_5	Ph	I^-	R^4	2	55% retention
						3	50% retention
						Ethanol : H_2O	
CH_3	Ph	*t*-C_4H_9	CH_2Ph	Br^-	R^4	99 : 1	58% retention
						95 : 5	68% retention
						75 : 25	68% retention
						50 : 50	65% retention
						25 : 75	57% retention
						10 : 90	54% retention
						5 : 95	61% inversion
						0 : 100	60% inversion
CH_3	Ph	CH_2Ph	Cyclohexyl	Br^-	R^3	1	92% inversion
						2	100% inversion
CH_3	Ph	CH_2Ph	$CH_2C(CH_3)_3$	I^-	R^3	1 or 2	100% inversion

[a] From (*125*).
[b] Reaction media: 1, 3 : 1 ethanol : water; 2, water; 3, 1 : 1 DMSO : water.
[c] Percentages based on phosphine oxide rotations.

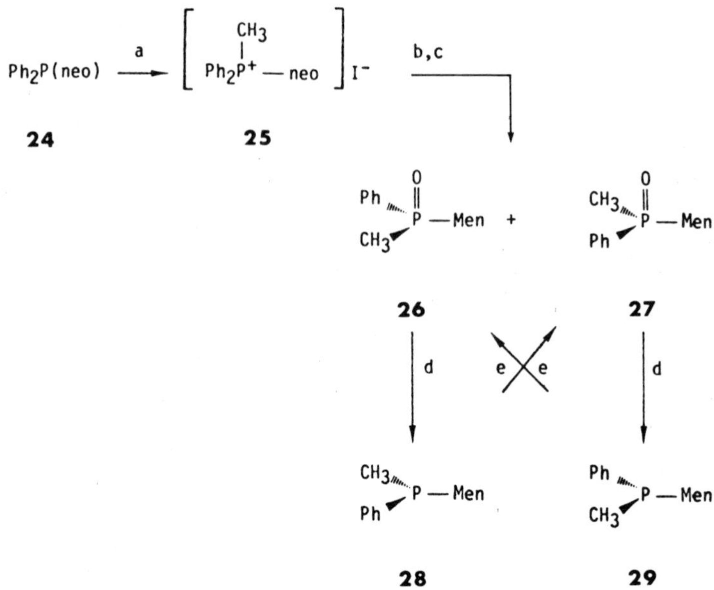

Scheme 10. Menthylmethylphenylphosphines and their phosphine oxides (*188*). Conditions: *a*, CH_3I, 23°C, 18 h, then reflux 2 h (95% yield). *b*, Reflux, NaOH/aqueous CH_3OH, 18 h (88%, 7/8 ratio ~1 : 1). *c*, Fractional crystallization. *d*, Si_2Cl_6, 80°C, 10 min, C_6H_6. *e*, 30% H_2O_2, 23°C, 18 h, C_6H_6.

Alkali-induced cleavage of quaternary phosphonium cations having chiral organic moieties ligated to achiral phosphorus has been used to prepare phosphines in which chiral organic moieties are ligated to chiral phosphorus (*188*) (see Scheme 10). The indicated synthesis affords *menthyl* derivatives from neomenthyldiphenylphosphonium cations. (*159, 188*), because **25** is rapidly epimerized by base at C-1, to give the more stable menthyl epimer. The structure of **26** was determined by X-ray crystallography, and the structures of **26–29** by chemical correlation. (*188*). Reaction (11) provides an alternative, but less convenient, route to

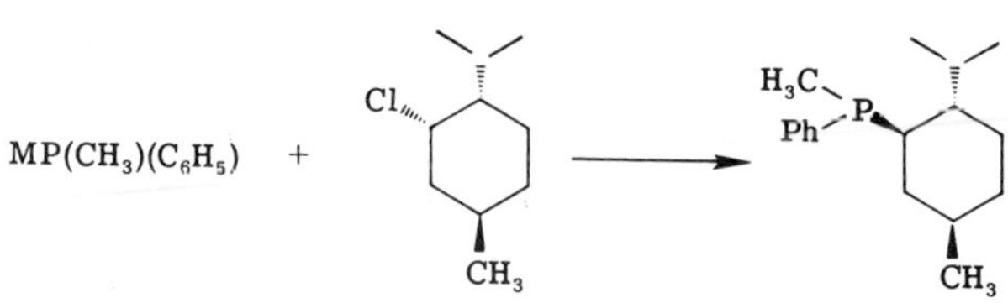

(11)

28 + **29**
(~1:1)

the phosphines **28** and **29** (*49*). The corresponding neomenthyl phosphines **32** and **33** were prepared following Scheme 11 (*49, 188*). An X-ray crystal-

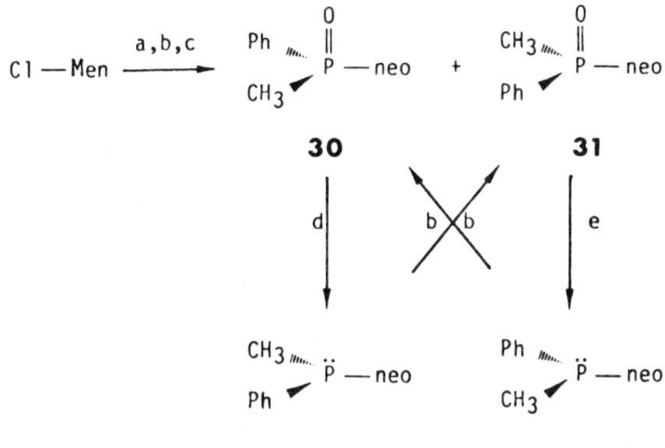

Scheme 11. Neomenthylmethylphenylphosphines and their phosphine oxides. Conditions: *a*, $KP(CH_3)(Ph)$, THF. *b*, 30% H_2O_2, C_6H_6. *c*, Fractional crystallization. *d*, Si_2Cl_6, 80°C, 2 h, C_6H_6. *e*, Si_2Cl_6, 80°C, 4 h, C_6H_6.

lographic structure determination on **31** allowed chemical correlation of all the structures, **30–33** (*188*). Both NMR studies (*204*) and conformational studies (*205*) have also been reported for menthyl- and neomenthylmethylphenylphosphines.

A different type of phosphine having both chiral phosphorus and chiral organic moiety has been prepared using reaction (12) (*101*). The chiral phosphines **34** and **35** were each obtained in low yields by fractional crystallization.

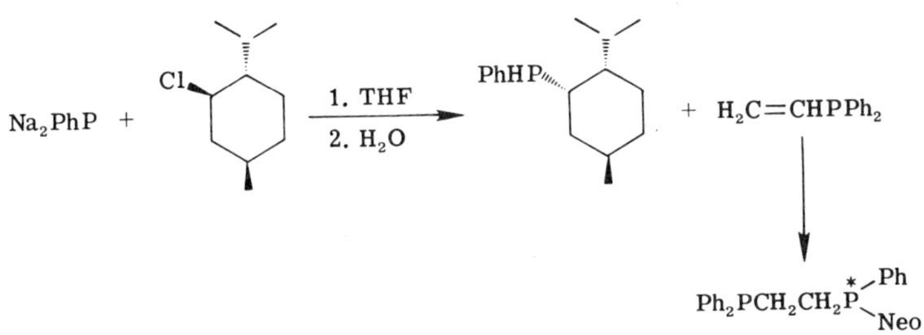

34 + **35**

(12)

3. Resolutions Using Diastereomeric Transition-Metal Complexes

The enantiomers of several P-chiral tertiary phosphines have been separated using diastereomeric transition-metal complexes containing other chiral ligands. Complex **36** reacts with racemic triarylphosphines to give a pair of diastereomers, indicated as **37**, which are separated by fractional crystallization. Treatment of the desired diastereomer with 1,2-bis-(diphenylphosphino)ethane then liberates the phosphine enantiomer [reaction (13)] (*168*).

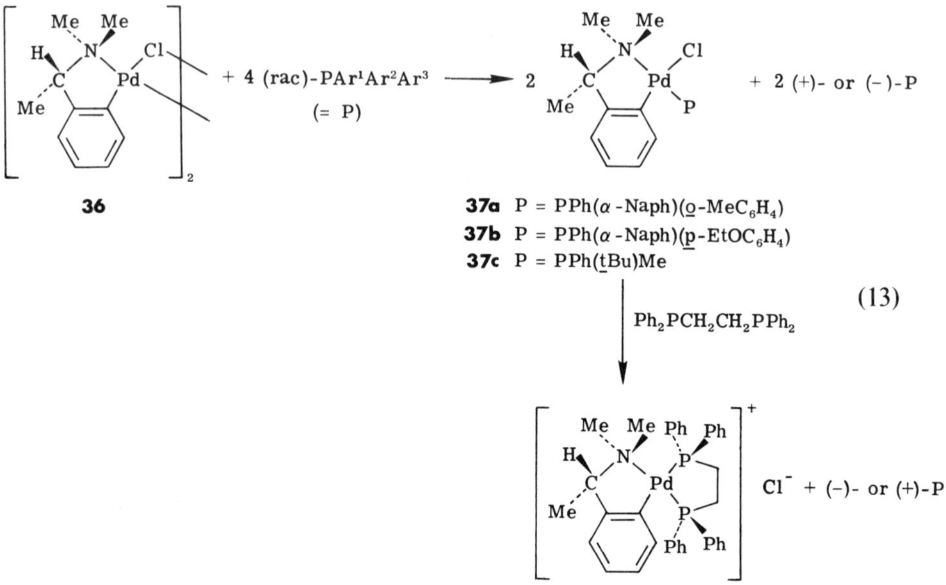

36

37a P = PPh(α-Naph)(o-MeC$_6$H$_4$)
37b P = PPh(α-Naph)(p-EtOC$_6$H$_4$)
37c P = PPh(tBu)Me

(13)

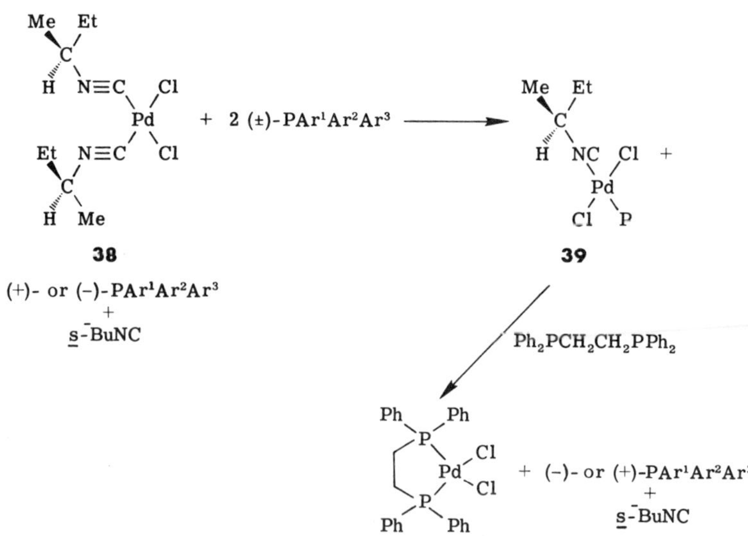

38 **39**

(14)

Use of **38** has been reported [reaction (14)] (*184*). For separations of aryldialkylphosphine enantiomers, however, complex **40** was preferred [reaction (15)] (*184*).

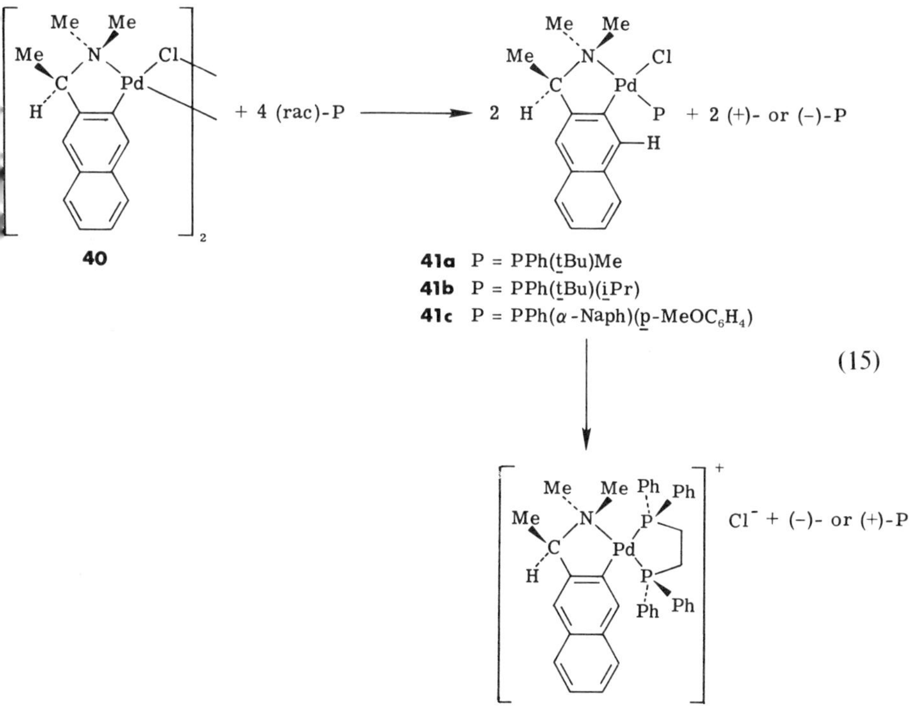

40

41a P = PPh(tBu)Me
41b P = PPh(tBu)(iPr)
41c P = PPh(α-Naph)(p-MeOC_6H_4)

(15)

Scheme 12 summarizes the preparation and resolution, using (*R*)-**36**, of the enantiomers of the chiral chelating bisphosphine **42** (*174*). Formation of metal complexes by **42** and the analogous arsenic compounds has been the subject of several papers (*46, 58, 73–75, 174–177, 201–203*). Similar studies have been reported of metal complexes of individual diastereomeric forms of 1,2-bis(methylphenylphosphino)ethane obtained by separation of phosphonium salts (*84–86*). Finally, separation of enantiomers of chiral phosphines containing achiral phosphorus has been reported (*158*).

4. Other Methods: P-Chiral Phosphinites, Phosphonites, and Related Compounds

In an emerging method of some promise, the enantiomers of P-chiral tertiary phosphines are obtained stereospecifically from attack of or-

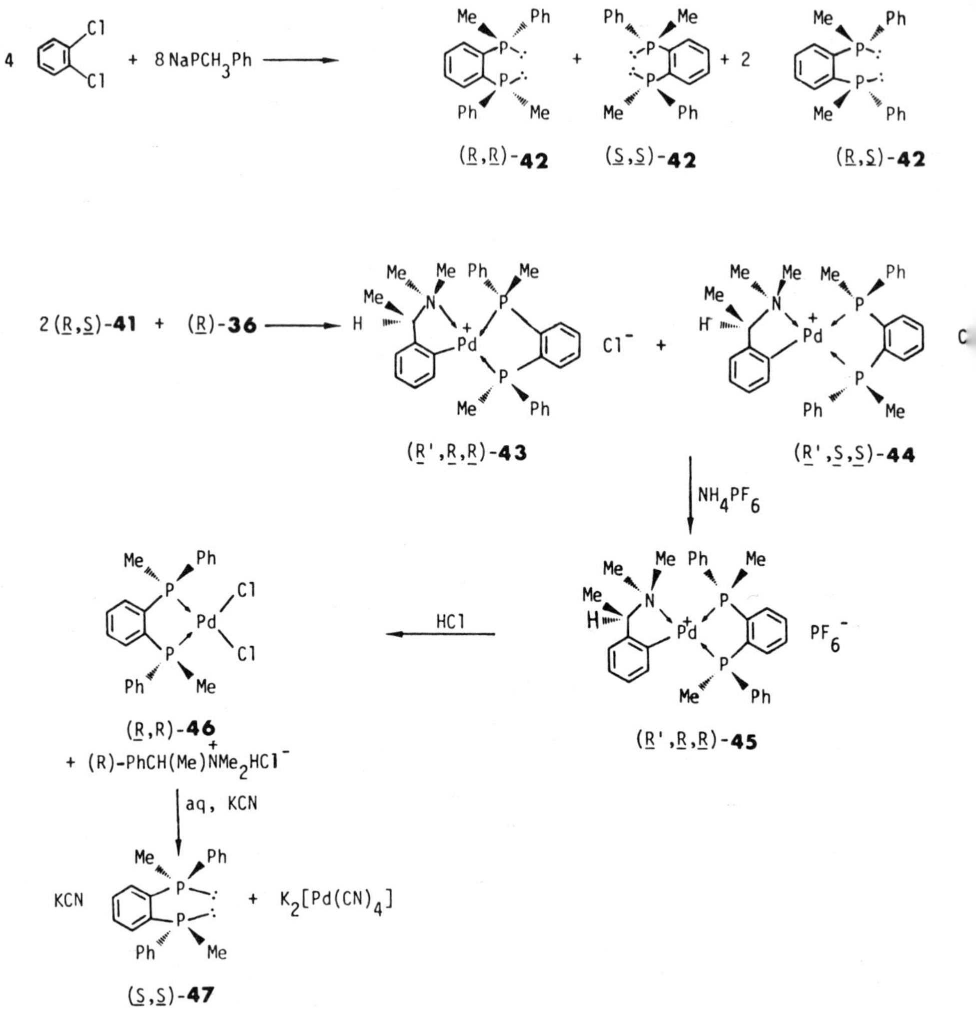

Scheme 12. Preparation and resolution of chiral chelating bisphosphines.

ganoalkali derivatives on enantiomers of P-chiral phosphonites R^1R^2POR. The P-chiral phosphonite enantiomers can be obtained in several ways.

Reactions of R^1ArPCl with achiral alcohols in the presence of chiral tertiary amines such (+)-*N*,*N*-dimethyl-(1-phenylethyl)amine afford the expected chiral R^1ArPOR^2 in modest enantiomeric excess (*156*). A second, more promising method to obtain chiral phosphinites is reaction of a chiral alcohol such as menthol (*156*) or cinchonine (*24, 25*) with R^1ArPCl to yield unequal amounts of the diastereomeric esters $R^1ArP(OR^*)$, which are then separated by conventional methods. A third method to obtain P-chiral phosphinites is outlined in Scheme 13. (*S*)-(−)-*tert*-Butylphenylphosphinoselenic acid (*112, 151*), obtained by methods described in Section III,C, was converted to the triflate **50**, whose reaction with ethanethiolate, $C_2H_5S^-$, proceeds with attack on selenium rather than phosphorus. This gives (*R*)-(+)-**51**, which is formed with clean retention of phosphorus configuration. The enantiomeric excess of (*R*)-(+)-**51** obtained in this way is ~85% (*156*). The menthyl ester **55** (*30*) can also be used (Scheme 14) (*156*); conversion of **55** to **57** is stereospecific. Another approach to P-chiral phosphinites has been described. Treatment of optically active *N,N*-diethylethylphenylphosphinoamidate with *p*-cresyl trifluoroacetate gives optically active *O*-*p*-cresylethylphenylphosphinite (*88*). Still another approach to chiral phosphinite esters is the reaction of alkoxy radicals with chiral tertiary phosphines (*206*). Thus, the reactions of (*S*)-(−)-$MePhPCH_2Ph$ and (*S*)-(+)-$PrMePCH_2Ph$ with ethoxy radicals

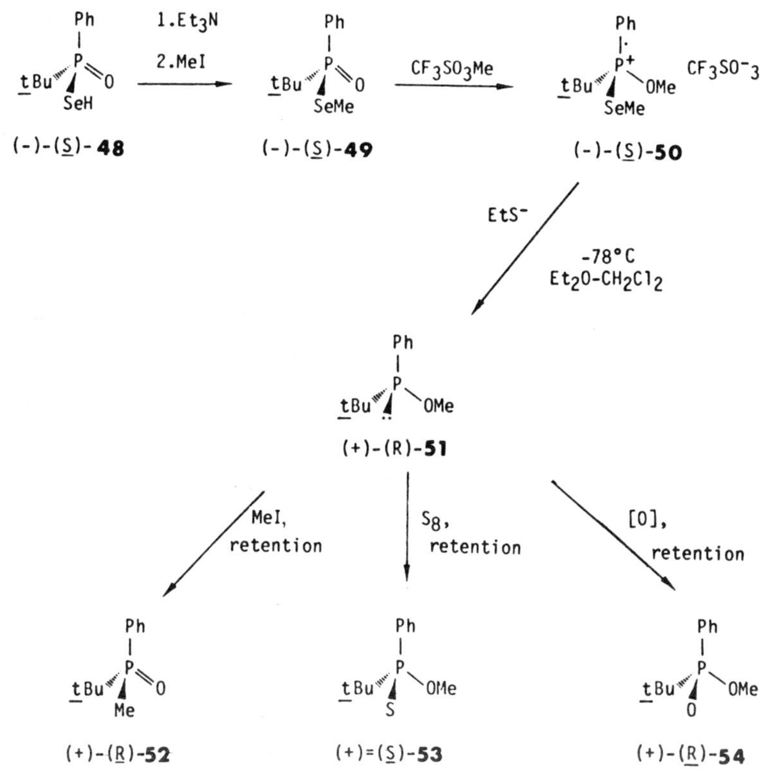

Scheme 13. Stereospecific synthesis of optically active *O*-methyl-*tert*-butylphenylphosphinite (**51**).

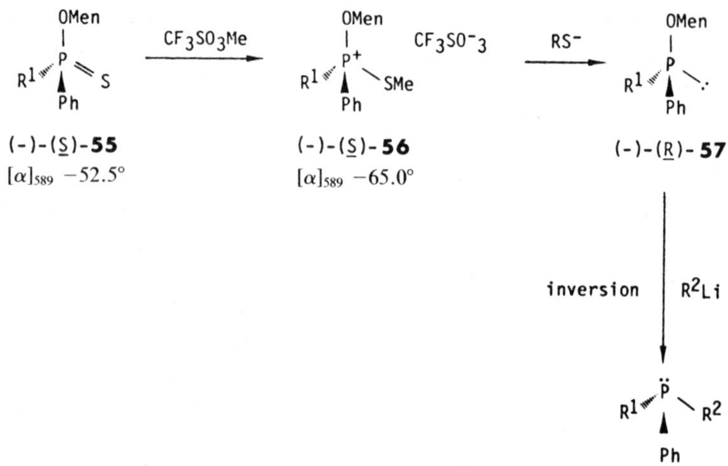

R = Me, R^1 = *n*Pr, $[\alpha]_{589}$ −18.4°

R = Me, R^1 = *t*Bu, $[\alpha]_{589}$ −20.6°

R = Et, R^1 = Me, $[\alpha]_{589}$ +3.4°

R = Et, R^1 = *n*Bu, $[\alpha]_{589}$ −10.8°

R = Et, R^1 = *t*Bu, $[\alpha]_{589}$ +109.4°

Scheme 14. Stereospecific synthesis of diastereoisomerically pure *O*-menthylmethylphenylphosphinite (**57**) and chiral tertiary phosphines.

yield the substitution products MePhPOEt and PrMePOEt, respectively, with net inversion of configuration at phosphorus. Synthetic utility appears to be limited by the subsequent racemization of the phosphinites due to attack on them by ethoxy radicals.

The P-chiral phosphinites have (as previously mentioned) lower epimeric stabilities than P-chiral tertiary phosphines. Racemization proceeds even at room temperature, especially in the presence of acids. P-Chiral phosphinites are nevertheless useful intermediates that can be converted stereospecifically to P-chiral phosphines, phosphine oxides and sulfides, phosphinates, and thiophosphinates, with the stereochemical results shown in Schemes 13 and 14. In addition to the transformations outlined in these schemes, some other ideas also appear promising.

Reaction (16) has been reported to give the phosphines **58** in up to 77% ee (*24, 25*). Still another idea is outlined in Scheme 15 (*14*). Finally, it is

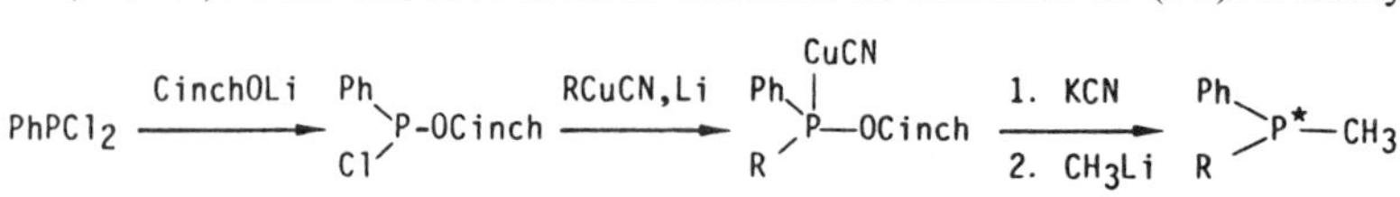

(16)

58

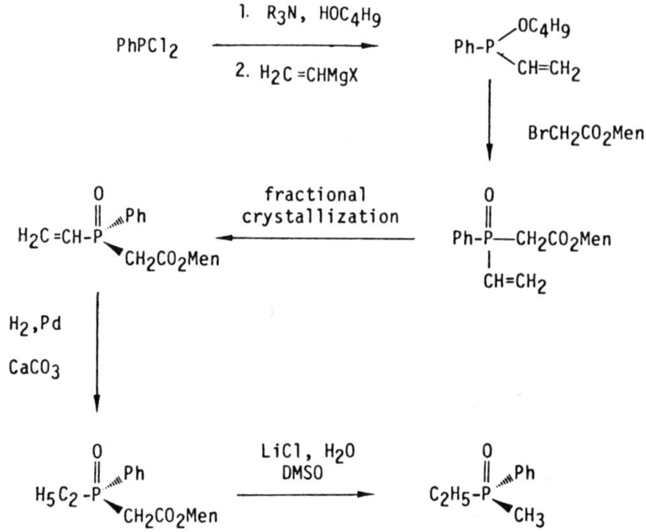

Scheme 15. Stereospecific synthesis and transformations of chiral phosphinites.

sometimes possible to convert one P-chiral tertiary phosphine into another one stereospecifically [reaction (17)] (*114, 116, 182*).

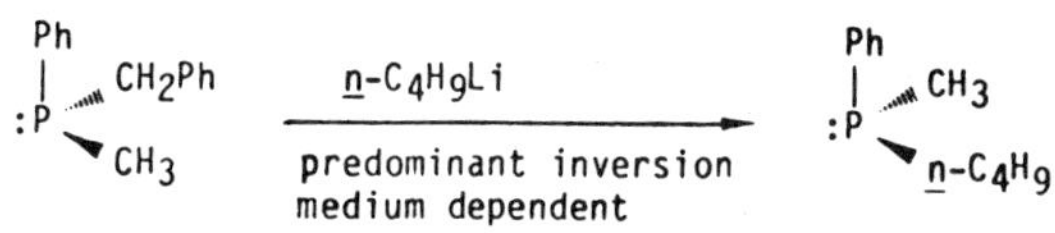

(17)

Various new approaches to P-chiral systems having P—OR or P—NR_2 moieties have been reported that are generally based on the use of phosphonium cations. In reaction (18), triflate chemistry is used to prepare chiral phosphonite **61** (*183*). Similar uses of triflate chemistry have been reported for P=O, P=S, and P=Se derivatives (*26, 165–166*), as well as for an $OSi(CH_3)_3$ analog of **61** (*189*).

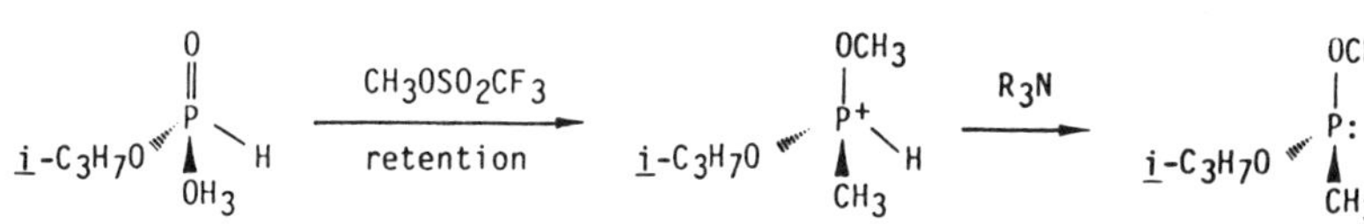

59 **60** **61**

(18)

Scheme 16 outlines some interesting uses of amidophosphonium salts in the preparation of P-chiral enantiomers (*79, 88, 90*). Compounds such as **66** can also be obtained from the corresponding phosphinate menthyl or

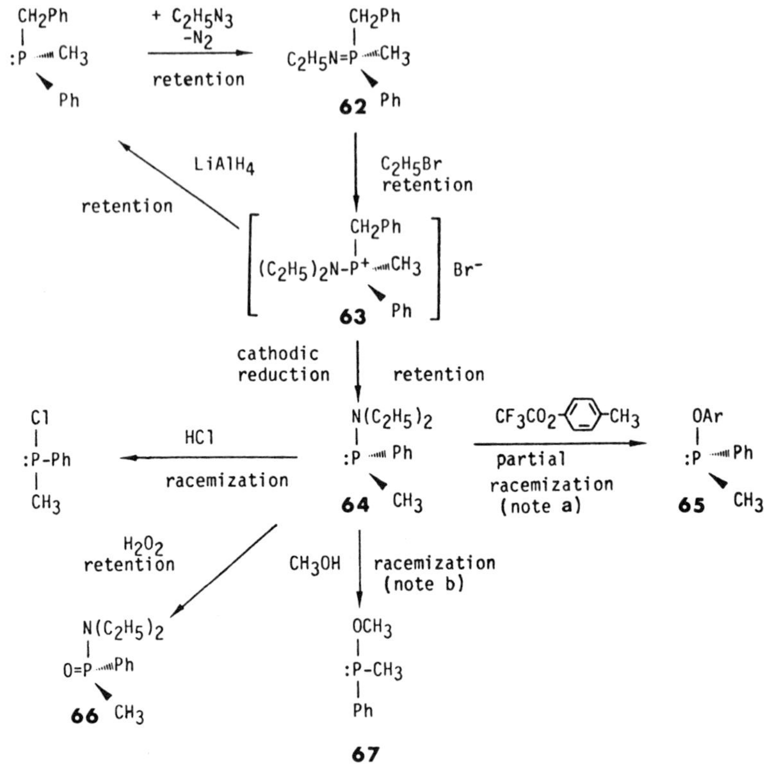

Scheme 16. Synthesis and transformations of P-chiral amidophosphonium salts. Notes: *a*, Actually reported (*88*) for diethylamidoethylphenylphosphine. *b*, The corresponding reaction with *p*-cresol has also been reported (*90*).

cholesteryl esters, whose reactions with $LiNR_2$ occur with predominant inversion of phosphorus configuration (*163*). A further possibility is shown in reaction (19). The phosphinamidate **68** can be converted to the corresponding phosphinimidite **69** with predominant retention of configuration (*37*).

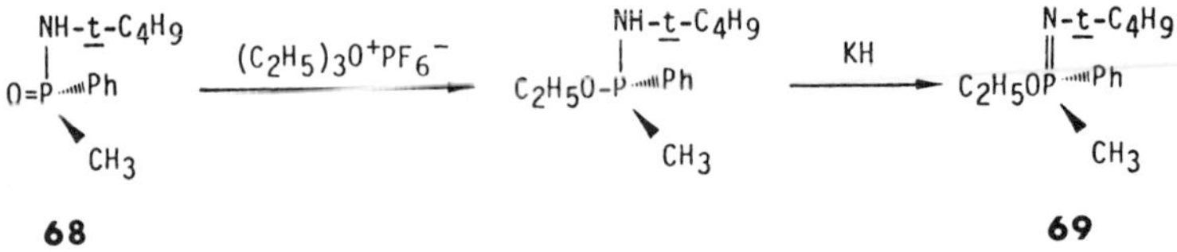 (19)

Other possibilities deserve mention. Some tertiary phosphine oxides are sufficiently acidic to form salts with camphorsulfonic acid. Salts of chiral phosphine oxides with camphorsulfonic acid are diastereomeric and can sometimes be separated by crystallization to effect resolution of the phosphine oxide (*20, 135–137*). Resolution of PAr[1]Ar[2]Ar[3] derivatives has been reported to be possible through separations involving diastereomeric salts of $HOCH_2PAr_3^+$ cations obtained by reaction of the phosphine with formaldehyde (*193*). Methods to obtain chiral phosphine sulfides and arsine sulfides have been reviewed (*135*). In the case of phosphine sulfides, the preferred method is S_8 oxidation, which occurs with retention.

5. Cyclic Phosphines and Related Substances

Several examples of phosphines are known in which a chiral phosphorus atom is incorporated into a ring system. Synthetic methods used are generally unremarkable. For example, **70** was obtained from addition

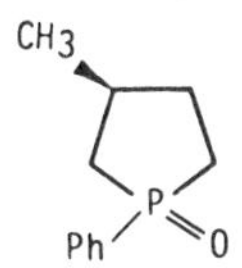

70

of $PhPCl_2$ to isoprene, while **72** was obtained using reaction (20) (*23*). The partial resolution of **71** was achieved using dibenzoyltartrate salts.

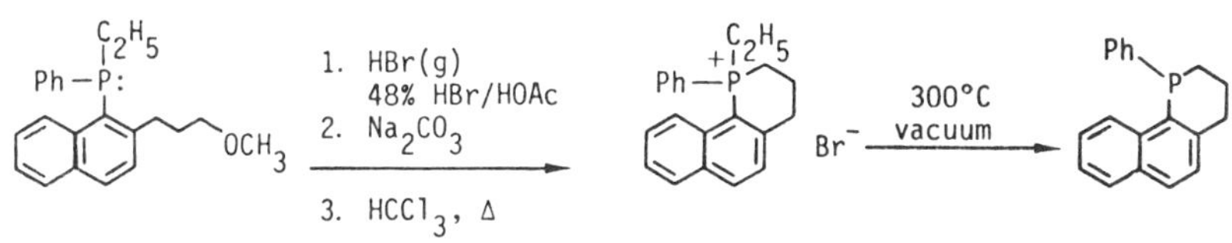

71 **72**

(20)

Stereochemical consequences of nucleophilic displacements at chiral phosphorus may be different in cyclic than in acyclic systems. Ring size, in particular, can affect stereochemical results. For example, the Si_2Cl_6 reductions of **73** (*31, 32*) [reaction (21)] and **76** (*42*) [reaction (22)] occur

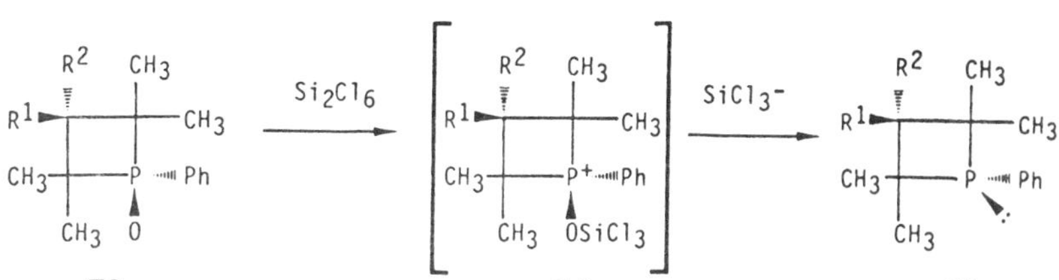

73 **74** **75**

overall retention

(21)

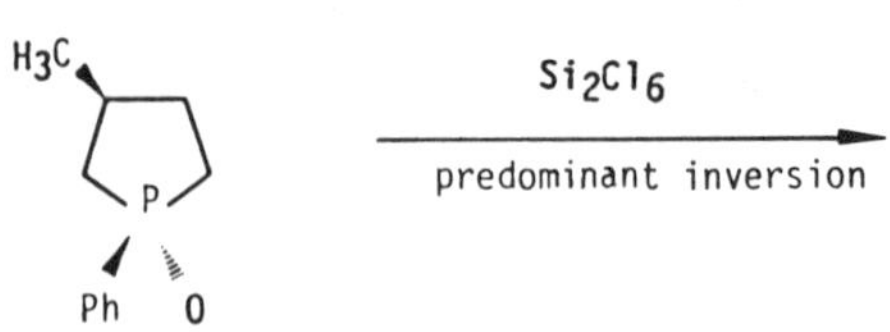

76 77 (22)

with predominant retention and predominant inversion, respectively. Similarly, alkaline hydrolysis of **78** (*42*) [reaction (23)] occurs with retention of configuration, while alkaline hydrolysis of **80** (*133*) [reaction (24)] occurs with partial inversion of configuration.

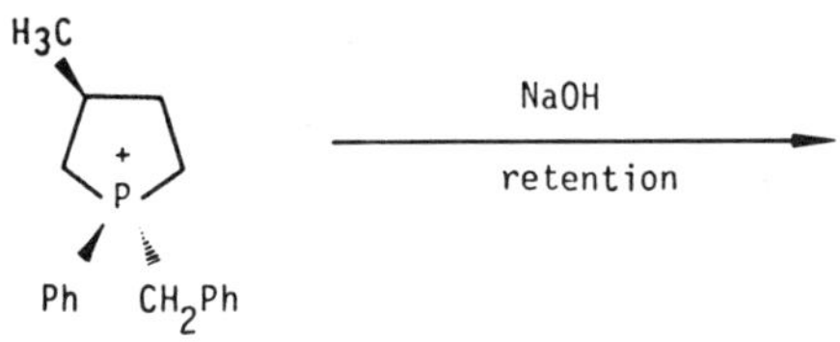

78 79 (23)

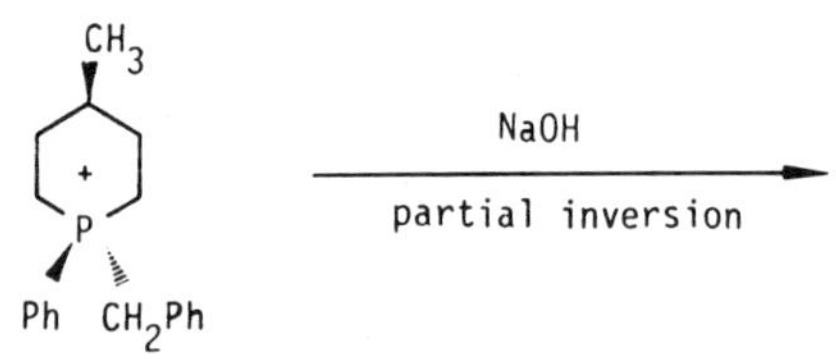

80 81 (24)

As already mentioned in Section II,B,2, stereochemical results such as those indicated in reactions (22)–(24) can be explained in terms of lifetime and preferred configurations of putative phosphorane intermediates, formed by attack of hydroxide in the case of phosphonium salt hydrolyses and by attack of $SiCl_3^-$ on **74** in the case of Si_2Cl_6 reduction of phosphine oxides. In the cyclic systems, ring strain and the electronic nature of the leaving group exert important influences, according to the detailed analysis given by Zon and Mislow (*197*).

C. Determinations of Enantiomeric Composition in Trivalent P-Chiral Systems

Optical rotations of tertiary phosphines and phosphinous acid derivatives are unreliable indicators of phosphorus epimer ratios. When derived from reduction of $R^1R^2R^3P{=}O$ of known epimeric purity, the epimeric purity of $R^1R^2R^3P$ can be estimated by reoxidation with H_2O_2 and measurement of the $R^1R^2R^3P{=}O$ rotation (*188*). Direct determinations of epimeric purity (*187*), however, are definitely preferable. One useful method involves quaternization of the phosphine, which is a mixture of P-epimers, with the bromide **82**. Analysis of the ^{1}H or ^{31}P NMR of the

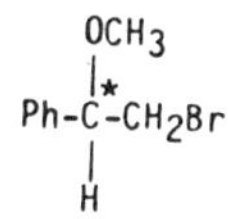

82

resulting phosphonium salts will then indicate the P-epimer ratio (*21*, *171*).

Enantiomeric excesses of P-chiral tertiary phosphines have also been determined by analysis of NMR spectra of diastereomeric transition-metal complexes containing both the phosphine enantiomers and other chiral ligands (*117*). Analysis of P-chiral phosphines that contain other chiral groups (such as chiral organic moieties or another chiral phosphorus center) is also often possible with NMR. Systems in which there are two or more chiral centers may also be accessible to HPLC analysis (*160*). Determination of enantiomeric purities of P-chiral organophosphorus compounds has been reviewed (*59*).

D. Lists of Chiral Tertiary Phosphines and Related Substances

Table 3 is a representative list of P-chiral enantiomers of tertiary phosphines, phosphine oxides, quaternary phosphonium salts, and some related substances. The list is intended only to illustrate possibilities since more detailed lists are available (*135*).

III. Four-Coordinate Phosphorus Thioacids and Related Compounds

Preparations of four-coordinate P-chiral compounds have been reviewed (*135*). Phosphine oxides, quaternary phosphonium salts, and phosphinates were considered in the previous section in connection with preparations of P-chiral tertiary phosphines. This section is concerned with preparations of other four-coordinate P-chiral compounds. Some of the subject compounds have potent biological activities, including Sarin (*189*) (**83**) and cyclophosphamide and related substances (**84–87**).

TABLE 3
Representative P-Chiral Tertiary Phosphines and Related Substances

Structure	Method	References
	Phosphines	
$:P(CH_3)(C_2H_5)(Ph)$	Phosphonium	*80, 87*
$:P(CH_3)(Ph)$(allyl)	Phosphinate	*7*
(*R*)-(−)-$:P(CH_3)(C_3H_7)(Ph)$[a]	Several	*13c, 16b, 21b, 182*
(*R*)-(+)-$:P(CH_3)(Ph)(t\text{-}C_4H_9)$	Phosphinate	*8, 9, 121*
(−)-$:P(CH_3)(Ph)(CH_2Ph)$	Phosphinate	*162*
	Phosphonium	*80*
(*R*)-(+)-$:(CH_3)(Ph)(2\text{-}CH_3OC_6H_4)$	Phosphinate	*21b*
(*R*)-(+)-$:P(CH_3)$(Cyclohexyl)-$(2\text{-}CH_3OC_6H_4)$[b]	Phosphinate	*103*
$(CH_3)(Ph)P$-Menthyl	Phosphonium, then separation of diastereomeric phosphine oxides	*188*
$(CH_3)(Ph)P$-Neomenthyl[c]	$LiP(CH_3)(Ph)$ + menthyl chloride then separation of phosphine oxides	*188*
$:P(CH_3)(Ph)(2\text{-}C_{10}H_7)$	Phosphinate	*162*
$:P(CH_3)(2\text{-}CH_3OC_6H_4)(2\text{-}C_{10}H_7)$	Resolution of the phosphine	*168*
	Resolution of the phosphine	*184*
$(CH_3)(Ph)PCH_2CH_2P(CH_3)(Ph)$[d]	Phosphonium	*13b*
$[(Ph)(2\text{-}CH_3OC_6H_4)PCH_2\text{—}]_2$[e]	Phosphinate	*191*
$[(Ph)(CH_3)M\text{—}CH_2\text{—}]$ M = As or P	Resolution of phosphine (arsine)	*175–177*
$Ph_2PCH_2CH_2P(Ph)$—Neomenthyl[f]	Separation of phosphine diastereomers	*101*
Polymer supported P-chiral phosphines	Phosphinate (to obtain monomers)	*10b*
Series of related P-chiral systems for hydrogenation studies	Phosphinate	*91*
	Quarternary phosphonium salts	
(*R*)-(−)-$P^+(CH_3)(Ph)(t\text{-}C_4H_9)$(allyl) Br^-	Resolution	*10b*
(*S*)-(+)-$P^+(CH_3)(C_3H_7)(C_4H_9)(Ph)$ Br^-	Resolution	*10b*
(*R*)-(−)-$P^+(CH_3)(C_3H_7)(t\text{-}C_4H_9)(Ph)$ Br^-	Resolution	*10c*
(−)-$P^+(CH_3)(C_4H_9)(CH_2CH_2CN)$-$(CH_2C_6H_5)$	Resolution	*113, 195b*

TABLE 3 (*Continued*)

Structure	Method	References
(S)-(−)-$P^+(CH_3)(C_3H_7)(Ph)$-(2-Cyclohex-1-enyl) Br^-	Resolution	*10b*
Both $P^+(CH_3)(CH_2CH{=}CH_2)(Ph)$-$(CH_2Ph)^+$ Br^-	Resolution	*10c*
(S)-(+)-$P^+(CH_3)(Ph)(CH_2Ph)(NR_2)$	Quaternization	*89*
	Phosphonous and phosphinous acid derivatives	
(R)-(−)-:$P(OC_2H_5)[OSi(CH_3)_3](Ph)$	Asymmetric synthesis	*189*
(R)-(+)-:$P(OCH_3)(O$-i-$C_3H_7)(CH_3)$	Asymmetric synthesis	*183*
:$P(CH_3)(Ph)(NR_2)$	Various	*79, 90, 92*
	Phosphinic acid derivatives	
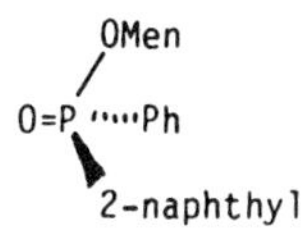	—	*108, 109*
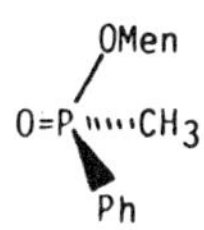	See footnote *g*	*108, 109*
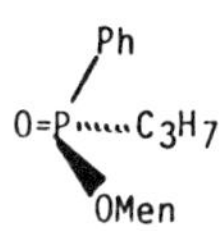	—	*108, 109*
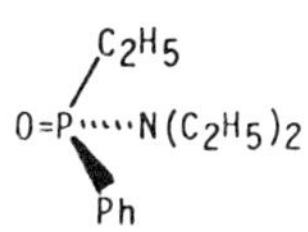	—	*92*
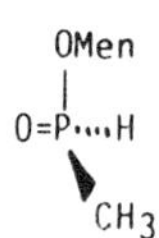	—	*12*
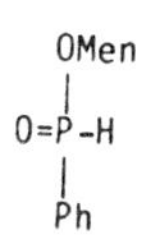	—	*47*

(*Continued*)

TABLE 3 (*Continued*)

[a] The (*S*)-(+) enantiomer has also been prepared (*151b*).
[b] CAMP.
[c] Both P-epimeric diastereomers and the corresponding phosphine oxides were isolated in pure form. Absolute configurations known.
[d] All diastereomers known (see text).
[e] DIPAMP. The (*R,R*)-(−) diastereomer is used to prepare *S* amino acids by catalytic asymmetric hydrogenations.
[f] Both diastereomers obtained in pure form.
[g] Both diastereomers. Absolute configuration known from X-ray crystallography and chemical correlations.

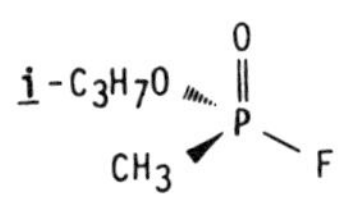

(S)-(-)-**83**

Sarin

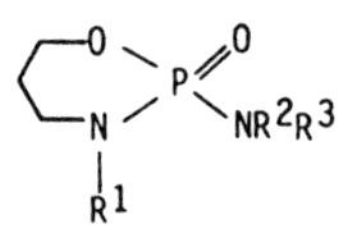

84 R^1=H; R^2=R^3 = CH_2CH_2Cl cyclophosphamide

85 R^2=H; R^1=R^3=CH_2CH_2Cl isophosphamide

86 R^1=R^2=R^3=CH_2CH_2Cl trophosphamide

87 R^1=CH_2CH_2Cl; R^2=H; R^3=$CH_2CH_2OSO_2CH_3$ sulphosphamide

Efficient preparations of **84–87** are clearly desirable. Asymmetric syntheses are gradually replacing older methods that are based on resolutions of diastereomeric salts of phosphorus thioacids. This section is concerned with the following preparative methods:

1. Resolutions involving diastereomeric salts of phosphorus thioacids
2. Asymmetric syntheses based on acyclic covalent diastereomers
3. Asymmetric syntheses based on cyclic covalent diastereomers

This classification has been used by Hall and Inch in their review (*59*), which supplements an earlier review by Mikolajczuk (*156b*) and the brief discussion given here.

As already mentioned (Scheme 2), many interconversions of four-coordinate P-chiral compounds can be achieved with high stereoselectivity under at least some conditions (*25, 50b, 59, 156b*). Transformation possibilities are summarized in Scheme 2. These transformations are synthetically useful and provide the basis for chemical correlation of absolute configurations of product with absolute configuration of starting material (*39*).

A. *Resolutions of P-Chiral Thioacids*

Anions of phosphinic acids $R_2PO_2^-$ are symmetric (as mentioned previously) and do not form diastereomeric salts with chiral cations (*44*). Anions of thiophosphinic acids, however, exist in the preferred form $R^1R^2P(S)O^-$ and can be resolved into enantiomeric forms through separations of diastereomeric salts. Many such resolutions have been effected (see Table 4). Classical resolutions of phosphorus thioacids are histori-

TABLE 4
Phosphorus Thio and Seleno Acids Resolved by Fractional Crystallization of Diastereomeric Salts

$R^1R^2P(X)OH$				
R^1	R^2	X	Resolving agent	References
C_2H_5	C_6H_5	S	Quinine	*57*
CH_3	C_6H_5	S	Quinine	*11*
t-C_4H_9	C_6H_5	S	Quinine	*30*
α-$C_{10}H_7$	C_6H_5	S	Quinine	*22*
t-C_4H_9	C_6H_5	Se	α-Phenylethylamine	*110*
C_2H_5	OC_2H_5	S	Quinine	*1*
CH_3	OCH_3	S	α-Phenylethylamine	*18*
CH_3	OC_2H_5	S	α-Phenylethylamine	*18*
CH_3	O-*i*-C_3H_7	S	α-Phenylethylamine	*18*
C_2H_5	OCH_3	S	Quinine	*2*
i-C_3H_7	OC_2H_5	S	Quinine	*2*
CH_3	O-*n*-C_4H_9	S	α-Phenylethylamine	*18*
t-C_4H_9	OCH_3	S	α-Phenylethylamine	*110*
i-C_4H_7	OCH_3	S	α-Phenylethylamine	*155*
CH_3	O-cyclo-C_5H_9	S	α-Phenylethylamine	*59*
CH_3	OC_2H_5	Se	α-Phenylethylamine	*59*
C_2H_5	OC_2H_5	Se	Brucine	*164*
n-C_4H_9O	OC_2H_5	S	Cinchonindine	*138*
C_5H_5O	O-*i*-C_3H_7	S	Quinine	*138*
α-$C_{10}H_7O$	OCH_3	S	(−)-Ephedrine	*5, 40*
CH_3O	SCH_3	S	Strychnine	*185*
CH_3O	NHC_6H_{11}	S	Quinine	*67b*
CH_3O	NC_4H_8O	S	Quinine	*52*
2,4-$Cl_2C_6H_3O$	NH-*i*-C_3H_7	S	Quinine	*180*
MeO	OC_6H_4-*p*-NO_2	S	Strychnine	*78*

[a] Dicyclohexylamine salt.
[b] Free acid.
[c] Lead salt.
[d] (+)-α-Phenylethylamine salt.
[e] (−)-α-Phenylethylamine salt.
[f] *S*-Methyl ester.
[g] Ammonium salt.

cally important (*156b*) and synthetically useful but suffer from the same limitations as other resolutions. Structure sensitivity can be severe; separations can be tedious and require racemates that are themselves not readily synthesized. Also, no information is directly obtainable concerning product configurations.

An example of the synthetic utility of resolved P-chiral thioacids is provided by the work of Moriyama and Bentrude (*210, 211*). The optically active phosphine oxides $MePhP(O)CH_2Ph$ and n-$PrMeP(O)CH_2Ph$ were prepared in high optical purity as precursors to the corresponding phosphines by two successive Grignard displacement reactions on the readily available, essentially optically pure *O*-isopropyl-*S*-alkyl methylphosphonothioate esters (where alkyl = Me or $EtOCH_2CH_2OCH_2CHO$—). Corresponding chiral tertiary phosphines were obtained by reduction with phenylsilane.

B. Asymmetric Syntheses Using Acyclic Covalent Diastereomers

Chiral dialkylphenylphosphates (*106*) and alkylphenylphosphonates (*105*) have been prepared using proline as a chiral template (*208, 209*) [reaction (25)]. This reaction has two potential limitations: the phenyl

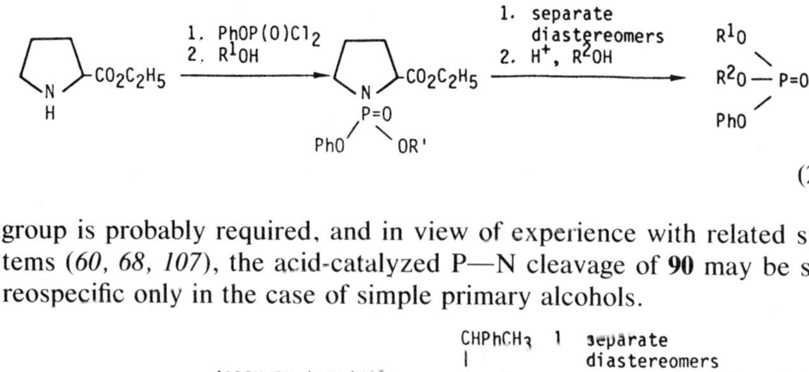

(25)

group is probably required, and in view of experience with related systems (*60, 68, 107*), the acid-catalyzed P—N cleavage of **90** may be stereospecific only in the case of simple primary alcohols.

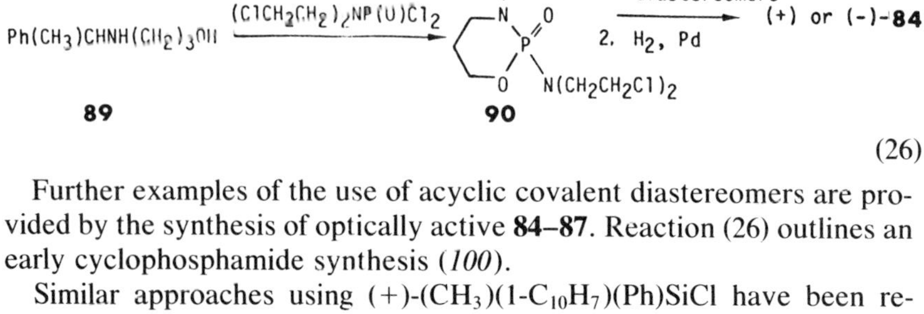

89 **90**

(26)

Further examples of the use of acyclic covalent diastereomers are provided by the synthesis of optically active **84–87**. Reaction (26) outlines an early cyclophosphamide synthesis (*100*).

Similar approaches using (+)-$(CH_3)(1\text{-}C_{10}H_7)(Ph)SiCl$ have been re-

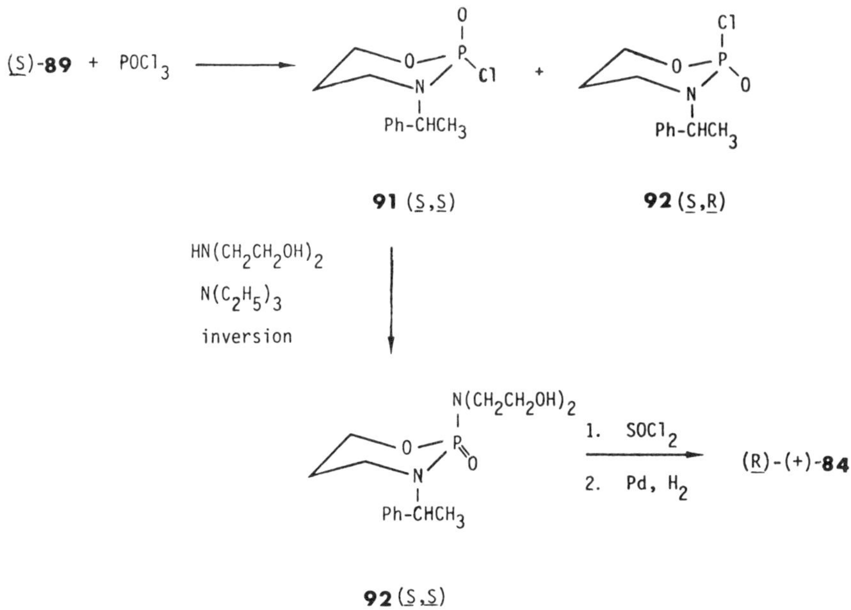

Scheme 17. An improved cyclophosphamide synthesis.

ported (*15, 99, 198, 199*). Enantiomers of **85** and **86** have been similarly prepared (*131*), and **86** has been resolved using a platinum complex (*194*). Absolute configurations of (*S*)-(−)-**84** (*4*) and (*R*)-(+)-**84** (*96*) have been determined. Improvement of reaction (26) has been reported (*179*) (see Scheme 17). The advantage of Scheme 17 over earlier syntheses is that reaction of **89** with $POCl_3$ (*169*) under carefully controlled conditions (*179*) affords **91** and **92** in ratios of approximately 10 : 1 to 12 : 1, respectively.

C. Asymmetric Syntheses Using Cyclic Covalent Diastereomers

Both carbohydrates and chiral β-aminoalcohols such as (−)-ephedrine have been used as templates in asymmetric synthesis of four-coordinate P-chiral compounds (*59*). The intermediates in these syntheses are cyclic covalent diastereomers in which chiral phosphorus atoms incorporated in the ring form geometric isomers according to the phosphorus configuration. Separations of the geometric isomers are relatively easy, after which ring opening and cleavage give rise to the chiral phosphorus product. The ring-opening processes display regiochemical and stereochemical diver-

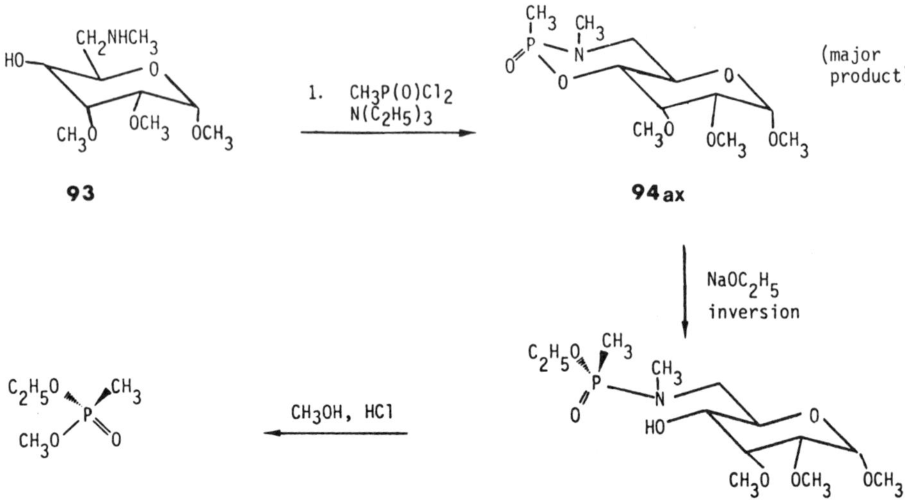

93 **94ax** **96** **95**

Scheme 18

sity not found in corresponding acyclic systems. This is apparently due to the influence of ring conformations on directions of attack by nucleophiles and on the geometry of possible transition states and intermediates. This subject has recently been reviewed (*62, 93*). Some examples of syntheses that use the cyclic covalent diastereomer approach are outlined below to illustrate potential, scope, and problems.

Scheme 18 summarizes the chemistry of the useful amino sugar **93** (*69*). Reactions (27) and (28) illustrate some other possibilities (*27, 69, 70*). Compounds **97** and **100** were obtained from **93** by treatment with $P(O)Cl_3$ or $P(S)Cl_3$, followed by two successive alkoxide displacements.

Some typical stereochemical results for ring-opening reactions of carbohydrate-derived cyclic covalent diastereomers are given in Tables 5 and 6. These tables are adapted from the review by Hall and Inch (*93*),

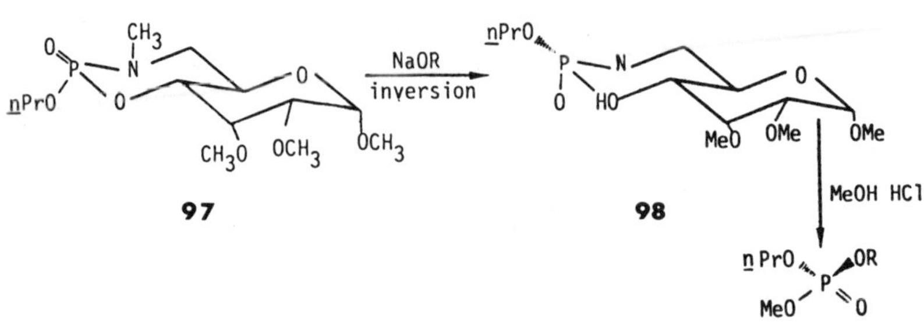

97 **98** **99**

(27)

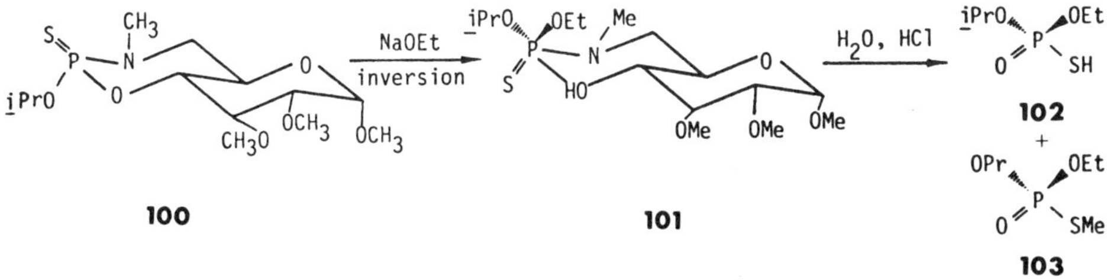

100 101 102 + 103

(28)

TABLE 5
Stereochemistry of Ring Opening of Cyclic Covalent Diastereomers: Leaving Group Axial[a]

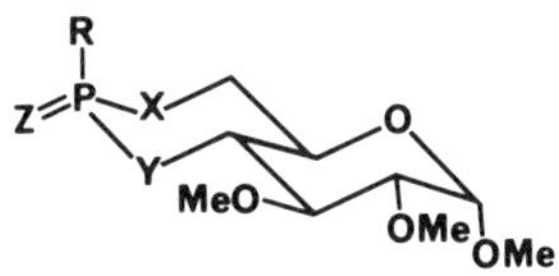

Ring heteroatoms X	Y	Z	Leaving group (R)	Nucleophile	Preponderant stereochemistry (% yields of major and minor isomers where isolated)
O	O	O	Cl	EtOH	Inversion (93 : 4)
S	O	O	Cl	EtOH	Inversion (78 : —)
O	S	O	Cl	EtOH	Inversion
O	O	S	Cl	EtOH	Inversion (81 : —)
O	O	O	Cl	NaOEt/EtOH	Retention (54 : 42)
S	O	O	Cl	NaOEt/EtOH	Retention (60 : —)
O	S	O	Cl	NaOEt/EtOH	Retention (28 : —)
O	O	S	Cl	NaOEt/EtOH	Retention (65 : 31)
NMe	O	O	Cl	NaOEt/EtOH	Inversion (80 : —)
NMe	O	S	Cl	NaOEt/EtOH	Inversion (87 : —)
O	O	O	Cl	Me_2N/PhH (*52*)	Inversion (87 : 9)
O	O	O	Cl	NaSPr/PhH (*53*)	Inversion (47 : 4)
O	O	O	Cl	NaOPh-*p*-NO_2 (1 equiv)	Inversion (60 : —)
S	O	O	Cl	NaOPh-*p*-NO_2 (1 equiv)	Inversion (53 : —)
NMe	O	O	Cl	NaOPh-*p*-NO_2 (1 equiv)	Inversion (56 : —)
O	O	O	SPr	NaOEt/EtOH (*54, 55*)	Retention (77 : —)
O	O	O	OPh-*p*-NO_2	NaOEt/EtOH	Retention (88 : —)
NMe	O	O	OPh-*p*-NO_2	NaOMe/MeOH	Inversion (68 : —)
O	O	O	F	NaOEt/EtOH (*56, 57*)	Retention (46 : 16)
O	O	O	Cl	MeMgI/Et_2O/PhH	Retention (32 : 6)
NMe	O	O	Cl	PhMgBr/Et_2O/PhH	Inversion (46 : 4)
NMe	O	S	Cl	MeMgI/Et_2O/PhH	Inversion (— : —)
O	O	O	Me_2N	EtOH/H	Inversion (>95 : —)

[a] From (*93*).

TABLE 6
Stereochemistry of Ring Opening of Cyclic Covalent Diastereomers: Leaving Group Equatorial[a]

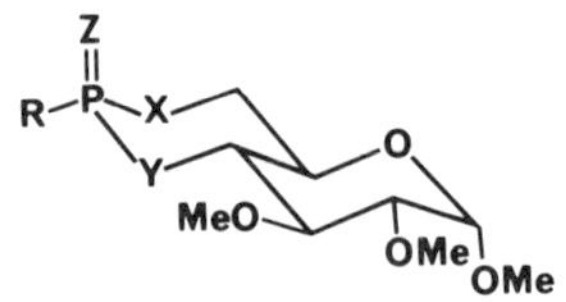

Ring heteroatoms X	Y	Z	Leaving group (R)	Nucleophile	Preponderant stereochemistry (% yields of major and minor isomers where isolated)
O	O	O	SPr	NaOEt/EtOH	Retention (67 : —)
O	O	O	OPh-*p*-NO_2	NaOEt/EtOH	Retention (75 : 15)
O	O	O	OPh-*p*-NO_2	OPh-*p*-NO_2/CH_3CN	Inversion (75 : —)
NMe	O	O	OPh-*p*-NO_2	NaOMe/MeOH	Inversion (67 : 9)
O	O	O	Me_2N	EtOH/H^+	Inversion (>95 : —)

[a] From (*93*).

who discussed possible theoretical and mechanistic interpretations. It is difficult to determine to what extent the differences between cyclic and acyclic systems are the result of (*a*) the inherent restrictions on nucleophilic attack resulting from phosphorus configuration and (*b*) the ability of the rings to assume nonchair configurations (*93*). Whatever the reasons, the results obtained by Hall and Inch on **105** and **106** are both difficult to explain and, in some cases, are useful in practical syntheses. According to Inch and Hall (*93*), compounds **105ax** and **105eq** react with phenylmagne-

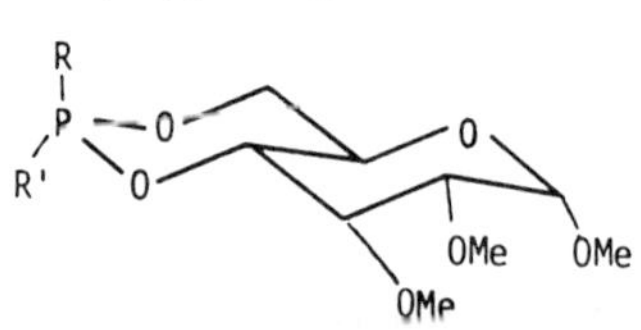

105ax R = Me, R^1 = O
105eq R = O, R^1 = Me
106ax R = Me, R^1 = S
106eq R = S, R^1 = Me

sium iodide with predominant cleavage of the P—O-6 bond, giving 50% product yields. From **106eq**–PhMgI, the main product (obtained in 61% yield) arose from P—O-6 bond cleavage with retention of phosphorus configuration. With **106ax**–PhMgI, both P—O-6 cleavage with inversion and P—O-4 cleavage with retention were observed. Suggested explanations for these results include effects of phosphorus configuration, ring conformation, and medium effects (*61, 93*).

Arbuzov reactions of enantiomeric cyclic phosphonites may also be used to generate chiral phosphine oxides. An example is provided by the work of Segi *et al.* (*212*). Arbuzov reactions of (2*S*,4*S*)-4-methyl-2-phenyl-1,3,2-dioxaphosphorinane with various alkyl halides gave the corresponding diastereomerically pure phosphinates with regioselective cleavage of the primary C—O bond. Reaction of the phosphinates with Grignard reagents then gave optically active phosphine oxides in high optical yields.

Because carbohydrate templates can be difficult to obtain, use of readily available β-amino alcohols such as (−)-ephedrine is inherently attractive. Many examples have been reported. Diastereomer separations are typically relatively easy. Ring-opening chemistry is similar to that observed with the carbohydrate templates. The examples that follow illustrate synthetic possibilities (see also Table 7) (*59*).

Reaction of $RP(S)Cl_2$ (R = alkyl or chloro) with (−)-ephedrine affords **107a** and **107b** (R = alkyl) or **108a** and **108b** in ratios depending on the preparative method. Both **a**,**b** pairs can be readily separated (*29*). Sele-

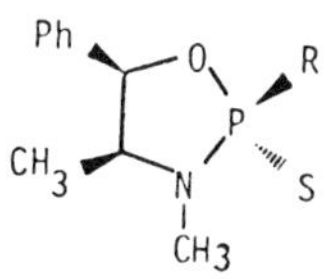

a

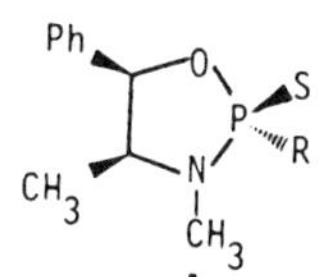

b

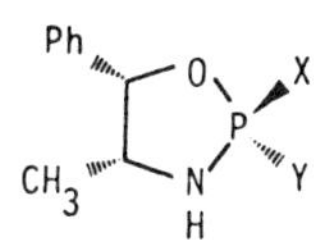

Derived from (+)-norephedrine

107 $R=CH_3$

108 R=Cl

109 R=OR

110 $R=NR_2$

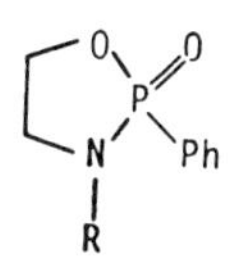

111

112a X=S $Y=CH_3$

112b $X=CH_3$ Y=S

113a X=O $Y=CH_3$

113b $X=CH_3$ Y=O

114a X=S $Y=OC_2H_5$

114b $X=OC_2H_5$ Y=S

nium analogs have also been prepared (*59b*). Alkoxide and amine both displace chloride from **108** with retention of phosphorus configuration to give **109** and **110**, respectively (*28b*). Compounds **111–114** are obtained in similar ways (*64*).

Ring-opening processes in **107–114** are influenced by the nature of X and Y, the nature of the nucleophilic reagent, and the reaction conditions. Ring opening may involve P—O or P—N bond cleavage, with different

TABLE 7
Absolute Configuration and Rotation Data for Organophosphorus Compounds Obtained in Enantiomeric Form Using (−)-Ephedrine as a Chiral Template[a]

R^1	R^2	R^3	X	Configuration
SH	OEt	OMe	O	(*R*)-(+), also (*S*)
SH	O-*i*Pr	OEt	O	(*R*)-(+), also (*S*)
SMe	OEt	OMe	O	(*R*)-(+), also (*S*)
SMe	O-*i*Pr	OEt	O	(*R*)-(+), also (*S*)
SMe	O-*i*Pr	OMe	O	(*R*)-(−), also (*S*)
SMe	Me	OEt	O	(*R*)-(+), also (*S*)
SeMe	Me	OEt	O	(*R*)-(+), also (*S*)
SeMe	O-*i*Pr	OMe	O	(*R*)-(−), also (*S*)
OMe	OEt	Me	O	(*R*)-(+), also (*S*)
OMe	OEt	O-*i*Pr	O	(*R*)-(−), also (*S*)
Cl	OMe	OEt	S	(*R*)-(+), also (*S*)
O-*i*Pr	OMe	OEt	S	(*R*)-(+), also (*S*)
NMe_2	OMe	OEt	S	(*R*)-(+)
NMe_2	O-*i*Pr	OEt	S	(*S*)-(+)
NH_2	OMe	OEt	S	(*R*)-(+)
NH_2	O-*i*Pr	OEt	S	(*S*)-(−)
SMe	NH_2	OEt	O	(*R*)-(+), also (*S*)
SMe	NH_2	OMe	O	(*R*)-(+)
SMe	NMe_2	OMe	O	(*R*)-(+)
SMe	NHMe	OMe	O	(*R*)-(+)
SMe_2	NMe_2	OEt	O	(*R*)-(+)
Cl	SMe	OMe	O	(*R*)
OMe	OEt	NMe_3	O	(*R*)-(−), also (*S*)
OMe	OEt	NH_2	O	(*R*)-(+), also (*S*)
O-*i*Pr	OMe	NHMe	O	(*S*)
NH_2	NMc_2	OMe	O	(*S*)-(−)
Cl	OMe	OEt	O	(*R*), also (*S*)
F	OEt	OMe	O	(*R*), also (*S*)
OPh	OEt	OMe	O	(*R*)-(+)
SH	Ph	Me	O	(*R*)-(+)
SH	*n*Bu	Me	O	(*R*)-(−)

[a] From (*3b, 28b, 59, 67a*).

stereochemical results. This may be either an advantage or a disadvantage from a synthetic viewpoint, but it is unexplained from a mechanistic, theoretical viewpoint. Some examples follow.

Substitution on nitrogen affects the ring opening of **111** during alkaline hydrolysis [reaction (29)] (*66*). Treatment of **109** with alkoxide results in

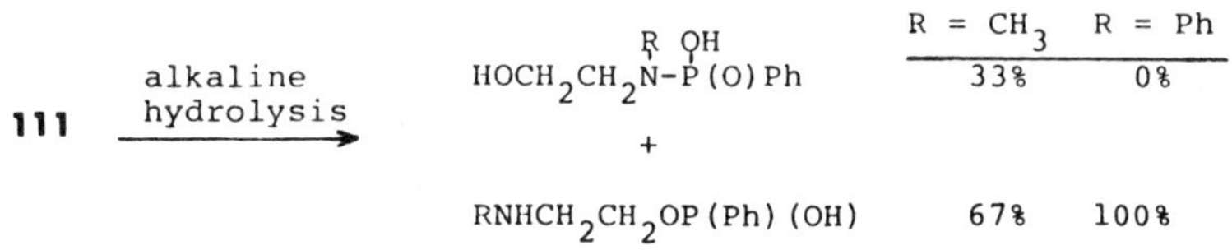

(29)

endocyclic P—N bond cleavage and inversion of phosphorus configuration. Treatment with hydroxide, however, results mainly in P—O bond cleavage, although some P—N bond cleavage is typically also observed (*63, 66*). Thus, reactions (30)–(32) have been reported. In reaction (32) the

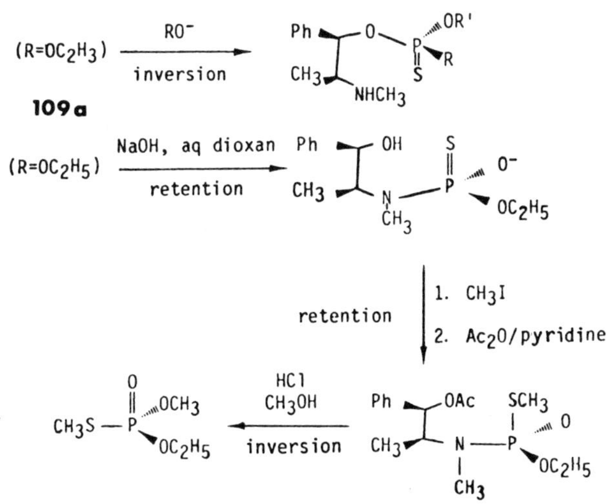

(30)

(31)

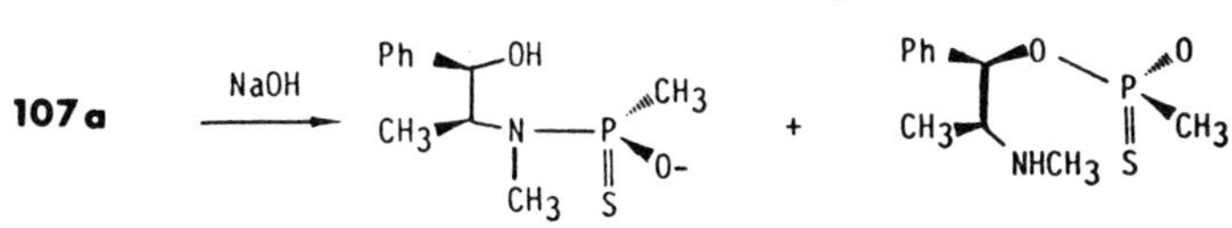

(32)

115 **116**

major product is **115** (P—O bond cleavage), which is formed with various stereochemistries depending on the solvent (e.g., 80% inversion in aqueous ethanol but 85% retention in acetonitrile). Compound **116** is obtained with 17% inversion of configuration in aqueous ethanol.

Compounds **112–114**, derived from (+)-norephedrine, exhibit slightly different behavior. For example, **112a** is cleaved by alkoxide with inver-

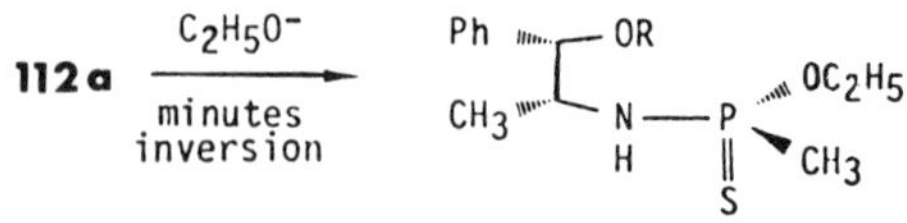 (33)

117 R = H

118 R = OAc

sion of configuration at phosphorus [reaction (33)] (*64*). Acetylation of **117** to **118** prevents further reactions. Treatment of **118** according to the sequence shown as reaction (34) then gives optically pure **120**.

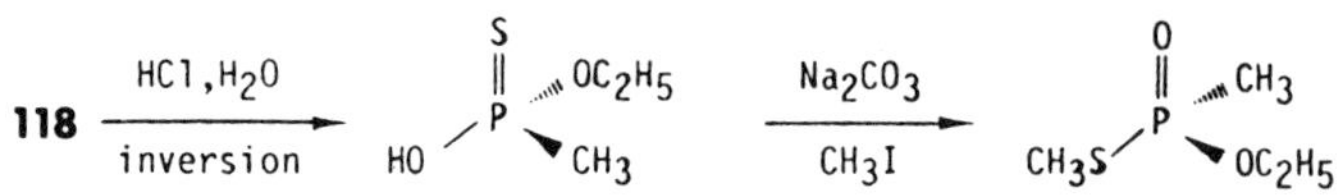 (34)

119 **120**

The conversion of **112a** to **120** occurs with net retention of phosphorus configuration. The stereospecific P—O bond cleavage of **112a** by ethoxide is distinct from stereospecific P—N bond cleavage of **107a** by ethoxide. Furthermore, when a **112a**–ethoxide solution is stored for several hours, the product mixture is that derived from P—N bond cleavage [reaction (35)]. Alkaline hydrolysis of **112a** is much faster than **107a** hydrolysis. It

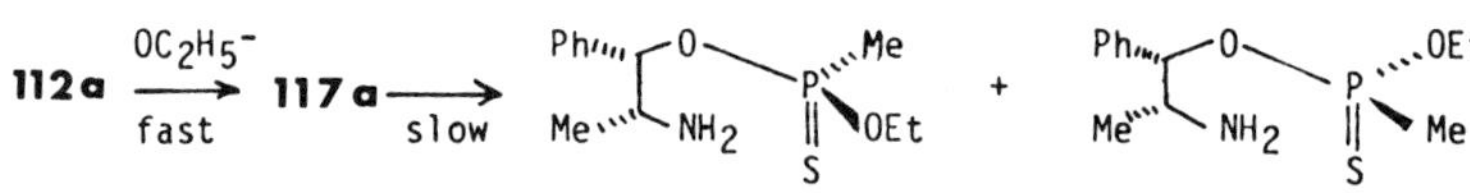 (35)

121 95% **122** 5%

has been suggested that **112a** hydrolysis involves elimination–addition with an intermediate such as **123** [reaction (36)]. Reaction (35) can be explained if **117a** is recyclized to **107a**, which then slowly hydrolyses by a nucleophilic displacement mechanism. Results obtained in the ring opening of **114b** by alkoxide were similar.

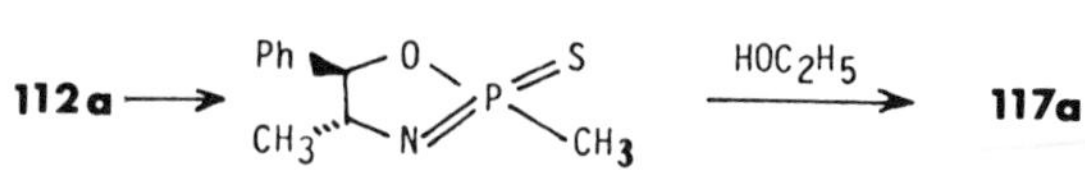 (36)

123

Several other applications of ephedrine as a template are noteworthy. Scheme 19 illustrates the synthesis of thiophosphates; Scheme 20 illustrates the synthesis of thiazaphospholidenes, as well as some of their ring-opening reactions (*65, 67*). Scheme 21 depicts the use of another chiral template, mandelic acid, to make chiral [^{16}O,^{17}O,^{18}O]phosphates (*29, 95, 122, 123*).

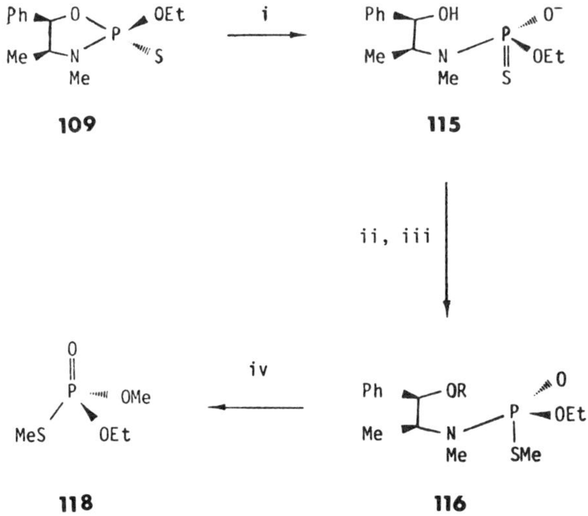

Scheme 19. Synthesis of thiophosphates (*66*). Reagents: *i*, HO—. *ii*, MeI. *iii*, Ac_2O–Pyridine. *iv*, MeOH–HCl.

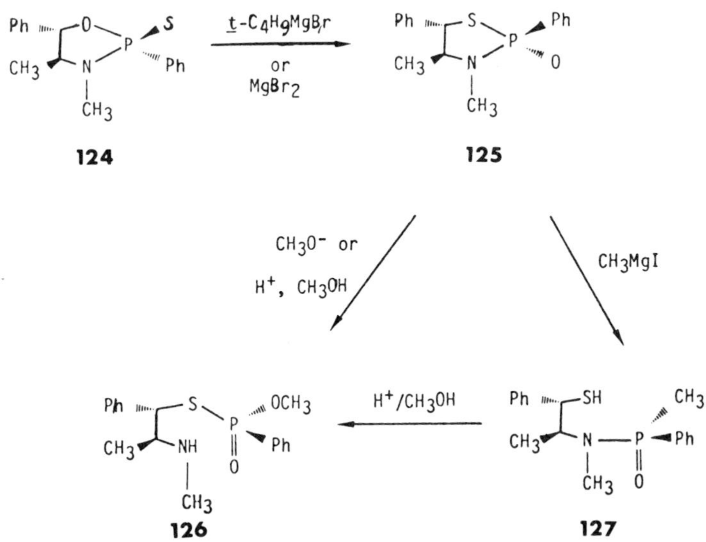

Scheme 20. Synthesis of thiazaphospholidenes (*65–67*).

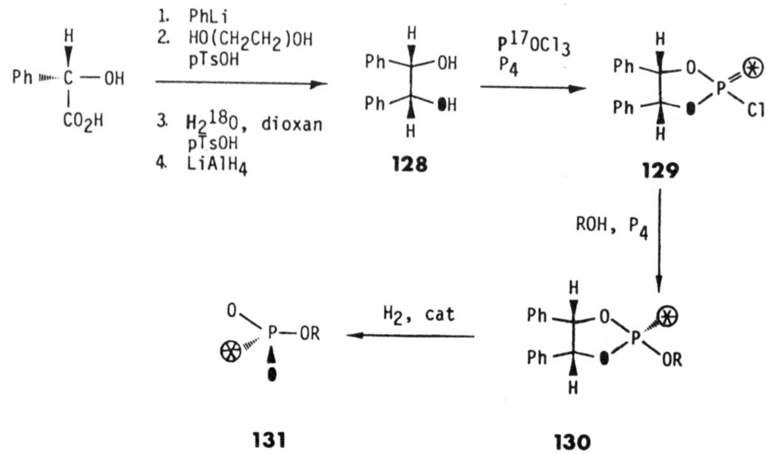

Scheme 21. Synthesis of chiral [^{16}O,^{17}O,^{18}O]phosphates from mandelic acid (*29, 95, 122, 123*). ●, ^{18}O; ⊛, ^{17}O.

D. Some Chemical Transformations of Phosphorus Thioacids and Their Derivatives

Many reactions of phosphorus acids have been summarized in Scheme 2 and are treated in detail in the review by Mikolajczuk and Leitloff *156b*. A few other reactions of phosphorus thioacid derivatives are described briefly here.

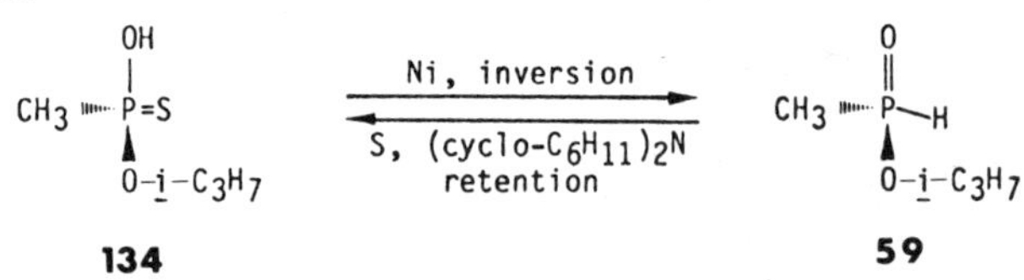

(37)

Desulfurization of **134** affords **135** with inversion of phosphorus configuration [reaction (37)]. Phosphorus thioacids can serve as phosphine oxide precursors via reaction (38) (*39*). There had been some initial confu-

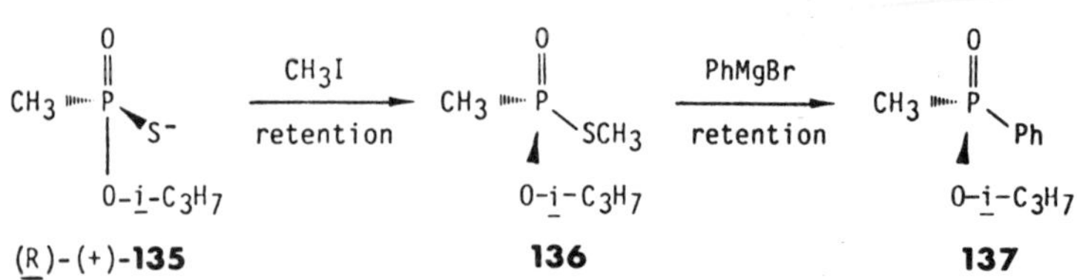

(38)

sion concerning the configuration of (R)-(+)-**135** because it had not been realized that reaction of phenylmagnesium bromide with **136** occurs with

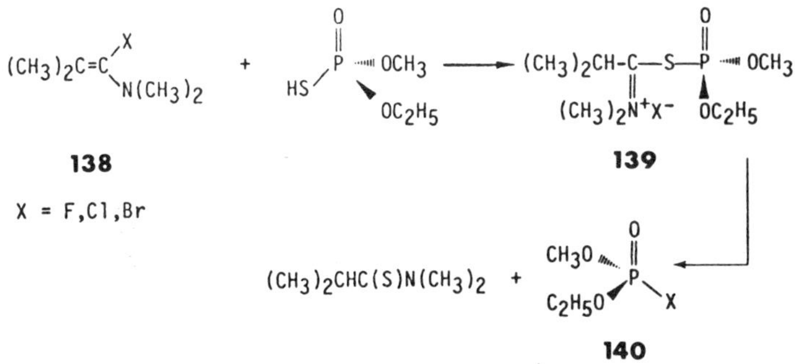

(39)

retention of configuration. The indicated absolute configuration was eventually confirmed by an X-ray crystallographic study (*189*).

Stereospecific halogenation of methylethylphosphorothioic acid (**138**) occurs with inversion of configuration [reaction (39)] (*66*). Fluoride, chloride, and probably bromide are displaced from **140** with inversion in reaction (40).

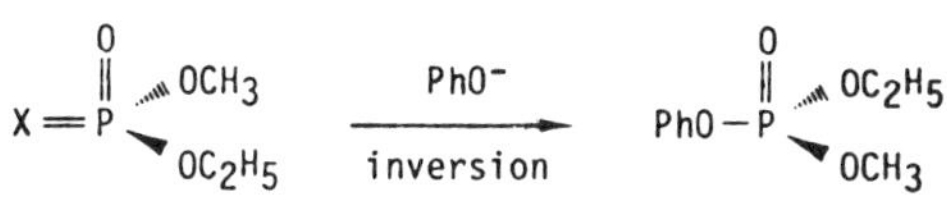

(40)

Substrate **141** gives a similar reaction with 4-nitrophenol [reaction (41)], but **143**, obtained from **141**, gives **144**, containing ~30% of the *R* enantiomer (*66*).

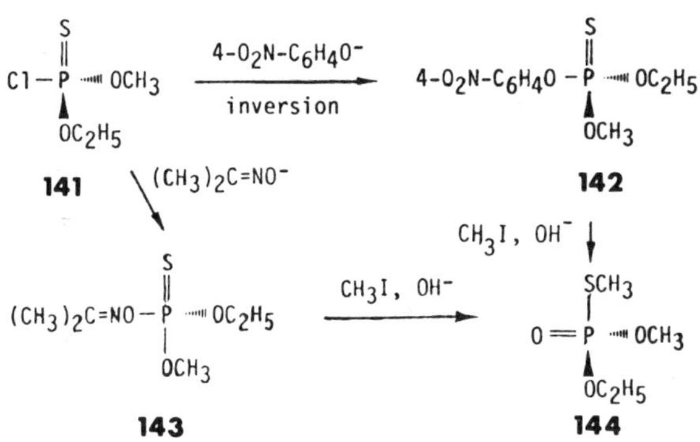

(41)

(42)

IV. Conclusions

The regiochemistry and stereochemistry of reactions of organophosphorus compounds have been thoroughly studied and are relatively well understood. The main reasons for making P-chiral organophosphorus compounds today are for applications in organic synthesis, biology, and medicine. Syntheses of P-chiral substances for applications typically must provide high enantiomeric excess, which is now usually achievable, and must be synthetically convenient (practical), which is often more difficult to achieve. Asymmetric syntheses are, at least in principle, more flexible and convenient than classical resolutions, and many methods have been developed that provide high enantiomeric excess. As with other types of asymmetric synthesis, choice of the right reagents and careful optimization to find the best conditions are both necessary. Syntheses of DIPAMP (**17**) and cyclophosphamide (**84**), described in this chapter, furnish examples of the value of choosing the right approach with careful optimization. Opportunities appear to exist to improve further the synthesis of P-chiral phosphines through use of phosphinates and phosphinothioates derived from template syntheses of the type described in Section III. It is also apparent that the range of compounds available from the template approach can be expanded significantly. Asymmetric synthesis of a considerable variety of P-chiral compounds now is a practical reality.

References

1. Aaron, H. S., Shryne, T. M., and Miller, J. I. (1958). *J. Am. Chem. Soc.* **80,** 107–110.

2. Aaron, H. S., Shryne, T. M., and Miller, J. I. (1960). *J. Am. Chem. Soc.* **82,** 596.

3. a. Aaron, H. S., Uyeda, R. I., Frack, H. F., and Miller, J. I. (1962). *J. Am. Chem. Soc.* **84,** 617; b. Abbot, S. J., Jones, J. R., Weinman, S. A., and Knowles, J. R. (1978). *J. Am. Chem. Soc.* **100,** 2558.

4. Adamiak, D. A., Kinas, R., Saenger, W., and Stec, W. J. (1977). *Angew. Chem. Int. Ed. Engl.* **16,** 330.

5. Akintonwa, D. A. A. (1978). *Tetrahedron* **34,** 959.

6. Anderson, K. K. (1962). *Tetrahedron Lett.* 93.

7. Baechler, R. D., Farnham, W. D., and Mislow, K. (1969). *J. Am. Chem. Soc.* **91,** 5686.

8. Baechler, R. D., and Mislow, K. (1970). *J. Am. Chem. Soc.* **92,** 3090.

9. Baechler, R. D., and Mislow, K. (1970). *J. Am. Chem. Soc.* **92,** 4758.

10. a. Baechler, R. D., and Mislow, K. (1971). *J. Am. Chem. Soc.* **93,** 773; b. Baker, G. L., Fritschel, S. J., and Stille, J. R. (1981). *J. Org. Chem.* **46,** 2954–2960; c. Balzer, W.-D. (1969). *Chem. Ber.* **102,** 3546.

11. Benschop, H. P., and van den Burg, G. R. (1968). *Rec. Trav. Chim. Pays Bas* **87,** 362.

12. Benschop, H. P., and Platenburg, D. H. J. M. (1970). *J. Chem. Soc. Chem. Commun.* 1098.

13. a. Berry, R. S. (1960). *J. Chem. Phys.* **32,** 933; b. Bercz, J. P., Horner, L., and Bercz, C. V. (1967). *Justus Liebigs Ann. Chem.* **703,** 17; c. Bestmann, H. J., and Tomoskozi, I. (1968). *Tetrahedron* **24,** 3299.

14. Bodalski, R., Rutkowska-Olma, E., and Petrusiewicz, K. M. (1980). *Tetrahedron* **36,** 2353–2355.

15. Bodor, N., Knutson, K. B., and Sato, T. (1980). *J. Am. Chem. Soc.* **102,** 3969.
16. a. Bone, S. A., Trippett, S., and Whittle, P. J. (1974). *J. Chem. Soc. Perkin Trans. 1* 2125; b. Borowitz, I. J., Kirby, Jr., K. C., Rusek, P. E., and Casper, E. W. (1971). *J. Org. Chem.* **36,** 88.
17. Boter, H. L., Ooms, A. J. J., van den Burg, G. R., and Van Dijk, C. (1966). *Rec. Trav. Chim Pays Bas,* **85,** 147.
18. Boter, H. L., and Platenburg, D. H. J. M. (1967). *Rec. Trav. Chim. Pays Bas,* **86,** 399.
19. Brierly, J., Trippett, S., and White, M. W. (1977). *J. Chem. Soc. Perkin Trans. 1* 273.
20. Brunner, H., Pieronczyk, W., Schoenhammer, B., Streng, K., Bernal, I., and Korp, J. *Chem. Ber.* **114,** 1137–1149.
21. a. Casey, J. P., Lewis, R. A., and Mislow, K. (1968). *J. Am. Chem. Soc.* **90,** 3869; b. Casey, J. P., Lewis, R. A., and Mislow, K. (1969). *J. Am. Chem. Soc.* **91,** 2789.
22. Chausov, V. A., and Lutsenko, I. F. (1973). *Zh. Obshch. Khim.* **43,** 69.
23. Chen, C. H., and Berlin, K. D. (1971). *J. Org. Chem.* **36,** 2791.
24. Chodkiewicz, W., and Guillerm, D. (1979). *Tetrahedron Lett.* 3573–3574.
25. Chodkiewicz, W., Jore, D., and Wodski, W. (1979). *Tetrahedron Lett.* 1069–1072.
26. Colle, K. S., and Lewis, E. S. (1978). *J. Org. Chem.* **43,** 571.
27. Cooper, D. B., Harrison, J. M., and Inch, T. D. (1974). *Tetrahedron Lett.* 2697.
28. a. Cooper, D. B., Inch, T. D., and Lewis, G. J. (1974). *J. Chem. Soc. Perkin Trans. 1* 1043 and references therein; b. Cooper, D. B., Hall, C. R., Harrison, J. M., and Inch, T. D. (1977). *J. Chem. Soc. Perkin Trans. 1* 1969; c. Christol, H., Cristau, H. J. (1971). *Ann. Chim.* **6,** 179, 191.
29. Cullis, P. M., Lowe, G. (1981). *J. Chem. Soc. Perkin Trans 1* 2317–2321.
30. De Ath, N. J., Ellis, K., Smith, D. J. H., and Trippett, S. (1971). *J. Chem. Soc. Chem. Commun.* 714.
31. De Bruin, K. E., Zon, G., Naumann, K., and Mislow, K. (1969). *J. Am. Chem. Soc.* **91,** 7027.
32. De Bruin, K. E., and Mislow, K. *J. Am. Chem. Soc.* **91,** 7393.
33. De Bruin, K. E., and Peterson, J. R. (1972). *J. Org. Chem.* **37,** 2772.
34. De Bruin, K. E., and Johnson, D. M. (1973). *J. Am. Chem. Soc.* **95,** 4675.
35. De Bruin, K. E., Padilla, A. G., and Campbell, M. I. (1973). *J. Am. Chem. Soc.* **95,** 4681.
36. De Bruin, K. E., and Perrin, E. D. (1975). *J. Org. Chem.* **40,** 1523.
37. De Bruin, K. E., and Thomas, L. L. (1977). *J. Chem. Soc. Chem. Commun.* 33.
38. Dennis, E. A., and Westheimer, F. H. (1966). *J. Am. Chem. Soc.* **88,** 3432.
39. Denohue, J., Mandel, N., Farnham, W. B., Murray, R. K., Mislow, K., and Benschop, H. P. (1971). *J. Am. Chem. Soc.* **93,** 3792 and references therein.
40. Donninger, C., and Hutson, D. H. (1968). *Tetrahedron Lett.* 4821.
41. Egan, W., Tang, R., Zon, G., and Mislow, K. (1970). *J. Am. Chem. Soc.* **92,** 1442.
42. Egan, W., Chauviere, G., Mislow, K., Clark, R. T., and Marsi, K. L. (1970). *J. Chem. Soc. Chem. Commun.* 733.
43. Emsley, J., and Hall, D. (1976). "The Chemistry of Phosphorus." Harper and Row, London.
44. Ephraim, F. (1911). *Ber. Dtsch. Chem. Ges.* **44,** 631.
45. Not cited.
46. Evans, D. J., Kane-Maguire, L. A. P., and Wild, S. B. (1982). *J. Organomet. Chem.* **232,** C9–C12.
47. Farnham, W. B., Murray, Jr., R. K., and Mislow, K. (1970). *J. Am. Chem. Soc.* **92,** 5809.
48. Farnham, W. B., Murray, Jr., R. K., and Mislow, K. (1971). *J. Chem. Soc. Chem. Commun.* 146.

49. Fisher, C., and Mosher, H. S. (1977). *Tetrahedron Lett.* 2487.

50. a. Galdecki, Z., Glowka, M. L., Musierowicz, S., and Michalski, J. (1980). *Pol. J. Chem.* **54,** 539–542; b. Gallagher, M. J., and Jenkins, I. D. (1968). *Top. Stereochem.* **3,** 1.

51. Gerrard, A. F., and Hamer, N. K. (1968). *J. Chem. Soc. B* 539.

52. Gerrard, A. F., and Hamer, N. K. (1967). *J. Chem. Soc. B* 1122.

53. Gillespie, P., Ramirez, F., Ugi, I., and Marquarding, D. (1973). *Angew. Chem. Int. Ed. Engl.* 12, 91.

54. Green, H., and Hudson, R. F. (1959). *Proc. Chem. Soc.* 227.

55. Green, H., and Hudson, R. F. (1963). *J. Chem. Soc.* 540.

56. Green, M., and Hudson, R. F. (1963). *J. Chem. Soc.* 3883.

57. Green, H., and Hudson, R. F. (1963). *J. Chem. Soc.* 540.

58. Grocott, S. C., and Wild, S. B. (1982). *Inorg. Chem.* **21,** 3526–3534.

59. a. Hall, C. R., and Inch, T. D. (1979). *Phosphorus Sulfur* **7,** 171–184; b. Hall, C. R., and Inch, T. D. (1976). *Tetrahedron Lett.* 3645.

60. Hall, C. R., and Inch, T. D. (1977). *Tetrahedron Lett.* 3765.

61. Hall, C. R., Inch, T. D., and Pottage, C. (1981). *Phosphorus Sulfur* **10,** 229–232.

62. Hall, C. R., and Inch, T. D. (1980). *Tetrahedron Lett.* **36,** 2059–2095.

63. Hall, C. R., and Inch, T. D. (1981). *Am. Chem. Soc. Symp Ser.* **171,** 89.

65. Hall, C. R., Inch, T. D., and Williams, N. E. (1982). *J. Chem. Soc. Perkin Trans. 1* 639.

65. Hall, C. R., and Williams, N. E. (1980). *Tetrahedron Lett.* 4959–4962.

66. a. Hall, C. R., and Inch, T. D. (1981). *J. Chem. Soc. Perkin Trans 1* 2368; b. Hall, C. R., and Williams, N. E. (1982). *Tetrahedron Lett.* 999–1002; c. Hall, C. R., Inch, T. D., and Lawston, I. W. (1979). *Tetrahedron Lett.* 2729–2732; d. Hall, C. R., and Inch, T. D. (1979). *J. Chem. Soc. Perkin Trans. 1* 1104.

67. a. Hall, C. R., and Williams, N. E. (1981). *J. Chem. Soc. Perkin Trans. 1* 2746–2750; b. Hamer, N. K. (1965). *J. Chem. Soc.* 2731.

68. Harger, M. J. P. (1977). *J. Chem. Soc. Perkin Trans. 1* 2057.

69. Harrison, J. M., Inch, T. D., and Lewis, G. J. (1975). *J. Chem. Soc. Perkin Trans. 1* 1892.

70. Harrison, J. M., Inch, T. D., Lewis, G. J., and Chittenden, R. A. (1975). *J. Chem. Soc. Chem. Commun.* 720.

71. Haufe, J. (1964). Diplomarbeit, Mainz.

72. Hays, H. R., Peterson, D. J. (1972). *In* "Organic Phosphorus Chemistry" (G. Kosalapoff and L. Maier, eds.), Vol. 3, p. 341. Wiley (Interscience), New York.

73. Henrick, K., and Wild, S. B. (1975). *J. Chem. Soc. Dalton Trans.* 1506.

74. Henrick, K., and Wild, S. B. (1974). *J. Chem. Soc. Dalton Trans.* 2500.

75. Henrick, K., and Wild, S. B. (1975). *J. Chem. Soc. Dalton Trans.* 2500.

76. Henson, P. D., Naumann, K., and Mislow, K. (1969). *J. Am. Chem. Soc.* **91,** 5465.

77. Herriott, A. (1971). *J. Am. Chem. Soc.* **93,** 3304.

78. Hilgetag, G., and Lehman, G. (1957). *Angew. Chem.* **69,** 506.

79. Horner, L. (1980). *Pure Appl. Chem.* **52,** 843.

80. Horner, L. (1964). *Pure Appl. Chem.* **9,** 225.

81. Horner, L., and Winkler, H. (1964). *Tetrahedron Lett.* 461.

82. Horner, L., and Mentrup, A. (1961). *Justus Liebig's Ann. Chem.* **646,** 65.

83. Horner, L., and Balzer, W. D. (1965). *Tetrahedron Lett.* 1157.

84. Horner, L., and Muller, E. (1972). *Phosphorus* **2,** 73.

85. Horner, L., and Muller, E. (1972). *Phosphorus* **2,** 77.

86. Horner, L., and Muller, E. (1972). *Phosphorus* **2,** 117.

87. Horner, L., Winkler, A., Mentrup, A., Hoffmann, H., and Beck, P. (1961). *Tetrahedron Lett.* 161.
88. Horner, L., and Jordan, M. (1979). *Phosphorus Sulfur* **6,** 491.
89. Horner, L., and Samaan, S. (1971). *Phosphorus* **1,** 207.
90. Horner, L., and Jordan, M. (1980). *Phosphorus Sulfur,* **8,** 235–242.
91. Horner, L., and Schlotthauer, B. (1978). *Phosphorus Sulfur* **4,** 155–156.
92. a. Horner, L., and Jordan, M. (1980). *Phosphorus Sulfur* **8,** 225–234; b. Horner, L., and Luckenbach, R. (1971). *Phosphorus* **1,** 73–76.
93. Inch, T. D., and Hall, C. R. (1981). *Am. Chem. Soc. Symp. Ser.* **171,** 83.
94. Jacobus, J., and Mislow, K. (1968). *J. Chem. Soc. Chem. Commun.* 253.
95. Jarvest, R. L., Lowe, G., and Potter, B. V. L. (1981). *J. Chem. Soc. Perkin Trans. 1* 3186–3195.
96. Karle, I. L., Karle, J. M., Egan, W., Zon, G., and Brandt, J. A. (1977). *J. Am. Chem. Soc.* **99,** 4803.
97. Kauffmann, T., and Berger, D. (1968). *Chem. Ber.* **101,** 3022.
98. Kauffman, T., Schonfelder, M., and Legler, J. (1970). *Justus Liebig's Ann. Chem.* **731,** 37, and references therein.
99. Kawashima, T., Kroshefsky, R. D., Kok, R. A., and Verkade, J. G. (1978). *J. Org. Chem.* **43,** 1111.
100. Kinas, R., Pankiewicz, K., and Stec, W. J. (1975). *Bull. Acad. Pol. Sci. Ser. Sci. Chim.* **23,** 981.
101. King, R. B., Bakos, J., Hoff, C. D., and Marko, L. (1979). *J. Org. Chem.* **44,** 3096–3100.
102. Knowles, W. S., Sabacky, M. J., Vineyard, B. D., and Weinkauff, D. J. (1975). *J. Am. Chem. Soc.* **97,** 2567.
103. Knowles, W. S., Sabacky, M. J., and Vineyard, B. D. (1974). *Adv. Chem. Ser.* **132,** 274–282.
104. Knowles, W. S. (1975). Seminar at Stanford University, Industrial Affiliates Symposium.
105. Koizumi, T., Amitani, H., and Yoshii, E. (1978). *Tetrahedron Lett.* 3741.
106. Koizumi, T., Kobayashi, Y., Amitani, H., and Yoshii, E. (1977). *J. Org. Chem.* **42,** 3459.
107. Koizumi, T., Kobayashi, Y., and Yoshii, E. (1974). *J. Chem. Soc. Chem. Commun.* 678.
108. Korpiun, O., and Mislow, K. (1967). *J. Am. Chem. Soc.* **89,** 4784.
109. Korpiun, Lewis, R. A., Chickos, J., and Mislow, K. (1968). *J. Am. Chem. Soc.* **90,** 4842, 4847.
110. Krawiecka, B., and Michalski, J. (1971). *Bull. Acad. Pol. Sci. Ser. Sci. Chim.* **9,** 373.
111. Krawiecka, B., Michalski, J., Mikolajczuk, J., Mikolajczuk, M., Omelanczuk, J., and Skowronska, A. (1974). *J. Chem. Soc. Chem. Commun.* 630.
112. Krawiecka, B., Skrzypczynski, Z., and Michalski, J. (1972). *Phosphorus* **3,** 177.
113. Kumli, K. F., Mc Ewen, W. E., and Vander Werf, C. A. (1959). *J. Am. Chem. Soc.* **81,** 248.
114. Kyba, E. P. (1975). *J. Am. Chem. Soc.* **97,** 2555.
115. Kyba, E. P. (1976). *J. Am. Chem. Soc.* **98,** 4805.
116. Kyba, E. P., and Hudson, C. W. (1975). *Tetrahedron Lett.* 1869.
117. Kyba, E. P., and Rines, S. P. (1982). *J. Am. Chem. Soc.* **104,** 4800–4801.
118. Lambert, J. B., Jackson, III, G. F., and Mueller, D. C. (1968). *J. Am. Chem. Soc.* **90,** 6401.
119. Lambert, J. B., Jackson, G. F., and Mueller, D. C. (1970). *J. Am. Chem. Soc.* **92,** 3093.

120. Lehn, J. M. (1970). *Forstschr. Chem. Forsch.* **15,** 311.
121. Lewis, R. A., and Mislow, K. *J. Am. Chem. Soc.* **91,** 7009; *idem* (1969). *J. Chem. Soc. Chem. Commun.* 1010.
122. Lowe, G., Potter, B. V. L., Sproat, B. S., and Hull, W. E. (1979). *J. Chem. Soc. Chem. Commun.* 733.
123. Lowe, G., Cullis, P. M., Jarvest, R. L., and Potter, B. V. L. (1981). *Am. Chem. Soc. Symp. Ser.* **171,** 103.
124. Luckenbach, R. (1973). "Dynamic Stereochemistry of Pentacoordinated Phosphorus and Related Elements." Thieme, Stuttgart.
125. Luckenbach, R. (1976). *Z. Naturforsch.* **31b,** 1127–1134.
126. Luckenbach, R. (1972). *Phosphorus* **1.** 223–228.
127. Luckenbach, R. (1975). *Chem. Ber.* **108,** 803.
128. a. Luckenbach, R. (1972). *Phosphorus* **1,** 229–236; b. Luckenbach, R. (1972). *Phosphorus* **1,** 293; c. Luckenbach, R. (1973). *Phosphorus* **3,** 117–123; d. Luckenbach, R. (1976). *Z. Naturforsch.* **30b,** 119; e. Luckenbach, R. (1971). *Phosphorus* **1,** 73–76; f. Luckenbach, R., and Endres, W. (1976). *Z. Naturforsch.* **31b,** 1011–1012.
129. Luckenbach, R. (1973). *Phosphorus* **2,** 293.
130. a. Luckenbach, R. (1976). *Z. Naturforsch.* **31b,** 1135–1137; b. Luckenbach, R., and Herweg, G. (1976). *Justus Liebig's Ann. Chem.* 2305–2315; c. Luckenbach, R., and Bechtolsheimer, H. H. (1977). *Z. Naturforsch.* **32b,** 589–593; d. Luckenbach, R., and Bechtolsheimer, H. H. (1977). *Z. Naturforsch.* **32b,** 584–588.
131. Ludeman, S. M., Bartlett, D. L., and Zon, G. J. (1979). *J. Org. Chem.* **44,** 1163.
132. Marsi, K. L. (1974). *J. Org. Chem.* **39,** 265.
133. Marsi, K. L., and Clark, R. T. (1970). *J. Am. Chem. Soc.* **92,** 3791.
134. Maryanoff, C. A., Maryanoff, B. E., Tang, R., and Mislow, K. (1973). *J. Am. Chem. Soc.* **95,** 5839.
135. Mc Ewen, W. E., Berlin, K. D. (eds.) (1975). "Organophosphorus Stereochemistry, Part I. Origins and P(III) and P(IV) Compounds: Part II. P(V) Compounds." Halsted.
136. Meisenheimer, J., Cooper, J., Horing, M., Lauter, W., Lichtenstadt, L., and Samuel, W. (1926). *Justus Liebig's Ann. Chem.* **449,** 213.
137. Meisenheimer, J., and Lichtenstadt, A. (1911). *Chem. Ber.* **44,** 356.
138. Michalski, J., and Ratajczak, A. (1963). *Rocz. Chem.* **37,** 1185.
139. Michalski, J., Mikolajczuk, M., Halpern, A., and Proszynska, K. (1969). *Tetrahedron Lett.* 1919.
140. Michalski, J., Mikolajczuk, M., and Pliska-Krawiecka, B. (1966). *Angew. Chem.* **73,** 716.
141. Michalski, J., Mikolajczuk, M., and Pliska-Krawiecka, B. (1968). *Bull. Acad. Pol. Sci. Ser. Sci. Chim.* **17,** 75.
142. Michalski, J., Mikolajczuk, M., and Ratajczak, A., (1963). *Bull. Acad. Pol. Sci. Ser. Sci. Chim.* **11,** 11.
143. Michalski, J., and Ratajczak, A. (1959). *Chem. Ind. London* 539.
144. Michalski, J., Mikolajczuk, M., and Omelanczuk, J. (1965). *Tetrahedron Lett.* 1779.
145. Michalski, J. (1967). *Bull. Soc. Chim. France* 1112.
146. Michalski, J., Mikolajczuk, M., and Ratajczuk, A. (1965). *Bull. Acad. Pol. Sci. Ser. Sci. Chim.* **13,** 277.
147. Michalski, J., and Mikolajczuk, M. (1966). *Tetrahedron* **22,** 3055.
148. Michalski, Mikolajczuk, M., Mlotkowska, B., and Omelanczuk, J. (1969). *Tetrahedron* **25,** 1743.
149. Michalski, J., and Musierowicz, S. (1967). *Angew. Chem.* **79,** 1070.
150. Michalski, J., Ratajczak, A., and Tulimowski, Z. (1963). *Bull. Acad. Pol. Sci. Ser. Sci. Chim.* **11,** 237.

151. a. Michalski, J., and Skryzpczynski, Z. (1975). *J. Organomet. Chem.* **97,** C3; b. Michalski, J., Okruszek, A., and Stec, W. (1970). *J. Chem. Soc. Chem. Commun.* 1495.
152. Mikolajczuk, M. (1966). *Chem. Ber.* **99,** 2083.
153. Mikolajczuk, M., Omelanczuk, J., and Para, M. (1972). *Tetrahedron* **28,** 385.
154. Mikolajczuk, M., Omelanczuk, J., and Michalski, J. (1969). *Bull. Acad. Pol. Sci. Ser. Sci. Chim.* **17,** 155.
155. Mikolajczuk, M., and Omelanczuk, J. (1973). *Phosphorus* **3,** 47.
156. a. Mikolajczuk, M. (1980). *Pure Appl. Chem.* **52,** 959–972; b. Mikolajczuk, M., and Leitloff, M. (1975). *Russ. Chem. Rev.* **44,** 670.
157. Mislow, K. (1973). *Trans. N. Y. Acad. Sci. Ser II,* **35,** 227–242.
158. Miyashita, A., Yasuda, A., Takaya, H., Toriumi, K., Ito, T., Souchi, T., and Noyori, R. (1980). *J. Am. Chem. Soc.* **102,** 7932–7934.
159. Morrison, J. D., and Masler, W. F. (1974). *J. Org. Chem.* **39,** 270.
160. Morrison, J. D. (1978). Private communication to D. Valentine, Jr.
161. Musierowicz, S., and Wroblewski, A. E. (1980). *Tetrahedron* **36,** 1375–1380.
162. Naumann, K., Zon, G., and Mislow, K. (1969). *J. Am. Chem. Soc.* **91,** 7012.
163. Nudelman, A., and Cram, D. J. (1971). *J. Org. Chem.* **36,** 335.
164. Nuretdinov, I. A., Buina, N. A., Bayandina, E. V., and Sibgatulina, F. G. (1974). *Isv. Akad. Nauk SSSR* 483.
165. Omelanczuk, J., and Mikolajczuk, M. (1979). *J. Am. Chem. Soc.* **101,** 7292.
166. Omelanczuk, J., Perlikowska, W., and Mikolajczuk, M. (1980). *J. Chem. Soc. Chem. Commun.* 24.
167. Oram, R. K., and Trippett, S. (1973). *J. Chem. Soc. Perkin Trans. I* 1300.
168. Otsuka, S., Nakamura, A., Kano, T., and Tani, K. (1971). *J. Am. Chem. Soc.* **93,** 430.
169. Pankiewicz, K., Kinas, R., Stec, W. J., Foster, A. B., Jarman, M., and Van Maanen, J. M. S. (1979). *J. Am. Chem. Soc.* **101,** 7712.
170. Peerdeman, A. F., Holst, J. P. C., Horner, L., and Winkler, H. (1965). *Tetrahedron Lett.* 811–815.
171. Raban, M. (1968). *In* "Topics in Stereochemistry" (N. L. Allinger and E. L. Eliel, eds.), Ch. 4. Wiley, New York.
172. Rauk, A., Allen, L. C., and Mislow, K. M. (1970). *Angew. Chem.* **82,** 453.
173. Rauk, A., Andose, J. D., Frick, W. R., Tang, R., and Mislow, K. (1971). *J. Am. Chem. Soc.* **93,** 6507.
174. Roberts, N. K., and Wild, S. B. (1979). *J. Am. Chem. Soc.* **101,** 6524.
175. Roberts, N. K., and Wild, S. B. (1981). *Inorg. Chem.* **20,** 1892.
176. Roberts, N. K., and Wild, S. B. (1981). *Inorg. Chem.* **20,** 1900.
177. Roberts, N. K., and Wild, S. B. *J. Chem. Soc. Dalton Trans.* 2015.
178. Sato, T. (1980). Japanese Patent No. 79 157581; *idem* (1980). *Chem. Abstr.* **93,** 26467.
179. Sato, T., Ueda, H., Nakagawa, K., and Bodor, N. (1983). *J. Org. Chem.* **47,** 98–101.
180. Seiber, J. N., and Tolkmith, H. (1969). *Tetrahedron* **25,** 381.
181. Seyferth, D., Welch, D. E., and Heeren, J. K. (1964). *J. Am. Chem. Soc.* **86,** 1100.
182. Smith, D. J. H., and Trippett, S. (1969). *J. Chem. Soc. Chem. Commun.* 855.
183. Szafraniec, L. J., Szafraniec, L. L., and Aaron, H. S. (1982). *J. Org. Chem.* **47,** 1936.
184. Tani, K., Brown, L. D., Ahmed, J., Ibers, J. A., Yokota, M., Nakamura, A., and Otsuka, S. (1977). *J. Am. Chem. Soc.* **99,** 7876.
185. Teichmann, H., and Lam, P. T. (1969). *Z. Chem.* **9,** 310.
186. Ugi, I., and Ramirez, F. (1972). *Chem Br.* **8,** 18.
187. Valentine, Jr., D., and Scott, J. W. (1978). *Synthesis* 329.
188. Valentine, Jr., D., Blount, J. F., and Toth, K. (1980). *J. Org. Chem.* **45,** 3691.

189. Van den Berg, G. R., Platenburg, D. H. J. M., and Benschop, H. P. (1972). *Rec. Trav. Chim. Pays Bas* **91,** 929–934.
190. Van den Berg, G. R., Platenburg, D. H. J. M., and Benschop, H. P. (1971). *J. Chem. Soc. Chem. Commun.* 606.
191. Vineyard, B. D., Knowles, W. S., Bachman, G. L., and Weinkauff, D. J. (1977). *J. Am. Chem. Soc.* **99,** 5946.
192. Westheimer, F. (1968). *Acc. Chem. Res.* **1,** 70.
193. Wittig, G., Christau, H. J., and Braun, H. (1967). *Angew. Chem. Int. Ed. Engl.* **6,** 700–701.
194. Wroblewsky, A. E., and Verkade, J. G. (1979). *J. Am. Chem. Soc.* **101,** 7719.
195. a. Wudl, F., and Lee, T. B. K. (1973). *J. Am. Chem. Soc.* **95,** 6349; b. Young, D. P., Mc Ewen, W. E., Velez, D. C., Johnson, J. W., and Vanderwerf, C. A., (1964). *Tetrahedron Lett.* 359.
196. Zon, G., De Bruin, K. E., Naumann, K., and Mislow, K. (1969). *J. Am. Chem. Soc.* **91,** 7023.
197. Zon, G., and Mislow, K. (1971). *Fortschr. Chem. Forsch.* **19,** 61.
198. Zon, G. (1975). *Tetrahedron Lett.* 3139.
199. Zon, G., Brandt, J., and Egan, W. J. (1977). *J. Natl. Cancer Inst.* **58,** 1117.
200. Bestmann, H. J., Lienert, J., and Heid, E. (1982). *Chem. Ber.* **115,** 3875–3879.
201. Palmer, J. A., and Wild, S. B. (1983). *Inorg. Chem.* **22,** 4054–4059.
202. Salem, G., and Wild, S. B. (1983). *Inorg. Chem.* **22,** 4049–4054.
203. Allen, D. G., Mc Laughlin, G. M., Robertson, G. B., Steffen, W. L., Salem, G., and Wild, S. B. (1982). *Inorg. Chem.* **21,** 1007–1014.
204. Shortt, A. B., Durham, L. J., and Mosher, H. J. (1983). *J. Org. Chem.* **48,** 3125–3126.
205. Shortt, A. B. (1982). *Diss. Abstr. Int. B* **43,** 1109.
206. Bentrude, W. G., Moriyama, M., Mueller, H. D., and Sopchik, A. E. (1983). *J. Am. Chem. Soc.* **105,** 6053–6061.
207. Guruswamy, N., and Berlin, K. D. (1982). *J. Am. Chem. Soc.* **104,** 3114–3119.
208. Gerit, J. A., Coderre, J. A., and Mehdi, S. (1983). *Adv. Enzymol. Relat. Areas Mol. Biol.* **55,** 291–380.
209. Stec, W. J. (1983). *Acc. Chem. Res.* **16,** 411–417.
210. Moriyama, M., and Bentrude, W. G. (1982). *Tetrahedron Lett.* 4547.
211. Moriyama, M., and Bentrude, W. G. (1983). *J. Am. Chem. Soc.* **105,** 4726–4731.
212. Segi, M., Nakamura, Y., Nakashima, T., and Suga, S. (1983). *Chem. Lett.* 913–916.

4

Synthesis and Utilization of Compounds with Chiral Nitrogen Centers

Franklin A. Davis
Department of Chemistry
Drexel University
Philadelphia, Pennsylvania

Robert H. Jenkins, Jr.
Texaco Research Center
Beacon, New York

I. Introduction

The stereochemical behavior of nitrogen in its compounds has been extensively investigated for nearly a century. Hantzsch and Werner were the first to suggest the possibility of asymmetric trivalent nitrogen (*1*). It is now known that the three substituents of an aliphatic amine are pyramidally arranged in a geometry around nitrogen similar to that of a tetravalent carbon. Consequently, if the three substituents are different, the nitrogen atom constitutes a chiral center, and in principle, such amines

ISBN 0-12-507704-1

are capable of existing as enantiomers. The ease of pyramidal inversion at nitrogen [Eq. (1)], however, makes the isolation of simple, optically ac-

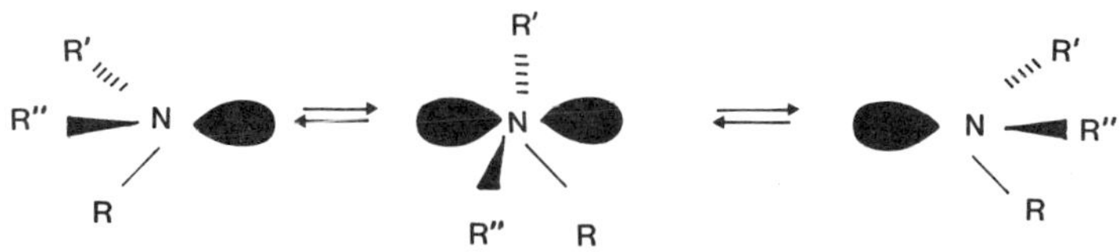

(1)

tive stereoisomers impossible. It is for this reason that most of the earlier attempts [reviewed by Shriner *et al.* (*2*)] to resolve amines failed.

The barrier to pyramidal inversion in trivalent amines is 6–7 kcal/mol at room temperature (*3*). For optically active amines containing a chiral nitrogen to exist at room temperature, the barrier to inversion [Eq. (1)] must be ≥23 kcal/mol (*4*). Experimentally, the influence of angle strain and electronic factors on these inversion barriers are well understood, and it is possible to increase the barrier [Eq. (1)] sufficiently to permit the isolation of optically active, chiral nitrogen-containing compounds (*3, 4, 5*). This has been accomplished by incorporating the nitrogen into three-membered rings and by having electronegative heteroatoms adjacent to nitrogen. The barrier to nitrogen inversion in quaternary ammonium salts and amine oxides is also high enough to permit the isolation of stereoisomers. The influence of vicinal n–σ^* interaction on the pyramidal stability of tricoordinated nitrogen has been discussed (*6*).

This chapter reviews the synthesis, reactions, and properties of nitrogen-containing compounds that derive their optical activity solely from an asymmetric nitrogen atom. Reference to earlier reviews of chiral nitrogen will be made as appropriate (*2, 7*).

II. Chiral Trivalent Nitrogen

A. Aziridines

Incorporation of a nitrogen atom into a three-membered ring substantially increases the barrier to pyramidal inversion [Eq. 1] (*3*). This barrier (18–20 kcal/mol) is still insufficient, however, to allow the isolation of optically active *N*-alkyl- and *N*-arylaziridines. To date the only reported examples of optically active aziridines are those where the aziridine nitrogen atom is attached to a highly electronegative atom, such as a halogen or alkoxy group.

Annunziata, Fornasier, and Montanari (*8*) were the first to report an optically active aziridine whose asymmetry results solely from a chiral nitrogen. The synthesis was accomplished by asymmetric chlorination of 2,2-diphenylaziridine (**1**) at −60 to −78°C with (−)-(1*R*,2*R*)-isobornylhypochlorite (**3**) and (+)-(*S*)-(*N*-chloro)methylphenylsulfoxamine (**4**) to give *N*-chloro-2,2-diphenylaziridine (**2**) ($[\alpha]_D$ +3.4° (*c* 0.4, acetone) and +4.3°

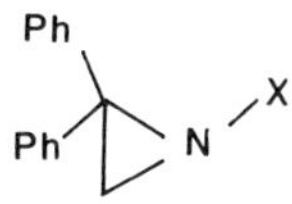

1 X = H

2 X = Cl

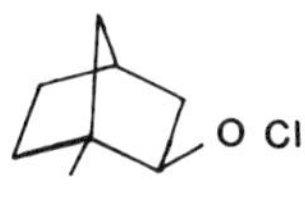

3

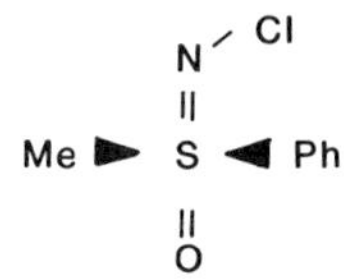

4

(*c* 0.4, acetone), respectively) (*8*). The higher asymmetric induction observed for **3** was attributed to the large groups attached to the chiral center and to the proximity of this center to the active site. After 4 days at 0°C, **2** was completely racemized.

Optically active aziridines **2**, having rotations of $[\alpha]_D$ −4.5 and +20.2°, have also been prepared by treatment of **1** with *tert*-butyl hypochlorite (*t*BuOCl) in the presence of such chiral solvents as *S*-(+)-2,2,2-trifluoro-1-phenylethanol and (*R*)-(−)-2,2,2-trifluoro-1-(1-naphthyl)ethanol, respectively (*9*). The absolute configuration of **2**, which is unknown, appears to be dependent on the chirality of the trifluorocarbinol. From the rate of racemization, the barrier to pyramidal nitrogen inversion was calculated to be 24.4 kcal/mol.

Attempts by Kostyanovskii and Rudchenko to carry out the asymmetric halogenation of 2,2-dimethyl- and 2,2-pentamethyleneaziridine with D-(+)-*N*-chloro- or bromocamphorimides at −70°C failed, as did enantiomeric enrichment of 1-bromo-2,2,-dimethylaziridine by partial decomposition with chiral amines (*10*). However, halogenation of (*S*)-**5** with *t*BuOCl and *N*-chlorosuccinimide (NCS) gave (+)-(*R*)-*N*-chloro-2,2-dimethylaziridine (**6**) ($[\alpha]_D$ +2.3° (*c* 5.6 *n*-hexane) and $[\alpha]_D$ +4.5° (*c* 2.8, *n*-

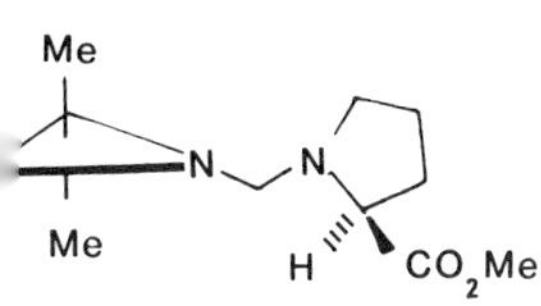

5

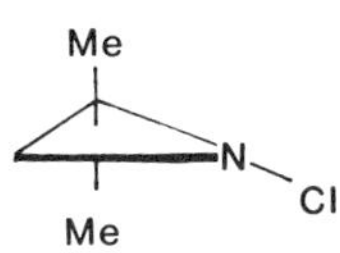

(+)-(R)-6

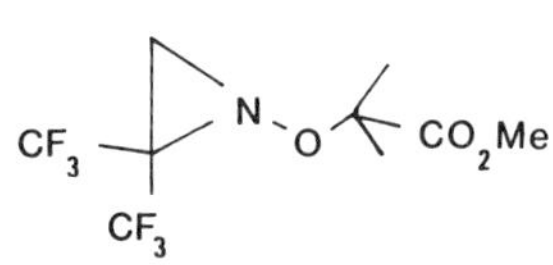

8

hexane), respectively), in 47 to 57% isolated yields (*11*). The configuration of **6** was assigned on the basis of its positive Cotton effect and a consideration of the quadrant rule. The half-life for racemization of (+)-(*R*)-**6** at 80°C was 45 min.

Kostyanovskii and Rudchenko isolated the enantiomers of (−)-*S*- and (+)-(*R*)-diethyl-1-methoxyaziridine 2,2-dicarboxylate (**7**) in high optical purity (*10*). This was accomplished by separating the diastereomeric (*R*)- and (*S*)-2-phenylethylamine salts of the monocarboxylic acid as outlined in Fig. 1 (*10*). The enantiomeric purity of (−)-(*S*)- and (+)-(*R*)-**7** was determined by NMR to be 93.4 and 96.2%, respectively, on the basis of the optical purity of the amine. The absolute configuration of the nitrogen atom in (+)-(*R*)-**7** was determined by X-ray crystallography of the α-(*p*-bromophenyl)ethylammonium salt of the monoacids of (+)-(*R*)-**7** (*12*).

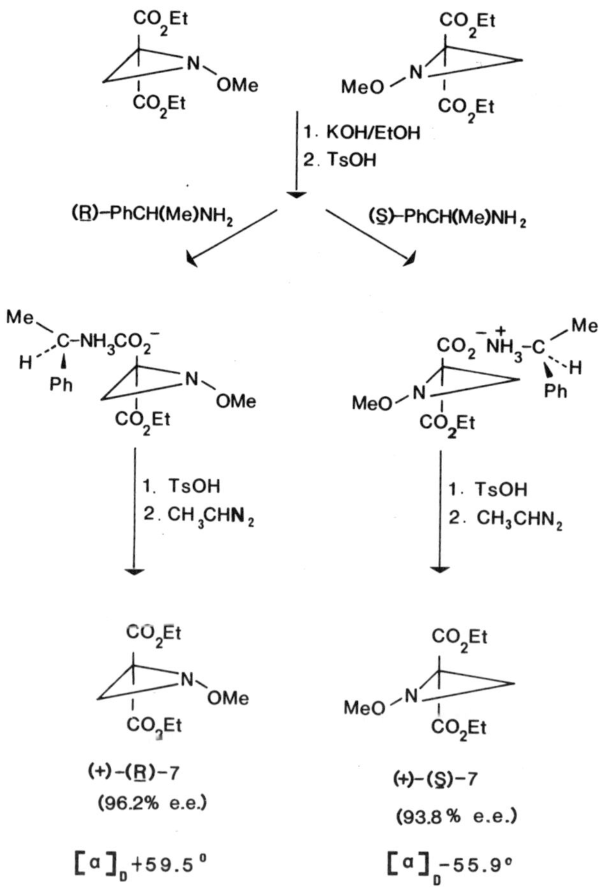

Fig. 1. Resolution of (−)-(*S*)- and (+)-(*R*)-diethyl-1-methoxyaziridine 2,2-dicarboxylate (**7**).

Treatment of (+)-(*R*)-**7** with ammonia gives the diamide in 66% yield with no loss of optical activity (*12*). At 90°C in C_2Cl_4, the half-life for racemization was 63 h ($\Delta G\ddagger$ = 31.0 kcal/mol). The inversion equilibrium of amide derivatives of aziridines, such as (+)-**7** (Et = NH_2, NHMe), can be displaced toward the (−)-(*S*) enantiomer by heating (70–100°C) with *l*-menthyl lactate (*13*).

Kostyanovskii *et al.* were unable to prepare optically active 1-alkoxy-2,2-bis(trifluoromethyl)aziridine (**8**) by asymmetric destruction (kinetic resolution) and/or crystallization from chiral solvents (*14*). The enantiomers of (−)-**8** (95% optically pure, $[\alpha]_D$ −16.4°) and (+)-**8** (85% optically pure, $[\alpha]_D$ +14.3°) can be separated using the procedure outlined in Fig. 1. Complete racemization of **8** occurs after 16 h at 100°C ($\Delta G\ddagger$ = 29.8 kcal/mol). The work of Kostyanovskii and his group on asymmetric nitrogen in aziridines has been summarized (*15*).

Ab initio SCF-CI methods were used by Rauk to study the lower electronic states and the oscillator and optical rotatory strengths of the lower electronic transitions of aziridines, *cis*- and *trans*-2-methylaziridine, diaziridines, and oxaziridines (*16*). The calculated and experimental results agree that the first transition of (1*S*,2*S*)-2-methylaziridine has a positive rotational strength, and the signs of the circular dichroism (CD) are determined principally by the configuration at nitrogen.

B. Diaziridines

The low basicity of the nitrogen in diaziridines precludes their separation into enantiomers via their diastereomeric salts of chiral acids. One approach taken by Kostyanovskii *et al.* for the preparation of chiral diaziridines was kinetic resolution. This procedure, outlined in Eq. (2),

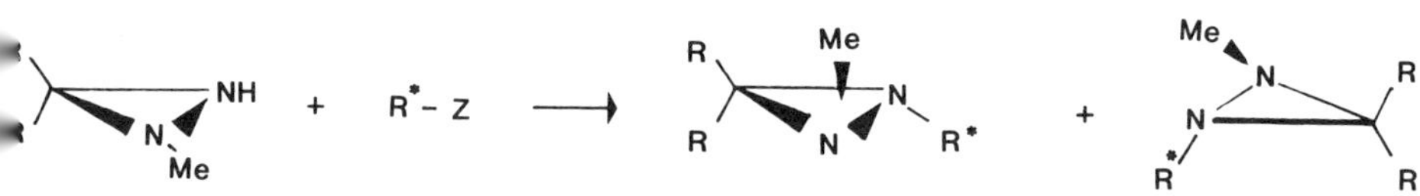

(±) - 9 10 a 10 b

(2)

involves reaction of an excess of the racemic diaziridine **9** with an optically active adjuvant (R*Z) affording a mixture of diaziridine diastereomers (**10ab**) (*17–22*). The unreacted diaziridine proved to be optically active. The optical purity of the chiral diaziridines prepared in this manner was determined by proton NMR and was based on the ratio of the diastereomers **10a,b** (*17*). These results are summarized in Table 1.

TABLE 1
Kinetic Resolution of Diaziridines

(±)-Diaziridine	Chiral adjuvant	% ee	Configuration	$[\alpha]_D$ (deg.)	Referenc
 **11**	(*S*) 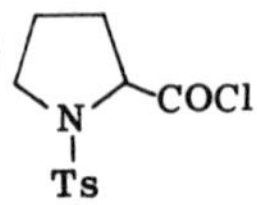	63.5	1*R*,2*R*	56.1	*19*
		44.3	1*R*,2*R*	—	*20*
	(*R*) 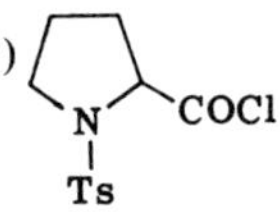	47.5	1*S*,2*S*	−42.2	*19*
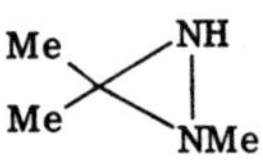 **12**	(*S*)-PhCH)Me)N=C=O	5.1	1*R*,2*R*	3.3	*17*
	(*S*)-PhCH(Me)N=C=O	4.4	1*R*,2*R*	3.3	*17*
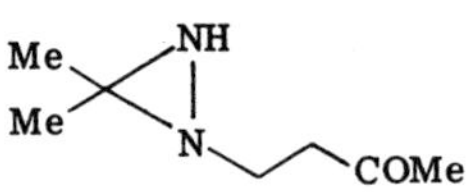 **13**	(*S*) 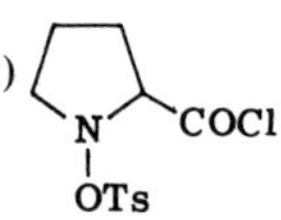	14		16.2	
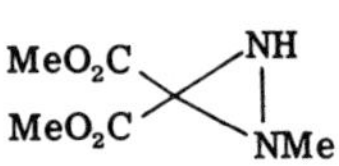 **14**	*d*-10-Camphorsulfonyl chloride		1*S*,2*S*	−0.95	*22*
	(−)-Ephedrine		1*R*,2*R*	7.39	*22*

The absolute configuration of (−)-(1*R*,2*R*)-**11** was determined by an X-ray crystal structure of the pure (*S*)-α-phenylethyl isocyanate diastereomer **15** (*18*). The antipodes of (+)-(1*R*,2*R*)- and (−)-(1*S*,2*S*)-**11** have

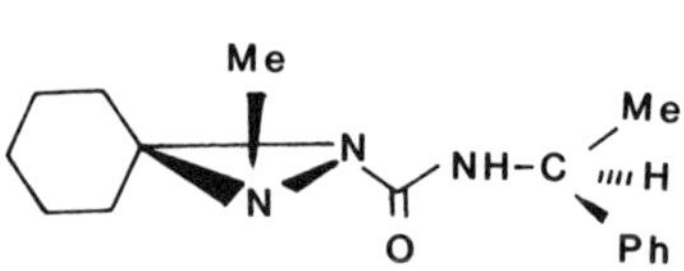

(−)-(1R,2S)-15

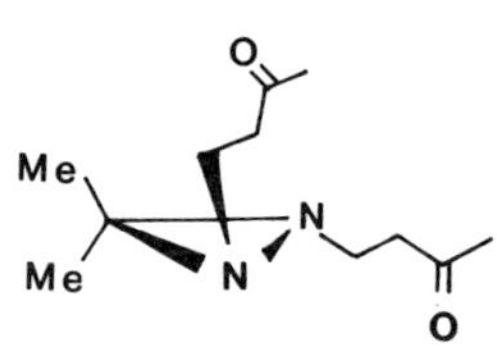

16

opposite and negative signs of the corresponding ORD and CD curves (*19*).

Treatment of (+)-**12** (22.3% optically pure, $[\alpha]_D$ +14.7° (*c* 4.3, *n*-heptane)) with chloral affords a racemic centrosymmetric dimer, which on reaction with base, gives racemic **12** and (+)-**12** of approximately three times the optical purity (68.3%, $[\alpha]_D$ +45.5° (*c* 4.2, n-heptane)) from the mother liquor (*20*). All attempts to remove the carbamoyl substituent from the pure (*S*)-α-phenylethyl isocyanate of **11** failed, affording only racemic **11** (*19*).

With methyl vinyl ketone (MVK), (+)-2-(3-butanone)-3,3-dimethyldiaziridine (**13**) gives **16** ($[\alpha]_D$ +2.5 (*c* 3.2, *n*-heptane)) (*21*). The fact that **16** was also optically active confirms the trans orientation of the diaziridine 1,2-substituents, a necessary condition for optical activity in symmetrically substituted diaziridines.

Kostyanovskii, Shustov, Mishchenko, and Markov described the preparation of a variety of optically active diaziridines **18** of unknown, but apparently low, optical purity by treatment of the *d*- and *l*-10-*O*-camphorsulfonylketoximes **17** with amines [Eq. (3)] (*22*). These results are summarized in Table 2. The fact that optically active diaziridines can be

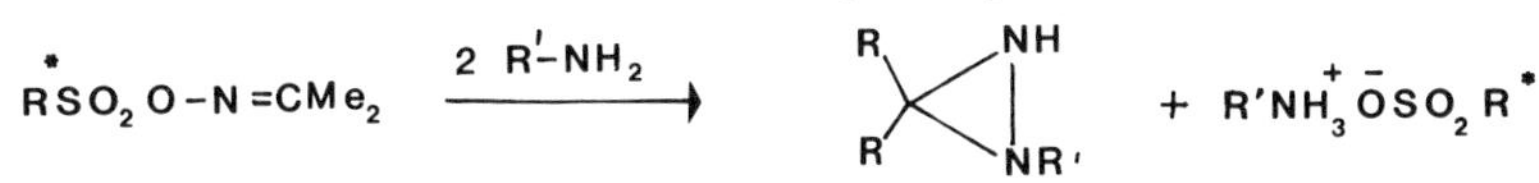

(3)

17 **18**

R* = *d*- or *l*-10-camphor

TABLE 2
Synthesis of Diaziridines from *l*-10-*O*-Camphorsulfonylketoximes[a]

Oxime **17** R	Amine	% Yield	Diaziridine	$[\alpha]_D$ (deg.)
Me	$MeNH_2$	22	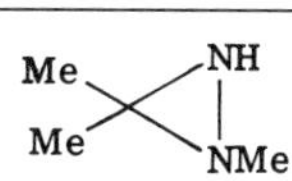	−4.64[b]
CF_3	$EtNH_2$	58	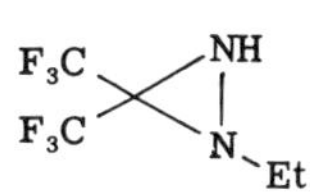	−1.3
CF_3	*i*$PrNH_2$	30	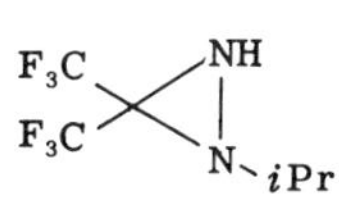	−0.13

[a] From (*22*).
[b] 1*S*,2*R* configuration; see Table 1.

formed in this manner argues against a nitrene mechanism; in other words, the chiral leaving group must take part in the rate-determining step.

When *d*-10-camphorsulfonylketoxime **17a** is treated with methyl amine at −16°C in methylene chloride, (+)-(1*R*,2*R*)-**11** is obtained in low optical purity (~0.2% ee). With *l*-10-camphorsulfonylketoxime **17b**, methylamine gives (−)-**12**. On the basis of this correlation, as well as the ORD and CD curves, the 1*S*,2*S* configuration was assigned to (−)-1-methyl-3,3-dimethyldiaziridine (**12**) (*19*).

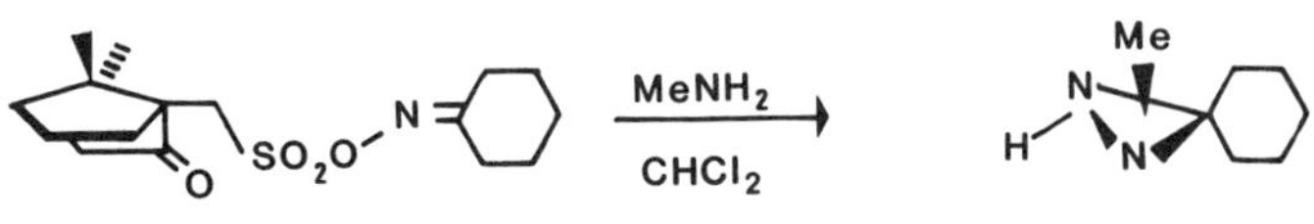

d-17a   −16°C   (+)-(1R,2R)-11

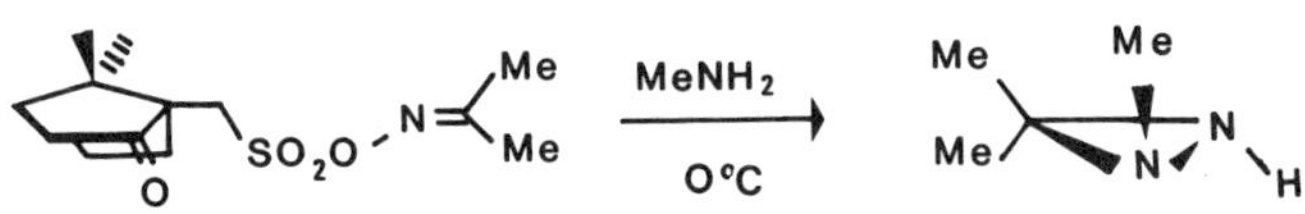

l-17b   (−)-(1S,2S)-12

Separation of the antipodes of (+)- and (−)-dimethyl-1-methyldiaziridine 3,3-dicarboxylate (**19**) has been accomplished in high optical purity by the general resolution procedure developed for aziridines (Fig. 1) (*23*). Treatment of the monoacid of racemic dimethyl-1-methyldiaziridine 3,3-dicarboxylate (**21**) with (+)- and (−)-ephedrine gave the corresponding

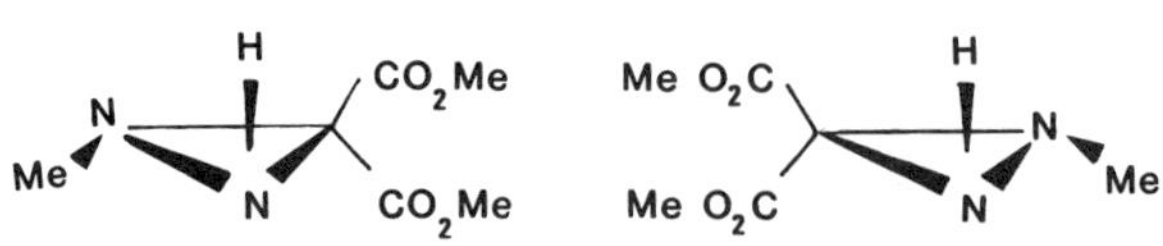

(+)-(1R, 2R) −19   (−)-(1S, 2S) −19

diastereomeric salts. Separation by crystallization of these salts followed by hydrolysis and reaction with diazomethane afforded (+)-(1*R*,2*R*)-**19** ($[\alpha]_D$ +72.9° (*c* 0.5, methanol)) and (−)-(1*S*,2*S*)-**19** ($[\alpha]_D$ −70.2 (*c* 0.5, methanol)). The optical purities of (+)-**19** and (−)-**19** were determined to be 98.0 and 94.3%, respectively, using an optically active shift reagent. The absolute configurations were established by an X-ray crystal structure of methyl-(−)-trans-1-methyl-3-*N*-[(*S*)-α-phenylethyl]carboxamide diaziridine 3-carboxylate (*24*).

(+)-(1*S*,2*S*)-1-[2-Dimethylaminoethyl-3,3-bis(trifluoromethyl)diaziridine (**20**) was isolated by separating the diastereomeric salts of *l*-10-camphorsulfonic acid (*25*). Crystallization below 0°C was necessary because of the low configurational stability of this diaziridine. The optical purity of the diastereomer salt was judged to be 85.5% by proton NMR. Treatment of the salt with KOH gave (+)-(1*S*,2*S*)-**20** ($[\alpha]_D$ +41.3° (*c* 3.1, *n*-hexane))

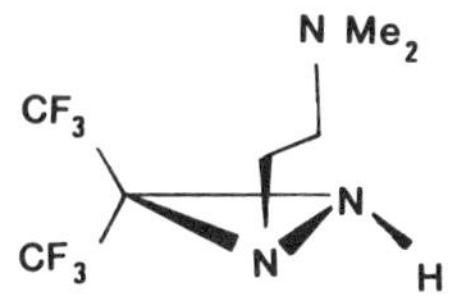

(+)-(1S,2S)-20

(85.5% ee). The absolute configuration of (+)-**20** was inferred from the CD spectrum and from the positive Cotton effect that was observed. With methyl iodide, (+)-**20** gave the optically active quaternary salt.

Hakli and Mannschreck described the separation of the enantiomers of 3-benzyl-1,2,3-trimethyldiaziridine (**21**) by column chromatography on triacetylcellulose (*26*). Using ethanol : water (96 : 4), the first 10% of the elutent contained predominantly (+)-**21** ($[\alpha]_{436}$ +85° (*c* 0.35, CCl_4)), the

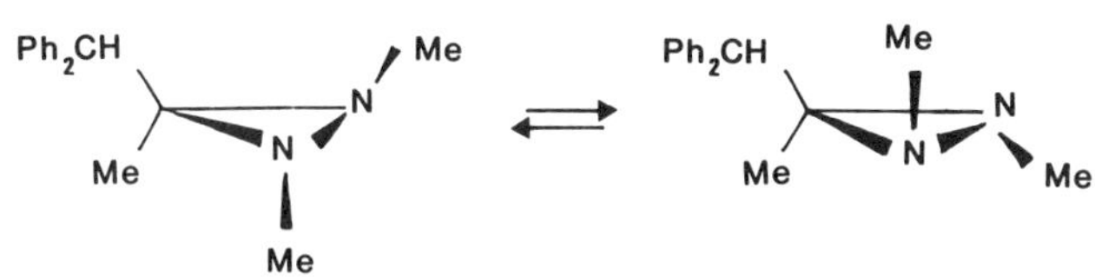

(1R, 2R)-21 (1S, 2S)-21

remaining elutent being practically pure (−)-**21**, ($[\alpha]_{436}$ −141° (*c* 0.05, CCl_4)). The rate of cis–trans isomerization of (+)-**21** was ascertained by following the rate of racemization in toluene at 89.9°C. This process was first order with a half-life of 56.9 min and $\Delta G‡ = 27.6$ kcal/mol.

With a slight excess of (*S*)-α-phenylethylamine, racemic diaziridine **19** gave a 1 : 1 mixture of only two trans diastereomers (**22** and **23**). The

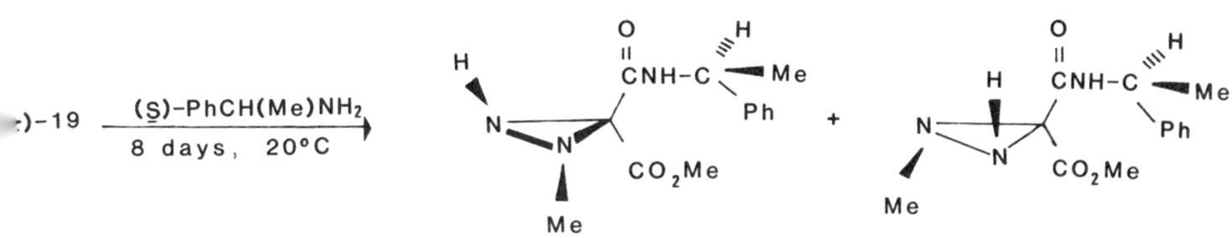

22 23

difference in reactivity of the two ester groups is apparently due to steric hindrance of the *cis*-$MeCO_2$ group by the substituent on nitrogen (*24*). One crystallization from benzene gave pure **22** (36% yield), and recrystallization three times from 4 : 1 carbon tetrachloride : hexane gave pure **23** (21% yield). The efficiency of this resolution may be due to the high asymmetry of the diastereomer molecules, which have four adjacent chiral centers.

A diastereomeric equilibrium mixture of 3,3-bis(trifluoromethyl)diaziridines **24/25** was obtained on treatment of hexafluoroactone-*O*-tosyl oxime with the methyl ester of (−)-*S*-alanine (*25*). The absolute configurations are unknown. On crystallization or removal of the solvent from **24/25**, a single diastereomer was obtained. The reverse process was observed on solution and indicates that crystallization is accompanied by inversional epimerization. Similar behavior was noted for other diaziridines related to **24/25** (*25*).

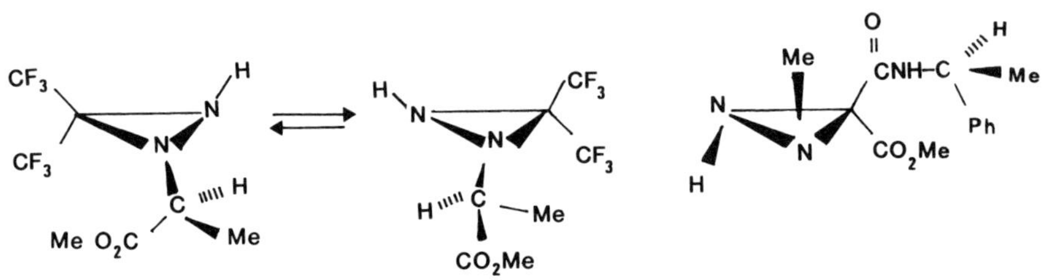

24 **25** **26**

The inversion parameters for the nitrogen atoms in diaziridines cannot be determined directly by NMR. These values were obtained by measuring the rate of racemization of (+)-**11** ($\Delta G\ddagger$ = 27.98 kcal/mol) and the epimerization rate of (−)-**11** at 70°C (*20*). The half-live for the cis to trans isomerization of **22** and **26** was determined by NMR to be 2.8 h ($\Delta G\ddagger$ = 27.3 kcal/mol) (*24*).

C. Oxaziridines

The most studied optically active chiral nitrogen-containing compounds are oxaziridines, compounds having carbon, nitrogen, and oxygen atoms in a three-membered ring. The calculated value for nitrogen inversion in these compounds is 32 kcal/mol (*27*), which agrees well with experimentally determined values of 24 to 31 kcal/mol (*28–30*).

Optically active oxaziridines are usually prepared by oxidation of achiral imines with chiral peracids, by oxidation of chiral imines with achiral peracids, and by oxidation of imines in chiral media. Other methods include treatment of oxaziridines with brucine, photolysis of nitrones in chiral solvents, and thermal isomerization of oxaziridines in optically active liquid crystals. These methods are reviewed below.

1. Chiral 2-Alkyloxaziridines

a. Oxidation of Achiral Imines with Chiral Peracids

Optically active oxaziridines with asymmetry due solely to the nitrogen atom were first reported by the groups of Boyd (*28, 31, 32*) and Montanari (*29*). Oxidation of imines ($R_2C{=}NR'$), in which both groups attached to the imino carbon are identical, with (1*S*)-(+)-monopercamphoric acid (MPCA) affords the corresponding optically active oxaziridines **27** (*31*)

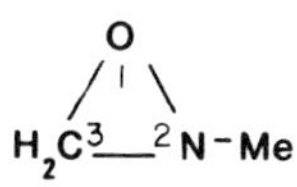

27

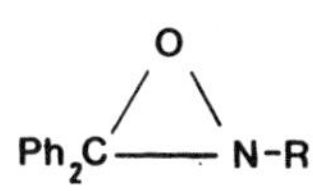

28 R=Me, Et, iPR, t Bu

and **28** (*29*) in apparently low optical yield (Table 3). It was possible to obtain 2-isopropyl and 2-*tert*-butyldiphenyloxaziridines (**28**) (R = *i*Pr and *t*Bu, respectively) in high optical purity by repeated crystallizations from ethyl ether (Table 3, entries 4 and 5).

In these oxidations the chirality of the peracid, and not the substituent on the nitrogen of the imine, determines the configuration of the oxaziridine obtained. For example, oxidation with (+)-(1*S*)-MPCA gives a series of oxaziridines (**28**) having negative rotations, whereas (−)-(*R*)-2-phenylpropionic acid gives the enantiomeric series (*29*). As observed in other oxidations using chiral peracids, lowering the temperature increases the enantioselectivity. The absolute configurations and enantiomeric purities of oxaziridines prepared in this manner are also solvent dependent. Solvents such as CCl_4 and $CHCl_3$ give higher enantioselectivity than alcohols. These effects are ascribed to variations in the "effective" bulk of the reactants by solvation of the ground or transition states (*31*).

Since unsymmetrically substituted imines can exist in both cis and trans forms (+)-MPCA oxidation of imine **29** affords *trans*- and *cis*-oxaziridines **30** and **31**, respectively, which have a diastereomeric relationship (Fig. 2) (*33*). The *trans*-oxaziridine **30** predominates when R = H. Multiple crystallizations of *trans*-2-*tert*-butyl-3-(*p*-nitrophenyl)oxaziridine (**30**, R = H,

TABLE 3
Properties of Oxaziridines Prepared by Oxidation of Imines with Monopercamphoric Acid

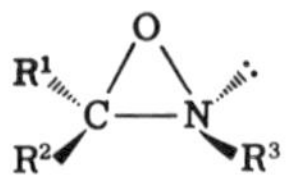

Entry	R^1	R^2	R^3	Temp. (°C)	% ee[a]	$[\alpha]_D$ (deg.)	$[\alpha]_{Dmax}$ (deg.)	Reference
1	H	H	*t*Bu	−78	0.6	−0.2	—	*31*
2	H	H	*t*Bu	−78	6.0	−2.07	−34.5	*35*
3	Ph	Ph	Me	−20	—	−42.2	—	*29*
4	Ph	Ph	*i*Pr	−20	10.3	−51.0	−192.0	*29*
5	Ph	Ph	*t*Bu	−20	29.8	−77.0	−258.0	*29*
6	—$(CH_2)_5$—		*t*Bu	3	—	−16.3	—	*28*
7	—$(CH_2)_5$—		Me	−78	14	5.4	38.5	*35*
8	Me	H	*t*Bu	−78	14	5.2	−37.1	*35*
9	Ph	H	*t*Bu	−78	21	−22.3	−85.6	*35*
10	H	Ph	*i*Pr	−78	35.0	—	—	*35*
11	Ph	H	Me	−78	40	—	—	*35*
12	*p*-BrPh	H	*t*Bu	−78	60	−44.6	−74.3	*35*
13	*p*-NO_2Ph	H	*t*Bu	−78	25	−30.9	−124.0	*35*
14	*p*-MeOPh	H	*t*Bu	−78	20	−17.8	−89.0	*35*
15	$PhCH_2$	H	*t*Bu	−78	20	−2.4	−12.2	*35*
16	*i*Pr	H	*t*Bu	−78	40.0	−13.1	−32.6	*35*
17	3-Cyclo-hexyl	H	*t*Bu	−78	58	−20.1	−34.7	*35*

[a] % ee determined by NMR; see (*35*).

R′ = *t*Bu) from *n*-pentane give a maximum rotation $[\alpha]_D$ −99° and of *cis*-2-*tert*-butyl-3-methyl-3-(*p*-nitrophenyl)oxaziridine (**31**, R = Me, R′ = *t*Bu), a maximum rotation $[\alpha]_D$ −160°. Thermal epimerization of *trans*/*cis*-**30**–**31** affords oxaziridines with different signs of rotation suggesting that the major enantiomers produced on oxidation have opposite absolute configurations at the ring carbon atoms. Furthermore, these results and those of Pirkle and Rinaldi (*35*) suggest that (+)-MPCA oxidation of imines produces oxaziridines having a bias in favor the *S* configuration at nitrogen (Table 3) (*35*).

Pirkle and Rinaldi prepared a large variety of optically active oxaziridines by oxidation of aldimines (RCH═NR′) with (+)-MPCA (Table 3, entries 2 and 7–17) (*34, 35*). The optical purity varied from a low of 6% ee to a high of 60% ee for 2-*tert*-butyl-3-(*p*-bromophenyl)oxaziridine. The extent of asymmetric induction appears to be related to the size of the group attached to the aldimine carbon. For R^3 = *tert*-butyl, the asymmet-

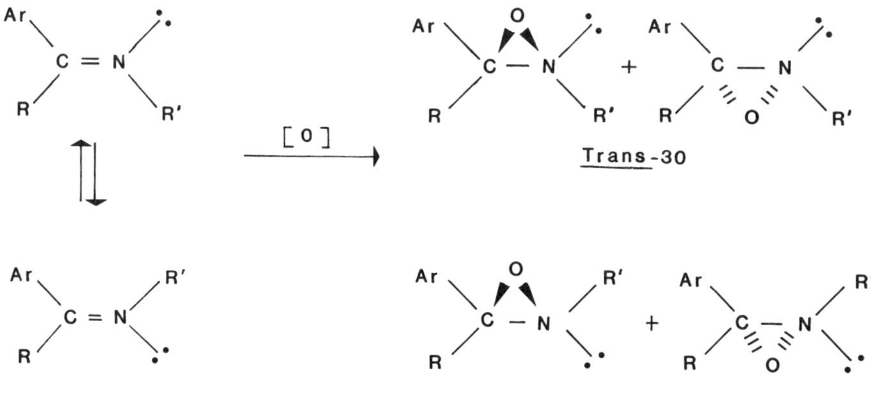

Ar = *p*-nitrophenyl
R = Me, H
R′ = alkyl

Fig. 2. Oxidation of cis and trans imines **29**.

ric induction increases for the series H < Me < benzyl < *i*Pr < 3-cyclohexyl = *p*-bromophenyl (6, 14, 20, 40, 58, and 60% ee, respectively) (Table 3, entries 2, 8, 15, 16, 17, and 12). The enantiomeric excesses were determined by NMR using chiral solvating agents (CSAs) (see below).

The older studies of Boyd and Montanari on asymmetric oxidation of imines with (+)-MPCA are not directly comparable with those of Pirkle and Rinaldi (*34–35*). The latter authors have shown that the usual method for preparation of this chiral peracid produces two isomers that give opposite stereochemical senses of asymmetric induction (*36*) (e.g., compare entries 1 and 2 in Table 3).

Bucciarelli, Forni, Mortetti, and Torre explored the oxidation of racemic (*R*,*S*)-*N*-diphenylmethylene-α-methylbezylamine (**32** and **33**, respectively) with chiral peracids under conditions of kinetic control (0.5 equiv of chiral peroxy acid) (*37*). With *S* peroxy acids a predominance of diastereomers **34** and **36** (88–99% yield) having negative rotations were obtained over **35** and **37**. The enantioselectivity (2–7% ee) was dependent

32 (α$\underline{R}$) 34 (2$\underline{S}$, α$\underline{R}$) 35 (2$\underline{R}$, α$\underline{R}$)

33 (α$\underline{S}$) 36 (2$\underline{R}$, α$\underline{S}$) 37 (2$\underline{S}$, α$\underline{S}$)

on the chirality of the peracid, the temperature, and the solvent. The difference in transition state energies between **36** and **37** was estimated to be 1.2 kcal/mol. The asymmetric bias observed in these oxidations is thought to be determined by the relative nonbonding interactions of the peracid and the imine carbon atoms in the transition state. The fact that asymmetric induction is dependent on the solvent and the temperature suggests a two-step Baeyer–Villiger type imine-to-oxaziridine oxidation mechanism rather than a concerted one-step mechanism.

b. Oxidation of Chiral Imines with Chiral and Achiral Peracids

Oxidation of symmetrically substituted ketimines prepared from (−)-(*S*)- or (+)-(*R*)-α-phenylethylamine [$R_2C{=}NCH(Me)Ph$] with *m*-chloroperbenzoic acid (MCPBA) produces unequal amounts (82–97 : 18–3) of oxaziridine diastereomers (*37, 38*). The composition of the reaction mixture was determined by NMR, and the oxaziridines were isolated by HPLC and chromatography. The absolute configuration of (−)-(2*S*)- and

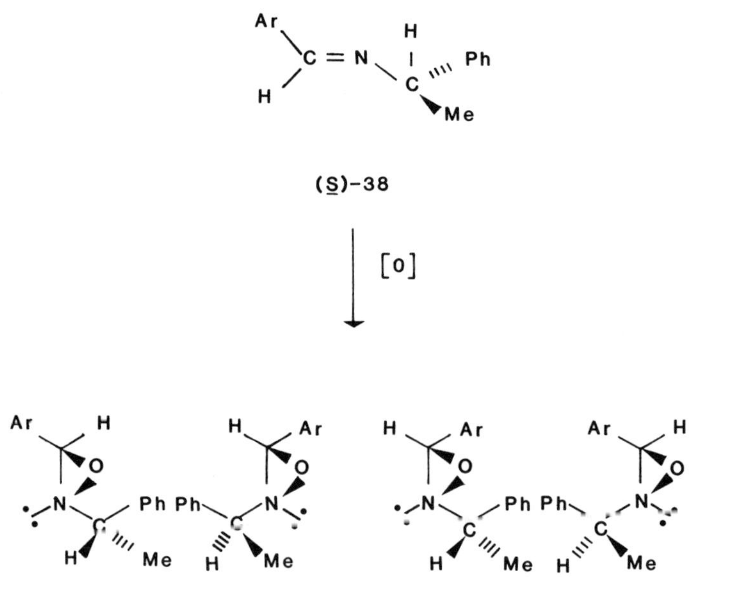

(R,R,αS)-39 (S,S,αS)-40 (R,S,αS)-41 (S,R,αS)-42

Fig. 3. Oxidation of (*S*)-4-bromo-*N*-benzylidiene-α-phenylethylamine (**38**) with MCPBA. Ar = 4-Bromophenyl.

(+)-(2*R*)-2-[(*R*)-α-methylbenzyl]-3,3-diphenyloxaziridine (**34**) (87%, $[\alpha]_D$ −97.0°) and **36** (13%, $[\alpha]_D$ +256.0°) was determined by an X-ray crystal structure of the former (*38*). The diastereoselectivity in these oxidations is dependent only on the temperature (*37*).

MCPBA oxidation of (*S*)-4-bromo-*N*-benzylidiene-α-phenylethylamine (**38**) gives all possible nonracemic diastereomers **39–42**, with those having the trans or *E* configuration being favored (Fig. 3, Table 4) (*35, 40*). An X-ray structure of the major diastereomer, **39**, established the absolute configuration as (2*R*,3*R*)-2-[(*S*)-1-phenylethyl]-3-*p*-bromophenyloxaziridine (*41*). Consequently, the configuration of **40** is 2*S*,3*S*,*S*. Since the thermal isomerization of oxaziridines involves only nitrogen inversion, the absolute configurations of **41** and **42** were ascertained by demonstrating that **42** isomerizes to **39** and that **41** isomerizes to **40** (*40*).

Similar results are reported for the MCPBA oxidation of optically active *N*-2-norbornylidine-α-phenylethylamine (*42*). Four oxaziridines are formed with the *E* isomers predominating. The asymmetric induction (trans : cis ratio) is 63 : 37 and is not observed on oxidation of the racemic imine.

Pirkle and Rinaldi investigated the oxidation of (−)-(*R*)- and (+)-(*S*)-**38** with both (+)-MPCA and MCPBA (Table 4) (*35*). As expected, very similar yields of diastereomers were obtained on oxidation of (−)-(*R*)- and (+)-(*S*)-**38** with MCPBA since the two experiments are mirror images of one other. A bias in favor of oxaziridines having a configuration at nitro-

TABLE 4
Oxidation of (−)-(*R*)- and (+)-(*S*)-4-Bromo-*N*-benzylidine-1-phenylethylamine (**38**) with MCPBA and (+)-MPCA[a]

Entry	Imine configuration	Peracid (temp., °C)	Diastereoisomers[b] (%)
1	(+)-(*S*)	(+)-MPCA (−78)	*R*,*R*,α*S* (54), *S*,*S*,α*S* (33), *R*,*S*,α*S* (9), *S*,*R*,α*S* (4)
2	(+)-(*S*)	MCPBA (0)	*R*,*R*,α*S* (61), *S*,*S*,α*S* (21), *R*,*S*,α*S* (13), *S*,*R*,α*S* (5)
3	(−)-(*R*)	(+)-MPCA (−78)	*S*,*S*,α*R* (61), *R*,*R*,α*R* (7), *S*,*R*,α*R* (28), *R*,*S*,α*R* (4)
4	(−)-(*R*)	MCPBA (0)	*S*,*S*,α*R* (58), *R*,*R*,α*R* (20), *S*,*R*,α*R* (16), *R*,*S*,α*R* (6)

[a] From (*35*).
[b] Determined by NMR on crude reaction mixture.

gen that is opposite to that of the group attached to nitrogen was noted by these authors as well as by Mostowicz and Belzecki (*40*). Similar trends were noted for chiral groups attached to the imine carbon (*40*).

Different amounts of diastereomers **39–42** were obtained on oxidation of (+)-(*S*)- and (−)-(*R*)-**38** with (+)-MPCA in contrast to the oxidations of these imines with MPCA (Table 3) (*35*). A bias in favor of *cis*- and *trans*-oxaziridines having the *S* configuration at nitrogen was observed. This trend may apply to other imine oxidations using (+)-MPCA (Table 1).

Only one oxaziridine diastereomer (**44**, >94%) is obtained on oxidation of (+)-camphorimines **43** with MCPBA (*35, 40, 43, 44*). Attack by peracid

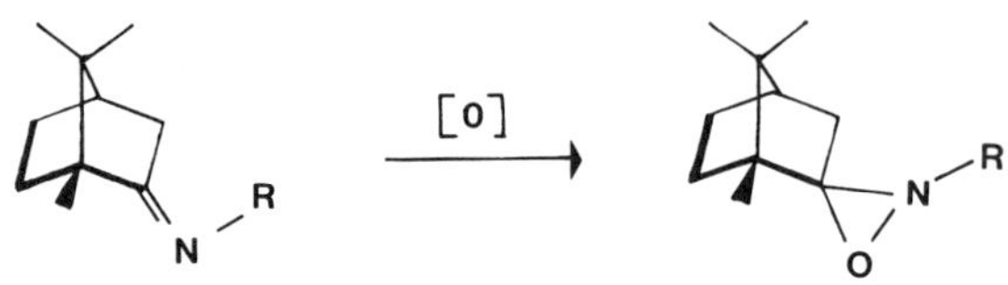

43 44

at the C—N bond from the exo direction gives oxaziridines having the (−)-*S*,*S* configuration, which is thought to predominate because hydride reagents reduce camphor from this direction.

c. Synthesis of Chiral Oxaziridines in Optically Active Media

Considerable decomposition occurs on heating racemic 2-*tert*-butyloxaziridine for 5 min at 148°C in the liquid-crystal cholesteryl benzoate (*45*). The residual oxaziridine, however, is optically active (20% ee) and has the (−)-(*S*) configuration. Preferential destruction of one of the enantiomers of 2-*n*-propyl-3-methyl-3-isobutyloxaziridine by brucine affords the $[\alpha]_D$ −3.94° oxaziridine (*46*). Rastetter *et al.* have shown that the reported oxygen transfer from this oxaziridine to form brucine *N*-oxide is incorrect (*47*).

Photolysis of nitrones **45** in the presence of (−)-(*R*)- or (+)-(*S*)-2,2,2-trifluoro-1-phenylethanol gave the corresponding optically active oxaziridine **46** (*48*). Maximum enantioselectivity (31%) was observed for ben-

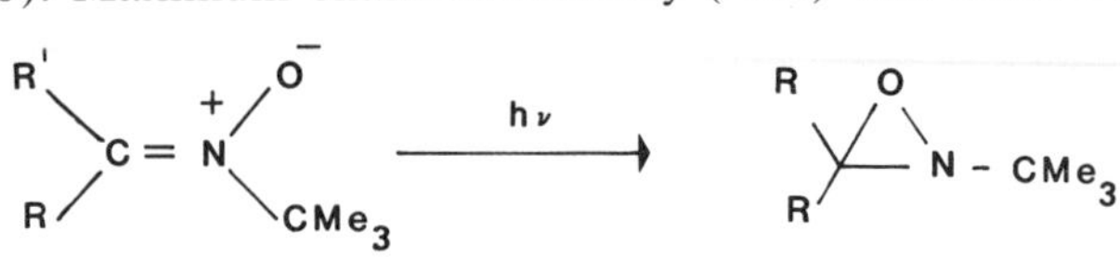

45 46

R = *p*-nitrophenyl
R′ = R = Ph; R′ = H

zophenone nitrones **45** (R = *t*Bu, R′ = Ph) when photolysis was carried out at −78°C. A higher asymmetric bias was observed for these nitrones (20–31%) as compared with the nitrone where R = *p*-nitrophenyl (3–6%). This effect was attributed by Boyd and Neill to interactions of the chiral alcohol with the nitro group.

The oxidation of benzophenoneimines ($Ph_2C{=}NR$, R = Me, *t*Bu) in the presence of chiral solvents such as (−)-(*R*)-menthol, (−)-(*R*)- and (+)-(*S*)-2-octanol, (−)-(*S*)-1-phenylethanol (**47a**), (+)-(*S*)-2,2,2-trifluoro-1-phenyethanol (**47b**), (+)-(*S*)-1-cyclohexylethanol (**48a**), (+)-(*S*)-1-cyclohexyl-2,2,2-trifluoroethanol (**48b**), and (−)-(*R*)-2,2,2-trifluoro-1-

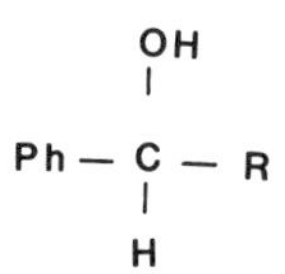

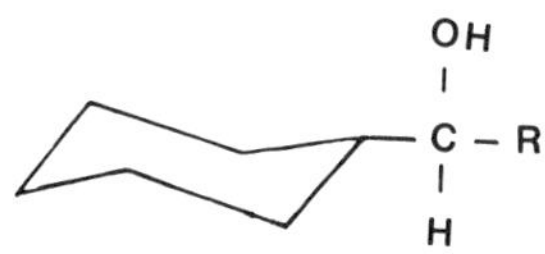

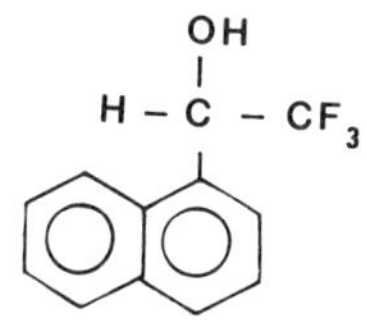

47a, R = Me
b, R = CF_3

48

49

(1-naphthyl)ethanol (**49**) at −40°C was investigated by Torre and coworkers (*49, 50*). Oxaziridines **28** (*R* = Me, *t*Bu) were obtained in 2 to 33% optical purity as summarized in Table 5 (*49, 50*). The enantioselectivity

TABLE 5
Oxidation of *N*-Methyl- and *N*-*tert*-Butylbenzophenoneimines in the Presence of Chiral Alcohols with MCPA at −40°C[a]

	Ph_2C—NR (O-bridged oxaziridine) % ee	
Chiral alcohol	R = Me	R = *t*Bu
(+)-(*R*)-**48b**	33.0	12.4
(+)-(*S*)-**48b**	2.6	2.0
(−)-(*S*)-**47a**	1.6	9.1
(+)-(*S*)-**47b**	22.2	19.2
(+)-(*S*)-**47b**	17.1	12.5
(+)-(*S*)-**47b**	13.6	7.5
(−)-(*R*)-**49**	25.3	27.8

[a] From (*49*).

increases with decreasing temperature and depends upon the structure of the imine and the structure and relative amounts of the chiral alcohol. Trifluoromethylcarbinols are 3 to 15 times more effective than methylcarbinols, and the stereochemistry of the product depends on the chirality of the solvent used. For example, oxaziridines with negative rotations are obtained with (*R*)-trifluoromethylcarbinols [or (*S*)-methylcarbinols] (Table 5). A fivefold excess of chiral alcohol **47b** results in higher enantioselectivity (22.2 versus 17.1% ee) for *N*-methylketimine oxidations. The asymmetric bias for *N*-methyloxaziridine was generally higher than for *N*-*tert*-butyloxaziridine (Table 5). Similar trends are observed for the oxidation of other imines, although the asymmetric bias is lower (*49*).

Torre *et al.* suggested that the optical yields and oxaziridine configuration are controlled by the strength and nature of the primary and secondary interactions between the chiral alcohol and the imine (*49*). Such interactions are stronger with the more acidic trifluoromethylcarbinol. The lower asymmetric bias noted for oxidation of the *N*-*tert*-butyl imines (Table 5) was interpreted in terms of steric inhibition of the imine–alcohol solvate by the bulky *tert*-butyl group on nitrogen.

The effect of chiral solvents on the oxidation of racemic α-methylbenzylimines [$R_2C{=}NCH(Me)Ph$, R = Me, Ph] by MCPBA, under conditions of kinetic control, was also investigated by these workers (*49*). As noted previously, the optical yields and absolute stereochemistry of the resulting oxaziridines depends on imine structure and the chirality of the carbinol. Best results were observed with (−)-(*R*)-2,2,2-trifluoro-1-(1-naphthyl)ethanol (**49**) and racemic imine **32/33** affording oxaziridine diastereomers **34** (9.6% ee) and **37** (14.2% ee). Oxidation of **32/33** in the presence of (*R*)-trifluorocarbinols **48b** and **49** gives oxaziridine diastereomers having the *S* configuration at nitrogen, whereas the 2-propanoneimine gives the opposite results. Solvate configurational correlation models **50** and **51**[1] were suggested by these authors to explain these

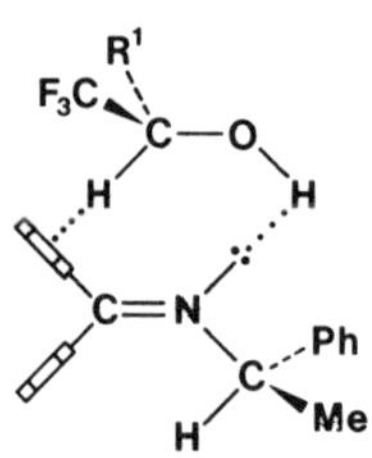

50

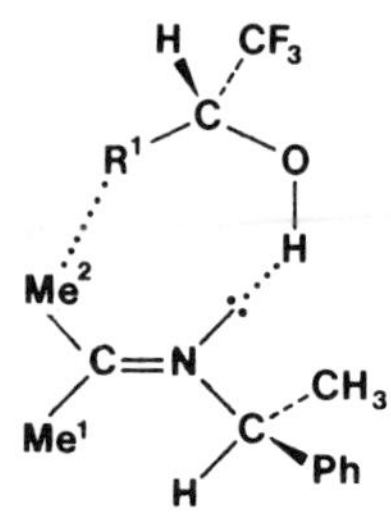

51

[1] Structures **50** and **51** from (*49*).

results. Attack by the peracid from the pro-*S* and pro-*R* directions affords oxaziridines having the *S* and *R* configurations, respectively, at the nitrogen atom and would be preferred on steric grounds.

d. Absolute Configuration and Enantiomeric Composition of Chiral Oxaziridines

Since there are no known stereoselective reactions of 2-alkyloxaziridines, numerous studies have been directed toward developing methods for ascertaining their absolute configuration and enantiomeric composition. Pirkle and Rinaldi used NMR and chiral arylperfluoroalkylcarbinols (chiral solvating agents, or CSAs) to determine the absolute configuration and enantiomeric composition of a variety of optically active oxaziridines (Table 3) (*34, 35*). These CSAs form chelate-like solvates with the oxaziridines. As a consequence of the diastereomeric nature of these solvent–oxaziridine complexes, they have nonidentical time-averaged NMR spectra because of the stereochemical dependence of the shielding effect on the oxaziridine groups by the aryl group of the CSA.

As shown in **52** and **53**,[2] oxaziridines have two basic sites, nitrogen and oxygen, with which to form primary hydrogen bonds with the CSA. Studies of oxaziridines with lanthanide shift reagents (LSRs), which select the same basic site as CSA, revealed that the site of coordination is very sensitive to steric effects. The primary binding site is nitrogen in unhindered oxaziridines **53**, but it changes to oxygen when bulky substituents are attached to the oxaziridine nitrogen or carbon atoms (*35*).

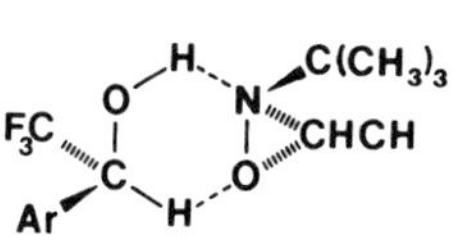

52

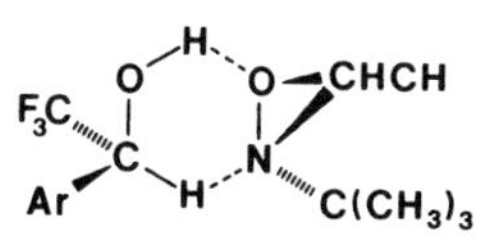

53

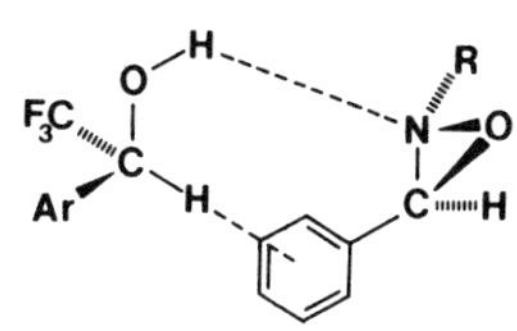

54

With (−)-(*S*,*S*)-2-*tert*-butyl-3-(1*R*,4*R*)-1,7,7-trimethylbicyclo[2.2.1]heptyloxaziridine (**44**, R = *t*Bu), an oxaziridine of known absolute configuration, the primary hydrogen-bonding site with (+)-(*S*)-2,2,2-trifluoro-1-(9-anthryl)ethanol is nitrogen. Opposite senses of nonequivalence for the *N*-*tert*-butyl (high field) and 10-methyl (low field) resonances were observed, as shown in **55**[3] (*35, 43*). It was argued that since CSAs are much less sensitive to steric effects than LSRs, the latter reagents are reliable for determining the absolute configurations of *N*-alkyloxaziridines (*35*).

[2] Structures **52–54** from (*35*). Copyright 1978 American Chemical Society.
[3] Structure **55** from (*35*). Copyright 1978 American Chemical Society.

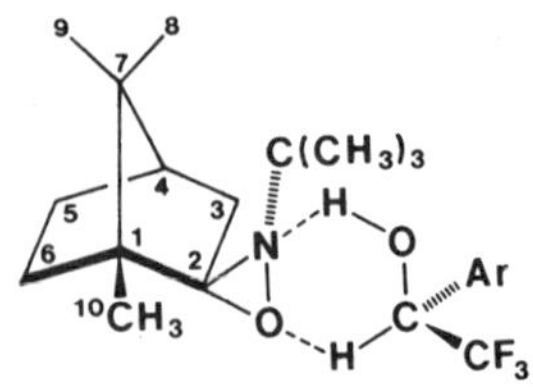

55

When an aromatic group is cis to the nitrogen lone pair, the primary hydrogen bonding of oxaziridines with CSA is at nitrogen with the secondary interaction at the aryl group, as shown in **54** (*35*). Such an interaction with the solvent is precluded for *cis*-3-aryloxaziridines, and solvation occurs as shown in **53** (*35*). Interpretation of the CSA–NMR data becomes less predictable when *C*-aryl groups are present because of additional basic sites for solvent complexation.

Mostowicz and Belzecki proposed a method for correlating the chirality of oxaziridines **39–42** with that of oxaziridines whose absolute configurations are known. This method is based on the difference in the molecular rotation $[\Delta M]_{586}$ between the oxaziridine and the imine (*40*).

e. Reactions and Properties of Chiral Oxaziridines

The configurational stability of nitrogen in oxaziridines has been studied using optically active oxaziridines. On heating, oxaziridines lose optical activity by racemization and by irreversible conversion to nitrones (*28, 29*). Both processes are first order and both occur for *N*-methyl-2,2-diphenyloxaziridine (R′ = *t*Bu, R = Ph). However, for the derivative where R′ = *t*Bu, only racemization is observed. The different entropies for thermal racemization ($\Delta S^\ddagger = -9.7$ eu) and for nitrone formation ($\Delta S^\ddagger = +5.0$ eu) suggest that the two processes are separate.

The thermal racemization of oxaziridines involves nitrogen inversion as demonstrated by the thermal epimerization of optically active *trans*-**30** and *cis*-**31** into *cis*-**31** and *trans*-**30**, respectively (*33*). The barriers to pyramidal nitrogen inversion in oxaziridines decrease as the size of substituents on nitrogen and carbon increases. This effect has been related to increases to nonbonded interactions that destabilize pyramidal nitrogen relative to the planar transition (*28*). A change from *N*-methyl to *N*-*tert*-butyl in **28**, for example, results in an 8×10^3 increase in rate (*29*). The lack of solvent effects on these barriers has been attributed to the low basicity of the nitrogen atom and to steric inhibition of solvation (*28, 29*). Photoracemization of oxaziridines involves C—O bond cleavage to form the nitrone RN(O)=CR_2, which rearranges back to the oxaziridine (*51*).

Photolysis of (2*R*,*S*)-6*a*(*e*)-*tert*-butyl-2-(α-methylbenzyl)-1,2-oxazaspiro[2.5.]octane (**56**) at 2537 Å gives the single lactam **57**, having the *S* configuration at the ring *tert*-butyl carbon. This result demonstrates that the C—C bond anti to the nitrogen lone pair is cleaved and supports the stereoelectronic theory. This theory attempts to explain regioselectivities observed in the photochemical and thermal rearrangements of oxaziridines as well as the photo-Beckmann rearrangement. Spirooxaziridine **56**

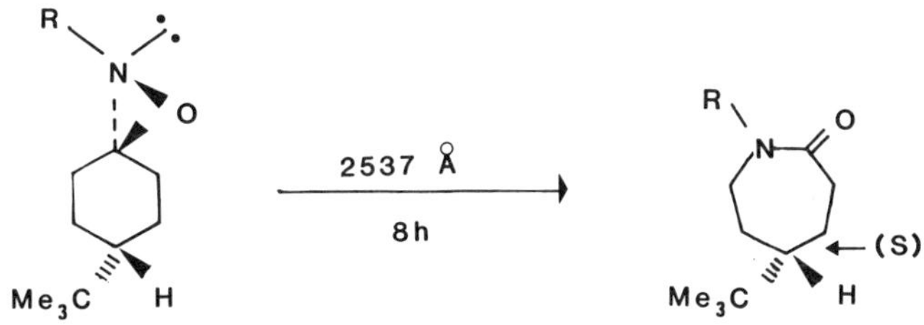

56 57

R = (−)-(*S*)-α-phenylamine

($[\alpha]_D$ −32.7°) was prepared by oxidation of the corresponding chiral imine, and its absolute configuration was determined by X-ray analysis (*52*).

The UV and CD spectra of camphoroxaziridines (−)-(*S*,*S*)-**44** (R = Me, Et, Me_2CH, Me_3C, $PhCH_2$) in ethanol and isooctane were determined (*44*). With the exception of **44** (R = Me_2CH), a positive pattern in the range 190–350 nm was observed. Ab initio SCF–CI calculations on (*R*)-oxaziridine, (2*R*,3*R*)-3-methyloxaziridine, and (2*S*,3*R*)-3-methyloxaziridine predict that these compounds have levorotatory first-transitions suggesting that the optical activity of oxaziridines may not be dominated by the configuration at nitrogen (*5*).

2. Chiral 2-Sulfonyloxaziridines

Optically active 2-sulfonyloxaziridines, a new class of oxaziridines discovered by Davis and co-workers (*53*), unlike 2-alkyloxaziridines have a highly electrophilic oxaziridine oxygen atom capable of undergoing a variety of oxygen-transfer reactions.

a. Preparation of Chiral 2-Sulfonyloxaziridines

The biphasic, basic oxidation of rapidly equilibrating cis–trans mixtures of sulfonimines **58** is highly stereospecific and affords only the *trans*-oxaziridine [Eq. (4)] (*53*). Davis and Stringer reported that oxidation of *N*-benzylidinebenzenesulfonamide (**58a**) with MCPBA in the presence of chiral-phase transfer reagents (−)-benzylquinidinium chloride and (+)-benzylchinchoninum chloride gives (−)- and (+)-2-benzenesulfonyl-3-phenyloxaziridine (**59a**) in low optical yield (1.4–3.1% ee) (*54*). The opti-

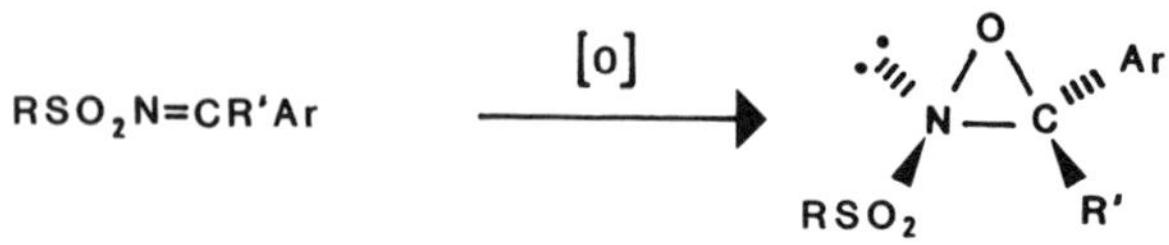

(4)

58 59

a R = Ar = Ph, R′ = H
b R = Me, Ar = Ph, R′ = H
c R = Ph, Ar = *p*-Tol, R′ = H
d R = Me, Ar = R′ = Ph

cal purity, determined using the chiral-shift reagent tris[3-(heptafluoropropyl)hydroxymethylene-*d*-camphorato]europium(III) derivative [Eu(hfc)$_3$], was improved from 1.4 to 10% ee by crystallization from ether/*n*-pentane.

Sulfonimines **58a–c** on oxidation with (+)-(1*S*)-MPCA under basic conditions (KOH, pH 9–10) in CH_2Cl_2–MeOH solution give the corresponding optically active 2-sulfonyloxaziridines **59a–c** in 17 to 19% ee with a predominance of the negative enantiomer (*55*). Multiple crystallizations of **59a–c** from ethyl ether give (−)-**59a–c** in greater than 95% optical purity. The lack of knowledge of the electronic transitions in these compounds, however, prevent determination of the absolute configuration based on the chiroptical properties (*55*).

All attempts to obtain optically active 2-methanesulfonyl-3,3-diphenyloxaziridine (**59d**), whose asymmetry would be due solely to the asymmetric nitrogen, by oxidation of **58d** with chiral peracids failed (*56*). Comparison of these results with those from the corresponding *N*-alkylimines lead Buciarelli *et al.* to conclude that the sulfonyl group acts to increase the lifetime of peracid–sulfonimine adducts **60a** and **b** (*56*). For *N*-alkylimines

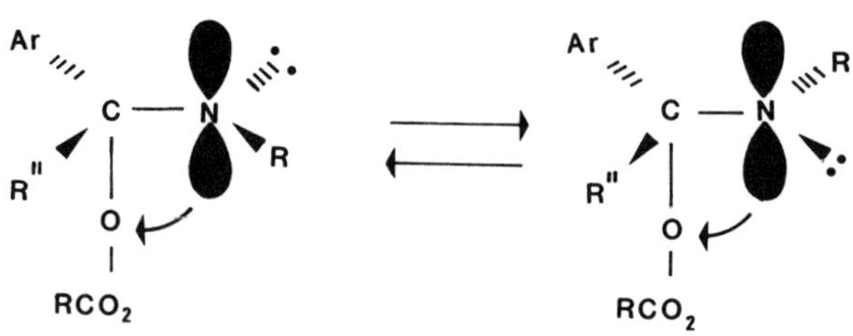

60a 60b

(R = alkyl), cyclization to the oxaziridine is faster than equilibration of **60a** and **b**. The sulfonyl group (R = SO_2R), in contrast, should increase the lifetime of **60a** and **b** so that rotation about the C—N bond is faster than cyclization, giving the more thermodynamically (*E*) oxaziridines **59a–c**.

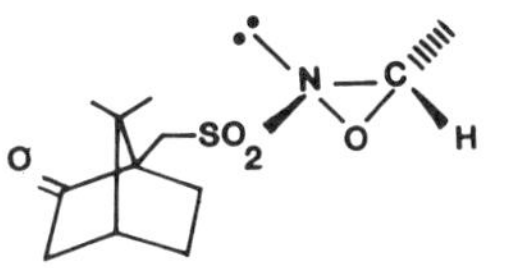

(−)-(S,S)-61

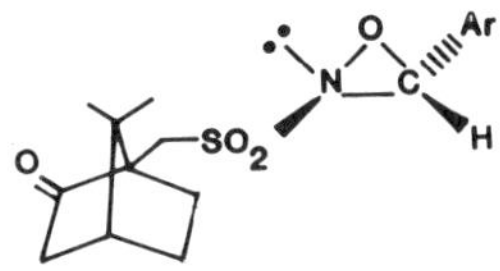

(+)-(R,R)-62

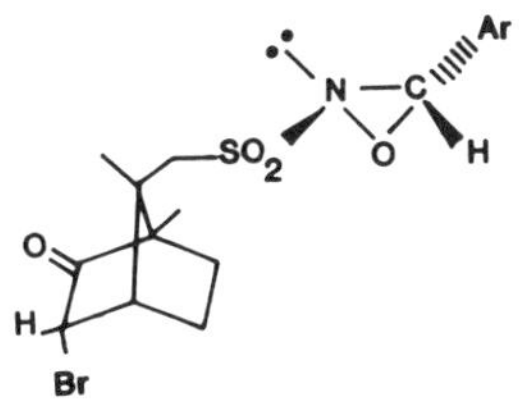

(−)-(S,S)-63

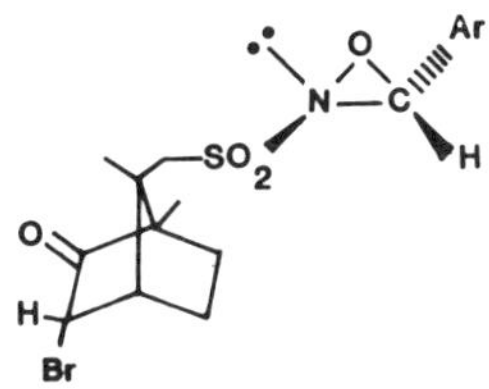

(+)-(R,R)-64

Oxidation of chiral sulfonimines ($R^*SO_2N{=}CHAr$) derived from *d*-10-camphorsulfonamide and *d*-α-bromo-π-camphorsulfonamide was observed by Davis and co-workers to give oxaziridine diastereomers **61/62** and **63/64**,[4] respectively, in excellent yield (*57*). The fact that oxaziridines **61/62**, derived from oxidation of the *d*-10-camphorsulfonimines, were obtained in a 65 : 35 ratio suggests that the carbonyl group plays a role in directing the peracid to a particular face of the C=N double bond. Only when the sulfonimine aryl group (Ar) is 2-chloro-5-nitrophenyl can the oxaziridine diastereomers be separated by crystallization (*57*).

An X-ray structure of (−)-**63** established that the absolute configuration of the oxaziridine three-membered ring is *S,S*, and therefore it is *R,R* in (+)-**64**. The configurations of *S,S* and *R,R* assigned to the three-membered rings of (−)-**61** and (+)-**62**, respectively, are based on the chiral recognition mechanism shown in Fig. 4 (*57*).

b. Asymmetric Oxidation of Sulfides and Disulfides

Oxidation of methyl and *tert*-butylphenyl sulfides at −50°C with (−)-2-sulfonyloxaziridines **60b** and **c** in methylene chloride gives optically active sulfoxides (Table 6) (*55*). The asymmetric induction was quite low (2–3.1% ee) and is both qualitatively and quantitatively similar to oxidations of the prochiral sulfides with chiral peracids (*55*).

[4] Structures **61–64** from (*57*). Copyright 1982 American Chemical Society.

TABLE 6
Asymmetric Oxidation of Prochiral Sulfide S Using Chiral 2-Sulfonyloxaziridines

$$ArSr + RSO_2N\text{—}CHAr' \rightarrow ArS(O)R + RSO_2N{=}CHAr'$$

Entry	Oxaziridine	Solvent	Temp. (°C)	Sulfoxide: Ar(O)Me Ar	Configuration	% ee	Sulfoxide: ArS(O)CMe$_3$ Ar	Configuration	% ee	Reference
1	(−)-**60a**	CH_2Cl_2	−50	Ph	*R*	3.1	Ph	*S*	2.0	55
2	(−)-**60b**	CH_2Cl_2	−50	Ph	—	0.0	Ph	*S*	2.8	55
3	(−)-**60c**	CH_2Cl_2	50	Ph	*R*	2.1	Ph	*S*	1.6	55
4	(−)-(*S*,*S*)-**61**	$CHCl_3$	25	*p*-Tol	*S*	17.0	*p*-Tol	*S*	11.0	57
5	(−)-(*S*,*S*)-**61**	$CHCl_3$	−50	*p*-Tol	*S*	31.0	*p*-Tol	*S*	9.2	57
6	(+)-(*R*,*R*)-**62**	$CHCl_3$	25	*p*-Tol	*R*	21.3	*p*-Tol	—	—	57
7	(−)-(*S*,*S*)-**63**	$CHCl_3$	25	*p*-Tol	*S*	10.3	*p*-Tol	*S*	11.0	57
8	(−)-(*S*,*S*)-**63**	$CHCl_3$	−50	*p*-Tol	*S*	21.4	*p*-Tol	*S*	12.5	57
9	(−)-(*S*,*S*)-**63**	PhH	25	*p*-Tol	—	—	*p*-Tol	*S*	9.9	57
10	(+)-(*R*,*R*)-**64**	$CHCl_3$	25	*p*-Tol	*R*	16.6	*p*-Tol	*R*	19.1	57
11	(+)-(*R*,*R*)-**64**	$CHCl_3$	−50	*p*-Tol	*R*	26.0	*p*-Tol	*R*	15.1	57

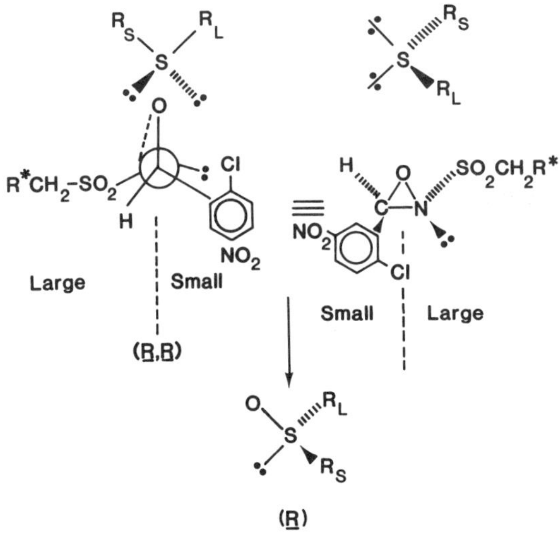

Fig. 4. Chiral recognition model. Ar = 2-Chloro-5-nitrophenyl.

The enantioselectivity for oxidation of methyl and *tert*-butyl-*p*-tolyl sulfides by oxaziridines **61–64**, in contrast, is five to eight times better than chiral peracids or **59a–c**. For example, the oxidation of methyl-*p*-tolyl sulfide with (−)-(*S*,*S*)-**61** at 25°C gives the sulfoxide in 17% optical yield. At −50°C the enantioselectivity increases to 31% (Table 6) (*57*). In all cases studied the sulfoxide configuration was controlled by the configuration of the oxaziridine three-membered ring and was independent of the solvent. Thus (−)-(*S*,*S*)-**61/63** and (+)-(*R*,*R*)-**62/64** give sulfoxides having the *S* and *R* configuration, respectively. These results are in marked contrast to oxidations with chiral peracids where both the absolute configuration and the asymmetric bias of the product are solvent dependent (*57*).

A chiral recognition mechanism based on a consideration of steric interactions in the diastereomeric transition states (Fig. 4) was proposed by Davis *et al.* to explain these results (*57*). In the region of the oxaziridine three-membered ring, the camphorsulfonyl group is considered to be larger than the 2-chloro-5-nitrophenyl group. Consequently, the preferred diastereomeric transition state for sulfide oxidation is the one in which the enantiotopic electron pair on sulfur attacks the electrophilic oxaziridine oxygen in such a way that the large (R_L) and small (R_S) groups of the substrate (R_L–S–R_S) face the small and large regions of the oxaziridine three-membered ring, respectively.

This model (Fig. 4) was also shown to work for the oxidation of disulfides to optically active thiosulfinates. (−)-(*S*)-*p*-Tolyl-*p*-toluenethiosulfinate [*p*-Tol—S(O)—S—Tol-*p*], was obtained in 2.1% ee on oxidation of

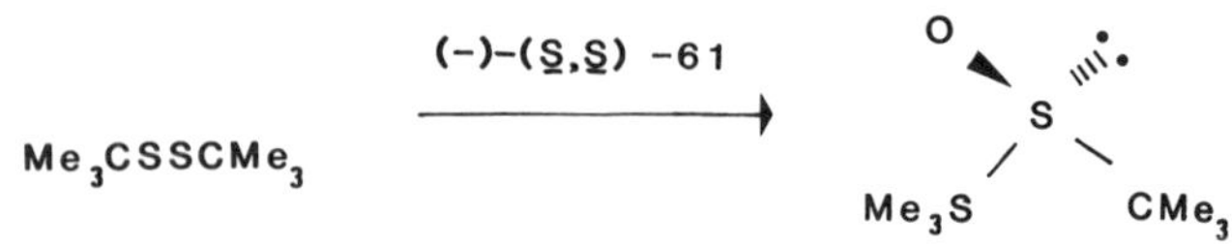

(−)-(S) 13.8%ee

the disulfide with (−)-(*S*,*S*)-**61** (*57*). Similarly, oxidation of di-*tert*-butyl disulfide gave (−)-*tert*-butyl-2-methyl-2-propanethiosulfinate in 13.8% ee, predicted to have the *S* configuration (*57*).

The differences in size of the groups attached to the oxaziridine three-membered ring and substrate, or the group size difference (GSD) effect, were shown to play an important role in determining the absolute configuration and the asymmetric bias. The results summarized in Table 7 show that the differences in size of the groups attached to the substrate increases as the enantioselectivity increases, from a minimum of 1.1% ee for methylbenzyl sulfide to a maximum of 46.0% ee for methyl-2,4,6-tri-*tert*-butylphenyl sulfide (*57*).

c. Oxidative Kinetic Resolution of Sulfoxides

The oxidative kinetic resolution of methyl-*p*-tolyl sulfoxide and *tert*-butylphenyl sulfoxide, using 0.5 equiv of diastereomeric oxaziridines **61** to **64**, was investigated by Davis and Billmers (*58*). In (−)-(*S*,*S*)-**63** and (+)-(*R*,*R*)-**64**, the configuration of the recovered sulfoxide is controlled by the configuration of the oxaziridine three-membered ring and can be pre-

TABLE 7
Effect of Group Size Difference (GSD) on the Oxidation of Sulfides to Sulfoxides by (−)-(*S*,*S*)-**61**[a]

Sulfide	% ee	Configuration
$PhCH_2SMe$	1.1	*S*
p-Tol-$SCMe_3$	11.1	*S*
p-Tol-SMe	17.0	*S*
p-Tol-$SCHMe_2$	20.8	*S*
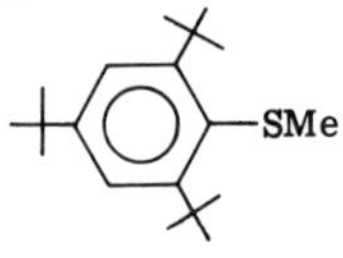	46.0	*S*

[a] From (*57*). Copyright 1982 American Chemical Society.

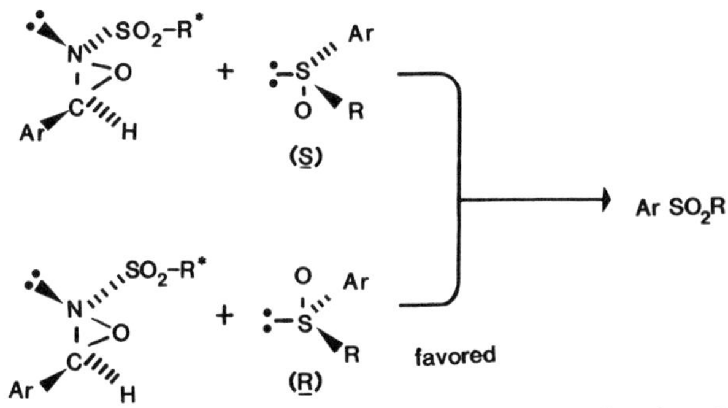

Fig. 5. Oxidative kinetic resolution of sulfoxides by chiral 2-sulfonyloxaziridines **61**–**64**.

dicted using steric arguments similar to those used for sulfide to sulfoxide asymmetric oxidations (Fig. 5). The (*S*,*S*)-oxaziridine preferentially oxidizes the (+)-(*R*)-sulfoxide enantiomer resulting in enrichment of the (−)-*S* enantiomer. Both the configuration and the asymmetric induction (17–22% ee) were insensitive to solvent polarity. Multistep kinetic resolution using (+)-(*R*,*R*)-**64** gave (+)-(*R*)-methyl-*p*-tolyl sulfoxide in 26% ee. This procedure involved oxidation of methyl-*p*-tolyl sulfide with 1 equiv of **64** followed by oxidation of the resulting sulfoxide with an additional 0.5 equiv.

The oxidative kinetic resolution of methyl-*p*-tolyl sulfoxide with oxaziridine (−)-(*S*,*S*)-**61**, in contrast to **63** and **64**, proved to be dependent on the polarity of the solvent (*58*). In nonpolar solvents such as benzene the (+)-(*R*)-sulfoxide enantiomer was preferentially formed, whereas in polar solvents the reverse is observed. A dipole–dipole interaction between the camphorcarbonyl and -sulfinyl group **65**[5] was proposed by Davis and

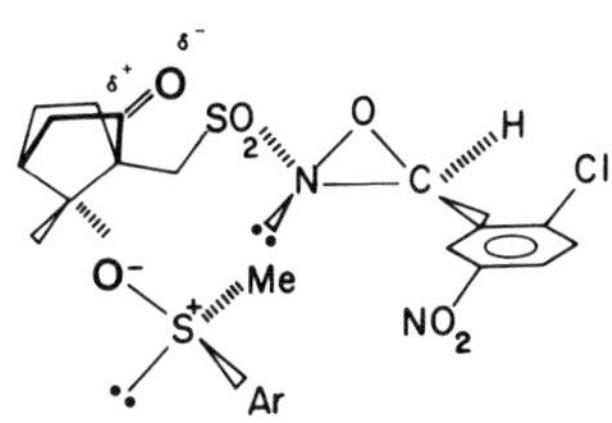

65

Billmers to explain these solvent effects. In nonpolar solvents the interaction shown in **65** governs the chiral recognition, but as the solvent polarity increases, this interaction becomes less important, and the chiral recognition is again controlled by steric factors (Fig. 4).

[5] Structure **65** from (*58*). Copyright 1983 American Chemical Society.

d. Asymmetric Oxidation of Selenides

The asymmetric oxidation of selenides using chiral 2-sulfonyloxaziridines were investigated by Davis *et al.* (*59*). Oxidation of methylphenyl selenide (**66**), under anhydrous conditions, by (−)-(*S*,*S*)-**61** and (+)-(*R*,*R*)-**62** afforded optically active (−)-(*S*)- and (+)-(*R*)-methylphenyl selenoxide (**67**) in 8 and 9% ee respectively (*59*). These selenoxides were isolated in low yield (20%) by sublimation, and the enantiomeric purity and absolute configuration were determined using (+)-2,2,2-trifluoro-1-(9-anthryl)ethanol. The absolute configuration of methylphenyl selenoxide is that predicted by the sulfoxide model (Fig. 4). The lower asymmetric induction observed for the oxidation of **66** compared to methyl-*p*-tolyl

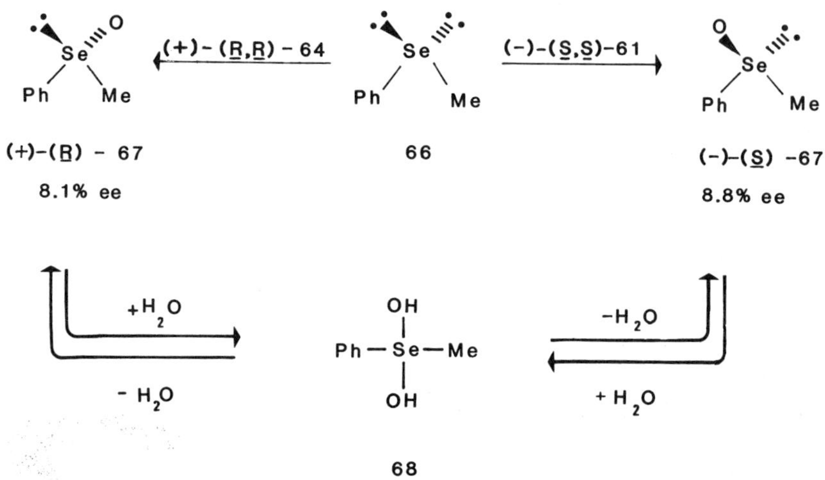

sulfide oxidations (Table 7) may be related to longer C—Se and Se—O bond lengths. Longer bond lengths may reduce the importance of the steric interactions in the diastereomeric transition states that govern the asymmetric induction.

The configurational lability of selenoxides was shown to be the result of acid-catalyzed hydrate formation (**68**) (*59*). (−)-(*S*)- and (+)-(*R*)-methylphenyl selenoxide (**67**) were completely racemized in less than 10 sec in the presence of 1 equiv of water. Difficulty in forming the tetracoordinate hydrate in (−)-(*R*)- and (+)-(*S*)-2,4,6-triisopropylphenyl selenoxide, prepared using chiral sulfonamides ($R^*SO_2NH_2$) and a kinetic resolution procedure, is thought to be the reason for its much greater stability in the presence of water ($t_{1/2}$ racemization = 30 h) (*60*).

(*E*)-Cinnamylphenyl selenide (**69**) gives (+)-(*S*)- and (−)-(*R*)-1-phenylallyl alcohol (**70**) in 8.7 and 12.8% optical yield on oxidation with (−)-(*S*,*S*)-**61** and (+)-(*R*,*R*)-**64**, respectively (*58*). A concerted or nearly concerted

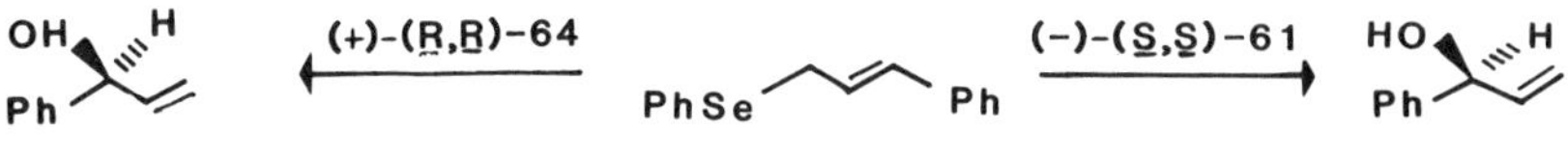

(R)-70 69 (S)-70

[2,3]sigmatropic rearrangement of the intermediate allylic selenoxide **71** through a five-membered cyclic transition of a doubly suprafacial migration is consistent with these results (Fig. 6). Note that the intermediate allylic selenoxide **71** has the configuration predicted by the oxaziridine–sulfoxide chiral recognition mechanism (Fig. 4).

e. Asymmetric Epoxidation of Alkenes

Diastereomeric 2-sulfonyloxaziridines (−)-(*S*,*S*)-**63** and (+)-(*R*,*R*)-**64** when heated with alkenes for 1 to 4 h at 60°C afford the corresponding optically active alkenes [Eq. (5)] (Table 8) (*61*). The enantioselectivity is two to five times better than similar oxidations using chiral peracids. For example, (1*R*,2*R*)-1,2-epoxy-1-phenylcyclohexane was obtained in 40%

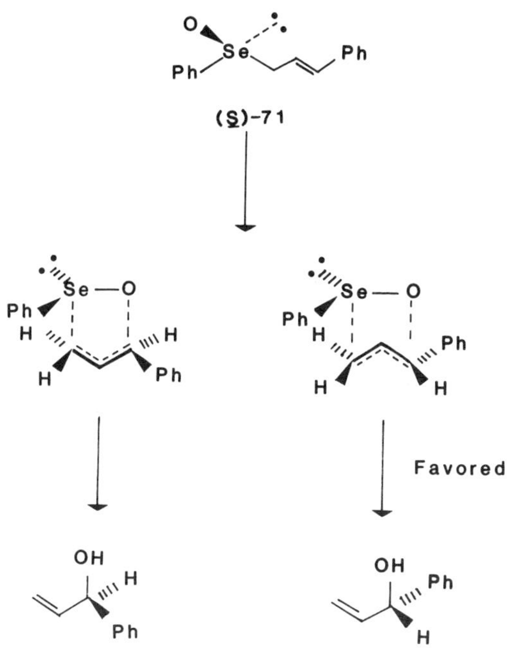

Fig. 6. [2,3]Sigmatropic rearrangement of allylic selenoxide (*S*)-**71**.

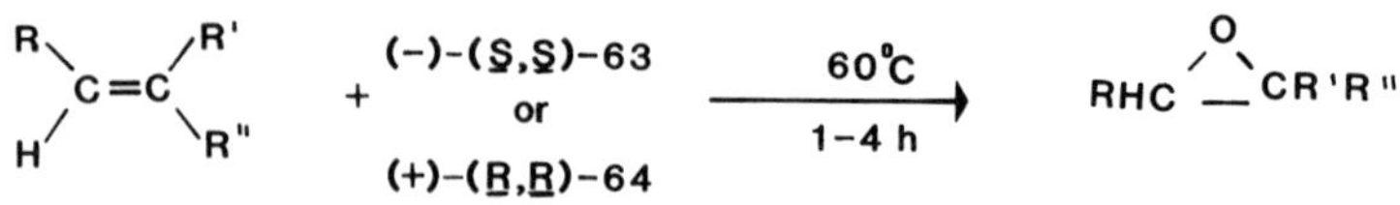

(5)

ee on oxidation of 1-phenylcyclohexene with (+)-(*R*,*R*)-**64**. As observed in other asymmetric oxidations using **61** to **64**, the configuration of the oxaziridine three-membered ring controls the stereochemistry of the product (Table 8).

Molecular orbital calculations (STO-3G/4-31G) by Bach and Wolber suggest that the planar transition state geometry shown in **72** is favored by

TABLE 8
Asymmetric Epoxidation of Olefins Using Chiral 2-Sulfonyloxaziridines (−)-(*S*,*S*)-**63** and (+)-(*R*,*R*)-**64** at 60°C in $CHCl_3$[a]

Oxaziridine[b]	Olefin	% Yield	Epoxide configuration	% ee
(−)-(*S*,*S*)-**63**	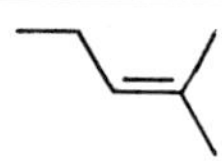	35	(+)-*S*	17.5
(+)-(*R*,*R*)-**64**		30	(−)-*R*	17.5
(−)-(*S*,*S*)-**63**	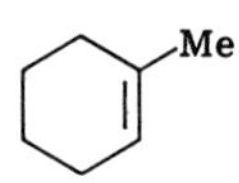	58	(+)-1*S*,2*R*	12.7
(+)-(*R*,*R*)-**64**		60	(−)-1*R*,2*S*	11.8
(+)-(*R*,*R*)-**64**	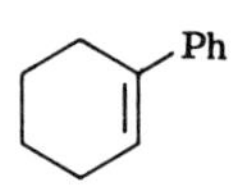	70	(+)-1*R*,2*R*	40.0
(−)-(*S*,*S*)-**63**	(+) 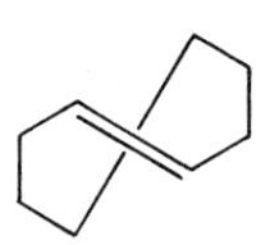	32	(+)-1*S*,8*S*	14.7
(+)-(*R*,*R*)-**64**		37	(−)-1*R*,8*R*	13.9
(−)-(*S*,*S*)-**63**	PhCH=CH_2	50	(−)-*S*	15.9
(+)-(*R*,*R*) **64**		57	(+)-*R*	15.9
		43[d]	(+)-*R*	15.7
(−)-(*S*,*S*)-**63**	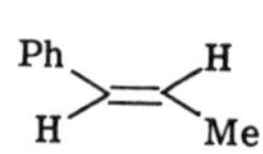	81	(−)-*S*,*S*	26.8
(+)-(*R*,*R*)-**64**		80	(+)-*R*,*R*	27.7
(−)-(*S*,*S*)-**63**	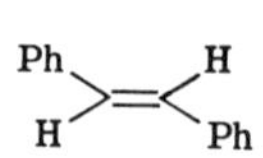	66	(−)-*S*,*S*	30.3
(+)-(*R*,*R*)-**64**		86	(+)-*R*,*R*	34.9
		62[d]	(+)-*R*,*R*	31.5

[a] From (*62*). Copyright 1983 American Chemical Society.
[b] Oxaziridines were greater than 95% optically pure.
[c] Isolated yields by TLC or gas chromatography.
[d] Benzene solvent.

3 kcal/mol over the spiro geometry in **73**[6] (*61, 62*). This can be verified experimentally using oxaziridines **63** and **64** (*62*). The reasonable assumption is made that the chiral recognition is largely controlled by minimizing nonbonded interactions in the transition state. If the geometry is planar (**72**), attack by the oxaziridine (−)-(*S*,*S*)-**63** is favored to occur on the *Si* and *Si*,*Si* enantiotopic faces of the C=C double bond to give epoxides having the *S* and *S*,*S* configurations. Conversely, if the transition state is spiro (**73**), (−)-(*S*,*S*)-**63** will afford epoxides having the *R* and *R*,*R* configu-

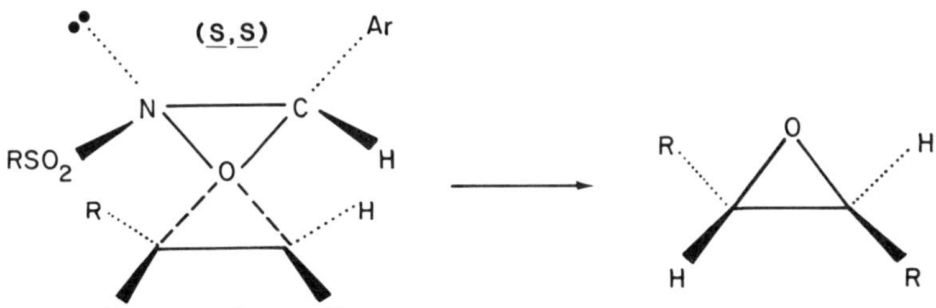

72

(S,S)

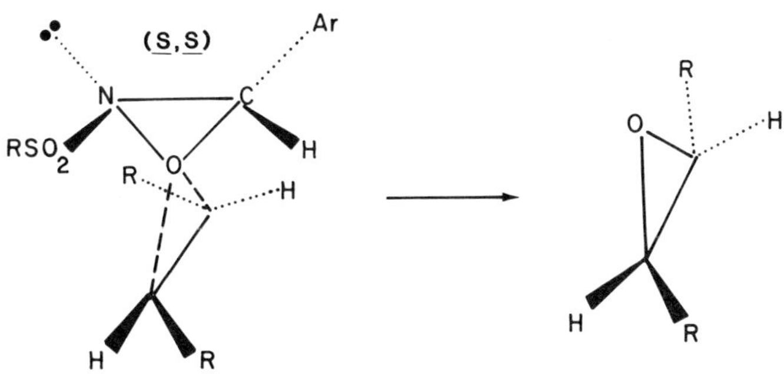

73

(R,R)

rations. Consistent with the theoretical predictions is that (*S*,*S*)-**63** and (*R*,*R*)-**64** give only epoxides having the *S*,*S* and *R*,*R* configurations, respectively (Table 8) (*62*).

The increase in asymmetric bias for the alkene series styrene, *trans*-α-methylstyrene, and *trans*-stilbene (15.8, 26.8, and 30.3% ee, respectively) can also be explained in terms of planar transition state **72**. As the sizes of the groups attached to the C=C double bond increase, attack of the

[6] Structures **72** and **73** from (*62*). Copyright 1983 American Chemical Society.

oxaziridine on the face of the C=C bond where these groups are eclipsed with the small oxaziridine C—H and lone pair is increasingly favored.

f. Asymmetric Oxidation of Enolates

2-Sulfonyloxaziridines **61** to **64** have also been used to asymmetrically oxidize enolates. Treatment of **74** with 4 equiv of KH at −78°C followed by addition of (−)-(*S*,*S*)-**61** or (+)-(*R*,*R*)-**62** gives (+)-(*R*)- and (−)-(*S*)-kjellmanianone (**75**) in 33.0 and 36.5% ee, respectively (*63*). The relationship of the camphorcarbonyl group to the active site in **61** and **62** appears to be important in establishing the transition state geometry for delivery of the hydroxyl group. Oxaziridines (−)-(*S*,*S*)-**63** and (+)-(*R*,*R*)-**64** give (+)- and (−)-**75** in much lower optical purity (8 and 12% ee, respectively).

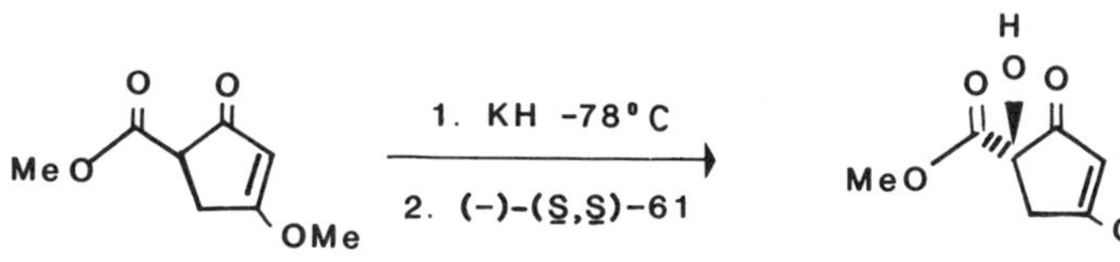

74 (+)-(R)-75

33% ee

In these oxaziridine diastereomers, the camphorcarbonyl group is apparently too far removed from the oxygen active site to play an important role in chiral recognition.

D. Chiral N-*Alkoxyamines*

In 1969 Muller and Eschenmoser demonstrated that chiral nitrogen attached to an electronegative oxygen can exist in five-membered ring systems (*64*). Two diastereomeric *N*-methoxy-3,3-dimethoxycarbonyl-5-cyano-1,2-oxazolidines **76** and **77** were obtained in 1,3-dipolar cycloaddition of nitronic ester **75** to acrylonitrile.

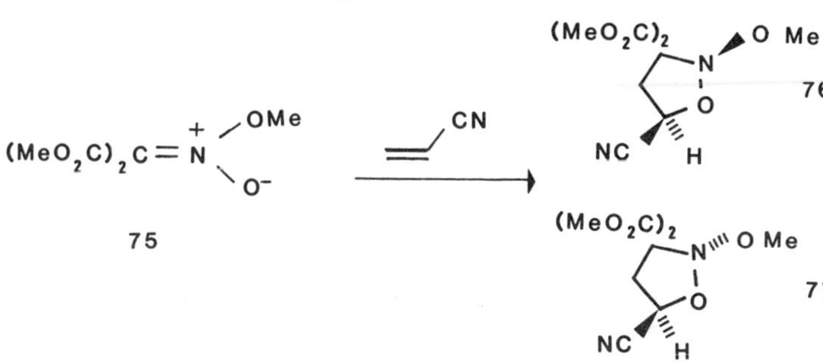

75 76 77

High-pressure (14,000 atm) induced cycloaddition of *Z* and *E* nitronic esters to 16-dehydro-20-oxo steroids **78** affords all four possible isomers of steroido[16α,17α-*d*]tetrahydro-1′,2′-oxazoles (**79a–d**) resulting from exo and endo attack of the dipolarophile. (*65, 66*). The structures of the cycloadducts were determined by PMR. It was shown that the thermal conversion of the less stable adducts **79b** to **79c** and **79d** to **79a** involved a

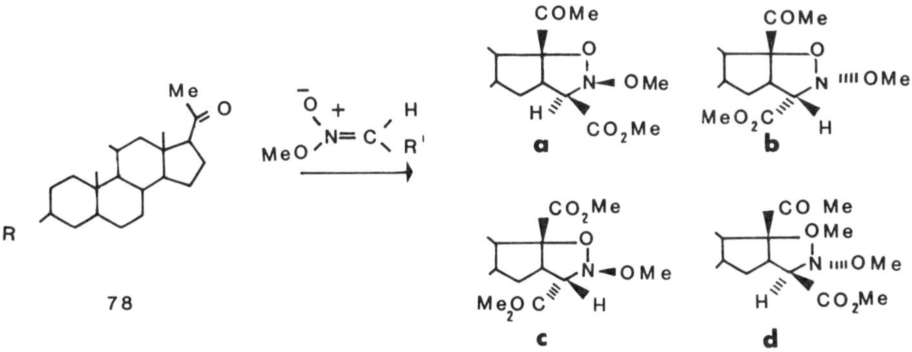

78

79

simultaneous nitrogen inversion and isoxazolidine cycle conformational change.

In 1979 Kostyanovsky and co-workers resolved the enantiomers of 1-alkoxyisoxazolidine-3,3-dicarboxylic ester (**80**), which contains only a nitrogen chiral center (*67*). A procedure similar to that previously used for *N*-methoxyaziridines **7** was utilized (Fig. 1). This involved resolution of the monoacid enantiomers via the (−)-(*S*)- and (+)-(*R*)-phenylethylammonium diastereomer salts, which were transformed into the bis amides (+)-(*S*)-**81** and (−)-(*R*)-**81**, respectively. Optical purities were determined by

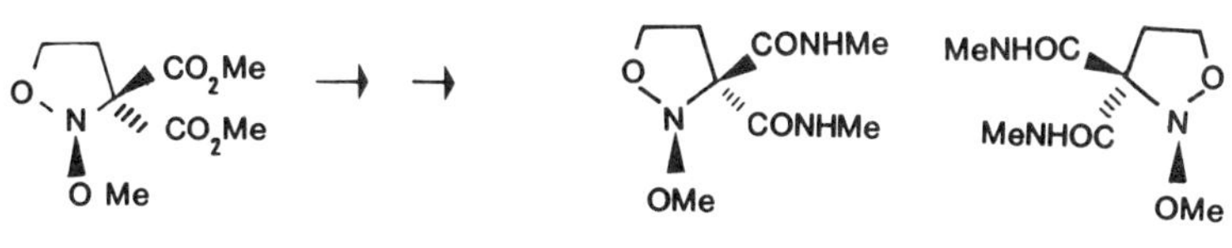

80 (+)-(2S)-81 (−)-(2R)-81

PMR, and the absolute configurations by an X-ray analysis of the (+)-(2*S*,3*R*)-phenylethylammonium salt.

The configurational stability of acyclic amines having two electronegative heteroatoms attached to nitrogen has been investigated by Kostyanovsky *et al.* (*68*). β-(*N*-Methoxy-*N*-benzoxyamino)isovaleric acid methyl ester **82** was prepared by treatment of the corresponding *N*-chloro derivative with benzyl alcohol in the presence of triethylamine. Partial resolution of **82** into enantiomers was accomplished via formation of diastereomeric salts with (*S*)- and (*R*)-phenylethylamine. NMR chiral shift reagent experiments proved **82** to be 16% optically pure. The half-life for

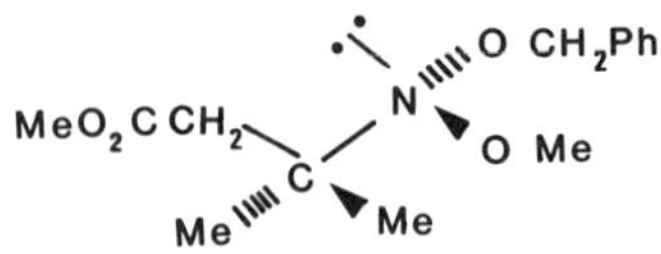

82

complete racemization of (+)- and (−)-**82** in methanol at 20°C was 5.18 h ($\Delta G\ddagger$ = 23.5 kcal/mol).

III. Chiral Tetravalent Nitrogen

A. *Chiral Quaternary Ammonium Salts*

1. Synthesis and Structure

Quaternary ammonium salts ($R_4N^+X^-$) are known to possess a tetrahedral structure in which four groups (R) are directly bonded to nitrogen (*2*). Electrostatic attraction is responsible for the association of the anion (X^-) with the cationic nitrogen core. One of the earliest observations of an optically active quaternary ammonium salt was made by Le Bel in 1891 (*69*). He found that *Penicillium glaucum* would interact with a solution of methylethylisopropylisobutylammonium chloride to preferentially destroy some of the dextrorotary isomer affording a solution with a specific rotation of −7 to −8°.

Since that time, many quaternary ammonium salts have been resolved, usually via crystallization of their *d*-10-camphorsulfonates or *d*-α-bromo-π-camphorsulfonates. This earlier work has been summarized by Shriner *et al*. (*2*).

The absolute configuration of (+)-(*R*)-benzylmethylphenylpropylammonium iodide was assigned by Horner *et al*. on the basis of the known configuration of (+)-(*S*)-benzylmethylphenylpropylphosphonium bromide using the quasiracemate method (*70*).

2. Reactions

a. The Stevens Rearrangement

Hill and Chan were the first to demonstrate the transfer of asymmetry from nitrogen to carbon in their study of the Stevens rearrangement of (+)-(*R*)-allylbenzylmethylphenylammonium iodide (**83**) (*71, 72*). When treated with potassium *tert*-butoxide in dimethylsulfoxide, **83** produced, in addition to *N*-methylaniline (35%) and 3-(*N*-methylanilino)-4-phenylbutene-1, a 15% yield of (−)-(*S*)-3-(*N*-methylanilino)-4-phenylbutene-1 (**84**,

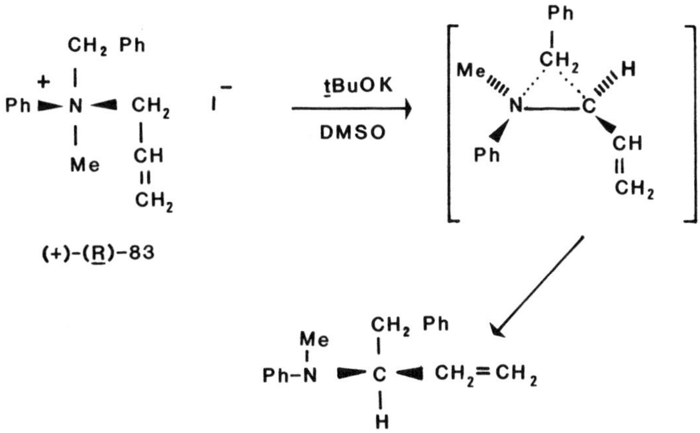

(+)-(R)-83

(−)-(S)-84

[α] −18.4°). The extent of asymmetric induction is unknown but probably substantial. Interestingly, the diastereomeric transition state leading to **84** would not be predicted on steric grounds. An electrostatic interaction between the cis phenyl and vinyl groups was suggested by the authors.

Brewster and Jones observed that heating (−)-4,4′-dimethoxy-1,1′,3,3′-tetrahydrospiro(isoindole-2,2′-isoindolium)bromide (**85**) with sodium hydroxide in diglyme gives (−)-**86** in low yield (7–11%) (*73*). The position of

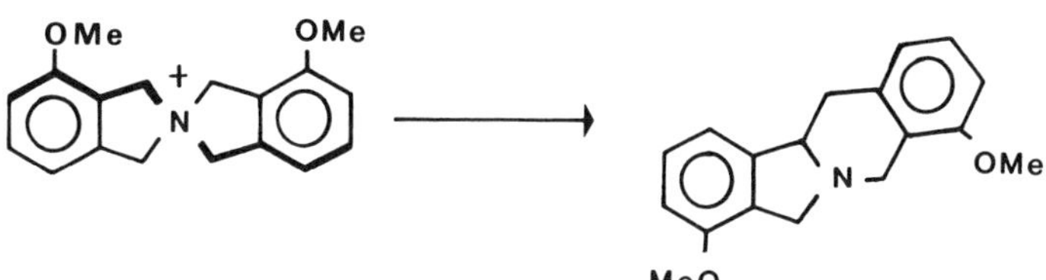

(−)-(R)-85 (−)-(S)-86

the methoxy groups was determined by NMR, and the absolute configuration of the asymmetric carbon center was shown to be *S* by ozonolysis of

86 to an aspartic acid derivative. Since the Stevens rearrangement occurs with retention of configuration, that is, bond breaking and bond making occur on the same side of the ring, the *R* configuration was assigned to salt **85**.

b. The Hofmann Elimination

The thermal decomposition of (−) and (+) quaternary hydroxides **87** at 55 to 80°C was shown by Cope *et al.* to gave *cis*- and *trans*-cyclooctenes (**88**) in a 60 : 40 ratio (*74*). Only the *trans*-(−)-(*R*)-cyclooctene (**88**) proved to be optically active (1.4% ee). The (+)-(*S*)-enantiomer **88** (2.4–2.7% ee) was produced when **87** was treated with KNH_2 at −40°C in liquid ammo-

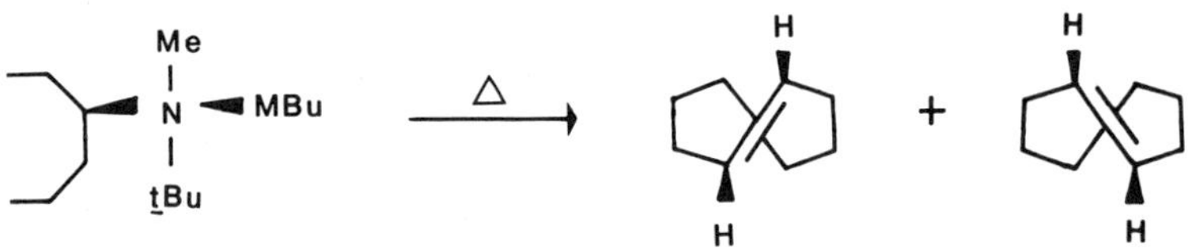

(−)-87 (−)-(R)-88 (+)-(S)-88

nia. Elimination probably occurs via an E_2-type mechanism, but the very low asymmetric induction precludes detailed analysis of the chiral recognition mechanism: the difference in energy between an asymmetric bias of −1.4 and +2.7 is only 0.1 kcal/mol (*75*).

B. Chiral Amine Oxides

1. Synthesis and Structure

Amine oxides ($R_3N^+O^-$), like quaternary ammonium salts, possess a tetrahedral structure in which an oxygen atom is bonded to nitrogen via a semipolar bond. Tertiary amine oxides with three different substituents can exist as configurationally stable enantiomers.

Optically active amine oxides are most commonly prepared by resolution. A large number of amine oxides were resolved by Meisenheimer using *d*-α-bromo-π-camphorsulfonic acid (*76–79*). The enantiomers of methylphenyl-4-methylcyclohexylamine *N*-oxide (*80*) and methylneopentyl-4-methylcyclohexylamine *N*-oxide (*81*) have also been resolved using (−)-dibenzoyltartaric acid.

Asymmetric oxidation of unsymmetric tertiary amines also gives optically active oxides, albeit in low optical yields. For example, when *N*-

methyl-1,2,3,4-tetrahydroquinoline (Kairoline) is treated with (+)-monopercamphoric acid at −70°C in chloroform ether, the *N*-oxide **89** was obtained (as the hydrochloride) with a specific rotation of $[\alpha]_{546}$ +0.27° (0.6% ee) (*82*). Similar oxidation at low temperatures of a series of *N*-alkyl-*N*-methylanilines [RN(Me)Ph; R = Et, *i*Pr, and *t*Bu] produced amine oxides with very small rotations ([α] +0.8° from 589 to 450 nm), whereas no rotations were observed for oxidations carried out at room temperature (*83*).

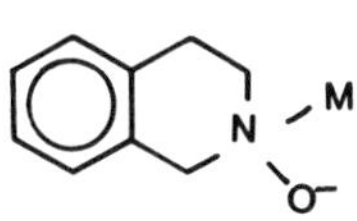

(−)-89

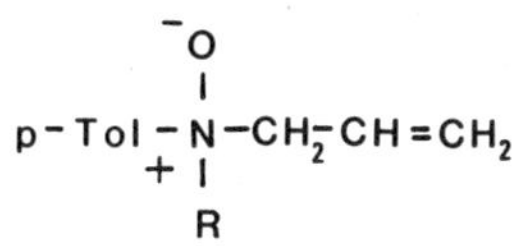

90 a, R = Et

b, R = Me

Inouye *et al.* obtained (+)-*N*-*trans*-but-2-enyl-*N*-ethyl-*p*-toluidine *N*-oxide (**90a**, $[\alpha]_D$ +5.5° (*c* = 9.65, MeOH)) and (+)-*N*-*trans*-crotyl-*N*-methyl-*p*-toluidine N-oxide (**90b**, $[\alpha]_D$ −76.3° (*c* 0.82, MeOH)) by oxidation of the corresponding tertiary amines with (−)-*O*,*O*-dibenzoyl-D-pertartaric acid (*84*). The extent of asymmetric induction was not determined.

The enantiomeric purity and absolute configuration of certain amine oxides can be determined using chiral solvents, such as (−)-2,2,2-trifluoro(α-naphthyl)ethanol and (+)-(*S*)-2,2,2-trifluorophenylethanol) (*85*, *86*). Pirkle and co-workers, for example, assigned the *S* configurations to (−)-*N*-methyl-*N*-ethyl-α-naphthylamine oxide and (−)-*N*-methyl-*N*-ethylaniline oxide based on NMR spectra nonequivalences in (+)-(*S*)-2,2,2-trifluorophenylethanol (*86*). These assignments were confirmed in the case of the latter amine oxide by an X-ray structure of the *d*-α-bromo-π-camphorsulfonic acid salt.

2. Reactions

Pyrolysis of optically active 4-methylcyclohexylamine *N*-oxides (**91**) has been investigated by Berti and Bellucci (*80*) and in more detail by Goldberg and Lam (*81*). At 110 to 120°C, **91** gave optically active 4-methylcyclohexene **92** (1.9–30.0% ee). The absolute configuration of **91** can be determined on the basis of the syn elimination mechanism for amine oxide pyrolysis and the known configuration of **92** by considering nonbonded steric interactions in the transition state. These results have been summarized by Morrison and Mosher (*87*).

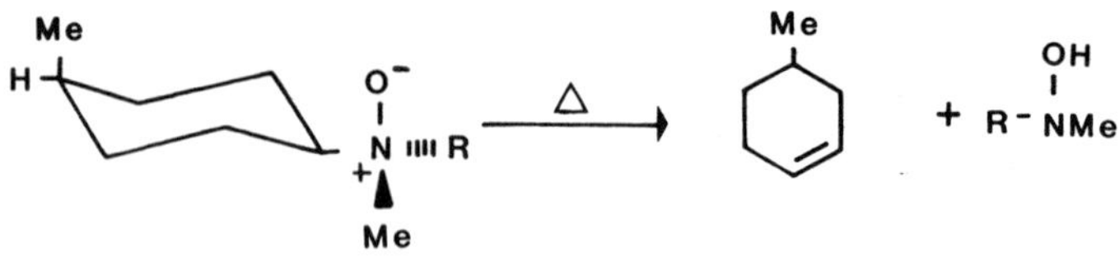

91 92

The [2,3]sigmatropic rearrangement of (+)-(*R*)-*N*-*trans*-crotyl-*N*-methyl-*p*-toluidine oxide (**93**) to (+)-(*S*)-*O*-methylvinylcarbinyl-*p*-tolylhydroxylamine (**95**) was explored by Moriwaki *et al.* (*88*). Oxidation of the corresponding amine with *O,O*-dibenzoyl-1-pertartaric acid gave, after several crystallizations from ethanol, the dibenzoyltartrate salt of **93** ($[\alpha]_D$ −78.4°). The configuration of the asymmetric nitrogen atom in **93**

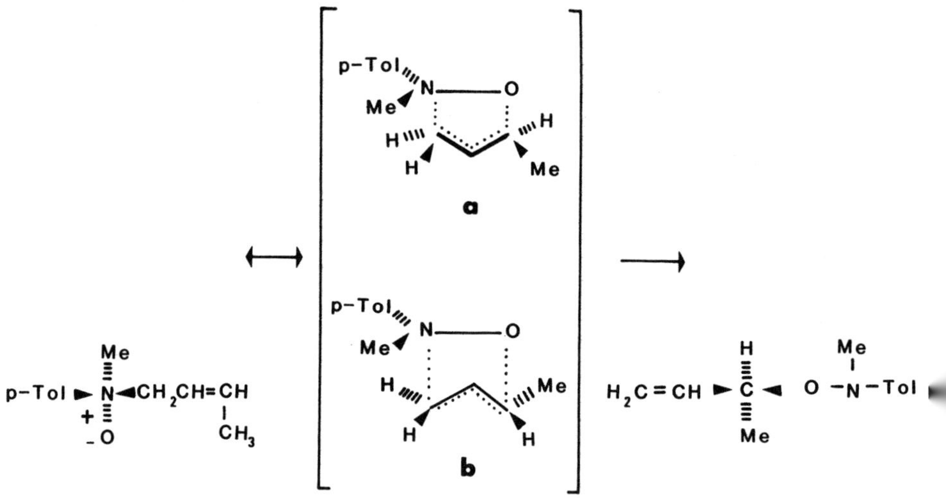

(+)-(R)-93 94 (+)-(S)-95

was assigned the *R* configuration, and the optical purity was determined to be 16% using (+)-(*S*)-2,2,2-trifluoro(α-naphthyl)ethanol and Pirkle's chiral NMR solute–chiral solvent interaction model. To avoid possible complications, however, the double bond in **93** was first reduced with potassium diazocarbonate to give (+)-*N*-methyl-*N*-*n*-butyl-*p*-toluidine, on which the NMR studies were actually carried out.

When (+)-(*R*)-**93** was refluxed with 10% aqueous sodium hydroxide, a 90% yield of (+)-(*S*)-**95** ($[\alpha]_D$ 2.38°) was obtained (*88*). The asymmetric bias and absolute configuration were determined by reduction to (+)-(*S*)-2-butanol, and the transfer of chirality was estimated to be at least 83%.

On the basis of these results, a mechanism involving a concerted [2,3]sigmatropic rearrangement via a five-membered cyclic transition state (**94**) was proposed. Suprafacial mode **94a** is thermodynamically more favorable than **94b** because interaction between the *N*-*p*-tolyl and methyl groups is minimized in **94a**.

Acknowledgment

Support during the writing of this chapter from the National Science Foundation is gratefully acknowledged.

References

1. Hantzsch, A., and Werner, A. (1890). *Ber. Dtsch. Chem. Ges.* **23,** 11.
2. Shriner, R. L., Adams, R., and Marvel, C. S. (1938). "Organic Chemistry, an Advanced Treatise" (H. Gilman, ed.), p. 328. Wiley, New York.
3. Lehn, J. M. (1970). "Topics in Current Chemistry," Vol. 15, p. 311. Springer-Verlag, Berlin.
4. Kessler, H. (1970). *Angew. Chem. Int. Ed. Engl.* **9,** 219.
5. Rauk, A., Allen, L. C., and Mislow, K. (1970). *Angew. Chem. Int. Ed. Engl.* **9,** 400.
6. Shustov, G. V., Zolotoi, A. B., and Kostyanovsky, R. G. (1983). *Tetrahedron* **39,** 2319.
7. Morrison, J. D., and Mosher, H. S. (1976). "Asymmetric Organic Reactions," p. 365. Am. Chem. Soc., Washington, D.C.
8. Annunziata, R., Fornasier, R., and Montanari, F. (1972). *J. Chem. Soc. Chem. Commun.* 1133.
9. Forni, A., Moretti, I., Proxyanki, A. V., and Torre, G. (1981). *J. Chem. Soc. Chem Commun.* 588.
10. Kostyanovskii, R. G., and Rudchenko, V. F. (1976). *Dokl. Akad. Nauk SSSR* **231,** 878.
11. Kostyanovskii, R. G., and Kadorkina, G. K. (1977). *Izv. Akad. Nauk SSSR Ser. Khim.* 1686.
12. Rudchenko, V. D., D'yachenko, O. A., Zolotoi, A. B., Atovmyan, L. O., and Kostyanovskii, R. G. (1979). *Dokl. Akad. Nauk SSSR* **246,** 1150.
13. Kostyanovskii, R. G., Rudchenko, V. F., and Shustov, G. W. (1977). *Izv. Akad. Nauk SSSR Ser. Khim.* 1687.
14. Kostyanovskii, R. G., Kadorkina, G. K., Chervin, I. I., Nasibow, Sh. S., and Varlamov, S. V. (1980). *Khim. Geterotsikl. Soedin.* 1495.
15. Rudchenko, V., D'yanchenko, O. A., Zolotol, A. B., Atovmyan, L. O., Chervin, I. I., and Kostyanovsky, R. G. (1982). *Tetrahedron* **38,** 961.
16. Rauk, A. (1981). *J. Am. Chem. Soc.* **103,** 1023.
17. Kostyanovskii, R. G., Polyankov, A. E., and Markov, V. I. (1974). *Izv. Akad. Nauk SSSR Ser Khim.* 1671

18. Dyachenko, O. A., Atovmyan, L. O., Aldoxhin, S. M., Polyakov, A. E., and Kostyanovskii, R. G. (1976). *J. Chem. Soc. Chem. Commun.* 50.
19. Kostyanovsky, R. G., Polyakov, A. E., and Shustov, G. V. (1976). *Tetrahedron Lett.* 2059.
20. Kostyanovskii, R. G., Polyakov, A. E., Shustov, G. V., Sakharov, K. S., and Markov, V. I. (1974). *Dokl. Akad. Nauk SSSR* **219,** 873.
21. Kostyanovskii, R. G., Plolyankov, A. E., and Markov, V. I. (1976). *Akad. Nauk SSSR Ser Khim.* 198.
22. Kostyanovskii, R. G., Shustov, G. B., Mishchenko, A. I., and Markov, V. I. (1976). *Akad. Nauk SSSR Ser. Khim.* 2026.
23. Kostyanovskii, R. G., and Shustov, V. (1977). *Dokl. Akad. Nauk SSSR* **232,** 1081.
24. Shushtov, G. V., D'yachenko, O. A., Aldoshin, S. M., Zolotoi, A. B., Isobaev, M. D., Chervin, I. I., Atovmyan, L. O., and Kostyanovskii, R. G. (1976). *Dokl. Akad. Nauk SSSR* **231,** 1174.
25. Kostyanovskii, R. G., Shustov, G. V., and Zaichenko, N. L. (1982). *Tetrahedron,* **38,** 949.
26. Hakli, H., and Mannschreck, A. (1977). *Angew. Chem. Intl. Ed. Engl.* **16,** 405.
27. Lehn, J. M., Munsch, B., Millie, P., and Veilland, A. (1969). *Theor. Chim. Acta* **13,** 313.
28. Bjorgo, J., and Boyd, D. R. (1973). *J. Chem. Soc. J.C.S. Perkin 2* 1575.
29. Montanari, F., Moretti, I., and Torre, G. (1973). *Gazz. Chim. Ital.* **103,** 681.
30. Forni, A., Garuti, G., Moretti, I., Torre, G., Andretti, G. D., Bocelli, G., and Sgarabotto, P. (1978). *J. Chem. Soc. Perkin 2* 401.
31. Boyd, R. R., and Graham, R. (1969). *J. Chem. Soc. C* 2648.
32. Boyd, D. R., Spratt, R., and Jerina, D. M. (1969). *J. Chem. Soc. C* 2650.
33. Bjorgo, J., Boyd, D. R., Campbell, R. M., Thompson, J. J., and Jennings, W. B. (1976). *J. Chem. Soc. Perkin 2* 606.
34. Pirkle, W. H., and Rinaldi, P. L. (1977). *J. Org. Chem.* **42,** 3217.
35. Pirkle, W. H., and Rinaldi, P. L. (1978). *J. Org. Chem.* **43,** 4475.
36. Pirkle, W. H., and Rinaldi, P. L. (1977). *J. Org. Chem.* **42,** 2080.
37. Bucciarelli, M., Forni, A., Moretti, I., and Torre, G. (1977). *J. Chem. Soc. Perkin 2* 1339.
38. Belzecki, C., and Mostowicz, C. (1975). *J. Org. Chem.* **40,** 3878.
39. Forni, A., Garuti, G., Moretti, I., Torre, G., Andreetti, G. D., Bocelli, G., and Sgarabotto, P. (1978). *J. Chem. Soc. Perkin 2* 401.
40. Mostowicz, D., and Belzecki, C. (1977). *J. Org. Chem.* **42,** 3917.
41. Bogucka-Ledochowska, M., Konitz, A., Hempel, A., Dauter, Z., Borowski, E., Belzecki, C., and Mostowicz, D. (1976). *Tetrahedron Lett.* 1025.
42. Berube, G., and Jankowiski, K. (1982). *Tetrahedron Lett.* 2857.
43. Forni, A., Moretti, I., and Torre, G. (1978). *Tetrahedron Lett.* 2941.
44. Forni, A., Moretti, I., Torre, G., and Vignudelli, E. (1979). *Tetrahedron Lett.* 907.
45. Pirkle, W. H., and Rinaldi, P. L. (1977). *J. Am. Chem. Soc.* **99,** 3510.
46. Emmons, W. D. (1957). *J. Am. Chem. Soc.* **79,** 5739.
47. Rastetter, W. H., Wagner, W. R., and Findeis, M. A. (1982). *J. Org. Chem.* **47,** 419.
48. Boyd, D. R., and Neill, D. C. (1977). *J. Chem. Soc. Chem. Commun.* 51.
49. Bucciarelli, M., Forni, A., Moretti, I., and Torre, G. (1980). *J. Chem. Soc. Perkin 1* 2152.
50. Forni, A., Moretti, I., and Torre, G. (1977). *J. Chem. Soc. Chem. Commun.*
51. Bjorgo, J., Boyd, D. R., Campbell, R., and Neill, D. C. (1976). *J. Chem. Soc. Chem. Commun.* 162.

52. Lattes, A., Oliveros, E., Riviere, M., Blzecki, C., Mostowics, D., Abramskj, W., Piccinni-Leopardi, C., Germain, G., and Van Merssche, M. (1982). *J. Am. Chem. Soc.* **104,** 3929.
53. Davis, F. A., Lamendola, Jr., J., Nadir, U., Kluger, E. W., Sedergran, T. C., Panunto, T. W., Billmers, R., Jenkins, Jr., R., Turchi, I. J., Watson, W. H., Chen, J. S., and Kimura, M. (1980). *J. Am. Chem. Soc.* **102,** 2000.
54. Davis, F. A., and Stringer, O. D. (1982). *J. Org. Chem.* **47,** 1774.
55. Bucciarelli, M., Forni, A., Marcaccioli, S., Moretti, I., and Torre, G. (1982). *Tetrahedron* **39,** 187.
56. Bucciarelli, M., Froni, A., Moretti, I., and Torre, G. (1983). *J. Chem. Soc. Perkin 2* 923.
57. Davis, F. A., Jenkins, Jr., R. H., Awad, S. B., Stringer, O. D., Watson, W. H., and Galloy, J. (1982). *J. Am. Chem. Soc.* 5412.
58. Davis, F. A., and Billmers, J. M. (1983). *J. Org. Chem.* **48,** 2672.
59. Davis, F. A., Stringer, O. D., and McCauley, Jr., J. P. *Tetrahedron* (in press).
60. Davis, F. A., Billmers, J. M., and Stringer, O. D. (1983). *Tetrahedron Lett.* 3191.
61. Bach, R. D., and Wolber, G. J. (1984). *J. Am. Chem. Soc.,* **106,** 1410.
62. Davis, F. A., Harakal, M. E., and Awad, S. B. (1983). *J. Am. Chem. Soc.* **105,** 3123.
63. Boschelli, D., Smith, III, A. B., Stringer, O. D., Jenkins, Jr., R. H., and Davis, F. A. (1981). *Tetrahedron Lett.* 4385.
64. Muller, K., and Eschenmoser, A. (1969). *Helv. Chim. Acta* **52,** 1823.
65. Kamernitzky, A. V., Levina, I. S., and Mortikova, E. I. (1975). *Tetrahedron Lett.* 3235.
66. Kamernitzky, A. V., Levina, I. S., Mortikova, E. I., Shitkin, V. M., and El'Yanov, B. S. (1977). *Tetrahedron* **33,** 2135
67. Kostyanovsky, R. G., Rudchenko, V. F., D'Yachenko, O. A., Chervin, I. I., Zolotoi, A. B., and Atovmyan, L. O. (1979). *Tetrahedron* **35,** 213.
68. Kostyanovsky, R. G., Rudchenko, V. F., Shtamburg, V. G., Chervin, I. I., and Nasibov, S. S. (1981). *Tetrahedron* **37,** 4243.
69. Le Bel (1891). *Compt. Rend.* **112,** 724.
70. Horner, L., Winkler, H., and Meyer, E. (1965). *Tetrahedron Lett.* 789.
71. Hill, R., and Chan, T. (1966). *J. Am. Chem. Soc.* **88,** 866.
72. Ref. *7,* page 380.
73. Brewster, J., and Jones, Jr., S. (1969). *J. Org. Chem.* **34,** 354.
74. Cope, A., Funke, W., and Jones, F. (1966). *J. Am. Chem. Soc.* **88,** 4693.
75. Reference 7; page 403.
76. Meisenheimer, J. (1980). *Chem. Ber.* **41,** 3966.
77. Meisenheimer, J. (1911). *Ann.* 117.
78. Meisenheimer, J., Glawe, H., Greeske, H., Schorning, A., and Vieweg, E. (1926). *Ann.* **449,** 188.
79. Meisenheimer, J. (1922). *Ann.* **428,** 252.
80. Berti, G., and Bellucci, G. (1964). *Tetrahedron Lett.* 3853.
81. Goldberg, S., and Lam, F-L. (1969). *J. Am. Chem. Soc.* **91,** 5113.
82. Morrison, J. D., and Long, K. P., unpublished results reported in ref. *7,* page 367.
83. Long, K. (1971). *Diss. Abstr. Int. B.* **31** (7), 3926–3927.
84. Moriwaki, M., Sawada, S., and Inouye, Y. (1970). *Chem. Commun.* 419.
85. Pirkle, W. H., Beare, S., and Muntz, R. (1969). *J. Am. Chem. Soc.* **91,** 4575.
86. Pirkle, W. H., Muntz, R. L., and Paul, I. C. (1971). *J. Am. Chem. Soc.* **93,** 2817.
87. For a discussion see ref. *7.*
88. Moriwaki, M., Yamamoto, Y., Oda, J., and Inouye, Y. (1976). *J. Org. Chem.* **41,** 300.

5

Synthesis and Utilization of Compounds with Chiral Silicon Centers

Cynthia A. Maryanoff
Bruce E. Maryanoff
McNeil Pharmaceutical
Spring House, Pennsylvania

I. Introduction

The fascination of organic silicon compounds greatly derives from silicon's position in the periodic table of the elements, in group IV, just below carbon. Since compounds of carbon have provided such rich and interesting chemistry, it is only natural that similarities and differences have been sought in the organosilicon area. Optically active carbon compounds have been present in nature since time immemorial. They are critical to the essence of life. More practically, they are important in the field of natural products and in the study of reaction mechanisms at carbon centers. From this backdrop, significant interest emerged in compounds with optical activity because of an asymmetric silicon center, work on which only began as late as the 1960s.

ISBN 0-12-507704-1

In this chapter, we present an overview of the synthesis and use of chiral silicon compounds (*1*). After an introductory discussion of mechanisms of reactions at silicon centers, we summarize the preparation of optically active organosilanes by classical resolution and asymmetric synthesis. Utilization of optically active organosilanes in organic synthesis is then reviewed.

The dominant method of preparation of optically active silicon compounds reported to date has been classical resolution. This has mainly been used in programs to elucidate reaction mechanisms at silicon centers. Although stereospecific reactions at silicon are common, and absolute configurations established, few practical asymmetric synthesis are known. Perhaps this is a consequence of the absence of chiral silicon centers in molecules in nature, such that the interest of inorganic chemists has vastly exceeded that of organic synthetic chemists.

II. Mechanism of Reactions at Silicon

Several excellent reviews of the stereochemistry and mechanisms of reactions at silicon are available. The first was a monograph by Sommer, who described the pioneering work conducted principally in his laboratory (*2*). Later, Sommer published a review addressing common reaction pathways (*3*). Corriu critically evaluated nucleophilic displacement reactions at silicon (*4*) and surveyed the chemical and stereochemical properties of silicon–transition-metal bonds (*5*). Corriu has also discussed stereochemical analogies between nucleophilic displacement reactions at silicon and phosphorus centers (*6, 7*).

Corriu's leading articles (*4*) on the stereochemistry of reactions at silicon began by establishing the mechanistic importance of nucleophilic substitution reactions. Silicenium ions are not believed to be intermediates for reactions in solution (*2–4*). Racemization of optically active halosilanes by solvents or salts has been reviewed (*8a*). Although the exact mechanism for racemization is under debate, there is general agreement that silicenium intermediates are not involved (*9*). In fact, vinyl and ferrocenyl silanes constructed to favor sp^2 cation formation by substituent conjugative stabilization did not lead to racemic solvolysis products (*9, 10*). Evidence ruling out one-electron transfer processes for nucleophilic substitution has been presented (*4, 11–13*). Thus, in synthesis, nucleophilic displacement reactions are of paramount interest.

Sommer originally described two nucleophilic substitution mechanisms relevant to monofunctional organosilicon compounds (*2, 3*): S_N2–Si and S_Ni–Si. In more recent reviews, however, Corriu has summarized evidence to support only a single mechanism, the former one (*4*). That is, reactions regarded as proceeding by an S_Ni–Si mechanism actually in-

S_N1–Si S_N2–Si

volve an S_N2–Si mechanism (assuming the reaction is first order in each reactant) with retention of configuration at silicon (see diagram).

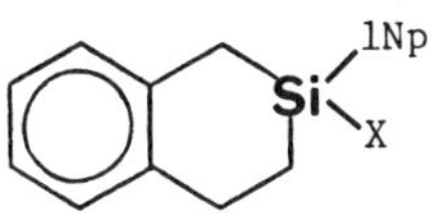

1

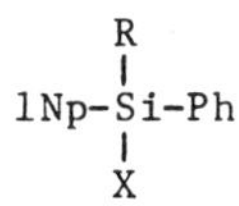

2

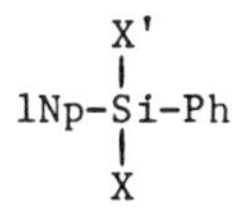

3

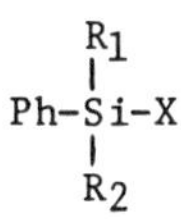

4

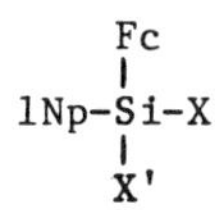

5

For example (*14*), reaction of **1** (X = F) with MeMgBr occurs with 81% inversion of configuration in ether, but with 93% retention in tetrahydrofuran (THF). In contrast, *n*PrMgBr gave predominant retention in both ether and dimethoxyethane (DME), although a large rate difference was evident between the two solvents ($k_{DME}/k_{ether} = 1800$). Reaction of **2** (R = Me, X = OMe) with *n*BuLi in a nonpolar solvent, wherein alkyllithium reagents are more aggregated, showed an increased amount of inversion. However, the chelators Kryptofix 211 (K-211) or tetramethylethylenediamine (TMEDA), deaggregating agents, enhanced retention (Table 1). If a mechanism involving electrophilic assistance of Si—X bond cleavage were operative, then a shift to inversion should occur with use of a basic solvent or additive, but this is not the case.

TABLE 1
Solvent Effect on Substitution by *n*-Butyllithium

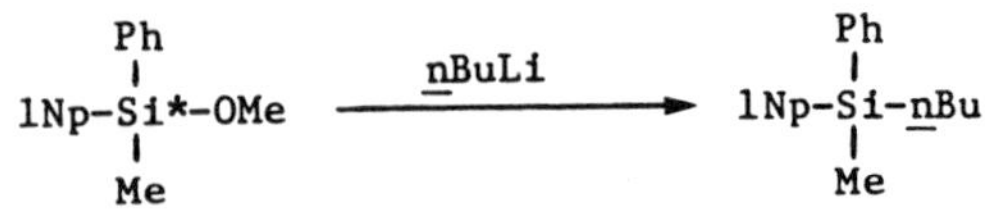

Conditions	Retention (%)
Heptane	68
Ether	86
TMEDA/ether	90
K-211/ether	100

Corriu has shown that nucleophilic substitution reactions at optically active silicon centers are stereoselective under proper reaction conditions. A wide variety of structural types, such as **1–5** (where 1Np = 1-naphthyl; X = a leaving group like halogen, alkoxy, or hydrogen; R = alkyl; Fc = ferrocenyl), were included. He concluded that the factors influencing the course of nucleophilic reaction strongly are the nature of the leaving group and the electronic character of the nucleophilic reagent. Silanes with easily displaced leaving groups (chloro, bromo, tosyloxy, acyloxy) underwent nucleophilic substitution with predominant inversion of configuration; silanes with difficultly displaced leaving groups (hydride) reacted with predominant retention of configuration. Intermediate stereochemical results were obtained with mercaptide, fluoride, and alkoxide leaving groups. For the same leaving group, stereochemistry was governed by the electronic character of the nucleophile: those with well-localized negative charge (hard reagents) led mainly to retention, whereas those with delocalized negative charge (soft reagents) led mainly to inversion. Corriu's reviews summarized experimental data up to 1981 (*4*). A discussion of the theoretical conclusions of Anh and Minot (*15*), in light of the experimental data, was also included.

In summary, Corriu has interpreted the experimental data for bimolecular substitution reactions in support of the intermediacy of a pentacoordinate silicon anion; kinetic (*16a,b,c*), ^{1}H-NMR (*16d*), and ^{29}Si-NMR (*16e*) data have been cited. In the case of substitution with inversion, the stereochemistry requires attack of the nucleophile opposite to the leaving

Inversion

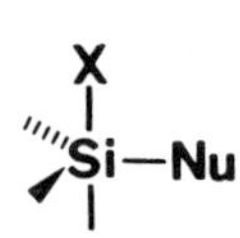

Retention

group, whereas in the case of retention it requires attack orthogonal to the leaving group (i.e., equatorial attack; see diagram). Such equatorial attack is supported by Martin's work on siliconates (*16f*).

Since a discussion of many important details relevant to the stereochemistry of reactions at silicon is beyond the scope of this chapter, the interested reader should consult Corriu's elaborate treatments of this field (*4*).

III. Preparation of Optically Active Organosilanes

Optically active organosilanes have been prepared by a variety of methods (*17*). Early work involved classical resolution, in which a bulk supply of a diastereomer was made and then transformed stereospecifically to silane products of choice. Asymmetric synthesis from achiral silicon starting materials has been achieved, but enantiomeric purity is generally low. Asymmetric synthesis from chiral silicon compounds (second-order asymmetric transformations through epimerization during reaction) have been reported. Partial second-order asymmetric synthesis of a diastereomer, followed by fractional crystallization, has afforded optically pure material. A kinetic resolution has also been published. In the succeeding section, we have collected representative examples from each of the above categories.

A. Classical Resolutions

The first compounds with optical activity due to asymmetric silicon, reported by Kipping, were resolved by fractional crystallization with considerable difficulty (*18–20*). Resolution of **6** was carried out with six differ-

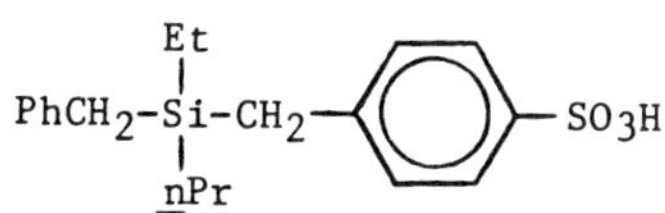

6

ent bases (*20*); optical purity was inferred by the fact that the same rotation for the free acid was obtained for each of the six procedures. The rotations were quite small, and enantiomeric purity was never actually established.

Later, Eaborn resolved *p*-carboxyphenylmethylethylsilane by fractional crystallization of quinine salts (*21*). The first really useful resolution, reported by Sommer's group, is depicted in Eq. (1) (*22, 23*). Chiral,

$$\mathrm{PhMeSi(OMe)_2} \xrightarrow{\mathrm{1NpMgBr}} (\pm)\text{-}\mathrm{1NpPhMeSiOMe} \xrightarrow[\mathrm{KOH}]{(-)\text{-menthol}} \quad (1)$$

$$(\pm)\text{-}\mathrm{1NpPhMeSi[(-)OMen]} \xrightarrow[\text{xtallization}]{\text{fractional}} (-)(-) \;+\; (+)(-) \text{ isomers}$$

racemic 1-naphthylphenylmethylmethoxysilane was reacted with the potassium salt of (−)-menthol, and the 1 : 1 mixture of diastereomers was fractionally crystallized to afford 46% of the (−),(−) diastereomer and 36% of the (+),(−) diastereomer. Absolute configuration was determined by X-ray crystallography (*23*). With these diastereomers in hand, Sommer inaugurated studies on the mechanism of reaction at chiral silicon centers (for summaries see refs. *1–3, 24* and Section IV).

Sommer found that diastereomerically pure 1-naphthylphenylmethylmenthoxysilane is converted stereospecifically to a variety of enantiomerically pure silanes (*24*). For example, lithium aluminum hydride (LAH) reduction yielded the corresponding Si—H compound with inversion of configuration.

In a similar manner, Corriu prepared optically active, bifunctional silanes [Eq. (2)]. Achiral 1-naphthylphenyldimethoxysilane was reacted

$$\mathrm{PhSi(OMe)_3} \xrightarrow{\mathrm{1NpMgBr}} \mathrm{1NpPhSi(OMe)_2} \xrightarrow[\text{cat. NaOMe}]{(-)\text{-menthol}} \quad (2)$$

$$(\pm)\text{-}\mathrm{1NpPhSiOMe[(-)OMen]} \xrightarrow[\text{xtallization}]{\text{fractional}} (-)(-) \;+\; (+)(-) \text{ isomers}$$

7

with (−)-menthol to give a 1 : 1 mixture of diastereomers in 74% yield. Fractional crystallization led to 35% (+),(−)-1-naphthylphenylmethoxymenthoxysilane and 18% of the (−),(−) diastereomer. Resolution of the

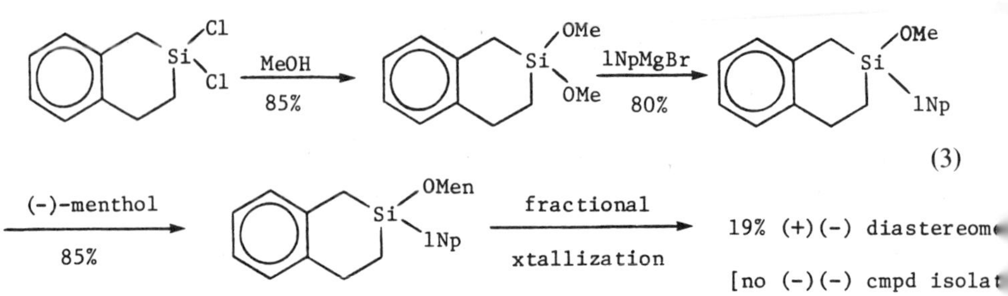

(3)

silatetralin system, shown in Eq. (3), yielded only 19% of the (+),(−) diastereomer (*26, 27*).

B. Asymmetric Synthesis

1. First-Order Asymmetric Transformations

By definition, first-order asymmetric transformations involve reactions of prochiral compounds to yield optically active organosilanes. Four types of prochiral silicon compounds have been reported in asymmetric syntheses: diamidosilanes, silyl ethers, silaethylenes, and silanes.

The first optically active, bifunctionally substituted silanes were prepared from 2-siloxazolidin-5-ones, which had been asymmetrically synthesized (*28, 29*). Reaction of prochiral bis(*N*-methylacetamido)methylphenylsilane (**8**) (*30*) with optically active amino acids resulted in formation of unequal amounts of 2-siloxazolidin-5-ones, such as **9** and **10** [Eq. (4)], and *N*-methylacetamide; thus, an asymmetric synthesis had

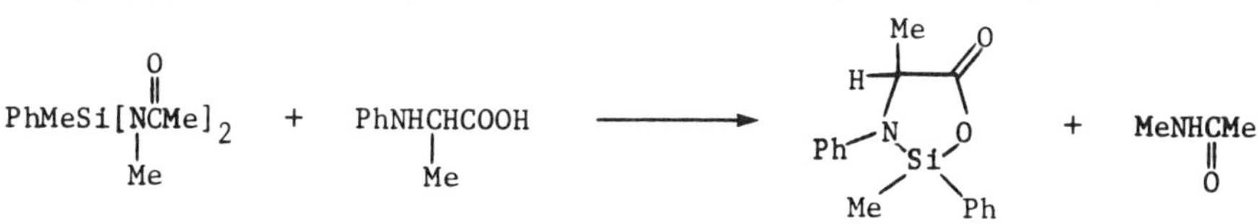

8

(4)

occurred. In solution, the *N*-methylacetamide catalyzed (*31*) an equilibrium between **9** and **10** [Eq. (5)]. Since only one diastereomer crystallized,

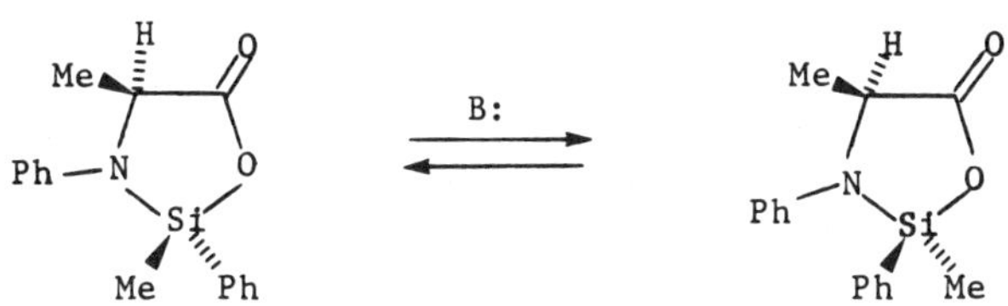

(5)

9 **10**

the equilibrium effected a second-order asymmetric transformation. The asymmetric synthesis of two *N*-alkyl-2-siloxazolidin-5-ones, **11** and **12**, were also reported. No base-catalyzed isomerization was observed with the *N*-alkyl derivatives.

Ring opening of the 2-siloxazolidin-5-ones with, for example, 1-naphthol was executed with ~10% racemization at silicon (^{1}H-NMR assay). Further reaction with methanol furnished the optically active, bi-

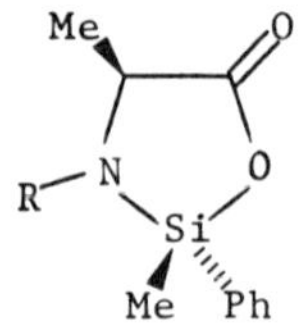

11 R = Me

12 R = iPr

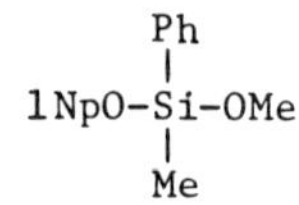

13

functionally substituted silane **13** with $[\alpha]_D$ 21.9° (c = 5, benzene) of unknown optical purity. Although this sequence was described as high yielding, no experimental yield was supplied (*28, 29*). The method appears to be experimentally facile and has been underutilized.

Homochirally substituted prochiral silyl ethers were reacted with ethyllithium to provide an assymmetric synthesis of silyl ethers with enantiomeric excesses of 15 to 69% (*32*). For instance, reaction of dimenthoxymethylphenylsilane with ethyllithium at −70°C gave 49% enantiomeric excess (ee), whereas at −30°C it gave 14% ee [Eq. (6)]. Reduction of **14**

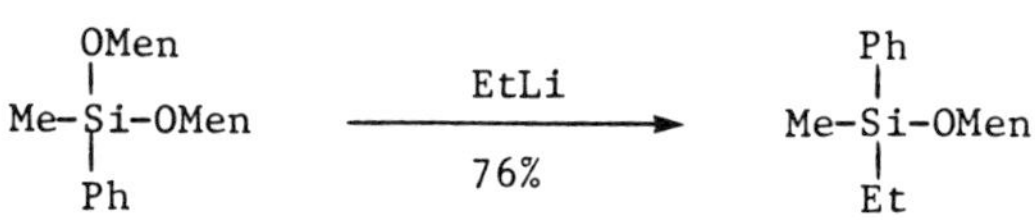

(14-49% ee) (6)

14

with LAH led to a 47% yield of the corresponding silane with $[\alpha]_D$ 3.4° (c = 17.9, hexane). Wide variability in the percentage asymmetric synthesis seemed to be dependent on concentration, solubility, temperature, and reaction time. Additional experimental work may make this method of more general applicability.

Photolysis of silacyclobutanes in the presence of chiral alcohols generated prochiral silaethylenes, trapping of which with chiral alcohols yielded unequal amounts of diastereomers [Eq. (7)] (*33*). However, the best asymmetric induction reported was only 30% ee.

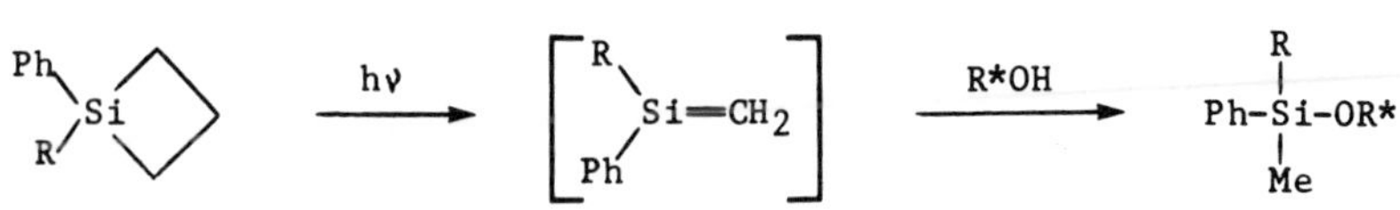

(7)

The use of prochiral silanes in hydrosilylation reactions has been reviewed (*1, 34–36*), two types of hydrosilylation reactions have been useful in the synthesis of optically active silicon compounds: (*a*) asymmetric alcoholysis of dihydrosilanes and (*b*) asymmetric hydrosilylation of ketones.

TABLE 2
Asymmetric Alcoholysis[a]

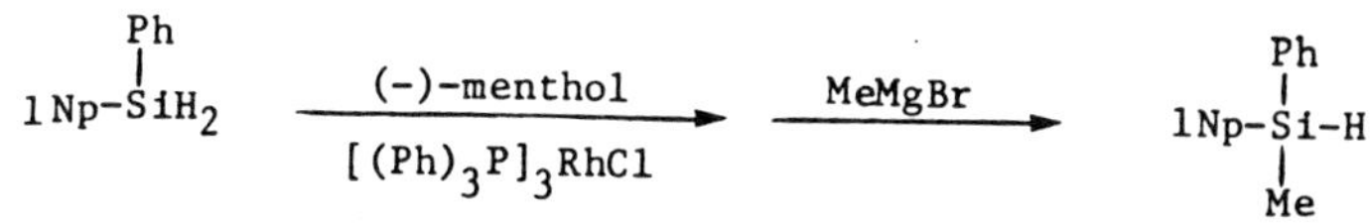

Conditions	% ee (sign of rotation)
(−)-Menthol, $(Ph_3P)_3RhCl$	48 (+)
(−)-Menthol, (−)-diop, $(COD)_2RhCl$	31 (+)
(−)-Menthol, (+)-diop, $(COD)_2RhCl$	49 (+)
MeOH, (+)-diop, $(COD)_2RhCl$	3 (+)
*i*PrOH, (+)-diop, $(COD)_2RhCl$	8 (+)
*t*BuOH, (−)-diop, $(COD)_2RhCl$	7 (−)

[a] From (*37*).

Corriu reported the first asymmetric alcoholysis of dihydrosilanes, promoted by Wilkinson-type catalysts (*37*). Optical purity of the products from a variety of dihydrosilanes was determined by reaction of the silyl ethers with a Grignard reagent to form known silanes [Eq. (8)]. Unfortu-

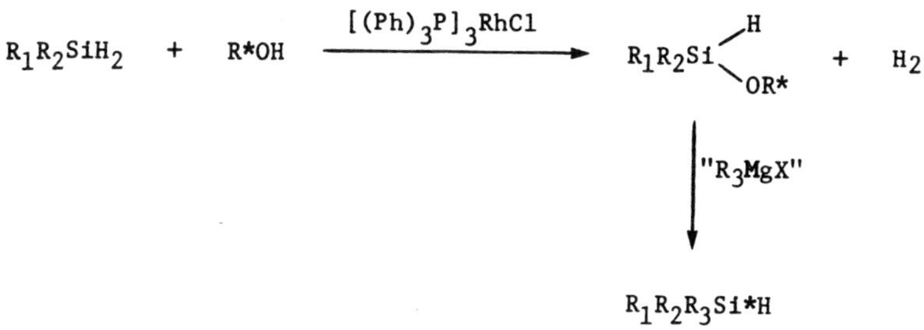

(8)

nately, reaction yields were not mentioned. To cite an exemplary reaction, 1-naphthylphenylsilane was reacted with (−)-menthol in the presence of Wilkinson's catalyst, tris(triphenylphosphine) rhodium(I) chloride, followed by reaction with methylmagnesium bromide, to give (+)-1-naphthylphenylmethylsilane (**15**) in 48% ee (Table 2). Use of a chiral ligand on rhodium, namely, (−)-diop (**16**), in the same reaction led to (+)-**15** in 31% ee, while use of (+)-diop led to (+)-**15** in 49% ee.

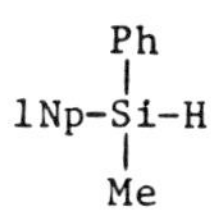

15

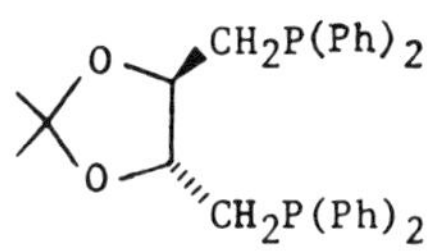

16

However, use of the diop ligand in conjunction with an achiral alcohol delivered poor enantiomeric excess (<10%).

In a similar manner, **15** was obtained with optical purity of 96%, in 51% yield, by asymmetric alcoholysis of 1-naphthylphenylsilane with ephedrine and Wilkinson's catalyst (*38*). This alcoholysis led to a 78 : 22 mixture of diastereomers, which on further reaction and recrystallization afforded either (+)- or (−)-**15** in high optical purity, depending on which ephedrine enantiomer was employed.

Corriu summarized all experimental work done on asymmetric alcoholysis of 1-naphthylphenylsilane and published conclusions regarding important aspects of the mechanism (*39*). He concluded that the structure of the organosilane and the alcohol were critical. The nature of the ligands on rhodium had little influence on the optical yields. This was attributed to direct nucleophilic attack of the alcohol on silicon. The method has been extended to the synthesis of novel ring compounds, such as **17**, in high optical purity [Eq. (9)] (*40, 41*).

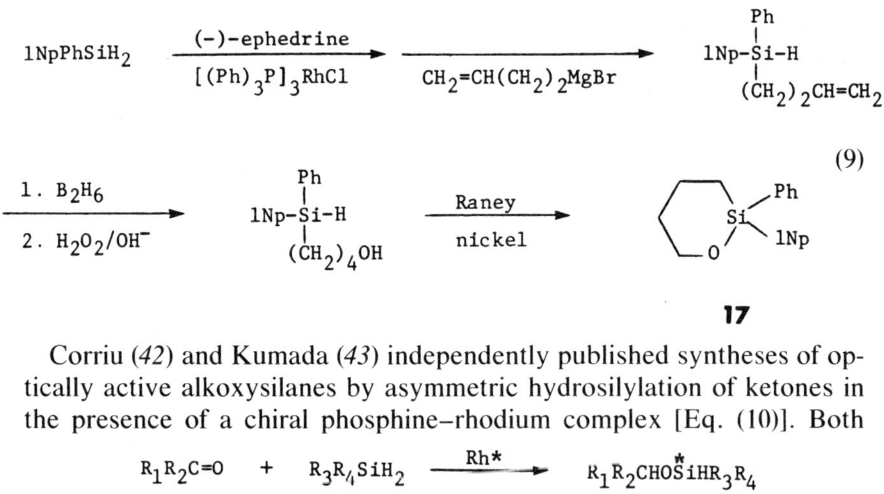

(9)

17

Corriu (*42*) and Kumada (*43*) independently published syntheses of optically active alkoxysilanes by asymmetric hydrosilylation of ketones in the presence of a chiral phosphine–rhodium complex [Eq. (10)]. Both

$$R_1R_2C{=}O \;+\; R_3R_4SiH_2 \xrightarrow{Rh^*} R_1R_2CHO\overset{*}{S}iHR_3R_4 \xrightarrow{R_5MgX} R_3R_4R_5Si^*H \qquad (10)$$

groups converted the silyl ether to a silane of known configuration. The reaction of a Grignard reagent was shown to occur with retention of configuration. Table 3 summarizes results that can be directly compared. From Table 3, it can be concluded that the more rigid diop ligand results in higher enantiomeric excess.

TABLE 3
Asymmetric Induction at Silicon Using 1-Naphthylphenylsilane Followed by Methyl Grignard

Ketone	Cat.[a]	Yield of 1-NpPhMeSiH (%)	Optical purity (%)	Configuration of product	Ref.
Acetone	1	76[b]	—	*S*	*43*
	2	89[b]	30	*R*	*42*
3-Pentanone	1	86[b]	7	*S*	*43*
	2	92	46	*R*	*42*
	3	90	44	*S*	*42*
Benzophenone	1	—	28	*S*	*43*
	2	93	31	*R*	*42*
2-Butanone	3	—	40	*S*	*42*
Acetophenone	3	—	32	*S*	*42*

[a] $[Rh[(R)\text{-benzylMePhP}]_2H_2S_2]^+$, where S = solvent; 2, (+)-diopRhCl; 3, (−)-diopRhCl.
[b] Yield of silyl ether.

Corriu went on to study the mechanism of hydrosilylation in more detail (*44*). Hydrosilylation of acetone with *R*-(+)-1-naphthylphenylmethylsilane in the presence of Wilkinson's catalyst occurred with 96% retention of configuration at silicon. This was proven by LAH reduction back to the starting silane (LAH reduction occurs with retention of configuration). Optically active alkoxysilanes with as much as 46% optical purity were obtained using the diop–rhodium catalyst, prochiral silanes, and ketones or aldehydes (*44*). Prochiral ketones led to different optical yields at silicon than carbon. The highest yields reported were obtained when a chiral ketone was hydrosilylated (*45*). For example, when 1-naphthylphenylsilane was reacted with (−)-menthone in the presence of Wilkinson's catalyst, followed by reaction with ethyl Grignard reagent, the silane was obtained in 67% ee with the *R* configuration. Analogous reaction of 1-naphthylphenylsilane with (−)-menthone with the (+)-diop catalyst produced silane in 82% ee with the *R* configuration. Conversely, reaction of 1-naphthylphenylsilane with (−)-menthone and the (−)-diop catalyst resulted in product silane in 46% ee with the *R* configuration. Further refinement resulted in 95% ee at silicon for the reaction of 1-naphthylphenylsilane with (−)-menthone (*46*). No clear-cut predictions were made based on mechanism (see Table 4). Rather, discussion of possible intermediates led to the conclusion that the structure of the silane and the nature of the ligand catalysts influenced optical purity and configuration. Thus, empirical observations were important to optimizing yields.

TABLE 4
Hydrosilylation of Chiral Ketones[a]

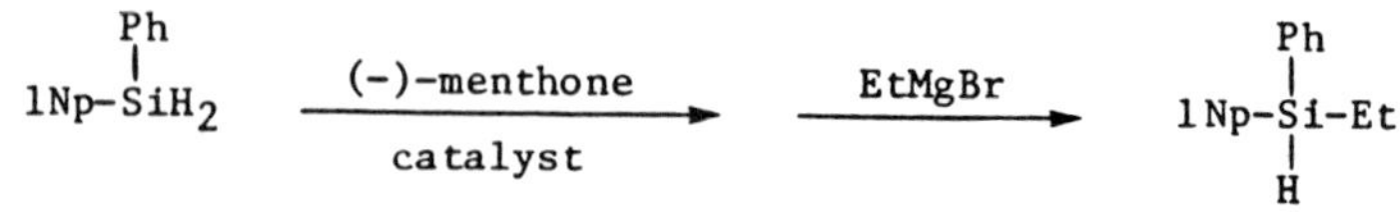

Catalyst	% ee	Configuration of product
$(Ph_3P)_3RhCl$	67	*R*
(+)-diopRhCl	82	*R*
(−)-diopRhCl	46	*R*
(+)-diopRhCl	95[b]	*R*

[a] From (*46*).
[b] Involved separation of diastereomeric silyl ethers; overall yield was 76%.

2. Second-Order Asymmetric Transformations

As mentioned above, base-catalyzed isomerization of *N*-phenyl-2-siloxazolidin-5-ones in solution, followed by crystallization of one diastereomer, exemplifies a second-order asymmetric transformation (*28, 29*).

Mentholysis of 1-naphthylphenylchlorosilane provided an experimentally facile method for obtaining (−)-1-naphthylphenyl-(−)-menthyloxysilane in high optical purity (*47, 48*). When equivalent molar amounts of (−)-menthol and 1-naphthylphenylchlorosilane were combined in pentane, the product obtained in 82% yield after distillation was enriched (~33% ee); crystallization then furnished one diastereomer (97% ee) in 58% overall yield. While this result was not as remarkable as the first example, it may constitute a second-order transformation, if enrichment was not the result of distillation.

3. Kinetic Resolutions

Two reports on kinetic resolutions of organosilanes have appeared, both from the Holt group. The first involved reduction of racemic 1-naphthylphenylmethylmethoxysilane with a 1 : 1 complex of quinidine and LAH in ether (*49*). The reaction was stopped when approximately one-third of the methoxysilane had been reduced. The product silane had an enantiomeric excess of ~6%, as did the recovered starting material (in the opposite direction). In an expansion of this work, Holt used a variety of LAH–alcohol complexes to partially reduce racemic methoxysilanes (*50*). The best enantiomeric excess obtained was 40%, while the average was less than 10%. Unreacted starting material and products always had opposite configurations (ORD, ^{1}H NMR, and CD studies).

In summary, the different approaches to the preparation of optically active organosilanes has been reviewed. The most useful experimental routes to optically pure materials turn out to be classical resolution and asymmetric synthesis through *N*-phenyl-2-siloxazolidin-5-ones, and asymmetric hydrosilylation.

IV. Utilization of Optically Active Organosilanes in Synthesis

The major utility of optically active organosilanes has been in probing the mechanism of reactions that take place at the silicon center. At the beginning of this chapter, we reviewed the mechanism of nucleophilic displacement reactions at silicon because of their importance in the synthesis of optically active organosilanes. In this section, their use in some other mechanistic inquiries is described. Also, the few examples that employ chiral silanes in synthesis are recounted.

Corriu established the stereoselectivity of hydrosilylation of ketones by the reaction shown in Eq. (11). *R*-(+)-1-Naphthylphenylmethylsilane was

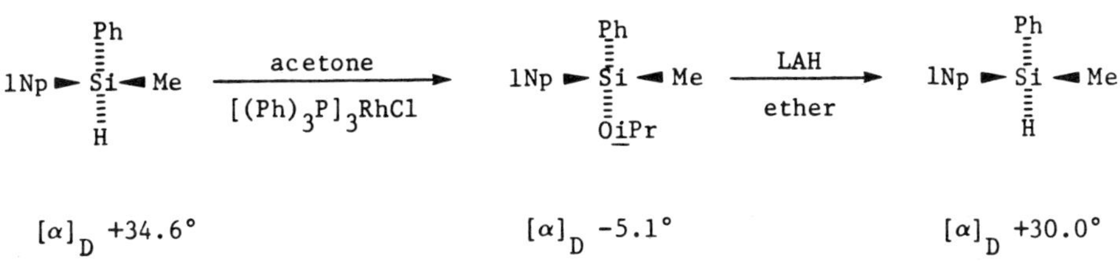

(11)

added to acetone by the assistance of Wilkinson's catalyst. The silyl ether product was reduced to starting silane and compared to the original sample. The hydrosilylation occurred with 96% retention of configuration at silicon.

Platinum-catalyzed hydrosilylation of 1-octene with (+)-1-naphthylphenylmethylsilane has been reported to be stereospecific (*51–53*). Chloroplatinic acid, 5% platinum on carbon, or $[(ethylene)PtCl_2]_2$ give retention of configuration (*51, 52*); however, chloroplatinic acid bound to the anion-exchange resin AV-17-8 gave inversion (*53*).

The optical stability of silyl radicals was addressed by Brook (*54*) and Kumada (*55*). Photolysis of (−)-acetyl-1-naphthylphenylmethylsilane in carbon tetrachloride, followed by LAH reduction of the crude product, yielded (+)-1-naphthylphenylmethylsilane in 64% optical purity [Eq. (12)]

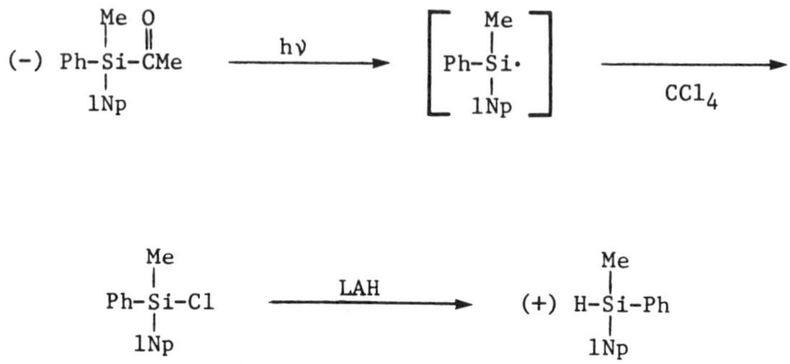

(12)

(*54*). Kumada (*55*) observed that when (+)-1-naphthylphenylmethylsilane was heated in CCl_4 with a catalytic amount of dibenzoylperoxide, followed by LAH reduction, (−)-silane was obtained in 94% yield and 88% optical purity. Thus, both authors (*54, 55*) showed that chloride abstraction occurred with retention of configuration and that LAH reduction occurred with inversion. Both demonstrated that the silyl radical has optical stability, but neither reported a value for the inversion barrier.

Insertion of dichloro- or dibromocarbene species, generated from the appropriate organomercury precursor, into the silicon–hydrogen bond of (+)-1-naphthylphenylmethylsilane proceeded with at least 95% retention of configuration [Eq. (13)] (*56, 57*).

$$\text{(+)-1NpPhMeSiH} \xrightarrow[\text{X = Br or Cl}]{PhHgCX_3} \text{(+)-1NpPhMeSiCHX}_2 \qquad (13)$$

Hydrogen–halogen exchange between silanes and triphenylmethyl halides had been suggested to involve carbenium-ion intermediates (*58*). Eaborn (*59*) found that treatment of (+)-1-naphthylphenylmethylsilane with triphenylmethyl chloride in refluxing benzene provided a 75% yield of (−)-chloromethyl-1-naphthylphenylsilane with complete retention of configuration. This result supported the proposal by Corey and West (*58*). Later work complicated the stereochemical implications by disclosure of high amounts of racemization (with net inversion) in chloroform and methylene chloride (*60*). More recently, Fry showed that reduction of 2-phenyl-2-butyl cation, from optically active or racemic 2-phenyl-2-butanol or 2-phenyl-1-butene, with optically active 1-naphthylphenylmethylsilane affords the hydrocarbon (2-phenylbutane) with ~2% ee (*61, 62*). No information on the stereochemistry at silicon was mentioned.

Jean and Lequan reported the generation of chiral allenes in a 60 : 40 ratio from the reaction of 3-bromobutyne with (+)-1-naphthylphenylmethylsilyllithium [Eq. (14)] (*63*).

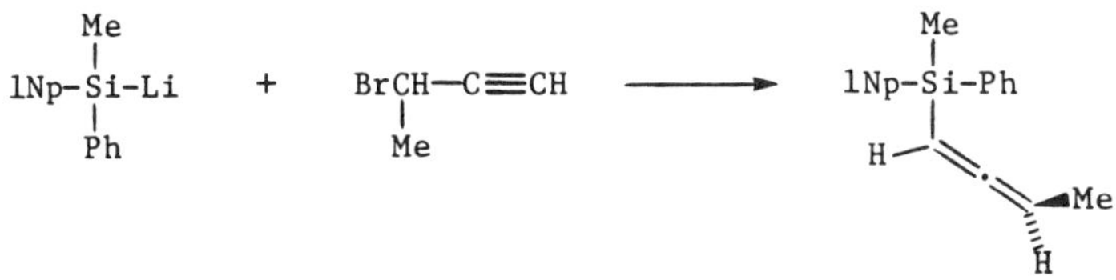

(14)

Optically active (1-naphthylphenylmethylsilyl)ketene (**18**) has been prepared (*64*). The only reactions of the ketone that were disclosed were those with water and alcohols. Reagents that would add to create a second chiral center may have use in future synthesis.

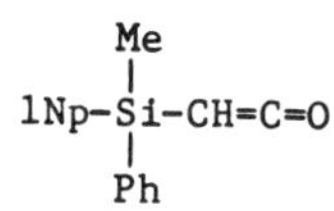

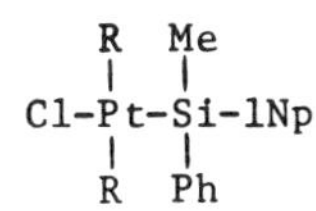

18 $[\alpha]_D$ +3° (+)-<u>trans</u>-**19** R = $PhPMe_2$

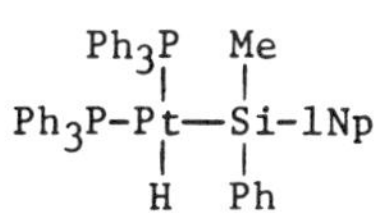

$Ph_2PCH_2CH_2SiHR_1R_2$

(−)-<u>cis</u>-**20** **21** R_1 = H; R_2 = Ph, Me

Two types of chiral polymers have been made. Anionic polymerization of (+)-1-naphthylphenylvinylsilane gave a solid material (*65*). Optically active polymers were obtained when chiral trimethylsiloxymethylepoxides were polymerized with optically active catalysts (*66*). In the latter case, a higher optical induction might be possible with the use of chiral epoxides bearing chiral siloxyalkyl substituents.

We described above reactions involving optically active silanes under rhodium (*44*) and platinum (*51–53*) catalysis. The observed overall stereospecificity implied an initial stereospecific insertion of the metal into the silicon–hydrogen bond. Eaborn (*67–69*) synthesized platinum complexes **19** and **20** without racemization of the silicon center. Using a phosphinoethylsilyl group, as in **21**, for bidentate chelation, Stobart achieved asymmetric induction at platinum (*70*). Use of chiral silicon ligands may provide new transition-metal-induced asymmetric syntheses in the future.

Paquette attempted to transfer chirality from silicon to carbon according to the scheme in Eq. (15); however, the reaction of (−)-**22** with singlet

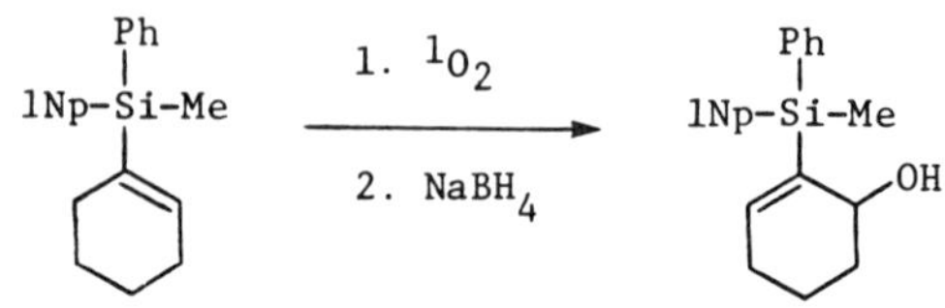

(−)-**22**

(15)

oxygen was unsuccessful (*71*). Furthermore, while epoxidation of (−)-**22** did occur with *m*-chloroperbenzoic acid, rearrangement to the desired allylic alcohol could not be induced.

Paquette also examined allyl group transfer from (−)-1-naphthylphenylmethylallylsilane to dimethyl acetals to yield homoallylic alcohols with carbon-centered asymmetry [Eq. (16)] (*75*). Both chemical and optical yields were unsatisfactory (<30% and ~5%, respectively).

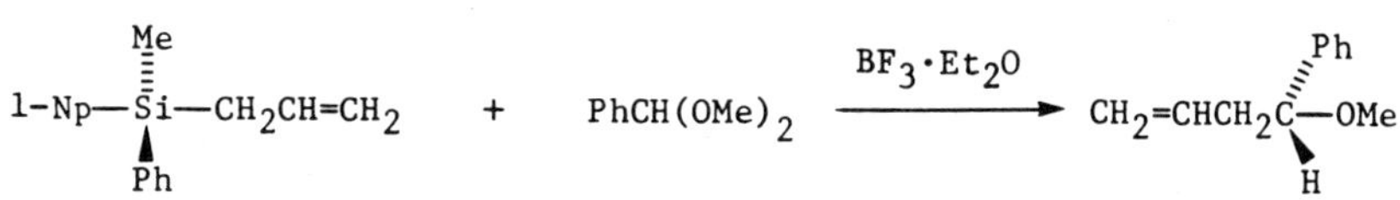

(16)

Another interesting reaction does not deal with optically active silicon centers but relies on the stereospecificity of reaction at silicon. Kumada reported that methoxymethyldisilanes could be converted to (chlorodimethylsilyl)trimethylsilylmethanes (*72*). Wetter extended the method to disilanylmethylsulfides by effecting the rearrangement with trimethyloxonium fluoborate [Eq. (17)] (*73*). Since the rearrangement occurred with

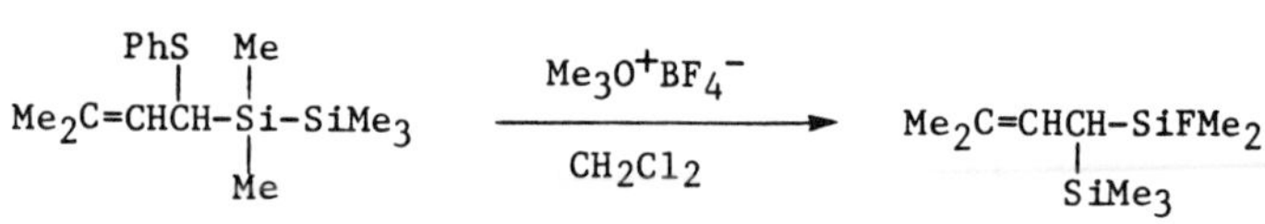

(17)

high configurational stability of the allylic double bond, the reaction was extended to optically active substrates [Eq. (18)] (*74*). Rearrangement of the α-(alkylthio)allyldisilane to the disilylalkene occurred with a minimum stereoselectivity of 92%, in 81% yield. Acylation furnished the vinylsilane with a stereoselectivity of ~94%, in 88% yield.

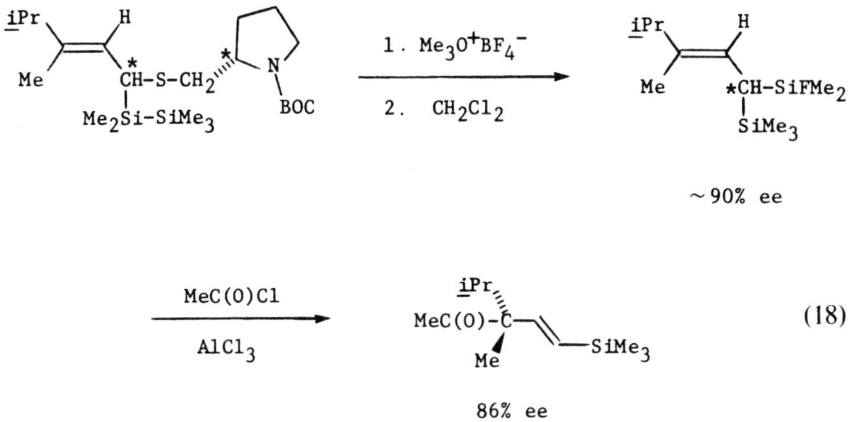

(18)

V. Concluding Remarks

Compounds with optical activity due to the presence of an asymmetric silicon center are relatively easy to obtain. Certain resolved organosilanes have played a pivotal role in the preparation of other silicon-centered asymmetric compounds, primarily by nucleophilic substitution reactions at silicon.

It is with this class of reaction that silicon chemistry clearly departs from typical carbon chemistry. Bimolecular nucleophilic displacement at saturated carbon proceeds with inversion of configuration and does not involve a pentacoordinate intermediate, whereas reaction at silicon occurs with retention or inversion and is best explained by the intermediacy of a pentacoordinate siliconate anion. Optically active silicon species have been invaluable in elucidating the mechanisms of silicon-substitution reactions.

Applications of optically active compounds containing a chiral silicon atom to organic synthesis have been rather sparse, especially considering the vast attention paid to asymmetric carbon chemistry in the past decade. Perhaps the only modest success experienced with asymmetric synthesis involving chiral silicon agents to date has discouraged many investigators from widely exploring this field. So, compounds with chiral silicon centers have only found a true purpose in mechanistic silicon chemistry; their practical aspects in organic synthesis are a matter for the future.

References

1. A review entitled "Synthesis and Reactions of Asymmetric Organosilicon Compounds" has appeared in Japanese: Matsumoto, H., and Nagai, Y. (1978). *Kagaku Kogyo* 619.
2. Sommer, L. H. (1965). "Stereochemistry, Mechanism and Silicon." McGraw-Hill, New York.
3. Sommer, L. H. (1973). *Intra-Science Chem. Rep.* **7,** 1.
4. a. Corriu, R. J. P., and Guérin, C. (1980). *J. Organomet. Chem.* **198,** 231–320; b. Corriu, R. J. P., and Guérin, C. (1982). *Adv. Organomet. Chem.* **20,** 265–312; c. Subsequent to submission of the present chapter, another review appeared: Corriu, R. J. P., Guérin, C., and Moreau, J. J. E. (1984). *Top. Stereochem.* **15,** 43–199.
5. Colomer, E., and Corriu, R. J. P. (1981). *Top. Curr. Chem.* **96,** 79.
6. Corriu, R. J. P., Dutheil, J. P., Lanneau, G. F., and Ould-Kada, S. (1979). *Tetrahedron* **35,** 2889.
7. Corriu, R. J. P., Dutheil, J. P., and Lanneau, G. (1981). *Tetrahedron* **37,** 3681.
8. a. Corriu, R., and Henner, M. (1974). *J. Organomet. Chem.* **74,** 1; b. Bassindale, A. R., and Stout, T. (1982). *J.Organomet. Chem.* **238,** C41.
9. Corriu, R. J. P., and Royo, G. (1968). *J. Organomet. Chem.* **14,** 291.
10. a. Corriu, R. J. P., and Royo, G. (1972). *Bull. Soc. Chim. Fr.* **4,** 1490; b. Corriu, R. J. P., Larcher, F., and Royo, G. (1976). *J. Organomet. Chem.* **104,** 161.
11. Krusic, P. J., Fagan, P. J., and San Filipo, J. (1977). *J. Am. Chem. Soc.* **99,** 250.
12. San Filipo, J., Silberman, J., and Fagan, P. J. (1978). *J. Am. Chem. Soc.* **100,** 4834.
13. a. Felkin, H., Knowles, P. J., and Munier, B. (1978). *J. Organomet. Chem.* **146,** 151; b. Corriu, R. J. P., Dabosi, G., and Martineau, M. (1979). *J. Chem. Soc. Chem. Commun.* 457.
14. a. Corriu, R., and Royo, G. (1972). *J. Organomet. Chem.* **40,** 229; b. Corriu, R., Massé, J., and Royo, G. (1971). *J. Chem. Soc. Chem. Commun.* 252; c. Corriu, R., Fernandez, J. M., and Guérin, G. (1978). *J. Organomet. Chem.* **152,** 21; Corriu, R., and Henner, B. (1974). *J. Organomet. Chem.* **71,** 393.
15. Anh, N. T., and Minot, C. (1980). *J. Am. Chem. Soc.* **102,** 103.
16. a. Corriu, R. J. P., and Henner, B. J. L. (1973). *J. Chem. Soc. Chem. Commun.* 116; b. Chauviere, G., Corriu, R. J. P., and Henner, B. J. L. (1975). *J. Organomet. Chem.* **86,** C1; c. Corriu, R. J. P., and Henner, B. J. L. (1975). *J. Organomet. Chem.* **102,** 407; d. Corriu, R. J. P., Royo, G., and DeSaxcè, A. (1980). *J. Chem. Soc. Chem. Commun.* 892; e. Helmer, B. J., West, R., Corriu, R. J. P., Poirer, M., Royo, G., and DeSaxcè, A. (1983). *J. Organomet. Chem.* **251,** 295; f. Perozzi, E. F., and Martin, J. C. (1979). *J. Am. Chem. Soc.* **101,** 1591.
17. For a review of principles of different methods for optical activation, see: Morrison, J. D., and Mosher, H. S. (1976). "Asymmetric Organic Reactions." Am. Chem. Soc., Washington, D. C.
18. Kipping, F. S. (1907). *J. Chem. Soc.* **91,** 209.
19. Kipping, F. S. (1908). *J. Chem. Soc.* **93,** 457.
20. Challenger, F., and Kipping, F. S. (1910). *J. Chem. Soc.* **97,** 755.
21. Eaborn, C., and Pitt, C. (1958). *Chem. Ind.* 830.
22. Sommer, L. H., and Frye, C. L. (1959). *J. Am. Chem. Soc.* **81,** 1013.
23. Sommer, L. H., Frye, C. L., Parker, G. A., and Michael, K. W. (1964). *J. Am. Chem. Soc.* **86,** 3271.

24. Belloli, R. (1969). *J. Chem. Educ.* **46,** 640.
25. Corriu, R. J. P., Lanneau, G. L., and Royo, G. L. (1972). *J. Organomet. Chem.* **35,** 35.
26. Corriu, R. J. P., and Massé, J. (1969). *Bull. Soc. Chim. Fr.* 3491.
27. Corriu, R. J. P., and Massé, J. (1968). *Tetrahedron Lett.* 5197.
28. Klebe, J. F., and Finkbeiner, H. *J. Am. Chem. Soc.* **88,** 4740.
29. Klebe, J. F., and Finkbeiner, H. (1968). *J. Am. Chem. Soc.* **90,** 7255.
30. Appropriate dichlorosilanes reacted in the presence of triethylamine to give siloxazolidinones, but the triethylamine hydrochloride presented experimental difficulties.
31. For a review of silyl proton exchange reactions, see: Klebe, J. F. (1970). *Acc. Chem. Res.* **3,** 299.
32. Richter, W. J. (1979). *J. Organomet. Chem.* **169,** 9.
33. Bertrand, G., Dubac, J., Mazerolles, P., and Ancelle, J. (1980). *J. Chem. Soc. Chem. Commun.* 382.
34. Ojima, I., Yamamoto, K., and Kumada, M. (1977). *Aspects Homogeneous Catal.* **3,** 186.
35. Ojima, I., *in* Vol. 5 of this treatise.
36. Speier, J. L. (1979). *Adv. Organomet. Chem.* **17,** 407.
37. Corriu, R. J. P., and Moreau, J. J. E. (1973). *Tetrahedron Lett.* 4469.
38. Corriu, R. J. P., and Moreau, J. J. E. (1975). *Bull. Soc. Chim. Fr.* 901.
39. Corriu, R. J. P., and Moreau, J. J. E. (1976). *J. Organomet. Chem.* **120,** 337.
40. Corriu, R., and Guérin, C. (1980). *J. Organomet. Chem.* **195,** 261.
41. Corriu, R., Guérin, C., and Massé, J. (1977). *J. Chem. Res. (M)* 1874.
42. Corriu, R. J. P., and Moreau, J. J. E. (1974). *J.Organomet. Chem.* **64,** C51.
43. Hayashi, T., Yamamoto, K., and Kumada, M. (1974). *Tetrahedron Lett.* 331.
44. Corriu, R. J. P., and Moreau, J. J. E. (1975). *J. Organomet. Chem.* **85,** 19.
45. Corriu, R. J. P., and Moreau, J. J. E. (1975). *J. Organomet. Chem.* **91,** C27.
46. Corriu, R. J. P., and Moreau, J. J. E. (1977). *Nouv. J. Chim.* **1,** 71.
47. Corriu, R. J. P., and Lanneau, G. F. (1971). *Tetrahedron Lett.* 2771.
48. Corriu, R. J. P., and Lanneau, G. F. (1974). *Organomet. Chem.* **64,** 63.
49. Holt, A., Jarvis, A. W. P., and Jarvis, G. J. (1968). *Tetrahedron Lett.* 4087.
50. Holt, A., Jarvis, A. W. P., and Jarvis, G. J. (1973). *J. Chem. Soc. Perkin Trans. 2* 114.
51. Sommer, L. H., Michael, K. W., and Fujimoto, H. (1967). *J. Am. Chem. Soc.* **89,** 1519.
52. Sommer, L. H., Lyons, J. E., and Fujimoto, H. (1969). *J. Am. Chem. Soc.* **91,** 7051.
53. Reikhsfel'd, V. O., Flerova, N. I., Filippov, N. A., and Zaslavskaya, T. N. (1981). *J. Gen. Chem. USSR* 1629.
54. Brook, A. G., and Duff, J. M. (1969). *J. Am. Chem. Soc.* **91,** 2118.
55. Sakurai, H., Murakami, M., and Kumada, M. (1969). *J. Am. Chem. Soc.* **91,** 519.
56. Sommer, L. H., Arlie-Ulland, L., and Ritter, A. (1968). *J. Am. Chem. Soc.* **90,** 4486.
57. Brook, A. G., Duff, J. M., and Anderson, D. G. (1970). *J. Am. Chem. Soc.* **92,** 7567.
58. Corey, J. Y., and West, R. (1963). *J. Am. Chem. Soc.* **85,** 2430.
59. Austin, J. D., and Eaborn, C. (1964). *J. Chem. Soc.* 2279.
60. Sommer, L. H., and Bauman, D. L. (1969). *J. Am. Chem. Soc.* **91,** 7076.
61. Fry, J. L. (1971). *J. Am. Chem. Soc.* **93,** 3558.
62. Fry, J. L., and Adlington, M. G. (1978). *J. Am. Chem. Soc.* **100,** 7641.
63. Jean, A., and Lequan, M. (1973). *J. Organomet. Chem.* **50,** 75.
64. Vodolazskaya, V. M., and Baukov, Y. I. *Zh. Obshch. Khim.* **43,** 2088.
65. Davydova, S. L., Purinson, Y. A., Lavrukhin, B. D., and Plate, N. A. (1965). *Izv. Akad. Nauk SSSR Ser. Khim.* 387.
66. Khanh, P. H., Koinuma, H., Inoue, S., and Tsuruta, T. (1970). *Makromolek. Chem.* **134,** 253.

67. Eaborn, C., Kapoor, P. N., Tune, D. J., Turpin, C. L., and Walton, D. R. M. (1972). *J. Organomet. Chem.* **34,** 153.
68. Eaborn, C., Tune, D. J., and Walton, D. R. M. (1973). *J. Chem. Soc. Dalton Trans.* 2255.
69. Eaborn, C., Tune, D. J., and Walton, D. R. M. (1972). *J. Chem. Soc. Chem. Commun.* 1223.
70. Holmes-Smith, R. D., Stobart, S. R., Cameron, T. S., and Jochem, K. (1981). *J. Chem. Soc. Chem. Commun.* 937.
71. Daniels, R. G., and Paquette, L. A. (1982). *Organometallics* **1,** 1449.
72. Tamao, K., and Kumada, M. (1971). *J. Organomet. Chem.* **30,** 329.
73. Wetter, H. (1978). *Helv. Chim. Acta* **61,** 3072.
74. Wetter, H., and Schweizer, W. B. (1979). *Helv. Chim. Acta* **62,** 1985.
75. Hathaway, S. J., and Paquette, L. A. (1983). *J. Org. Chem.* **48,** 3351.

Index

P

Q

R

S

T

V

X

Y